Photosynthesis Bibliography

volume 13 1982

References no. 48410–52388 / AAR–ZWE

Editors Z. Šesták & J. Čatský

1985 SPRINGER-SCIENCE+BUSINESS MEDIA, B.V.

Contributors:
Z. Šesták
J. Čatský
I. Tichá
J. Pospíšilová
J. Solárová
D. Hodáňová
J. Zima
J. Kutík

ISBN 978-90-6193-533-9   ISBN 978-94-017-2634-4 (eBook)
DOI 10.1007/978-94-017-2634-4

PREFACE

The bibliography includes papers in all fields of photosynthesis research - from studies of model biochemical and biophysical systems of the photosynthetic mechanism to primary production studied by the so-called growth analysis. In addition to papers devoted entirely to photosynthesis, papers on other topics are included if they contain data on photosynthetic activity, photorespiration, chloroplast structure, chlorophyll and carotenoid synthesis and destruction, *etc.*, or if they contain valuable methodological information (measurement of selected environmental factors, leaf area, *etc.*). In many branches it has been difficult to define the limits of interest for photosynthesis researchers. This problem has arisen *e.g.* in topics dealing with the tranfer of gases, where - in addition to the papers on carbon dioxide transfer - some papers on water vapour transfer are included, these being of general application or bringing new approaches. On the other hand, many papers dealing with the anatomy and physiology of stomata have been omitted, if the aspect of carbon dioxide or water vapour exchange has not been discussed.

This volume contains references to papers published in the year 1982, and similarly to preceding volumes also addenda including references published in the preceding period (*i.e.* 1966 to 1981). The numbers of the additional references are labelled with an asterisk (*) in the list of references.

To maximize the value of the bibliography the references are arranged alphabetically by authors' names, and each volume is provided with three indexes. The Authors' Index contains all names of authors, co-authors and editors. The Subject Index covers primary items chosen according to their interest for photosynthesis researchers. Starting with Volume 6, the Subject Index has been newly arranged and enlarged. It contains more details on the electron transport chain, carbon fixation pathways, gas exchange on leaf and canopy level, *etc.*, and also on internal and environmental factors affecting photosynthesis and related processes. In the Plant Index, the most important crop plants and selected plant types and groups are indexed.

Cumulative indexes accompany Volumes 1, 5, and then every fifth volume, *i.e.* Volumes 10, 15, *etc.*

We have tried to cover fully the relevant papers which have appeared in the most important scientific periodicals and books. Articles published in local journals, mimeographed booklets, *etc.*, were chosen mostly from reprints received directly from the authors. Abstracts of papers contributed to scientific meetings, theses and dissertations are usually not included.

Since more than 4000 relevant papers are currently published every year and included in this bibliography, and since almost all citations have been checked with the originals, collecting and preparing for publication of such a large amount of material would have been impossible without the collaboration of the authors of the relevant publications. The courtesy of those authors who have already supplied us with reprints is highly appreciated.

We acknowledge with thanks the cooperation of our colleagues from the Institute of Experimental Botany of the Czechoslovak Academy of Sciences in Prague, especially Mrs. **LUDMILA HÁVOVÁ,** Mrs. **JINDŘIŠKA SRBOVÁ,** and Mrs. **LENKA KOLČABOVÁ** who helped in preparing the card material. The librarian of the Institute, Mrs. **ZORA ZAWOYSKÁ** helped us with checking the references.

Dr. **Z. ŠESTÁK** and Dr. **J. ČATSKÝ**

Institute of Experimental Botany
Czechoslovak Academy of Sciences

Flemingovo nám. 2
CS-160 00 PRAHA 6
Czechoslovakia

INSTRUCTIONS FOR USE

All references are arranged alphabetically according to the authors' names and the year of publication. They are numbered and these numbers are used in the indexes. In case of a book title, the number is preceded by B. An asterisk (*) preceding the number denotes the reference published in the preceding period (1966 - 1981).

The references contain the original unshortened title of the paper (book). English, French and German titles are cited in the original language. Titles in other languages are supplemented with a translation in English (sometimes using the title of the respective English abstract or a shortened title with omitted deadweight words). Titles of Japanese, Chinese, etc. papers are given in English translation only. The journals' names are abbreviated mainly according to the "Style Manual for Biological Journals" (2nd Ed., Amer. Inst. Biol. Sci., Washington, D.C. 1964), e.g.:

Abhandlungen	Chromatography	inorganic	quarterly
Abstracts	Commission	Institute etc.	Rabota
Abteilung	Communication	international	Radiation
Academy etc.	comparative	Investigation	Radiobiology etc.
Acta	Comptes rendus	italian etc.	Rastenie
Africa	Conference	Izvestiya	Recherche
agricultural	Congress	Jahrbuch	Report
Agriculture	Contribution	japanese	Research
Agronomy	Cytochemistry	Japan	Review
Akademie (-emiya)	Cytology etc.	Journal	royal
Algology	czechoslovak	Klasse	russian
allgemeine	Dendrology	Laboratory etc.	russkiĭ
american	Department	Landwirtschaft	scandinavian
America	Deutschland	Letters	Science
analytical etc.	Disease	Limnology	Section
Anatomy etc.	Dissertation	Magazine	Series (-iya) etc.
angewandte	Doklady	marine	Society etc.
Annals etc.	Dopovidi	Mathematics	sovetskiĭ
annual	Ecology etc.	Microbiology etc.	soviet
anorganische	Education	miscellaneous	special
applied etc.	Embryology	molecular	SSSR
Arbeit	Encyclopedia	Monograph	Station
Archives	Engineer	moskovskiĭ	Supplement
Atmosphere	environmental	Mycology	Survey
atomic	Enzymology etc.	national	Symposium
Australia	european	natural	technical etc.
Beihefte	experimental etc.	Naturforschung	Technics etc.
Belgique	Experiment etc.	neerlandicus	Technology
Berichte	Faculty etc.	Netherland	Tijdschrift
biochemical etc.	Federation	New Zealand	Transactions
Biochemistry etc.	Fizika	nuclear etc.	Travaux
biokhimicheskiĭ	Fiziologiya	Oceanography etc.	tropical
Biokhimiya	Forestry	Optics etc.	Trudy
biological etc.	Forschung	organic etc.	ukrainian
Biology etc.	Foundation	original	UK
biophysical etc.	France	Otdelenie	US, USA
Biophysics etc.	Gazette	Pathology	USSR
Bodenkunde	general	Pflanzen-	University
bolgarskiĭ	genetical etc.	Philosophy	végétal
botanical etc.	Genetics etc.	physical etc.	Virology
Botany etc.	Gesellscháft	Physics etc.	Virusforschung
british	Giornale	physiological etc.	Volume
Bulletin	helveticus	Physiology etc.	Weekblad
Canada	Histochemistry	Phytopathology etc.	Wetenschappen
cellular etc.	Histology etc.	Plant (-arum)	Wissenschaft
central	Horticulture	polish	Zeitschrift
chemical etc.	hungarian etc.	Proceedings	Zeitung
Chemistry etc.	Husbandry	Publication	Zentralblatt
chimique etc.	imperial	Publishers	Zeszity
chinese	industrial	quantitative	Zhurnal

The numbers at the end of each reference of a journal article denote: volume (issue) : first page - last page, year of publication. The number of issue is given only in the journal where each issue is paginated separately.

Book titles are cited according to the title page, not to the book jacket or cover (if the names of the editors are not given on the title page, they are not cited in the reference). The publisher, place and year of publication are included.

Brackets at the end of the reference give bibliographic details and explanations to the contents, not given in the original. The following abbreviations are used most frequently:

ab	abstract	IRGA	infra-red gas analyser
Arm.	Armenian	Ital.	Italian
Belorus.	Belorussian	Jap.	Japanese
Bil	biliproteins	Latv.	Latvian
Bulg.	Bulgarian	Lithu.	Lithuanian
Car	carotenoids	Norweg.	Norwegian
CC	column chromatography	PC	paper chromatography
Chin.	Chinese	PhAR	photosynthetically active radiation
Chl	chlorophyll	Pol.	Polish
Croat.	Croatian	Ps	photosynthesis
Cyt	cytochromes	R	Russian
Dan.	Danish	Roum.	Roumanian
E	English	Serb.	Serbian
Est.	Estonian	Span.	Spanish
F	French	Swed.	Swedish
G	German	TLC	thin-layer chromatography
GC	gas chromatography	Tr	transpiration
Georg.	Georgian	Ukr.	Ukrainian
Hung.	Hungarian	Uz.	Uzbeg

The transliteration of cyrillic characters is in accordance with the BSI-ASA/SC--Z39 draft table, *i.e.*

Translit.	Cyrill.	Translit.	Cyrill.	Translit.	Cyrill.	Translit.	Cyrill.
a	а	i	и	r	р	ya	я
b	б	ï	й	s	с	yu	ю
ch	ч	k	к	sh	ш	z	з
d	д	kh	х	shch	щ	zh	ж
e	е	l	л	t	т	"	ъ
ė	э	m	м	ts	ц	'	ь
ë	ё	n	н	u	у		
f	ф	o	о	v	в		
g	г	p	п	y	ы		

Several exceptions apply for Ukrainian, Belorussian and Serbian:

Ukrainian:		Belorussian:		Serbian:			
y	и	ŭ	ў	c	ц	j	ј
i	і	i	і	č	h	lj	љ
ï	ï			ǧ	ч	nj	њ
				đ	b	š	ш
				dž	џ	ž	ж
				h	х		

Authors' names are presented in the spelling used in the original paper. If this spelling does not correspond to the original spelling used by the author (*e.g.* Russian papers of English authors), one spelling is referred to the other in the Authors' index.

Printers' errors in the original papers are marked by underlining the respective word (letters).

ERRATA

Reference no./page	For	Read
Volume 4		
15802 / p.25	carboxylation	carboxylation and decarboxylation
Volume 9		
35186 / p.156	L'organizacione	L'organicazione
Volume 11		
41350 / p.58	Kohlprotein	Rohprotein
Volume 12		
47743 / p.214	tserevichni	tsarevichni
47802 / p.218	WOODHOUSE, R.U.	WOODHOUSE, R.M.

48410 - **AARESKJOLD, K., LIAAEN-JENSEN, S.** : Determination of enantiometric composition of partly racemized carotenols. - Acta chem. scand. *B 36* : 499 - - 504, 1982.

48411 - **AARONSON, S., DUBINSKY, Z.** : Mass production of microalgae. - Experientia *38* : 36 - 40, 1982.

48412 - **ABBOTT, M.R., RICHERSON, P.J., POWELL, T.M.** : *In situ* response of phytoplankton fluorescence to rapid variations in light. - Limnol. Oceanogr. *27* : 218 - 225, 1982.

48413 - **ABDULAEV, N.G., OVCHINNIKOV, Yu.A.** : Some approaches to determining the primary structure of membrane proteins. - In : COLOWICK, S.P., KAPLAN, N.O. (ed.) : Methods in Enzymology. Vol.88. Pp. 723 - 729. Academic Press, New York - London - San Diego - San Francisco - São Paulo - Sydney - Tokyo - - Toronto 1982. [Bacteriorhodopsin.]

*48414 - **ABDULLAEV,A.A.,AKHMEDOV, Yu.D.** : Ontogeneticheskie izmeneniya konformatsii spin-mechennoĭ ribulozo-1,5-difosfatkarboksilazy. [Ontogenetic changes in the conformation of spin-labelled ribulose 1,5-bisphosphate carboxylase.] - Dokl. Akad. Nauk tadzh. SSR *24* : 692 - 695, 1981. [In R, ab : Tajik.]

48415 - **ABDULLAEV, Kh.A., KASPAROVA, I.S., NASYROV, Yu.S.** : Vliyanie solevogo stressa na intensivnost' fotosinteza i fotodykhanie khlopchatnika. [Effect of salt stress on photosynthetic rate and photorespiration in cotton.] - Dokl. Akad. Nauk tadzh. SSR *25* : 306 - 308, 1982. [In R, ab : Tajik.]

48416 - **ABER, J.D., PASTOR, J., MELILLO, J.M.** : Changes in forest canopy structure along a site quality gradient in southern Wisconsin. - Amer. Midland Naturalist *108* : 256 - 265, 1982. [Foliage distribution.]

48417 - **ABRAHAM, R.J., SMITH, K.M., GOFF, D.A., LAI, J.J.** : NMR spectra of porphyrins. 18. A ring-current model for chlorophyll derivatives. - J. amer. chem. Soc. *104* : 4332 - 4337, 1982.

48418 - **ABRAHAMSON, W.G., CASWELL, H.** : On the comparative allocation of biomass, energy, and nutrients in plants. - Ecology *63* : 982 - 991, 1982.

48419 - **ABRAMCHIK, L.M., CHAĬKA, M.T., VOLODARSKIĬ, A.D.** : Immunokhimicheskoe issledovanie belkovykh komponentov ètioplastnykh i khloroplastnykh membran. [Immunochemical investigation of protein components of etioplast and chloroplast membranes.] - Fiziol. Rast. *29* : 171 - 173, 1982. [In R.]

48420 - **ABRAMCHIK, L.M., VOLODARSKIĬ, A.D., MIKHAĬLOVA, S.A., PROSVIRNINA, E.N., CHAĬKA, M.T.** : Issledovanie immunokhimicheskikh svoĭstv khlorofillovykh pigmentov khloroplastnykh membran. [Immunochemical properties of chlorophyll pigments in chloroplast membranes.] - Dokl. Akad. Nauk belorus. SSR *26* : 947 - 950, 1982. [In R, ab : E.]

*48421 - **ABROSOV, N.S., GUBANOV, V.G., KOVROV, B.G.** : Teoretiko-èksperimental'nyĭ analiz krugovorota veshchestva v zamknutoĭ mikroèkosisteme. 1. Postroenie matematicheskoĭ modeli. [A theoretical and experimental analysis of substances turnover in a closed microecosystem. 1. The building of a mathematical model.] - Izv. sibir. Otd. Akad. Nauk SSSR *1981*, Ser. biol. Nauk 3 : 68 - 75, 1981. [In R, ab : E.]

48422 - **ABROSOV, N.S., GUBANOV, V.G., KOVROV, B.G.** : Teoretiko-èksperimental'nyĭ analiz krugovorota veshchestva v zamknutoĭ mikroèkosisteme. II. Statsionarnye sostoyaniya i limitiruyushchie faktory krugovorota. [A theoretical and experimental analysis of substances turnover in a closed microecosystem. II. The stable steady states and limiting factors of the turnover.] - Izv. sibir. Otd. Akad. Nauk SSSR *1982*(10), Ser. biol. Nauk 2 : 57 - 64, 1982. [Canopy productivity; in R, ab : E.]

48423 - **ACKER, S., LAGOUTTE, B., PICAUD, A., DURANTON, J.** : Protein identification of purified particles isolated from spinach thylakoids by deoxycholate electrophoresis. - Photosynthesis Res. *3* : 215 - 225, 1982.

48424 - **ACKERSON, R.C.** : Synthesis and movement of abscisic acid in water-stressed cotton leaves. - Plant Physiol. *69* : 609 - 613, 1982. [Stomatal resistance.]

48425 – ADAMS, M.W. : Plant architecture and yield breeding. – Iowa Sta. J. Res.
       56 : 225 – 254, 1982. [Growth analysis.]

48426 – ADAMSON, H. : Chloroplast development in green barley leaves transferred
       to darkness. – In : AKOYUNOGLOU, G., EVANGELOPOULOS, A.E., GEORGATSOS, J.,
       PALAIOLOGOS, G., TRAKATELLIS, A., TSIGANOS, C.P. (ed.) .: Cell Function and
       Differentiation, Part B. Biogenesis of Energy Transducing Membranes and
       Membrane and Protein Energetics. Pp. 189 – 199. Alan R. Liss, New York 1982.

48427 – ADAMSON, H. : Evidence for a light-independent protochlorophyllide reducta-
       se in green barley leaves. – In : AKOYUNOGLOU, G., EVANGELOPOULOS, A.E.,
       GEORGATSOS, J., PALAIOLOGOS, G., TRAKATELLIS, A., TSIGANOS, C.P. (ed.) :
       Cell Function and Differentiation, Part B. Biogenesis of Energy Transducing
       Membranes and Membrane and Protein Energetics. Pp. 33 – 41. Alan R. Liss,
       New York 1982.

48428 – ADLERCREUTZ, P., HOLST, O., MATTIASSON, B. : Oxygen supply to immobilized
       cells: 2. Studies on a coimmobilized algae-bacteria preparation with *in si-
       tu* oxygen generation. – Enzyme microb. Technol. *4* : 395 – 400, 1982.

48429 – ADLERCREUTZ, P., MATTIASSON, B. : Oxygen supply to immobilized cells: 1.
       Oxygen production by immobilized *Chlorella pyrenoidosa.* – Enzyme microb.
       Technol. *4* : 332 – 336, 1982.

48430 – ADMON, A., SHAHAK, Y., AVRON, M. : Adenosine triphosphate-generated trans-
       membrane electric potential in chloroplasts. – Biochim. biophys. Acta *681* :
       405 – 411, 1982.

*48431 – ADYGEZALOV, V.F., GRODZINSKIĬ, D.M. : Membrannyĭ potentsial i ego fotoin-
       dutsiruemye izmeneniya u kletok palisadnoĭ parenkhimy list'ev pshenitsy i
       kukuruzy. [Membrane potential and its photoinduced oscillations in palisa-
       de parenchyma cells of wheat and maize leaves.] – In : Svetozavisimaya
       Bioêlektricheskaya Aktivnost' List'ev Rasteniĭ. Pp. 99 – 104, 119. Ural'.
       Gos. Univ., Sverdlovsk 1977. [In R, ab : E.]

48432 – AFLALO, C., SHAVIT, N. : Source of rapidly labeled ATP tightly bound to non-
       -catalytic sites on the chloroplast ATP synthetase. – Europe. J. Biochem.
       126 : 61 – 68, 1982.

*48433 – AFRIA, B.S., MUKHERJEE, D. : Biochemical changes in leaves of certain C_3
       and C_4 plants under prolonged darkness and on exposure to light. – Plant
       biochem. J. *8* : 1 – 12, 1981.

48434 – AGATA, W., TAKEDA, T. : Studies on matter production in sweet potato plants
       1. The characteristics of dry matter and yield production under field con-
       ditions. – J. Fac. Agr. Kyushu Univ. *27* : 65 – 73, 1982.

48435 – AGATA, W., TAKEDA, T. : Studies on matter production in sweet potato plants
       2. Changes of gross and net photosyntheses, dark respiration and solar
       energy utilization with growth under field conditions. – J. Fac. Agr. Kyu-
       shu Univ. *27* : 75 – 82, 1982.

48436 – AGUIRRE, R., HATCHIKIAN, E.C., MONSAN, P. : Use of immobilized hydrogenase
       for hydrogen production. – In : HALL, D.O., PALZ, W. (ed.) : Photochemical,
       Photoelectrochemical and Photobiological Processes. Pp. 208 – 212. D. Reidel
       Publ. Co., Dordrecht – Boston – London 1982.

48437 – AGUIRRE, R., HATCHIKIAN, E.C., MONSAN, P., COCQUEMPOT, M.F., LISSOLO, T. :
       Utilization of free and immobilized *Desulfovibrio* hydrogenase in hydrogen
       photoproduction: coupling efficiency of cytochrome C_3, ferredoxin and flavo-
       doxin. – Biotechnol. Lett. *4* : 297 – 302, 1982.

48438 – AHL, P.L., CONE, R.A. : Photodichroism and the rotational motions of rhodo-
       psin and bacteriorhodospin. – In : COLOWICK, S.P., KAPLAN, N.O. (ed.) :
       Methods in Enzymology. Vol.88. Pp. 741 – 750. Academic Press, New York –
       – London – Paris – San Diego – San Francisco – São Paulo – Sydney – Tokyo –
       – Toronto 1982.

*48439 - AHMED, A.M., HEIKAL, M.M., ZIDAN, M.A. : Photosynthesis and some other growth parameters of some leguminous plants as affected by salinization treatments. - Nat. Monspel. Sér. Bot. *33* : 1 - 10, 1980.

48440 - AHMED, J., BONHAM, C.D. : Optimum allocation in multivariate double sampling for biomass estimation. - J. Range Manage. *35* : 777 - 779, 1982.

48441 - AIBA, S. : Growth kinetics of photosynthetic microorganisms. - Adv. biochem. Eng. *23* (Microbial Reactions) : 85 - 156, 1982. [Ps.]

48442 - AKAO, S., TSUKAHARA, S., YAMAGATA, M. : [A difference between leaves on main stem and branch regarding translocation patterns of ^{14}C-photosynthates.] - Nippon Dojo Hiryogaku Zasshi *53* : 319 - 326, 1982. [In Jap., ab : E.]

48443 - AKAZAWA, T., ASAMI, S., TAKABE, T. : Photorespiration in photosynthetic bacteria. - In : ZABORSKY, O.R. : CRC Handbook of Biosolar Resources. Vol. I/1. Pp. 317 - 322. CRC Press, Boca Raton 1982.

48444 - AKAZAWA, T., NISHIMURA, M. : Photorespiration and various types of CO_2 fixation. a.Photorespiration. - Recent Progr. nat. Sci. Jap. *7* : 53 - 55, 1982.

48445 - ÅKERLUND, H.-E., JANSSON, C., ANDERSSON, B. : Reconstitution of photosynthetic water splitting in inside-out thylakoid vesicles and identification of a participating polypeptide. - Biochim. biophys. Acta *681* : 1 - 10, 1982.

48446 - AKINYEMIJU, O.A., DICKMANN, D.I. : Contrasting effects of simazine on the photosynthetic physiology and leaf morphology of two *Populus* clones. - Physiol. Plant. *55* : 402 - 406, 1982.

48447 - AKOYUNOGLOU, G. : Reorganization of thylakoid components during chloroplast development in higher plants. - In : AKOYUNOGLOU, G., EVANGELOPOULOS, A.E., GEORGATSOS, J., PALAIOLOGOS, G., TRAKATELLIS, A., TSIGANOS, C.P. (ed.) : Cell Function and Differentiation, Part B. Biogenesis of Energy Transducing Membranes and Membrane and Protein Energetics. Pp. 171 - 188. Alan R. Liss, New York 1982.

B48448 - AKOYUNOGLOU, G., EVANGELOPOULOS, A.E., GEORGATSOS, J., PALAIOLOGOS, G., TRAKATELLIS, A., TSIGANOS, C.P. (ed.) : Cell Function and Differentiation, Part B. Biogenesis of Energy Transducing Membranes and Membrane and Protein Energetics. (Progress in Clinical and Biological Research. Vol. 102 B.) - Alan R. Liss Inc., New York 1982.

48449 - ALABACK, P.B. : Dynamics of understory biomass in Sitka spruce-western hemlock forests of southwestern Alaska. - Ecology *63* : 1932 - 1948, 1982.

48450 - ALAM, S.M. : Assimilation of $^{14}CO_2$ in rice plants with and without nitrogen application. - Pak. J. Bot. *14* : 99 - 101, 1982.

*48451 - ALAM, S.M., ADAMS, W.A. : Effect of lime on the nutrient composition and yield of barley and oats grown on acid soils. - Pak. J. sci. Res. *32* : 239 - 247, 1980. [Photosynthates.]

48452 - ALBAN, D.H., LAIDLY, P.R. : Generalized biomass equations for jack and red pine in the Lake States. - Can. J. Forest Res. *12* : 913 - 921, 1982.

48453 - ALBERGONI, F.G., BASSO, B., BRUSA, T., RECALCATI, L.M. : On the measurement of photosynthetic potentiality. - Maydica *27* : 97 - 105, 1982.

48454 - ALBERTSSON, P.-A. : Interaction between the lumenal sides of the thylakoid membrane. - FEBS Lett. *149* : 186 - 190, 1982.

48455 - ALBRACHT, S.P.J., ALBRECHT-ELLMER, K.J., SCHMEDDING, D.J.M., SLATER, E.C. : On the active site of hydrogenase from *Chromatium vinosum*. - Biochim. biophys. Acta *681* : 330 - 334, 1982.

48456 - ALEMDAG, I.S. : Aboveground dry matter of jack pine, black spruce, white spruce and balsam fir trees at two localities in Ontario. - Forest. Chron. *58* : 26 - 30, 1982.

48457 - **ALEMDAG, I.S.** : Biomass of the merchantable and unmerchantable portions of the stem. - Petawawa nat. Forest. Inst. Inform. Rep. *PI-X-20* : 1 - 20, 1982.

48458 - **ALEMDAG, I.S.** : Methods of estimating forest biomass from stand volumes: A case study with Ontario jack pine. - Pulp Paper Can. *83* : 41 - 43, 1982.

48459 - **ALEMDAG, I.S., STIELL, W.M.** : Spacing and age effects on biomass production in red pine plantations. - Forest. Chron. *58* : 220 - 224, 1982.

48460 - **ALEXANDRE, D.Y.** : Aspects de la régénération naturelle en forêt dense de Côte-d'Ivoire. - Candollea *37* : 579 - 588, 1982. [Radiation in canopy.]

48461 - **ALEXANDRE, D.Y.** : Étude de l'éclairement du sous-bois d'une forêt dense humide sempervirente (Taï, Côte-d'Ivoire). - Acta oecol.-Oecol. gen. *3* : 407 - 447, 1982. [Radiation in canopy.]

48462 - **ALEXANDRE, D.Y.** : Pénétration de la lumière au niveau du sous-bois d'une forêt dense tropicale. - Ann. Sci. forest. *39* : 419 - 438, 1982.

48463 - **ALFANO, A.J., FONG, F.K.** : Elementary reconstitution of the water splitting light reaction in photosynthesis. 2. Optical double resonance study of $(Chl\ a \cdot 2H_2O)_n$ two-photon interactions in nonpolar solutions. - J. amer. chem. Soc. *104*: 2767 - 2773, 1982.

48464 - **ALI, S.M., NAIDU, A.P.** : Screening for drought tolerance in maize. - Indian J. Genet. *42* : 381 - 388, 1982. [Chl.]

48465 - **ALIEV, D.A., ASADOV, A.A., KURBANOV, K.B., AKHMEDOV, G.A.** : Soderzhanie i fotokhimicheskaya aktivnost' reaktsionnykh tsentrov fotosistem 1 i 2 v khloroplastakh list'ev pshenitsy s razlichnoĭ fotosinteticheskoĭ funk-tsieĭ i urozhaĭnost'yu. [Contents and photochemical activities of reaction centres of photosystems 1 and 2 in chloroplasts of wheat leaves with different photosynthetic functions and yield capacities.] - Dokl. Akad. Nauk azerb. SSR *38* (7) : 3 - 7, 1982. [In R, ab : Azerb., E.]

48466 - **ALIEV, D.A., MOSKALENKO, A.A., SULEĬMANOV, S.Yu., EROKHIN, Yu.E.** : Nekoto-rye kharakteristiki tilakoidov gran i stromy iz khloroplastov pshenitsy. [Some characteristics of granal and stroma thylakoids from wheat chloro-plasts.] - Dokl. Akad. Nauk azerb. SSR *38* (4) : 47 - 51, 1982. [In R, ab : Azerb., E.]

48467 - **ALIEV, D.A., RAGIMOV, V.I., ADYGEZALOV, V.F.** : Ustanovka dlya odnovremennoĭ registratsii fotosinteza i fotoindutsirovannogo biopotentsiala u list'ev vysshikh rasteniĭ. [Device for simultaneous recording of photosynthesis and photoinduced biopotential in leaves of higher plants.] - Dokl. Akad. Nauk azerb. SSR *38*(8) : 40 - 42, 1982. [In R, ab : Azerb., E.]

48468 - **ALINA, B.A., KLYSHEV, L.K.** : Ribosomy khloroplastov gorokha v usloviyakh khloridnogo zasoleniya. [Ribosomes of pea chloroplasts under chloride sali-nity.] - Fiziol. Biokhim. kul't. Rast. *14* : 342 - 345, 1982. [In R, ab : E.]

48469 - **ALKEMA, J., SEAGER, S.L.** : The chemical pigments of plants. - J. chem.Educ. *59* : 133 - 136, 1982. [Chl, Car.]

48470 - **ALLDREDGE, A.L., COX, J.L.** : Primary productivity and chemical composition of marine snow in surface waters of the Southern California Bight. - J. mar. Res. *40* : 517 - 527, 1982.

48471 - **ALLDREDGE, A.L., SILVER, M.W.** : Abundance and production rates of floating diatom mats (*Rhizosolenia castracanei* and *R. imbricata* var. *shrubsolei*) in the eastern Pacific Ocean. - Mar. Biol. *66* : 83 - 88, 1982.

48472 - **ALLEWELDT, G., EIBACH, R., RÜHL, E.** : Untersuchungen zum Gaswechsel der Re-be I. Einfluß von Temperatur, Blattalter und Tageszeit auf Nettophoto-synthese und Transpiration. - Vitis *21* : 93 - 100, 1982.

48473 - **ALLEWELDT, G., RÜHL, E.** : Untersuchungen zum Gaswechsel der Rebe II. Ein-fluß langanhaltender Bodentrockenheit auf die Leistungsfähigkeit verschie-dener Rebsorten. - Vitis *21* : 313 - 324, 1982.

48474 - ALMON, H., BÖHME, H. : Photophosphorylation in isolated heterocysts from the blue-green alga *Nostoc muscorum*. - Biochim. biophys. Acta *679* : 279 - - 286, 1982.

48475 - ALONI, B., ROSENSHTEIN, G. : Effect of flooding on tomato cultivars: The relationship between proline accumulation and other morphological and physiological changes. - Physiol. Plant. *56* : 513 - 517, 1982. [Stomatal resistance.]

48476 - ALONSO, A., RESTALL, C.J., TURNER, M., GOMEZ-FERNANDEZ, J.C., GOÑI, F.M., CHAPMAN, D. : Protein-lipid interactions and differential scanning calorimetric studies of bacteriorhodopsin reconstituted lipid-water systems. - Biochim. biophys. Acta *689* : 283 - 289, 1982.

48477 - ALSCHER-HERMAN, R. : Chloroplast alkaline fructose 1,6-bisphosphatase exists in a membrane-bound form. - Plant Physiol. *70* : 728 - 734, 1982.

48478 - ALSCHER-HERMAN, R. : The effect of sulphite on light activation of chloroplast fructose 1,6-bisphosphatase in two cultivars of soybean. - Environm. Pollut. Ser.A *27* : 83 - 96, 1982.

48479 - ALTMAN, A. : Retardation of radish leaf senescence by polyamines. - Physiol. Plant. *54* : 189 - 193, 1982. [Chl.]

48480 - ALTUS, D.P., CANNY, M.J. : Loading of assimilates in wheat leaves. I The specialization of vein types for separate activities. - Aust. J.Plant Physiol. *9* : 571 - 581, 1982.

48481 - AMAGASA, T. : Change in properties of phosphoenolpyruvate carboxylase in *Kalanchoe daigremontiana* with leaf age. - Plant Cell Physiol. *23* : 1471 - - 1474, 1982.

48482 - AMAGASA, T. : The influence of leaf age on the diurnal change of malate and starch in the CAM plant *Kalanchoe daigremontiana* HAMET et PERR. - Z. Pflanzenphysiol. *108* : 93 - 96, 1982.

48483 - AMBERGER, A., GUTSER, R., WUNSCH, A. : Iron chlorosis induced by high copper and manganese supply. - J. Plant Nutr. *5* : 715 - 720, 1982.

48484 - AMBLARD, C. : Interêts du dosage des adénosines 5'-phosphate pour l'étude de la dynamique des populations phytoplanctoniques lacustres (le Pavin - - France). - Hydrobiologia *85* : 257 - 270, 1981.

48485 - AMINUDDIN, bin Haji M. : Light requirements of *Dyera costulata* seedlings. - Malays. Forest. *45* : 203 - 208, 1982.

48486 - AMINUDDIN, bin M., NG, F.S.P. : Influence of light on germination of *Pinus caribaea, Gmelina arborea, Sapium baccatum* and *Vitis pinnata*. - Malays. Forest. *45* : 62 - 68, 1982. [Dry-matter accumulation.]

48487 - AMIRDZHANOV, A.G. : Ob otsenke sortov vinograda po priznaku produktivnosti. I. Kriterii produktivnosti. [Evaluation of grape vine cultivars with respect to productivity. I. Productivity criteria.] - Sadovodstvo Vinogradarstvo Vinodelie Mold. *37* (8) : 25 - 28, 1982. [Energy utilization; in R.]

48488 - ANAN'EV, G.M., ZAKRZHEVSKIĬ, D.A. : Izbiratel'noe podavlenie ingibitorami fotosinteza otdel'nykh faz kinetiki vydeleniya O_2 kletkami khlorelly pri impul'snom osveshchenii. [Selective suppression by photosynthetic inhibitors of separate phases of O_2 evolution kinetics in *Chlorella* cells upon pulsed illumination.] - Fiziol. Rast. *29* : 1114 - 1119, 1982. [In R, ab : E.]

48489 - ANDERSON, C.A., LADIGES, P.Y. : Lime-chlorosis and the effect of fire on the growth of three seedling populations of *Eucalyptus obliqua* L'HÉRIT. - Aust. J. Bot. *30* : 47 - 66, 1982. [Chl, Car.]

48490 - ANDERSON, D.M., KOTHMANN, M.M. : A two-step sampling technique for estimating standing crop of herbaceous vegetation. - J. Range Manage. *35* : 675 - - 677, 1982.

48491 - ANDERSON, J.E. : Factors controlling transpiration and photosynthesis in *Tamarix chinensis* LOUR. - Ecology *63* : 48 - 56, 1982.

48492 - **ANDERSON, J.J.** : The nitrite-oxygen interface at the top of the oxygen minimum zone in the eastern tropical North Pacific. - Deep-Sea Res. *29* : 1193 -
- 1201, 1982. [Chl.]

48493 - **ANDERSON, J.J., OKUBO, A.** : Resolution of chemical properties with a vertical profiling pump. - Deep-Sea Res. *29* : 1013 - 1019, 1982.

48494 - **ANDERSON, J.M.** : Distribution of the cytochromes of spinach chloroplasts
between the appressed membranes of grana stacks and stroma-exposed thylakoid regions. - FEBS Lett. *138* : 62 - 66, 1982.

48495 - **ANDERSON, J.M.** : The role of chlorophyll-protein complexes in the function
and structure of chloroplast thylakoids. - Mol. cell. Biochem. *46* : 161 -
- 172, 1982.

48496 - **ANDERSON, J.M.** : The significance of grana stacking in chlorophyll *b*-con-
taining chloroplasts. - Photobiochem. Photobiophys. *3* : 225 - 241, 1982.

48497 - **ANDERSON, J.M., ANDERSSON, B.** : The architecture of photosynthetic membranes: lateral and transversal organization. - Trends biochem. Sci. *7* : 288 -
- 292, 1982.

48498 - **ANDERSON, J.M., MALKIN, R.** : An EPR study of the lateral organization of
electron carriers in chloroplast thylakoids. - FEBS Lett. *148* : 293 - 296,
1982.

48499 - **ANDERSON, L.E.** : Ribulose 1,5-bisphosphate carboxylase from plants and photosynthetic bacteria. - In : **ZABORSKY, O.R.** (ed.) : CRC Handbook of Biosolar Resources. Vol.I/1. Pp. 195 - 206. CRC Press, Boca Raton 1982.

48500 - **ANDERSON, L.E., ASHTON, A.R., MOHAMED, A.H., SCHEIBE, R.** : Light/dark modulation of enzyme activity in photosynthesis. - BioScience *32* : 103 - 107,
1982.

48501 - **ANDERSSON, B., ÅKERLUND, H.-E., JERGIL, B., LARSSON, C.** : Differential phosphorylation of light-harvesting chlorophyll-protein complex in appressed
and non-appressed regions of the thylakoid membrane. - FEBS Lett. *149* :
181 - 185, 1982.

48502 - **ANDERSSON, B., ANDERSON, J.M., RYRIE, I.J.** : Transbilayer organization of
the chlorophyll-proteins of spinach thylakoids. - Europe. J. Biochem. *123* :
465 - 472, 1982.

48503 - **ANDERSSON, B., HAEHNEL, W.** : Location of photosystem I and photosystem II
reaction centers in different thylakoid regions of stacked chloroplasts. -
FEBS Lett. *146* : 13 - 17, 1982.

48504 - **ANDO, T.** : Occurrence of two different modes of photosynthesis in *Dendro-
bium* cultivars. - Sci. Hort. (Amsterdam) *17* : 169 - 175, 1982.

48505 - **ANDRÉ, M., MASSIMINO, J., DAGUENET, A., MASSIMINO, D., THIERY, J.** : The
effect of a day at low irradiance of a maize crop. II. Photosynthesis,
transpiration and respiration. - Physiol. Plant. *54* : 283 - 288, 1982.

48506 - **ANDREEV, V.P., MASLOV, Yu.I., NASONOVA, E.A.** : Perestroĭka fotosinteticheс-
kogo apparata *Anabaena variabilis* pod deĭstviem diurona. [Diuron effects
on photosynthetic apparatus of *Anabaena variabilis*.] - Vestnik leningrad.
Univ. *1982*[21 (Biol.4)] : 65 - 69, 123, 1982. [In R, ab : E.]

48507 - **ANDREEVA, A.** : Chlorophyll fluorescence and EPR signal I kinetics during
dark-light transition in whole leaves of higher plants. - Photobiochem.
Photobiophys. *4* : 17 - 23, 1982.

48508 - **ANDREEVA, T.F.** : Fotosintez i azotnyĭ obmen rasteniya. [Photosynthesis and
nitrogen metabolism of plant.] - In : Fiziologiya Fotosinteza. Pp. 89 -
- 104. Nauka, Moskva 1982. [In R.]

48509 - **ANDREEVA, T.F., MAEVSKAYA, S.N., STEPANENKO, S.Yu., STROGANOVA, L.E.,
MURASHOV, I.N.** : Aktivnost' ribulozobisfosfatkarboksilazy-oksigenazy pri
dlitel'nom vozdeĭstvii na rastenie sveta i CO_2. [Ribulose 1,5-bisphosphate
carboxylase-oxygenase activity in plants under prolonged action of light
and CO_2.] - Fiziol. Rast. *29* : 1203 - 1206, 1982. [In R, ab : E.]

48510 - ANDREO, C.S., PATRIE, W.J., McCARTY, R.E. : Effect of ATPase activation and the δ subunit of coupling factor 1 on reconstitution of photophosphorylation. - J. biol. Chem. *257* : 9968 - 9975, 1982.

48511 - ANDRIETTI, F., PEZZOTTA, R. : Diffusion along intercellular spaces: an analysis of a transient situation. - Bull. math. Biol. *44* : 879 - 891, 1982.

48512 - ANGELOV, M.N., STANEV, V.P., POPOV, G.S. : Effect of potassium deficiency on photoassymilated ^{14}C metabolism in sunflower. - Dokl. bolg. Akad. Nauk *35* : 383 - 386, 1982.

48513 - ANTIA, N.J., CHENG, J.Y. : The keto-carotenoids of two marine coccoid members of the *Eustigmatophyceae*. - Brit. phycol. J. *17* : 39 - 50, 1982.

48514 - ANTICA, M., WRISCHER, M. : Reversible light-dependent transformation of mutant plastids in variegated leaves of *Euonymus fortunei* (TURCZ.) var. *radicans* (MIQ.) REHD. - Acta bot. croat. *41* : 19 - 27, 1982.

*48515 - ANTONIELLI, M., LUPATTELLI, M., VENANZI, G. : Studio comparato di diversi parenchimi clorofilliani di alcune piante C_4. [Comparative study of different chlorophyllous parenchymas in some C_4 plants.] - Ann. Fac. agr. Univ. Studi Perugia *34* : 400 - 418, 1980. [In Ital., ab : E.]

48516 - ANTONYUK, L.P., PUSHKIN, A.V., VOROBYEVA, L.M., SOLOVJEVA, N.A., EVSTIG-NEEVA, Z.G., KRETOVICH, W.L. : Multiple molecular forms of glutamine synthetase in pea seeds. - Mol. cell. Biochem. *47* : 55 - 57, 1982. [Chl.]

48517 - ANTOSZEWSKI, R., LENZ, F. : Regulation of transport between leaf and fruit. - In : Proceedings 21st International Horticultural Congress. Vol.I. Pp. 91 - 106. International Society for Horticultural Science, Wageningen 1982. [Photosynthates.]

48518 - AOKI, M., KATOH, S. : Oxidation and reduction of plastoquinone by photosynthetic and respiratory electron transport in a cyanobacterium *Synechococcus* sp. - Biochim. biophys. Acta *682* : 307 - 314, 1982.

48519 - AOKI, S. : Changes in fraction-1 protein and Hill reaction of overwintered tea leaves during development of new shoots. - Chagyo Gijutsu Kenkyu *63* : 19 - 22, 1982.

48520 - AOKI, S. : Changes in photosynthetic capacities of the overwintered tea leaves induced by shoot plucking. - Jap. J. Crop Sci. *51* : 439 - 444, 1982.

*48521 - APEL, K. : The protochlorophyllide holochrome of barley (*Hordeum vulgare* L.). - Phytochrome-induced decrease of translatable mRNA coding for the NADPH : protochlorophyllide oxidoreductase. - Europe. J. Biochem. *120* : 89 - 93, 1981.

48522 - ARANA, J.L., VALLEJOS, R.H. : Involvement of sulfhydryl groups in the activation mechanism of the ATPase activity of chloroplast coupling factor 1. - J. biol. Chem. *257* : 1125 - 1127, 1982.

*48523 - ARCHER, S., TIESZEN, L.L. : Growth and physiological responses of tundra plants to defoliation. - Arct. alp. Res. *12* : 531 - 552, 1980. [Ps.]

48524 - ARGADE, P.V., ROTHSCHILD, K.L. : Kinetic resonance Raman spectroscopy of purple membrane using rotating sample. - In : COLOWICK, S.P., KAPLAN, N.O. (ed.) : Methods in Enzymology. Vol.88. Pp. 643 - 648. Academic Press, New York - London - Paris - San Diego - San Francisco - São Paulo - Sydney - - Tokyo - Toronto 1982.

48525 - ARGYROUDI-AKOYUNOGLOU, J.H. : Polypeptide composition of the pigment-protein complexes of *Phaseolus vulgaris* thylakoids. Cation-induced disaggregation of oligomeric to monomeric forms correlates with the increase in the ratio F685/F730 in their fluorescence spectra at -196°. - In : AKOYUNOGLOU, G., EVANGELOPOULOS, A.E., GEORGATSOS, J., PALAIOLOGOS, G., TRAKATELLIS, A., TSIGANOS, C.P. (ed.) : Cell Function and Differentiation, Part B. Biogenesis of Energy Transducing Membranes and Membrane and Protein Energetics. Pp. 277 - 289. Alan R. Liss, New York 1982.

48526 - ARGYROUDI-AKOYUNOGLOU, J.H., AKOYUNOGLOU, A., KALOSAKAS, K., AKOYUNOGLOU, G. : Reorganization of the photosystem II unit in developing thylakoids of higher plants after transfer to darkness. Changes in chlorophyll b, light-harvesting chlorophyll protein content, and grana stacking. - Plant Physiol. *70* : 1242 - 1248, 1982.

*48527 - ARIANOUTSOU-FARAGGITAKI, M., MARGARIS, N.S. : Producers and the fire cycle in a phryganic ecosystem. - In : MARGARIS, N.S., MOONEY, H.A. (ed.) : Components of Productivity of Mediterranean-Climate Regions. Basic and Applied Aspects. Pp. 181 - 190. Dr W. Junk Publ., The Hague - Boston - London 1981. [Chl.]

48528 - ARMITAGE, A.M., VINES, H.M. : Net photosynthesis, diffusive resistance, and chlorophyll content of shade- and sun-tolerant plants grown under different light regimes. - HortScience *17* : 342 - 343, 1982.

48529 - ARNOLD, C.-G. : Zur phylogenetischen Herkunft der Mitochondrien und Plastiden. - Biol. Zentralbl. *101* : 365 - 374, 1982.

48530 - ARNOLD, J.T., THOMPSON, L.F. : Chlorosis in blueberries: A soil-plant investigation. - J. Plant Nutr. *5* : 747 - 753, 1982.

48531 - ARNON, D.I. : Sunlight, Earth life. The grand design of photosynthesis. - Sciences *1982*(10) : 22 - 27, 1982.

*48532 - ARNON, D.I., TSUJIMOTO, H.Y., TANG, G.M.-S. : Contrasts between oxygenic and anoxygenic photoreduction of ferredoxin: Incompatibilities with prevailing concepts of photosynthetic electron transport. - Proc. nat. Acad. Sci. USA *77* : 2676 - 2680, 1980.

48533 - ARNON, D.I., TSUJIMOTO, H.Y., TANG, G.M.-S. : Different roles of plastoquinone in the photoreduction of ferredoxin and of membrane-bound iron-sulfur centers of chloroplasts. - Biochem. biophys. Res. Commun. *106* : 450 - 457, 1982.

48534 - ARNTZEN, C.J., PFISTER, K., STEINBACK, K.E. : The mechanism of chloroplast triazine resistance: Alterations in the herbicide site of action. - In : LeBARON, H., GRESSEL, J. (ed.) : Herbicide Resistance in Plants. Pp. 185 - - 214. Wiley, New York 1982.

48535 - ARO, E.-M. : A comparison of the chlorophyll-protein composition and chloroplast ultrastructure in two bryophytes and two higher plants. - Z. Pflanzenphysiol. *108* : 97 - 105, 1982.

48536 - ARO, E.-M. : Polypeptide patterns of the thylakoid membranes of bryophytes. Plant Sci. Lett. *24* : 335 - 345, 1982.

48537 - ARTECA, R.N. : Calcium infiltration inhibits greening in "Katahdin" potatoes. - HortScience *17* : 79 - 80, 1982. [Chl.]

48538 - ARTECA, R.N. : Effect of root applications of kinetin and gibberellic acid on transplanting shock in tomato plants. - HortScience *17* : 633 - 634, 1982. [Growth analysis.]

48539 - ARTECA, R.N., POOVAIAH, B.W. : Changes in phosphoenolpyruvate carboxylase and ribulose-1,5-bisphosphate carboxylase in *Solanum tuberosum* L. as affected by root zone applications of CO_2. - HortScience *17* : 396 - 398, 1982.

48540 - ARTECA, R.N., POOVAIAH, B.W. : Absorption of $^{14}CO_2$ by potato roots and its subsequent translocation. - J. amer. Soc. hort. Sci. *107* : 398 - 401, 1982. [Ps.]

48541 - ARUGA, Y. : Primary productivity of macroalgae in Japanese regions. - In : ZABORSKY, O.R. (ed.) : CRC Handbook of Biosolar Resources. Vol.I/2. Pp. 455 - 466. CRC Press, Boca Raton 1982.

*48542 - ASADA, K. : Biological carboxylations. - In : INOUE, S., YAMAZAKI, N. (ed.): Organic and Bio-organic Chemistry of Carbon Dioxide. Pp. 185 - 251. Kodansha Ltd., Tokyo 1981. [Ps.]

*48543 - ASADA, K., KANEMATSU, S., OKAKA, S., HAYAKAWA, T. : Phylogenic distribution of three types of superoxide dismutase in organisms and in cell organelles. - In : BANNISTER, J.V., HILL, H.A.O. (ed.) : Chemical and Biochemical Aspects of Superoxide and Superoxide Dismutase. Pp. 136 - 153. Elsevier/ /North-Holland, Amsterdam 1980. [Chloroplast.]

48544 - ASADA, K., KATOH, S. : Carriers of photosynthetic electron transport in higher plants and algae. - Recent Progr. nat. Sci. Jap. 7 : 25 - 31, 1982.

48545 - ASADA, K., TAKAHASHI, M., NAKANO, Y., SHIMADA, S. : Active oxygen in chloroplasts: suppression of singlet oxygen formation and scavenging of hydrogen peroxide. - In : NOZAKI, M., YAMAMOTO, S., ISHIMURA, Y., COON, M.J., ERNSTER, L., ESTABROOK, R.W. (ed.) : Oxygenases and Oxygen Metabolism: A Symposium in Honor of Osamo Hayaishi. Pp. 255 - 266. Academic Press, New York 1982.

48546 - ASAHI, Y., TOYAMA, S. : Some factors affecting the chloroplast replication in the moss *Plagiomnium trichomanes*. - Protoplasma *112* : 9 - 16, 1982.

*48547 - ASCENÇÃO, L., PAIS, M.S.S. : Ultrastructural aspects of secretory trichomes in *Cistus monspeliensis*. - In : MARGARIS, N.S., MOONEY, H.A. (ed.) : Components of Productivity of Mediterranean-Climate Regions. Basic and Applied Aspects. Pp. 27 - 38. Dr.W.Junk Publ., The Hague - Boston - London 1981. [Chloroplast.]

48548 - ASHLEY, D.A. : Soybean. - In : TEARE, I.D., PEET, M.M. (ed.) : Crop-Water Relations. Pp. 389 - 422. John Wiley & Sons, New York - Chichester - Brisbane - Toronto - Singapore 1982. [Ps.]

48549 - ASHTON, A.R. : A role for ribulose-1,5-bisphosphate carboxylase as a metabolite buffer. - FEBS Lett. *145* : 1 - 7, 1982.

48550 - ASMUS, R. : Field measurements on seasonal variation of the activity of primary producers on a sandy tidal flat in the northern Wadden Sea. - Neth. J. Sea Res. *16* : 389 - 402, 1982.

48551 - ATANASIU, L., SPIRESCU, I., POLESCU, L. : L'influence de la kinétine sur la croissance de l'algue *Chlorella*. - Rev. roum. Biol. Sér. Biol. vég. *27* : 17 - 22, 1982. [Chl.]

48552 - ATTIWILL, P.M., CROMER, R.N. : Photosynthesis and transpiration of *Pinus radiata* D. DON under plantation conditions in southern Australia. I. Response to irrigation with waste water. - Aust. J. Plant Physiol. *9* : 749 - - 760, 1982.

48553 - ATTIWILL, P.M., SQUIRE, R.O., NEALES, T.F. : Photosynthesis and transpiration of *Pinus radiata* D. DON under plantation conditions in southern Australia. II. First-year seedlings and 5-year-old trees on aeolian sands at Rennick (south-western Victoria). - Aust. J. Plant Physiol. *9* : 761 - - 771, 1982.

48554 - AUCLAIR, J.C., DEMERS, S., FRÉCHETTE, M., LEGENDRE, L., TRUMP, C.L. : High frequency endogenous periodicities of chlorophyll synthesis in estuarine phytoplankton. - Limnol. Oceanogr. *27* : 348 - 352, 1982.

48555 - AUSTIN, R.B. : A combined genetic and chemical approach to increasing and stabilizing wheat yields. - In : HAWKINS, A.F., JEFFCOAT, B. (ed.) : Opportunities for Manipulation of Cereal Productivity. Monograph 7. Pp. 193 - - 203. Brit. Plant Growth Regulator Group, Wantage 1982. [Ps.]

48556 - AUSTIN, R.B., MORGAN, C.L., FORD, M.A., BHAGWAT, S.G. : Flag leaf photosynthesis of *Triticum aestivum* and related diploid and tetraploid species. - Ann. Bot. *49* : 177 - 189, 1982.

48557 - AVAKYAN, A.B., VENEDIKTOV, P.S., DOBRETSOV, G.E., RUBIN, A.B. : Vzaimodeĭstvie fluorestsentnogo zonda 1-anilinonaftalin-8-sul'fonata s khloroplastami. [Interaction between fluorescent probe 1-anilino-naphthalene-8-sulphonate and chloroplasts.] - Biofizika *27* : 415 - 419, 1982. [In R, ab : E.]

48558 - **AVAKYAN, A.B., VENEDIKTOV, P.S., DOBRETSOV, G.E., RUBIN, A.B.** : Vliyanie te-
plovoǐ obrabotki khloroplastov gorokha na ikh vzaimodeǐstvie s 1-anilino-
naftalin-8-sul'fonatom. [Effect of heat treatment on pea chloroplasts
interaction with 1-anilinonaphthalene-8-sulphonate.] - Biol. Nauki *1982*
(11) : 30 - 34, 1982. [In R.]

48559 - **AVARMAA, R.A.** : Tonkostrukturnye spektry i kinetika lyuminestsentsii khlorofilla i ego proizvodnykh. [Fine-structure spectra and kinetics of luminescence of chlorophyll and its derivatives.] - Eesti NSV Tead. Akad. Toim.,
Füüs., Mat. *31* : 133 - 138, 1982. [In R, ab : E, Est.]

*B48560 - **AVRATOVSHCHUKOVA, N.** : Genetika Fotosinteza. [Genetics of Photosynthesis.]
- Kolos, Moskva 1980. [In R.]

48561 - **AXELSSON, L., DAHLIN, C., RYBERG, H.** : The function of carotenoids during
chloroplast development. V. Correlation between carotenoid content, ultrastructure and chlorophyll *b* to chlorophyll *a* ratio. - Physiol. Plant. *55* :
111 - 116, 1982.

48562 - **AXLER, R.P., GERSBERG, R.M., GOLDMAN, C.R.** : Inorganic nitrogen assimilation
in a subalpine lake. - Limnol. Oceanogr. *27* : 53 - 65, 1982. [Productivity.]

48563 - **AYAZLOO, M., GARSED, S.G., BELL, J.N.B.** : Studies on the tolerance to sulphur dioxide of grass populations in polluted areas. II. Morphological and
physiological investigations. - New Phytol. *90* : 109 - 126, 1982.

48564 - **AYRES, P.G.** : Water stress modifies the influence of powdery mildew on root
growth and assimilate import in barley. - Physiol. Plant Pathol. *21* : 283 -
- 293, 1982.

48565 - **AZOV, Y., GOLDMAN, J.C.** : Free ammonia inhibition of algal photosynthesis
on intensive cultures. - Appl. environm. Microbiol. *43* : 735 - 739, 1982.

48566 - **AZOV, Y., SHELEF, G., MORAINE, R.** : Carbon limitation of biomass production in high-rate oxidation ponds. - Biotechnol. Bioeng. *24* : 579 - 594,
1982.

*48567 - **BABU RAO, M., MULEY, E.V., MUKHOPATHYAY, S.K.** : Seasonal fluctuations of
primary production in two freshwater ponds at Wagholi, Poona, India. - Geobios *8* : 207 - 210, 1981.

48568 - **BACH, T.J., LICHTENTHALER, H.K.** : Inhibition of mevalonate biosynthesis
and of plant growth by the fungal metabolite mevinolin. - In : WINTERMANS,
J.F.G.M., KUIPER, P.J.C. (ed.) : Biochemistry and Metabolism of Plant Lipids. Pp. 515 - 521. Elsevier Biomedical Press, Amsterdam 1982. [Chl, Car.]

48569 - **BACHOFEN, R., SNOZZI, M.** : General characteristics of photosynthetic bacteria: *Rhodospirillum rubrum*. - In : ZABORSKY, O.R. (ed.) : CRC Handbook
of Biosolar Resources. Vol.I/2. Pp. 3 - 18. CRC Press, Boca Raton 1982.
[Ps.]

48570 - **BACON, C.W., LUTTRELL, E.S.** : Competition between ergots of *Claviceps pur-
purea* and rye seed for photosynthates. - Phytopathology *72* : 1332 - 1336,
1982.

48571 - **BADGER, M.R., ANDREWS, T.J.** : Photosynthesis and inorganic carbon usage
by the marine cyanobacterium, *Synechococcus* sp. - Plant Physiol. *70* : 517 -
- 523, 1982.

48572 - **BADGER, M.R., BJÖRKMAN, O., ARMOND, P.A.** : An analysis of photosynthetic
response and adaptation to temperature in higher plants: temperature acclimation in the desert evergreen *Nerium oleander* L. - Plant Cell Environm.
5 : 85 - 99, 1982.

48573 - **BADJI, M., FEYEN, J., BASSTANIE, L.** : Effet du déficit en eau du sol sur
l'évapotranspiration et la production de féveroles: une évaluation de modèles. - Agronomie *2* : 213 - 218, 1982.

*48574 - BÅGANDER, L.E. : Bacterial cycling of sulfur in a Baltic sediment: An *in situ* study in closed systems. - Geomicrobiol. J. *2* : 141 - 159, 1980. [Ps bacteria.]

48575 - BAGLEY, K., DOLLINGER, G., EISENSTEIN, L., SINGH, A.K., ZIMĀNYI, L. : Fourier transform infrared difference spectroscopy of bacteriorhodospin and its photoproducts. - Proc. nat. Acad. Sci. *79* : 4972 - 4976, 1982.

48576 - BAGYINKA, C., KOVÁCS, K.L., RAK, E. : Localization of hydrogenase in *Thiocapsa roseopersicina* photosynthetic membrane. - Biochem. J. *202* : 255 - 258, 1982.

48577 - BAHL, J.R., GUPTA, P.K. : Chlorophyll mutations in mungbean (*Vigna radiata* (L.) WILCZEK). - Theor. appl. Genet. *63* : 23 - 26, 1982.

48578 - BAILEY, C., LARSON, D.W. : Water quality and pH effects on *Umbilicaria mammulata* (ACH.) TUCK. - Bryologist *85* : 431 - 437, 1982. [Ps.]

48579 - BAILEY, W.G., STEWART, R.B. : A method for assessing leaf area. - Can. J. Plant Sci. *62* : 211 - 214, 1982.

48580 - BAKER, G.M., BHATNAGAR, D., DILLEY, R.A. : Site-specific interaction of ATP-ase-pumped protons with photosystem II in chloroplast thylakoid membranes. - J. Bioenerg. Biomembr. *14* : 249 - 264, 1982.

48581 - BAKER, N.R., FERNYHOUGH, P., MEEK, I.T. : Light-dependent inhibition of photosynthetic electron transport by zinc. - Physiol. Plant. *56* : 217 - 222, 1982.

48582 - BAKER, N.R., MARKWELL, J.P., THORNBER, J.P. : Adenine nucleotide inhibition of phosphorylation of the light-harvesting chlorophyll *a/b*-protein complex. - Photobiochem. Photobiophys. *4* : 211 - 217, 1982.

48583 - BAKER, N.R., STRASSER, R.J. : Development and spectral characteristics of excitation energy transfer between the photosystems during chloroplast biogenesis. - Photobiochem. Photobiophys. *4* : 265 - 273, 1982.

48584 - BAKER, R.J., GEBEYEHOU, G. : Comparative growth analysis of two spring wheats and one spring barley. - Crop Sci. *22* : 1225 - 1229, 1982.

48585 - BAKKER, E.P., CAPLAN, S.R. : Phospholipid substitution of the purple membrane. - In : COLOWICK, S.P., KAPLAN, N.O. (ed.) : Methods in Enzymology. Vol.88. Pp. 26 - 30. Academic Press, New York - London - Paris - San Diego - - San Francisco - São Paulo - Sydney - Tokyo - Toronto 1982.

*48586 - BAKKER-GRUNWALD, T., HESS, B. : Interactions of bacteriorhodopsin-containing membrane systems with polyelectrolytes. - J. Membrane Biol. *60* : 45 - - 49, 1981.

48587 - BALANDREAU, J., GUCKERT, A., WEINHARD, P. : Vers une modélisation de la fixation d'azote. - Acta oecol.-Oecol. gen. *3* : 91 - 110, 1982. [Ps.]

48588 - BALASUBRAMANIAN, D., MOHAN RAO, C. : Biological application of photoacoustic spectroscopy. - Curr. Sci. *51* : 111 - 117, 1982. [Ps, Chl.]

48589 - BALDY, P. : Glyoxylate aminotransférases de peroxysomes foliaires de *Fagopyrum esculentum*. - Physiol. vég. *20* : 477 - 486, 1982.

48590 - BALL, J.T., BERRY, J.A. : The C_i/C_a ratio: a basis for predicting stomatal control of photosynthesis. - Carnegie Inst. Washington Year Book *81* : 88 - - 92, 1982.

48591 - BALL, M.C., CRITCHLEY, C. : Photosynthetic responses to irradiance by the grey mangrove, *Avicennia marina*, grown under different light regimes. - Plant Physiol. *70* : 1101 - 1106, 1982.

48592 - BALLONI, W., MATERASSI, R., DE ZARLO, S., PELOSI, E., SILI, C. : Outdoor mass culture of algae in southern Italy utilizing seawater enriched with algal digested sludge. - In : GRASSI, G., PALZ, W. (ed.) : Energy from Biomass. Vol.3. Pp. 107 - 113. D. Reidel Publ. Co., Dordrecht - Boston - - London 1982.

48593 - BALOGH-NAIR, V., NAKANISHI, K. : Synthetic analogs of retinal, bacterio-
rhodopsin, and bovine rhodopsin. - In : COLOWICK, S.P., KAPLAN, N.O. (ed.):
Methods in Enzymology. Vol.88. Pp. 496 - 506. Academic Press, New York -
- London - Paris - San Diego - San Francisco - São Paulo - Sydney - Tokyo -
- Toronto 1982.

48594 - BALTSCHEFFSKY, M., BALTSCHEFFSKY, H., BOORK, J. : Evolutionary and mechani-
stic aspects on coupling and phosphorylation in photosynthetic bacteria. -
In : BARBER, J. (ed.) : Electron Transport and Photophosphorylation. Pp.
249 - 272. Elsevier Biomedical Press, Amsterdam - New York - Oxford 1982.

48595 - BAMBERGER, E.S., KING, D., ERBES, D.L., GIBBS, M. : H_2 and CO_2 evolution by
anaerobically adapted *Chlamydomonas reinhardtii* F-60. - Plant Physiol. *69* :
1268 - 1273, 1982.

48596 - BANNIKOVA, I.A., IZMAĬLOVA, N.N., MAKSIMOVICH, S.V. : Vodnyĭ balans i produktivnost' raznotravno-zlakovoĭ Karaganovoĭ stepi v Vostochnom Khangae
(MNR). [Water relations and productivity of Karagan steppe in Eastern
Changai (Mongolia) with different grass species.] - Ėkologiya *1982* (3) :
6 - 12, 1982. [In R.]

48597 - BARAKAT, R., STREKAS, T.C. : pH variation of midpoint potential for three
photosynthetic bacterial cytochromes c'. A link between physical and func-
ional properties. - Biochim. biophys. Acta *679* : 393 - 399, 1982.

48598 - BARANYAI, M., MATUS, Z., SZABOLCS, J. : Determination, by HPLC, of carotenoids in paprika products. - Acta aliment. *11* : 309 - 323, 1982.

*48599 - BARBAROSH, M.N. : Intensivnost' fotosinteza i dykhaniya list'ev yabloni
sorta Dzhonatan v zavisimosti ot podvoya i doz predplantazhnogo udobreniya.
[Photosynthetic and respiration rates of leaves of apple tree cultivar
Jonathan as affected by rootstock and pre-planting fertilization.] - In :
Voprosy Intensifikatsii Plodovodstva. Pp. 47 - 49. Kishinev. sel'skokhoz.
Institut, Kishinev 1978. [In R.]

B48600 - BARBER, J. (ed.) : Electron Transport and Photophosphorylation. - Elsevier
Biomedical Press, Amsterdam - New York - Oxford 1982.

48601 - BARBER, J. : Influence of surface charges on thylakoid structure and function. - Annu. Rev. Plant Physiol. *33* : 261 - 295, 1982.

48602 - BARBER, J., HORLER, D.N.H., CHAPMAN, D.J. : Photosynthetic pigments and
efficiency in relation to the spectral quality of absorbed light. - In :
SMITH, H. (ed.) : Plants and the Daylight Spectrum. Pp. 341 - 354. Academic
Press, London - Ney York 1982.

*48603 - BARBOTIN, J.N., THOMASSET, B. : Immobilized organelles and whole cells into
protein foam structures: Scanning and transmission electron microscope
observations. - Biochimie *62* : 359 - 365, 1980.

48604 - BARDELL, D. : Bacterial photosynthesis without chlorophyll. - Amer. Biol.
Teach. *44* : 278 - 279, 313, 1982.

48605 - BARKO, J.W., HARDIN, D.G., MATTHEWS, M.S. : Growth and morphology of submersed freshwater macrophytes in relation to light and temperature. - Can.
J. Bot. *60* : 877 - 887, 1982.

48606 - BARR, R., CRANE, F.L. : Ca^{2+} and calmodulin antagonists inhibit the proton
gradients associated with non-cyclic and cyclic photophosphorylation in
spinach chloroplasts. - Biochem. biophys. Res. Commun. *109* : 1215 - 1221,
1982.

48607 - BARR, R., CRANE, F.L. : Sulfhydryl reagents inhibit electron transport in
photosystem II of spinach chloroplasts. - Biochim. biophys. Acta *681* : 139 -
- 142, 1982.

48608 - BARR, R., TROXEL, K.S., CRANE, F.L. : Calmodulin antagonists inhibit electron transport in photosystem II of spinach chloroplasts. - Biochem. biophys. Res. Commun. *104* : 1182 - 1188, 1982.

48609 - **BARROS, R.S., MAESTRI, M., MOREIRA, R.C.** : Sources of assimilates for expanding flower buds of coffee. - Turrialba *32* : 371 - 377, 1982.

48610 - **BARROW, S.R., COCKBURN, W.** : Effects of light quantity and quality on the
decarboxylation of malic acid in Crassulacean acid metabolism photosynthesis. - Plant Physiol. *69* : 568 - 571, 1982.

48611 - **BARSKIĬ, E.L., KONDRASHIN, A.A., SAMUILOV, V.D.** : Obrazovanie raznosti
ělektricheskikh potentsialov kompleksami reaktsionnykh tsentrov *Rhodospirillum rubrum*, lishennymi tyazheloĭ sub"edinitsy. [Generation of electric potential differences by *Rhodospirillum rubrum* reaction centres complexes de-
void of the heavy subunit.] - Biokhimiya *47* : 1755 - 1758, 1982. [In R, ab
: E.]

✻48612 - **BARSKY, E.L., DANCSHAZY, Z., DRACHEV, L.A., IL'INA, M.D., JASAITIS, A.A.,
KONDRASHIN, A.A., SAMUILOV, V.D., SKULACHEV, V.P.** : Reconstitution of biological molecular generators of electric current. Bacteriochlorophyll and
plant chlorophyll complexes. - J. biol. Chem. *251* : 7066 - 7071, 1976.

48613 - **BARSKY, E.L., GUSEV, M.V., KONDRASHIN, A.A., SAMUILOV, V.D.** : Reconstitution of electrogenic function in isolated pigment-protein complexes of *Anabaena variabilis*. - Biochim. biophys. Acta *680 :* 304 - 309, 1982.

48614 - **BARTH, R.C., KLEMMEDSON, J.O.** : Amount and distribution of dry matter, nitrogen, and organic carbon in soil-plant systems of mesquite and palo verde. - J. Range Manage. *35* : 412 - 418, 1982.

48615 - **BARTLETT, S.G., GROSSMAN, A.R., CHUA, N.-H.** : *In vitro* synthesis and uptake
of cytoplasmically-synthesized chloroplast proteins. - In : EDELMAN,M.,HALLICK,
R.B., CHUA, N.-H. (ed.) : Methods in Chloroplast Molecular Biology. Pp. 1081 -
- 1091. Elsevier Biomedical Press, Amsterdam - New York - Oxford 1982.

48616 - **BARTSCH, M., KIMURA, M., SUBRAMANIAN, A.-R.** : Purification, primary structure, and homology relationships of a chloroplast ribosomal protein. - Proc.
nat. Acad. Sci. USA *79* : 6871 - 6875, 1982.

48617 - **BARUA, D.N., SARMA, P.C.** : Effect of leaf-pose and shade on the yield of
cultivated tea. - Indian J. agr. Sci. *52* : 653 - 656, 1982.

48618 - **BARUCH, Z.** : Patterns of energy content in plants from the Venezuelan pa-
ramos. - Oecologia *55* : 47 - 52, 1982.

48619 - **BAR-ZVI, D., SHAVIT, N.** : Modulation of the chloroplast ATPase by tight
binding of nucleotides. - Biochim. biophys. Acta *681* : 451 - 458, 1982.

48620 - **BASSHAM, J.A.** : Paths of carbon and their regulation. - In : SCHIFF, J.A.,
LYMAN, H. (ed.) : On the Origins of Chloroplasts. Pp. 117 - 130. Elsevier/
/North-Holland, New York - Amsterdam - Oxford 1982.

48621 - **BASSHAM, J.A.** : Photosynthetic carbon dioxide assimilation via the reducti-
ve pentose phosphate cycle (C_3 cycle). - In : ZABORSKY, O.R. (ed.) : CRC
Handbook of Biosolar Resources. Vol.I/1. Pp. 181 - 183. CRC Press, Boca
Raton 1982.

48622 - **BASSHAM, J.A., BUCHANAN, B.B.** : Carbon dioxide fixation pathways in plants
and bacteria. - In : GOVINDJEE (ed.) : Photosynthesis. Vol.II. Pp. 141 -
- 189. Academic Press, New York - London - Paris - San Diego - San Francisco - São Paulo - Sydney - Toronto 1982.

48623 - **BASSI, R., PASSERA, C.** : Effect of growth conditions on carboxylating enzymes of *Zea mays* plants. - Photosynthesis Res. *3* : 53 - 58, 1982.

48624 - **BASSMAN, J.H., DICKMANN, D.I.** : Effects of defoliation in the developing
leaf zone on young *Populus × euramericana* plants. I. Photosynthetic physiology, growth, and dry weight partitioning. - Forest Sci. *28* : 599 - 612,
1982.

48625 - **BASZYŃSKI, T.** : Niektóre czynniki wpływające na syntezę i działanie plas-
tocjaniny u roślin. [Some factors affecting synthesis and function of plastocyanin in plants.] - Zesz. nauk. Uniw. jagiell. *689* (Prace Biol. mol.9):
47 - 59, 1982. [In Pol., ab : E.]

48626 - BASZYŃSKI, T., KRÓL, M., KRUPA, Z., RUSZKOWSKA, M., WOJCIESKA, U., WOLIŃ-SKA, D. : Photosynthetic apparatus of spinach exposed to excess copper. - Z. Pflanzenphysiol. *108* : 385 - 395, 1982.

48627 - BATES, L.M., HALL, A.E. : Diurnal and seasonal responses of stomatal conductance for cowpea plants subjected to different levels of environmental drought. - Oecologia *54* : 304 - 308, 1982.

48628 - BATSCHAUER, A., SANTEL, H.-J., APEL, K. : The presence and synthesis of the NADPH-protochlorophyllide oxidoreductase in barley leaves with a high temperature-induced deficiency of plastid ribosomes. - Planta *154* : 459 - 464, 1982.

*48629 - BAUDO, R., GALANTI, G., GUILIZZONI, P., MERLINI, L., VARINI, P.G. : Relationships between heavy metals and aquatic organisms in Lake Mezzola hydrographic system (Northern Italy). 5. Net photosynthesis of the submersed macrophytes *Potamogeton crispus* L. and *Potamogeton perfoliatus* L. - Mem. Ist. ital. Idrobiol. "Dott. Marco de Marchi" (Palanza) *39* : 227 - 242, 1981.

48630 - BAYLEY, H., HÖJEBERG, B., HUANG, K.-S., KHORANA, H.G., LIAO, M.-J., LIND, C., LONDON, E. : Delipidation, renaturation, and reconstitution of bacteriorhodopsin. - In : COLOWICK, S.P., KAPLAN, N.O. (ed.) : Methods in Enzymology. Vol.88. Pp. 74 - 81. Academic Press, New York - London - Paris - - San Diego - San Francisco - São Paulo - Sydney - Tokyo - Toronto 1982.

*48631 - BAYLEY, H., RADHAKRISHNAN, R., HUANG, K.-S., KHORANA, H.G. : Light-driven proton translocation by bacteriorhodopsin reconstituted with the phenyl analog of retinal. - J. biol. Chem. *256* : 3797 - 3801, 1981.

48632 - BAZZAZ, F.A., CARLSON, R.W. : Photosynthetic acclimation to variability in the light environment of early and late successional plants. - Oecologia *54* : 313 - 316, 1982.

*48633 - BAZZAZ, M.B. : New chlorophyll chromophores isolated from a chlorophyll--deficient mutant of maize. - Photobiochem. Photobiophys. *2* : 199 - 207, 1981.

48634 - BAZZAZ, M.B., BRADLEY, C.V., BRERETON, R.G. : 4-Vinyl-4-desethyl chlorophyll *a*: characterisation of a new naturally occurring chlorophyll using fast atom bombardment, field desorption and "in bean" electron impact mass spectroscopy. - Tetrahedron Lett. *23* : 1211 - 1214, 1982.

48635 - BAZZAZ, M.B., BRERETON, R.G. : 4-Vinyl-4-desethyl chlorophyll *a*: A new naturally occurring chlorophyll. - FEBS Lett. *138* : 104 - 108, 1982.

48636 - B"CHVAROV, P., PAVLOVA, A., K"DREV, T. : Vliyanie na razlichnite nachini za poluchavane na khidroponna biomasa v"rkhu s"d"rzhanieto na belt"k, pigmenti i stepenta na ispolzuvane na zapasnite khranitelni veshchestva na semenata. [Effect of various methods of producing hydroponic biomass on protein and pigment contents and the extent of utilizing nutrients stored in the seeds.] - Fiziol. Rast. (Sofiya) *8* (3) : 74 - 79, 1982. [In Bulg., ab : E, R.]

48637 - BEADLE, C.L. : Plant-growth analysis. - In : COOMBS, J., HALL, D.O. (ed.) : Techniques in Bioproductivity and Photosynthesis. Pp. 20 - 25. Pergamon Press, Oxford - New York - Toronto - Sydney - Paris - Frankfurt 1982.

48638 - BEADLE, C.L., TALBOT, H., JARVIS, P.G. : Canopy structure and leaf area index in a mature Scots pine forest. - Forestry *55* : 105 - 123, 1982.

48639 - BEALE, S.I., FOLEY, T. : Induction of δ-aminolevulinic acid synthase activity and inhibition of heme synthesis in *Euglena gracilis* by N-methyl mesoporphyrin IX. - Plant Physiol. *69* : 1331 - 1333, 1982. [Chl.]

48640 - BEARDALL, J., GRIFFITHS, H., RAVEN, J.A. : Carbon isotope discrimination and the CO_2 accumulating mechanism in *Chlorella emersonii*. - J. exp. Bot. *33* : 729 - 737, 1982.

*48641 - BECACOS-KONTOS, T. : Investigations of primary production in Euboicos Gulf. - Rapp. Commun. int. Mer médit. *27* (7) : 73 - 74, 1981.

*48642 - BECACOS-KONTOS, T. : The effect of effluent discharge on primary production. Rapp. Commun. int. Mer médit. *27* (7) : 75 - 76, 1981.

*48643 - BECACOS-KONTOS, T. : Primary productivity and pollution in marine ecosystems in Greek waters. - In : Proceedings. International Conference on Environmental Pollution. Pp. 590 - 595. Thessaloniki 1981.

48644 - BECHER, B. : Monitoring of protein conformation changes during photocycle. - In : COLOWICK, S.P., KAPLAN, N.O. (ed.) : Methods in Enzymology. Vol.88. Pp. 265 - 267. Academic Press, New York - London - Paris - San Diego - - San Francisco - São Paulo - Sydney - Tokyo - Toronto 1982. [Bacteriorhodopsin.]

48645 - BECHTHOLD, P.S., KOHL, K.-D., SPERLING, W. : Low temperature photoacoustic spectroscopy of the purple membrane of *Halobacterium halobium*. - Appl. Opt. *21* : 127 - 132, 1982.

48646 - BECK, E., HOPF, H. : Carbohydrate metabolism. - Progr. Bot. *44* : 132 - 153, 1982.

48647 - BECKER, D.W., BRAND, J.J. : An *in vivo* requirement for calcium in photosystem II of *Anacystis nidulans*. - Biochem. biophys. Res. Commun. *109* : 1134 - 1139, 1982.

48648 - BECKER, E.W. : Ernährung, Wachstumsleistung, Ertrag und Konkurenzfähigkeit von Algenmassenkulturen. - Kali-Briefe (Büntehof) *16* : 249 - 269, 1982.

48649 - BECKER, E.W. : Physiological studies on Antarctic *Prasiola crispa* and *Nostoc commune* at low temperatures. - Polar Biol. *1* : 99 - 104, 1982. [Ps.]

B48650 - BECKER, E.W., VENKATARAMAN, L.V. : Biotechnology and Exploitation of Algae. The Indian Approach. - Deutsche Gesellschaft für Technische Zusammenarbeit (GTZ), Eschborn 1982.

48651 - BECKER, W. : Chancen und Grenzen der Mikroalgen. - Umschau Wiss. Tech. *82* : 494 - 498, 1982.

48652 - BECKWITH, A.C., JURSINIC, P.A. : An alternative mathematical approach to the analysis of photosynthetic oxygen evolution. - J. theor. Biol. *97* : 251 - 265, 1981.

*48653 - BEDBROOK, J.R., SMITH, S.M., ELLIS, R.J. : Molecular cloning and sequencing of cDNA encoding the precursor to the small subunit of chloroplast ribulose-1,5-bisphosphate carboxylase. - Nature *287* : 692 - 697, 1980.

48654 - BEDDARD, G.S., COGDELL, R.J. : Structure and excitation dynamics of light--harvesting protein complexes. - In : FONG, F.K. (ed.) : Light Reaction Path of Photosynthesis. Pp. 46 - 79. Springer-Verlag, Berlin - Heidelberg - - New York 1982.

48655 - BEDENKO, V.P., USHAROVA, G.P., INTYKBAEVA, B.B., FEDYUSHIN, A.A., SHARIPOV, K.A. : Fotosinteticheskaya deyatel'nost' u razlichnykh po produktivnosti sortov ozimoĭ pshenitsy. [Photosynthetic activity in winter wheat cultivars differing in productivity.] - Izv. Akad. Nauk kazakh. SSR, Ser. biol. *1982* (2) : 11 - 16, 1982. [In R, ab : Azerb.]

48656 - BEER, S., STEWART, A.J., WETZEL, R.G. : Measuring chlorophyll a and ^{14}C-labeled photosynthate in aquatic angiosperms by the use of a tissue solubilizer. - Plant Physiol. *69* : 54 - 57, 1982.

48657 - BEER, S., WETZEL, R.G. : Photosynthesis in submersed macrophytes of a temperate lake. - Plant Physiol. *70* : 488 - 492, 1982.

48658 - BEHRENDS, W., RAUSCH, U., LÖFFLER, H.-G., KINDL, H. : Purification of glycollate oxidase from greening cucumber cotyledons. - Planta *156* : 566 - - 571, 1982.

48659 - BEHRENS, P.W., MARSHO, T.V., RADMER, R.J. : Photosynthetic O_2 exchange kinetics in isolated soybean cells. - Plant Physiol. *70* : 179 - 185, 1982.

48660 - BELANGER, F.C., DUGGAN, J.X., REBEIZ, C.A. : Chloroplast biogenesis. Identification of chlorophyllide a (E458 F674) as a divinyl chlorophyllide a. - J. biol. Chem. *257* : 4849 - 4858, 1982.

48661 - BELANGER, F.C., REBEIZ, C.A. : Chloroplast biogenesis. Detection of monovinyl magnesium-protoporphyrin monoester and other monovinyl magnesium-por-
phyrins in higher plants. - J. biol. Chem. *257* : 1360 - 1371, 1982.

*48662 - BELBACHIR, O., MATRINGE, M., TISSUT, M., CHEVALLIER, D. : Physiological
actions of dinoterb, a phenol derivate 1. Physiological effects on the
whole plant and on tissue fragments of pea. - Pestic. Biochem. Physiol.
14 : 303 - 308, 1980. [Chl.]

48663 - BELKNAP, W.R., TOGASAKI, R.K. : The effects of cyanide and azide on the
photoreduction of 3-phosphoglycerate and oxaloacetate by wild type and
two reductive pentose phosphate cycle mutants of *Chlamydomonas reinhardtii*.
- Plant Physiol. *70* : 469 - 475, 1982.

48664 - BELL, C.J. : A model of stomatal control. - Photosynthetica *16* : 486 - 495,
1982.

48665 - BELL, C.J., INCOLL, L.D. : Translocation from the flag leaf of winter
wheat in the field. - J. exp. Bot. *33* : 896 - 909, 1982. [Photosynthates.]

48666 - BELL, J.N.B. : Sulphur dioxide and the growth of grasses. - In : UNSWORTH,
M.H., ORMROD, D.P. (ed.) : Effects of Gaseous Air Pollution in Agriculture
and Horticulture. Pp. 225 - 246. Butterworth Scientific, London - Boston -
- Sydney - Wellington - Durban - Toronto 1982. [Growth analysis.]

48667 - BELLAMY, N., RISK, M.J. : Coral gas: Oxygen production in *Millepora* on the
Great Barrier Reef. - Science *215*: 1618 - 1619, 1982.

48668 - BELLEMARE, G., BARTLETT, S.G., CHUA, N.-H. : Biosynthesis of chlorophyll
a/b-binding polypeptides in wild type and the chlorina f2 mutant of barley.
- J. biol. Chem. *257* : 7762 - 7767, 1982.

48669 - BELYAEVA, E.V., IVANOVA, M.A., DOMAN, N.G. : Izoělektricheskie tochki ri-
bulozodifosfatkarboksilazy-oksigenazy i ee sub"edinits iz list'ev kukuruzy
i bobov. [Isoelectric points of ribulose bisphosphate carboxylase/oxygenase
and its subunits from maize and bean leaves.] - Fiziol. Rast. *29* : 173 -
- 176, 1982. [In R.]

*48670 - BELYAEVA, E.V., TEREKHOVA, I.V., DOMAN, N.G. : Vliyanie substratov na ak-
tivnost' ribulozodifosfat-karboksilazy iz C_3 i C_4 rasteniĭ. [Effect of
substrates on the activity of ribulosebisphosphate carboxylase from C_3 and
C_4 plants.] - Prikl. Biokhim. Mikrobiol. *17* : 422 - 429, 1981. [In R, ab :
E.]

*B48671 - BELYANIN, V.N., SID'KO, F.Ya., TRENKENSHU, A.P. : Ėnergetika Fotosintezi-
ruyushcheĭ Kul'tury Mikrovodorosleĭ. [Energetics of the Photosynthesizing
Culture of Microalgae.] - Nauka, Sibirskoe Otdelenie, Novosibirsk 1980.
[In R.]

48672 - BEN AMOTZ, A., AVRON, M. : The potential use of *Dunaliella* for the production of glycerol, β-carotene and high-protein feed. - In : SAN PIETRO, A.
(ed.) : Biosaline Research: A Look to the Future. Pp. 207 - 214. Plenum
Publ. Corp., New York 1982.

*48673 - BEN-AMOTZ, A., GRUNWALD, T. : Osmoregulation in the halotolerant alga *Aste-
romonas gracilis*. - Plant Physiol. *67* : 613 - 616, 1981. [Ps.]

48674 - BEN-AMOTZ, A., KATZ, A., AVRON, M. : Accumulation of β-carotene in halotolerant algae: purification and characterization of β-carotene-rich globu-
les from *Dunaliella bardawil (Chlorophyceae)*. - J. Phycol. *18* : 529 - 537,
1982.

48675 - BEN-AMOTZ, A., SUSSMAN, I., AVRON, M. : Glycerol production by *Dunaliella*. -
Experientia *38* : 49 - 52, 1982. [Chl.]

48676 - BENDALL, D.S. : Chloroplast cytochromes of higher plants. - In : ZABORSKY,
O.R. (ed.) : CRC Handbook of Biosolar Resources. Vol.I/1. Pp. 153 - 160.
CRC Press, Boca Raton 1982.

48677 - BENDALL, D.S. : Photosynthetic cytochromes of oxygenic organisms. - Biochim.
biophys. Acta *683* : 119 - 151, 1982.

*48678 - BENEMANN, J.R., MIYAMOTO, K., HALLENBECK, P.C. : Bioengineering aspects of
          biophotolysis. - Enzyme Microb. Technol. 2 : 103 - 111, 1980.

 48679 - BENNETT, K.J., McPHERSON, H.G., WARRINGTON, I.J. : Effect of pretreatment
          temperature on response of photosynthesis rate in maize to current tempe-
          rature. - Aust. J. Plant Physiol. 9 : 773 - 781, 1982.

 48680 - BENNOUN, P. : A respiratory chain in the thylakoid membrane of Chlamydomo-
          nas reinhardtii. - In : AKOYUNOGLOU, G., EVANGELOPOULOS, A.E., GEORGATSOS,
          J., PALAIOLOGOS, G., TRAKATELLIS, A., TSIGANOS, C.P. (ed.) : Cell Function
          and Differentiation, Part B. Biogenesis of Energy Transducing Membranes
          and Membrane and Protein Energetics. Pp. 291 - 298. Alan R. Liss, New York
          1982.

 48681 - BENNOUN, P. : Evidence for a respiratory chain in the chloroplast. - Proc.
          nat. Acad. Sci. USA 79 : 4352 - 4356, 1982. [Chl.]

*48682 - BENZ, J., RÜDIGER, W. : Chlorophyll biosynthesis: Various chlorophyllides
          as exogenous substrates for chlorophyll synthetase. - Z. Naturforsch. 36C :
          51 - 57, 1981.

*48683 - BERENDSE, F. : Competition between plant populations with different rooting
          depths II. Pot experiments. - Oecologia 48 : 334 - 341, 1981. [Dry-matter
          accumulation.]

 48684 - BERGSTRÖM, J., VÄNNGÅRD, T. : EPR signals and orientation of cytochromes
          in the spinach chloroplast thylakoid membrane. - Biochim. biophys. Acta
          682 : 452 - 456, 1982.

 48685 - BERGTROM, G., SCHALLER, M., EICKMEIER, W.G. : Ultrastructural and biochemi-
          cal bases of resurrection in the drought-tolerant vascular plant, Selagi-
          nella lepidophylla. - J. Ultrastruct. Res. 78 : 269 - 282, 1982. [Ps.]

 48686 - BERHOW, M.A., SALUJA, A., McFADDEN, B.A. : Rapid purification of D-ribulo-
          se 1,5-bisphosphate carboxylase by vertical sedimentation in a reoriented
          gradient. - Plant Sci. Lett. 27 : 51 - 57, 1982.

 48687 - BERISH, C.W. : Root biomass and surface area in three successional tropical
          forests. - Can. J. Forest Res. 12 : 699 - 704, 1982. [Root area index.]

 48688 - BERKOWITZ, G.A., GIBBS, M. : Effect of osmotic stress on photosynthesis
          studied with the isolated spinach chloroplast : Generation and use of re-
          ducing power. - Plant Physiol. 70 : 1143 - 1148, 1982.

 48689 - BERKOWITZ, G.A., GIBBS, M. : Effect of osmotic stress on photosynthesis
          studied with the isolated chloroplast : Site-specific inhibition of the
          photosynthetic carbon reduction cycle. - Plant Physiol. 70 : 1535 - 1540,
          1982.

 48690 - BERLIN, J., QUISENBERRY, J.E., BAILEY, F., WOODWORTH, M., McMICHAEL, B.L.:
          Effect of water stress on cotton leaves I. An electron microscopic ste-
          reological study of the palisade cells. - Plant Physiol. 70 : 238 - 243,
          1982. [Chloroplast.]

 48691 - BERRIE, G.K., WEBSTER, P.M. : Ultrastructure of plastids and mitochondria
          in gemmae of Marchantia polymorpha L. - Ann. Bot. 50 : 199 - 206, 1982.

 48692 - BERRY, J.A., DOWNTON, W.J.S. : Environmental regulation of photosynthesis.-
          In : GOVINDJEE (ed.) : Photosynthesis. Vol.II. Pp. 263 - 343. Academic
          Press, New York - London - Paris - San Diego - San Francisco - São Paulo -
          - Sydney - Tokyo - Toronto 1982.

*48693 - BERRY, J.A., RAISON, J.K. : Responses of macrophytes to temperature. - In :
          LANGE, O.L., NOBEL, P.S., OSMOND, C.B., ZIEGLER, H. (ed.) : Physiological
          Plant Ecology I. Pp. 277 - 338. Springer-Verlag, Berlin - Heidelberg -
          - New York 1981. [Ps.]

 48694 - BERTRAND, P., GAYDA, J.-P. : Contribution of the exchange interactions to
          the redox properties of the [2Fe-2S] ferredoxins. - Biochim. biophys. Acta
          680 : 331 - 335, 1982.

48695 - **BERVILLE, A., CHARBONNIER, M.** : Mitochondria and chloroplasts: speculations and reflections on the molecular mechanism of heterosis. - In : SHERIDAN, W.F. (ed.) : Maize for Biological Research. Pp. 267 - 274. Plant Molecular Biology Association, Charlottesville 1982.

48696 - **BESSIERE, J., MONTIEL, A.** : Méthode rapide de dosage sélectif des chlorophylles *a* et *b* utilisation de la séparation par chromatographie liquide haute pression. A selective and rapid methode of chlorophylls *a* and *b* separative determination by high-pressure liquid chromatography. - Water Res. *16* : 987 - 993, 1982.

48697 - **BETHENOD, O., BOUSQUET, J.F., LAFFRAY, D., LOUGUET, P.** : Rééxamen des modalités d'action de l'ochracine sur la conductance stomatique des feuilles de plantules de blé, *Triticum aestivum* L., cv "Etoile de Choisy". - Agronomie *2* : 99 - 102, 1982.

48698 - **BETHENOD, O., JACOB, C., RODE, J.-C., MOROT-GAUDRY, J.-F.** : Influence de l'âge sur les caractéristiques photosynthétiques de la feuille de maïs, *Zea mays* L. - Agronomie *2* : 159 - 166, 1982.

48699 - **BETTI, J.A., BLANKENSHIP, R.E., NATARAJAN, L.V., DICKINSON, L.C., FULLER, R.C.** : Antenna organization and evidence for the function of a new antenna pigment species in the green photosynthetic bacterium *Chloroflexus aurantiacus*. - Biochim. biophys. Acta *680* : 194 - 201, 1982.

48700 - **BEURET, E.** : Le résistance chloroplastique aux triazines: Définition et situation en Suisse. - Rev. suisse Vitic. Arboric. Hortic. *14* : 297 - 302, 1982.

B48701 - **BEWLEY, J.D., BLACK, M.** : Physiology and Biochemistry of Seeds in Relation to Germination. Vol.2. Viability, Dormancy, and Environmental Control. - Springer-Verlag, Berlin - Heidelberg - New York 1982. [Chl.]

48702 - **BEWLEY, J.D., KROCHKO, J.E.** : Desiccation-tolerance. - In : LANGE, O.L., NOBEL, P.S., OSMOND, C.B., ZIEGLER, H. (ed.) : Physiological Plant Ecology II. Water Relations and Carbon Assimilation. Pp. 325 - 378. Springer-Verlag, Berlin - Heidelberg - New York 1982. [Ps.]

48703 - **BEZDENEZHNYKH, V.A.** : K voprosu o formalizatsii protsessa adaptatsii rasteniĭ. [On formalization of plant adaptation process.] - In : Vliyanie Faktorov Vneshneĭ Sredy i Fiziologicheski Aktivnykh Veshchestv na Termorezistentnost' i Produktivnost' Rasteniĭ. Pp. 94 - 103. Inst. Biol., Karel'. Filial Akad. Nauk SSSR, Petrozavodsk 1982. [Ps; in R.]

48704 - **BHAGWAT, A.S.** : Activation and inhibition of spinach ribulose 1,5-bisphosphate carboxylase by 2-phosphoglycolate. - Phytochemistry *21* : 285 - 289, 1982.

48705 - **BHAGWAT, A.S.** : Modification of essential tyrosine residues of spinach ribulose 1,5-bisphosphate carboxylase. - Plant Sci. Lett. *27* : 345 - 353, 1982.

48706 - **BHARATI, P.A.L., BAULAIGUE, R., MATHERON, R.** : Degradation of cellulose by mixed cultures of fermentative bacteria and anaerobic sulfur bacteria. - Zentralbl. Bakt. Hyg. I. Abt. Orig. *C3* : 466 - 474, 1982. [Chl.]

48707 - **BHARDWAJ, R., BURKEY, K.O., TRIPATHY, B.C., GROSS, E.L.** : Effect of trypsin treatment of photosystem I particles on the electron donation to P700. - Plant Physiol. *70* : 424 - 429, 1982.

*48708 - **BHARDWAJ, R., PAN, R.L., GROSS, E.L.** : Photopotential generation by photosystem I-mediated and proflavine-catalyzed photoreduction of methyl viologen. - Photobiochem. Photobiophys. *3* : 19 - 30, 1981.

48709 - **BHATIA, D.S., GUPTA, S., MALIK, C.P.** : Effect of some phenols on stomatal aperture and some enzymes in isolated epidermal peelings of *Gaillardia pulchella*. - Acta bot. Indica *10* : 1 -3, 1982.

*48710 - **BHATTATHIRI, P.M.A., DEVASSY, V.P.** : Effect of salinity on pigment concentrations of some tropical phytoplankters. - Indian J. Fish. *22* : 107 - 112, 1975.

48711 - BHOWMIK, P.C., DOLL, J.D. : Corn and soybean response to allelopathic effects of weed and crop residues. - Agron. J. *74* : 601 - 606, 1982.

48712 - BHUNIA, A.B., CHOUDHURY, A. : Primary production of the estuarine water around Sagar Island, Sunderbans. - Indian J. mar. Sci. *11* : 87 - 89, 1982.

48713 - BIAŁEK-BYLKA, G.E., SHKUROPATOV, A.Ya., KADOSHNIKOV, S.I., FRĄCKOWIAK, D.: Excitation energy transfer between β-carotene and chlorophyll-*a* in various systems. - Photosynthesis Res. *3* : 241 - 254, 1982.

48714 - BIELAWSKI, W., NIZIOŁEK, S., NALBORCZYK, E. : Photosynthesis in detached rye leaves at normal and low oxygen concentration. I. Incorporation of $^{14}CO_2$ into 2-oxo acids. - Acta biochim. pol. *29* : 331 - 338, 1982.

48715 - BIGGINS, J. : Thylakoid conformational changes accompanying membrane protein phosphorylation. - Biochim. biophys. Acta *679* : 479 - 482, 1982.

48716 - BILLINGS, W.D., LUKEN, J.O., MORTENSEN, D.A., PETERSON, K.M. : Arctic tundra: A source or sink for atmospheric carbon dioxide in a changing environment? - Oecologia *53* : 7 - 11, 1982.

48717 - BINDER, A. : Respiration and photosynthesis in energy-transducing membranes of cyanobacteria. - J. Bioenerg. Biomembrane *14* : 271 - 286, 1982.

48718 - BÎNDIU, C., MIHALCIUC, V., FIDANOFF, F. : Réaction physiologique des arbres au stress d'endommagement de la couronne causé par la neige. - Rev. roum. Biol. *27* : 9 - 15, 1982. [Ps.]

48719 - BINET, P., THAMMAVONG, B. : Production primaire et accumulation des bioéléments au niveau d'une population pure d'*Atriplex hastata* L. des rives de l'estuaire de la Seine (France). - Acta oecol.-Oecol.Plant. *3* : 219 - - 230, 1982. [Dry-matter accumulation.]

48720 - BINKLEY, D. : Nitrogen fixation and net primary production in a young Sitka alder stand. - Can. J. Bot. *60* : 281 - 284, 1982.

*48721 - BIRAN, I., BRAVDO, B., BUSHKIN-HARAV, I., RAWITZ, E. : Water consumption and growth rate of 11 turfgrasses as affected by mowing height, irrigation frequency, and soil moisture. - Agron. J. *73* : 85 - 90, 1981. [Ps.]

48722 - BIRD, I.F., CORNELIUS, M.J., KEYS, A.J. : Affinity of RuBP carboxylases for carbon dioxide and inhibition of the enzymes by oxygen. - J. exp. Bot. *33* : 1004 - 1013, 1982.

48723 - BIRMINGHAM, B.C., COLEMAN, J.R., COLMAN, B. : Measurement of photorespiration in algae. - Plant Physiol. *69* : 259 - 262, 1982.

48724 - BISHOP, D.G., KENRICK, J.R., CODDINGTON, J.M., JOHNS, S.R., WILLING, R.I.: The role of membrane fluidity in the maintenance of chloroplast function. - In : WINTERMANS, J.F.G.M., KUIPER, P.J.C. (ed.) : Biochemistry and Metabolism of Plant Lipids. Pp. 339 - 344. Elsevier Biomedical Press, Amsterdam 1982.

48725 - BISWAL, U.C., KASEMIR, H., MOHR, H. : Phytochrome control of degreening of attached cotyledons and primary leaves of mustard (*Sinapis alba* L.) seedlings. - Photochem. Photobiol. *35* : 237 - 241, 1982. [Chl.]

48726 - BJÖRKMAN, O., POWLES, S.B. : High light and water stress effects on photosynthesis in *Nerium oleander*. II. Inhibition of photosynthetic reactions under water stress: Interaction with light level. - Carnegie Inst. Washington Year Book *81* : 76 - 77, 1982.

48727 - BJØRNLAND, T. : Chlorophylls and carotenoids of the marine alga *Eutreptiella gymnastica*. - Phytochemistry *21* : 1715 - 1719, 1982.

48728 - BLAAUBOER, M.C.I., VAN KEULEN, R., CAPPENBERG, T.E. : Extracellular release of photosynthetic products by freshwater phytoplankton populations, with special reference to the algal species involved. - Freshwater Biol. *12* : 559 - 572, 1982.

48729 - BLACK, C.C. : Primary productivity of terrestrial plants.- In : ZABORSKY, O.R. (ed.) : CRC Handbook of Biosolar Resources. Vol.I/2. Pp. 487 - 492. CRC Press, Boca Raton 1982.

48730 - **BLACK, C.C.** : Carbon dioxide in terrestrial environments. - In : ZABORSKY,
O.R. (ed.) : CRC Handbook of Biosolar Resources. Vol.I/2. Pp. 513 - 515.
CRC Press, Boca Raton 1982. [Ps.]

48731 - **BLACK, C.C.,Jr.** : Biosolar resources: Fundamental biological processes. -
In : ZABORSKY, O.R. (ed.) : CRC Handbook of Biosolar Resources. Vol.I/1.
Pp. 3 - 9. CRC Press, Boca Raton 1982. [Energy utilization.]

48732 - **BLACK, C.C.,Jr.** : Crassulacean acid metabolism. - In : ZABORSKY, O.R. (ed.):
CRC Handbook of Biosolar Resources. Vol.I/1. Pp. 191 - 193. CRC Press,
Boca Raton 1982.

48733 - **BLACK, V.J.** : Effects of sulphur dioxide on physiological processes in
plants. - In : UNSWORTH, M.H., ORMROD, D.P. (ed.) : Effects of Gaseous Air
Pollution in Agriculture and Horticulture. Pp. 67 - 91. Butterworth Scientific, London - Boston - Sydney - Wellington - Durban - Toronto 1982. [Ps,
resistances.]

48734 - **BLACK, V.J., ORMROD, D.P., UNSWORTH, M.H.** : Effects of low concentration
of ozone, singly, and in combination with sulphur dioxide on net photosynthesis rates of *Vicia faba* L. - J. exp. Bot. *33* : 1302 - 1311, 1982.

48735 - **BLACKWELL, M.F.** : Electron paramagnetic resonance studies of triplet states
in photosystem I of chloroplasts. - Rep. Lawrence Berkeley Lab. *LBL-15012* :
1 - 196, 1982.

48736 - **BLAD, B.L.** : Atmospheric demand for water. - In : TEARE, I.D., PEET, M.M.
(ed.) : Crop-Water Relations. Pp. 1 - 44. John Wiley & Sons, New York -
- Chichester - Brisbane - Toronto - Singapore 1982. [Growth analysis.]

48737 - **BLAIR, R.M.** : Growth and nonstructural carbohydrate content of southern
browse species as influenced by light intensity. - J. Range Manage. *35* :
756 - 760, 1982. [Dry-matter accumulation.]

48738 - **BLANCHET, R., MERRIEN, A., GELFI, N., CAVALIE, G., COURTIADE, B., PUECH,J.:**
Estimation et évolution comparée de l'assimilation nette de couverts de
maïs (*Zea mays* L.), tournesol (*Helianthus annuus* L.) et soja (*Glycine max*
(L.) MERRILL), au cours de leurs cycles de développement. - Agronomie *2* :
149 - 154, 1982.

48739 - **BLANPIED, G.D.** : Studies of "Delicious" apple maturation and ripening as a
chronometric response to light-dark cycles. - J. amer. Soc. hort. Sci. *107* :
116 - 118, 1982. [Chl.]

48740 - **BLASIE, J.K., PACHENCE, J.M., TAVORMINA, A., ERECINSKA, M., DUTTON, P.L.,
STAMATOFF, J., EISENBERGER, P., BROWN, G.** : The location of redox centers
in biological membranes determined by resonance X-ray diffraction. II.
Analysis of the resonance diffraction data. - Biochim. biophys. Acta *679* :
188 - 197, 1982. [Photosynthetic reaction centre - cytochrome *c* complex.]

48741 - **BLAUROCK, A.E.** : Analysis of bacteriorhodopsin structure by X-ray diffraction. - In : COLOWICK, S.P., KAPLAN, N.O. (ed.) : Methods in Enzymology.
Vol.88. Pp. 124 - 132. Academic Press, New York - London - Paris - San Diego - San Francisco - São Paulo - Sydney - Tokyo - Toronto 1982.

*48742 - **BLEIN, J.P., DUCRUET, J.M., GAUVRIT, C.** : Effets des urées substituées sur
cellules végétales isolées, mitochondries et liposomes. - Trav. Soc. pharm.
Montpellier *41* : 158 - 160, 1981. [Ps.]

48743 - **BLICHARSKI, J.S., HARAŃCZYK, H., JAEGER, G., JEWKO, W., STRZAŁKA, K.** :
Proton spin-lattice relaxation in thylakoid membranes of chloroplasts II.
Relaxation at different concentrations of the membranes. - Acta phys. pol.
A62 : 151 - 156, 1982.

*48744 - **BLICHARSKI, J.S., HARAŃCZYK, H., STRZAŁKA, K.** : Proton spin-lattice relaxation in thylakoid membranes of chloroplasts. Proton dilution studies. - Poly-
mer Bull. *5* : 285 - 289, 1981.

48745 - **BLUMWALD, E., TEL-OR, E.** : Osmoregulation and cell composition in salt-
-adaptation of *Nostoc muscorum*. - Arch. Microbiol. *132* : 168 - 172, 1982.
[Ps.]

48746 - BÖCHER, T.W. : A developmental analysis of the photosynthesizing organs in
        Prosopis kuntzei. - Biol. Skrifter kongel. dan. Videnskab. Selskab. *23* (4):
        1 - 50, 1982.

48747 - BOCKHOLT, R. : Zur Leistungsfähigkeit von Saatgrasland auf einem Bodden-
        standort mit Brackwassereinstau. - Arch. Acker- Pflanzenbau Bodenk. *26* :
        789 - 797, 1982.

48748 - BODMER, S., BACHOFEN, R. : Analysis of nucleotides bound to the ATP synthe-
        tase from spinach chloroplasts isolated under different conditions. - Expe-
        rientia *38* : 1181 - 1184, 1982.

48749 - BOEHLER-KOHLER, B.A., LÄPPLE, G., HELLMANN, V., BÖGER, P. : Paraquat-induc-
        ed production of hydrocarbon gases. - Pestic. Sci. *13* : 323 - 329, 1982.
        [Ps.]

48750 - BOFFEY, S.A., LEECH, R.M. : Chloroplast DNA levels and the control of chlo-
        roplast division in light-grown wheat leaves. - Plant Physiol. *69* : 1387 -
        - 1391, 1982.

48751 - BOGENRIEDER, A. : Action spectra for the depression of photosynthesis by
        UV irradiation in *Lactuca sativa* L. and *Rumex alpinus* L. - In : BAUER, H.,
        CALDWELL, M.M., TEVINI, M., WORREST, R.C. (ed.) : Biological Effects of
        UV-B Radiation. Pp. 132 - 139. Gesellsch. Strahlen-Umweltforsch., München
        1982.

48752 - BOGENRIEDER, A., DOUTÉ, Y. : The effect of UV on photosynthesis and growth
        in dependence of mineral nutrition (*Lactuca sativa* L. and *Rumex alpinus*
        L.). - In : BAUER, H., CALDWELL, M.M., TEVINI, M., WORREST, R.C. (ed.) :
        Biological Effects of UV-B Radiation. Pp. 164 - 168. Gesellsch. Strahlen-
        -Umweltforsch., München 1982.

48753 - BOGENRIEDER, A., KLEIN, R. : Preliminary results regarding the spectral
        efficiency of UV on the depression of photosynthesis in higher plants. -
        In : CALKINS, J. (ed.) : The Role of Solar Ultraviolet Radiation in Marine
        Ecosystems. Pp. 617 - 620. Plenum Press, New York - London 1982.

48754 - BOGENRIEDER, A., KLEIN, R. : Possible errors in photosynthetic measurements
        arising from the use of UV-absorbing cuvettes: some examples in higher
        plants. - In : CALKINS, J. (ed.) : The Role of Solar Ultraviolet Radiation
        in Marine Ecosystems. Pp. 621 - 628. Plenum Press, New York - London 1982.

48755 - BOGENRIEDER, A., KLEIN, R. : Does solar UV influence the competitive relat-
        ionship in higher plants? - In : CALKINS, J. (ed.) : The Role of Solar
        Ultraviolet Radiation in Marine Ecosystems. Pp. 641 - 649. Plenum Press,
        New York - London 1982. [Chl.]

48756 - BÖGER, P. : Photosynthese und Nutzung der Sonnenenergie. - In : HEBER, U.
        (ed.) : Photosynthese. Ergebnisse eines Rundgesprächs in Würzburg 1981.
        Pp. 191 - 198. Deut. Forschungsgemeinschaft, Bonn 1982.

48757 - BÖGER, P. : Replacement of photosynthetic electron transport inhibitors
        by silicomolybdate. - Physiol. Plant. *54* : 221 - 224, 1982.

48758 - BÖGER, P., BÖHME, H. : Cytochromes of algae. - In : ZABORSKY, O.R. (ed.) :
        CRC Handbook of Biosolar Resources. Vol.I/1. Pp. 147 - 152. CRC Press,
        Boca Raton 1982.

48759 - BOGOMOLNI, R.A., SPUDICH, J.L. : Identification of a third rhodopsin-like
        pigment in phototactic *Halobacterium halobium*. - Proc. nat. Acad. Sci. USA
        79 : 6250 - 6254, 1982.

48760 - BOGOMOLNI, R.A., WEBER, H.J. : Assay of pigment P_{588} and its discrimination
        from bacteriorhodopsin by flash spectroscopy techniques. - In : COLOWICK,
        S.P., KAPLAN, N.O. (ed.) : Methods in Enzymology. Vol.88. Pp. 434 - 439.
        Academic Press, New York - London - Paris - San Diego - San Francisco -
        - São Paulo - Sydney - Tokyo - Toronto 1982.

48761 - BOGORAD, L. : Regulation of intracellular gene flow in the evolution of
        eukaryotic genomes. - In : SCHIFF, J.A., LYMAN, H. (ed.) : On the Origins
        of Chloroplasts. Pp. 277 - 295. Elsevier/North-Holland, New York - Amster-
        dam - Oxford 1982.

48762 - BOHLER, M.-C., BINDER, A., BACHOFEN, R. : Reconstitution of photophosphorylation with the coupling factor AF_1 of a thermophilic cyanobacterium. -
FEMS Microbiol. Lett. *15* : 115 - 118, 1982.

48763 - BÖHME, H., BÖGER, P. : Ferredoxin. - In : ZABORSKY, O.R. (ed.) : CRC Handbook of Biosolar Resources. Vol.I/1. Pp. 119 - 124. CRC Press, Boca Raton
1982.

48764 - BÖHME, H., PELZER, B. : Comparative immunological characterization of various photosynthetic cytochromes *c* from pro- and eucaryotic algae. - Arch.
Microbiol. *131* : 356 - 359, 1982.

48765 - BÖHMOVÁ, B. : Comparison of the effect of an equitoxical concentration of
NMU and NEU on spring barley (*Hordeum vulgare* L.). - Acta Fac. Rer. nat.
Univ. Comen., Genet. *13* : 1 - 10, 1982. [Chl.]

48766 - BOHNERT, H.J., CROUSE, E.J., SCHMITT, J.M. : Organization and expression
of plastid genomes. - In : PARTHIER, B., BOULTER, D. (ed.) : Nucleic Acids
and Proteins in Plants II. Pp. 475 - 530. Springer-Verlag, Berlin - Heidelberg - New York 1982. [Ps.]

48767 - BOĬCHENKO, E.A., UDEL'NOVA, T.M. : Uchastie metalloflavoproteidov v ėvo-
lyutsii avtotrofnoĭ assimilyatsii uglekisloty. [Participation of the me-
talflavoproteins in the evolution of autotrophic assimilation of carbon
dioxide.] - Izv. Akad. Nauk SSSR, Ser. biol. *1982* : 333 - 340, 1982. [In
R, ab : E.]

48768 - BOLHÀR-NORDENKAMPF, H.R. : Shoot morphology and leaf anatomy in relation
to photosynthetic efficiency. - In : COOMBS, J., HALL, D.O. (ed.) : Techniques in Bioproductivity and Photosynthesis. Pp. 58 - 65. Pergamon Press,
Oxford - New York - Toronto - Sydney - Paris - Frankfurt 1982. [Hill reaction.]

48769 - BOLOGA, A.S., FRANGOPOL, P.T. : Data on the vertical distribution of planktonic primary productivity in the offshore zone of Constanţa (The Black
Sea). - Rev. roum. Biol.,Sér. Biol. vég. *27* : 141 - 146, 1982. [Ps.]

*48770 - BOLOGA, A.S., USURELU, M., FRANGOPOL, P.T., FRANGOPOL, M. : La productivi-
té primaire planctonique superficielle dans le secteur de Constantza sur
le littoral roumain de la Mer Noire en 1978-1979. - Rapp. Commun. int. Mer
médit. *27* (7) : 77 - 78, 1981.

48771 - BONHOMME, R., RUGET, F., DERIEUX, M., VINCOURT, P. : Relations entre production de matière sèche aérienne et énergie interceptée chez différentes
génotypes de maïs. - Compt. rend. Acad. Sci. Paris, Sér. III *294* : 393 -
- 398, 1982.

48772 - BONIN, D.J., ANTIA, N.J., PELAEZ-HUDLET, J. : Influence of temperature and
light intensity on the utilization of glycine as nitrogen source for phototrophic growth of a marine unicellular cyanophyte (cyanobacterium). - Bot.
mar. *25* : 493 - 499, 1982.

48773 - BONNERJEA, J., EVANS, M.C.W. : Identification of multiple components in the
intermediary electron carrier complex of photosystem I. - FEBS Lett. *148* :
313 - 316, 1982.

48774 - BONOMI, F., KURTZ, D.M.,Jr. : Kinetics and equilibria of active site core
extrusion from spinach ferredoxin in aqueous N,N-dimethylformamide/Triton
X-100 solutions. - Biochemistry *21* : 6838 - 6843, 1982.

48775 - BOOTE, K.J. : Peanut. - In : TEARE, I.D., PEET, M.M. (ed.) : Crop-Water
Relations. Pp. 255 - 286. John Wiley & Sons, New York - Chichester - Brisbane - Toronto - Singapore 1982. [Ps.]

48776 - BOOTE, K.J., BENNETT, J.M. : Teaching techniques in crop water relations,
gas exchange, and growth analysis. - J. agron. Educ. *11* : 13 - 16, 1982.

48777 - BORISOV, A.Yu.: Photoelectric conversion of solar energy in photosynthetic
membranes. - In : AKOYUNOGLOU, G., EVANGELOPOULOS, A.E., GEORGATSOS, J.,
PALAIOLOGOS, G., TRAKATELLIS, A., TSIGANOS, C.P. (ed.) : Cell Function
and Differentiation, Part B. Biogenesis of Energy Transducing Membranes
and Membrane and Protein Energetics. Pp. 449 - 459. Alan R.Liss, New York
1982.

48778 - BORISOV, A.Yu., GADONAS, R.A., DANELYUS, R.V., PISKARSKAS, A.S., RAZZHIVIN, A.P. : Kinetika i mekhanizmy migratsii énergii pri fotosinteze. [Kinetics and energy migration in photosynthesis.] - Eesti NSV Tead. Akad. Toim., Füüs., Mat. *31* : 208 - 214, 1982. [In R, ab : E, Est.]

48779 - BORISOV, A.Yu., GADONAS, R.A., DANIELIUS, R.V., PISKARSKAS, A.S., RAZJIVIN, A.P. : Minor component B-905 of light-harvesting antenna in *Rhodospirillum rubrum* chromatophores and the mechanism of singlet-singlet annihilation as studied by difference selective picosecond spectroscopy. - FEBS Lett. *138* : 25 - 28, 1982.

48780 - BORISOVA, I.V., ANISIMOVA, K.I., BESPALOVA, Z.G., BOBROVSKAYA, N.I., KAZANTSEVA, T.I., POPOVA, T.A., SLEMNEV, N.N. : *Salsola passerina (Chenopodiaceae)* v severnoĭ Gobi (MNR). [*Salsola passerina (Chenopodiaceae)* in the northern Gobi (Mongolia).] - Bot. Zh. *67* : 1196 - 1206, 1982. [Ps; in R, ab : E.]

48781 - BÖRNER, T. : Zusammenwirken von Kerngenom und Chloroplastengenom (Plastom) beim Aufbau des Photosyntheseapparates. - In : HOFFMANN, P., HIEKE, B.(ed.): Photosynthese: Regulation und Evolution. (Colloquia Pflanzenphysiologie Nr.5.) Pp. 49 - 60. Humboldt-Universität, Berlin 1982.

*48782 - BÖRNER, T., FÖRSTER, H. : Zum Einfluß des Plastoms auf Plastidenzahl und Zellform bei der Sorte "Mrs. Parker" von *Pelargonium zonale* hort. - Wiss. Z. M.-Luther-Univ. Halle-Wittenberg, math.-naturwiss. R. *30* (3) : 77 - 81, 1981.

48783 - BOSE, S. : Chlorophyll fluorescence in green plants and energy transfer pathways in photosynthesis. - Photochem. Photobiol. *36* : 725 - 731, 1982.

48784 - BOSE, S. : Coloured derivatives of plastoquinone-A & dibromothymoquinone: Possible role in photosynthesis. - Indian J. Biochem. Biophys. *19* : 44 - - 48, 1982.

*48785 - BOSE, S., HOCH, G.E. : P_{700} sensitization by low concentration of DCMU in isolated pea chloroplasts. - Biochem. biophys. Res. Commun. *98* : 541 - 547, 1981.

48786 - BOTHE, H. : Hydrogen production by algae. - Experientia *38* : 59 - 64, 1982.

48787 - BÖTTCHER, U., BRANDT, P., MÜLLER, B., TISCHNER, R. : Physiologische Charakterisierung der Endocyanelle *Cyanocyta korschikoffiana* HALL & CLAUS. I. Photosynthetische und N-assimilatorische Eigenschaften in der symbiotischen Assoziation *Cyanophora-Cyanocyta*. - Z. Pflanzenphysiol. *106* : 167 - 172, 1982.

48788 - BOTTOMLEY, W., BOHNERT, H.J. : The biosynthesis of chloroplast proteins. - In : PARTHIER, B., BOULTER, D. (ed.) : Nucleic Acids and Proteins in Plants II. Structure, Biochemistry and Physiology of Nucleic Acids. Pp. 531 - 596. Springer-Verlag, Berlin - Heidelberg - New York 1982.

48789 - BOUGES-BOCQUET, B. : The electrogenic loop in green algae and higher plants: carriers involved, relation to the plastoquinone pool, coupling to the transfer chain. - In : TRUMPOWER, B.L. (ed.) : Function of Quinones in Energy Conserving Systems. Pp. 409 - 423. Academic Press, New York 1982.

48790 - BOURDU, R. : I.-Structure et photosynthèse: des complexes moléculaires des chloroplastes au couvert végétal. - Compt. rend. Séances Acad. Agr. Fr. *68* : 816 - 825, 1982.

48791 - BOURQUE, D.P., CAPEL, M.S. : Isolation and purification of tobacco chloroplast ribosomes. - In : EDELMAN, M., HALLICK, R.B., CHUA, N.-H.(ed.) : Methods in Chloroplast Molecular Biology. Pp. 617 - 628. Elsevier Biomedical Press, Amsterdam - New York - Oxford 1982.

48792 - BOUSSAC, A., ETIENNE, A.-L. : Heterogeneity of system II secondary electron acceptors in Tris-washed chloroplasts. - Biochim. biophys. Acta *682* : 281 - - 288, 1982.

48793 - **BOUSSAC, A., ETIENNE, A.L.** : Oxido-reduction kinetics of signal II slow in tris-washed chloroplasts. - Biochem. biophys. Res. Commun. *109* : 1200 - - 1205, 1982.

48794 - **BOUSSAC, A., ETIENNE, A.-L.** : Spectral and kinetic pH-dependence of fast and slow signal II in tris-washed chloroplasts. - FEBS Lett. *148* : 113 - - 116, 1982.

48795 - **BOUTHYETTE, P.-Y., JAGENDORF, A.T.** : Oligomycin effects on ATPase and photophosphorylation of pea chloroplast thylakoid membranes. - Plant Physiol. *69* : 888 - 896, 1982.

48796 - **BOWES, J.M., HORTON, P.** : The effect of redox potential on the kinetics of fluorescence induction in photosystem II particles from *Phormidium laminosum*. Sigmoidicity, energy transfer and the slow phase. - Biochim. biophys. Acta *680* : 127 - 133, 1982.

*48797 - **BOWES, J.M., HORTON, P., BENDALL, D.S.** : Does the acceptor Q_2 fulfil an indispensable function in the primary reactions of photosystem II? - FEBS Lett. *135* : 261 - 264, 1981.

48798 - **BOWMAN, C.M., DYER, T.A.** : Purification and analysis of DNA from wheat chloroplasts isolated in nonaqueous media. - Anal. Biochem. *122* : 108 - - 118, 1982.

48799 - **BOWMAN, M.K., NORRIS, J.R.** : Picosecond events in magnetic resonance spectroscopy of the bacteriochlorophyll special pair cation. - J. amer. chem. Soc. *104* : 1512 - 1515, 1982.

48800 - **BOWN, A.W.** : An investigation into the roles of photosynthesis and respiration in H^+ efflux from aerated suspensions of *Asparagus* mesophyll cells. - Plant Physiol. *70* : 803 - 810, 1982.

*48801 - **BOXER, S.G., BUCKS, R.R.** : Chlorophyll-amino acid interactions in synthetic models. - Isr. J. Chem. *21* : 259 - 264, 1981.

48802 - **BOXER, S.G., CHIDSEY, C.E.D., ROELOFS, M.G.** : Anisotropic magnesium interactions in the primary radical ion-pair of photosynthetic reaction centers. - Proc. nat. Acad. Sci. USA *79* : 4632 - 4636, 1982.

48803 - **BOXER, S.G., CHIDSEY, C.E.D., ROELOFS, M.G.** : Use of large magnetic fields to probe photoinduced electron-transfer reactions: an example from photosynthetic reaction centers. - J. amer. chem. Soc. *104* : 1452 - 1454, 1982.

48804 - **BOXER, S.G., CHIDSEY, C.E.D., ROELOFS, M.G.** : Dependence of the yield of a radical-pair reaction in the solid state on orientation in a magnetic field. - J. amer. chem. Soc. *104* : 2674 - 2675, 1982. [Ps.]

48805 - **BOXER, S.G., KUKI, A., WRIGHT, K.A., KATZ, B.A., XUONG, N.H.** : Oriented properties of the chlorophylls: Electronic absorption spectroscopy of orthorhombic pyrochlorophyllide *a*-apomyoglobin single crystals. - Proc. nat. Acad. Sci. USA *79* : 1121 - 1125, 1982.

48806 - **BOYER, J.S.** : Plant productivity and environment. - Science *218* : 443 - - 448, 1982. [Ps.]

48807 - **BOYLE, F.A., KEYS, A.J.** : Regulation of RuBP carboxylase activity associated with photoinhibition of wheat. - Photosynthesis Res. *3* : 105 - 111, 1982.

48808 - **BOYNTON, W.R., KEMP, W.M., KEEFE, C.W.** : A comparative analysis of nutrients and other factors influencing estuarine phytoplankton production. - In : KENNEDY, V.S. (ed.) : Estuarine Comparisons. Pp. 69 - 90. Academic Press, New York 1982.

48809 - **BOZARTH, C.S., KENNEDY, R.A., SCHEKEL, K.A.** : The effects of leaf age on photosynthesis in rose. - J. amer. Soc. hort. Sci. *107* : 707 - 712, 1982.

48810 - **BOZHENKO, V.K., VASIL'EV, I.R., NIKOLAEV, G.M., MATORIN, D.N., RUBIN, A.B.:** Sostoyanie vody i funktsionirovanie reaktsionnogo tsentra fotosistemy II v digitonin-tritonovykh subkhloroplastnykh chastitsakh. [Water state and functional activity of photosystem II reaction centre in digitonin-triton chloroplast particles.] - Fiziol. Rast. *29* : 1120 - 1125, 1982. [In R, ab : E.]

48811 - BOZHOK, G.V., KARPILOVA, I.F., GOSTIMSKIĬ,S.A. : Fotokhimicheskaya i fiziolo-
go-biokhimicheskaya kharakteristika zhelto-zelenykh mutantov gorokha s ne-
aktivnymi fotosistemami I i II. [Photochemical, physiological and biochemical characteristics of pea yellow-green mutants with inactive photosystems
I and II.] - Fiziol. Rast. *29* : 705 - 712, 1982. [In R, ab : E.]

*48812 - BRABEC, E., KOVÁŘ, P., DRÁBKOVÁ, A. : Particle deposition in three vegetat-
ion stands: a seasonal change. - Atmos. Environm. *15* : 583 - 587, 1981.
[Growth analysis.]

48813 - BRADFORD, K.J. : Regulation of shoot responses to root stress by ethylene,
abscisic acid, and cytokinin. - In : WAREING, P.F. (ed.) : Plant Growth
Substances. Pp. 599 - 608. Academic Press, London 1982. [Ps.]

48814 - BRADFORD, K.J., HSIAO, T.C. : Physiological responses to moderate water
stress. - In : LANGE, O.L., NOBEL, P.S., OSMOND, C.B., ZIEGLER, H. (ed.) :
Physiological Plant Ecology II. Water Relations and Carbon Assimilation.
Pp. 263 - 324. Springer-Verlag, Berlin - Heidelberg - New York 1982. [Ps.]

48815 - BRADFORD, K.J., HSIAO, T.C. : Stomatal behavior and water relations of
waterlogged tomato plants. - Plant Physiol. *70* : 1508 - 1513, 1982.

48816 - BRAIMAN, M., MATHIES, R. : Resonance Raman spectra of bacteriorhodopsin's
primary photoproduct: Evidence for a distorted 13-*cis* retinal chromophore.
- Proc. nat. Acad. Sci. USA *79* : 403 - 407, 1982.

48817 - BRAIMAN, M., MATHIES, R. : Spinning sample Raman spectroscopy at 77°K: bacteriorhodopsin's primary photoproduct. - In : COLOWICK, S.P., KAPLAN, N.P.
(ed.) : Methods in Enzymology. Vol.88. Pp. 648 - 659. Academic Press,
New York - London - Paris - San Diego - San Francisco - São Paulo - Sydney -
- Tokyo - Toronto 1982.

48818 - BRAINERD, K.E., FUCHIGAMI, L.H. : Stomatal functioning of *in vitro* and green
greenhouse apple leaves in darkness, mannitol, ABA, and CO_2. - J. exp. Bot.
33 : 388 - 392, 1982.

48819 - BRAND, L.E. : Persistent diel rhythms in the chlorophyll fluorescence of
marine phytoplankton species. - Mar. Biol. *69* : 253 - 262, 1982.

48820 - BRANDLEY, B.K. : Light-dependent particle movement in the outer envelope
of chloroplasts from *Codium australicum* (SILVA). - Protoplasma *110* : 15 -
- 19, 1982.

*48821 - BRANDLMEIER, T., SCHEER, H., RÜDIGER, W. : Chromophore content and molar
absorptivity of phytochrome in the P_r form. - Z. Naturforsch. *36C* : 431 -
- 439, 1981. [Bil.]

48822 - BRANDT, P., ERNST, D. : Diversification of isoelectric points from isolated
chloroplasts during the cell cycle of *Euglena gracilis,* strain z. - Electrophoresis *3* : 174 - 175, 1982.

48823 - BRANDT, P., KAISER-JARRY, K., WIESSNER, W. : Chlorophyll-protein complexes.
Variability of CPI, and the existence of two distinct forms of LHCP and one
low-molecular-weight chlorophyll *a* protein. - Biochim. biophys. Acta *679* :
404 - 409, 1982.

48824 - BRAUMANN, T., WEBER, G., GRIMME, L.H. : Carotenoid and chlorophyll composition of light-harvesting and reaction centre proteins of the thylakoid
membrane. - Photobiochem. Photobiophys. *4* : 1 - 8, 1982.

48825 - BRAUN, J.G., REAL, F., DE ARMAS, J.D. : Production studies in Canary Island
waters. - Rapp. Proc.-verb. Réun. Cons. int. Explorat. Mer *180* : 219 - 220,
1982. [Chl.]

48826 - BRAVDO, B., PALLAS, J.E.,Jr. : Photosynthesis, photorespiration and RuBP
carboxylase/oxygenase activity in selected peanut genotypes. - Photosynthetica *16* : 36 - 42, 1982.

48827 - BRAYMAN, A.A., SCHAEDLE, M. : Photosynthesis and respiration of developing
Populus tremuloides internodes. - Plant Physiol. *69* : 911 - 915, 1982.

48828 - **BRECHT, E.** : Analyse der optischen Spektren des lichtsammelnden Chlorophyll *a/b*-Protein-Komplexes aus *Vicia faba*. - In : HOFFMANN, P., HIEKE, B. (ed.): Photosynthese: Regulation und Evolution. (Colloquia Pflanzenphysiologie Nr. 5.) Pp. 148 - 153. Humboldt-Universität, Berlin 1982.

*48829 - **BRESLER, E., DASBERG, S., RUSSO, D., DAGAN, G.** : Spacial variability of crop yield as a stochastic soil process. - Soil Sci. Soc. Amer. J. *45* : 600 - 605, 1981.

48830 - **BRETON, J.** : The 695 nm fluorescence (F_{695}) of chloroplasts at low temperature is emitted from the primary acceptor of photosystem II. - FEBS Lett. *147* : 16 - 20, 1982.

48831 - **BRETON, J., VERMEGLIO, A.** : Orientation of photosynthetic pigments *in vivo*. - In : GOVINDJEE (ed.) : Photosynthesis. Vol.1. Pp. 153 - 194. Academic Press, New York - London - Paris - San Diego - San.Francisco - São Paulo - - Sydney - Tokyo - Toronto 1982.

48832 - **BRIANTAIS, J.-M., VERNOTTE, C., MAISON, B.** : Influence of stacking on the distribution of light energy in the photosynthetic apparatus. - Physiol. vég. *20* : 111 - 122, 1982.

48833 - **BRICAGE, P.** : Pigmentation and soluble peroxidase isozyme patterns of leaves of *Pedilanthus tithymaloides* L. *variegatus* as a result of daily temperature differences. - Plant Physiol. *69*: 668 - 671, 1982. [Chl.]

48834 - **BRICKER, T.M., NEWMAN, D.W.** : Changes in the chlorophyll-proteins and electron transport activities of soybean (*Glycine max* L., cv. Wayne) cotyledon chloroplasts during senescence. - Photosynthetica *16* : 239 - 244, 1982.

48835 - **BRICKER, T.M., SHERMAN, L.A.** : Triton X-114 phase-fractionation of maize thylakoid membranes in the investigation of thylakoid protein topology. - FEBS Lett. *149* : 197 - 202, 1982.

48836 - **BRIGGS, S.P., HAUG, A.R., SCHEFFER, R.P.** : Interaction of nitroxide spin labels with chloroplasts. - Plant Physiol. *70* : 668 - 670, 1982. [Ps.]

48837 - **BRIMBERG, U.I.** : Kinetics of bleaching of vegetable oils. - J. amer. Oil Chem. Soc. *59* : 74 - 78, 1982. [Chl.]

48838 - **BRINGFELT, B.** : A forest evapotranspiration model using synoptic data. - SMHI Rep. Meteorol. Climatol. *36* : 1 - 62, 1982. [Stomatal resistance.]

*48839 - **BRINSON, M.M., LUGO, A.E., BROWN, S.** : Primary productivity, decomposition and consumer activity in freshwater wetlands. - Annu. Rev. Ecol. Syst. *12* : 123 - 161, 1981.

48840 - **BRITTON, G.** : Carotenoid biosynthesis in higher plants. - Physiol. vég. *20* : 735 - 755, 1982.

B48841 - **BRITTON, G., GOODWIN, T.W.** (ed.) : Carotenoid Chemistry and Biochemistry. - Pergamon Press, Oxford - New York - Toronto - Sydney - Paris - Frankfurt 1982.

48842 - **BROCKMANN, U.H., KATTNER, G., DAHL, E.** : Plankton spring development in a South Norwegian fjord. - In : GRICE, G.D., REEVE, M.R. (ed.) : Marine Mesocosms. Biological and Chemical Research in Experimental Ecosystems. Pp. 195 - 204. Springer-Verlag, New York - Heidelberg - Berlin 1982.

48843 - **BRODA, E.** : Grundlagen der Photosynthese. (Im Zusammenhang mit der Gewinnung der Biomass.). - Österr. Chem. Z. *83* (2) : 21 - 26, 1982.

48844 - **BRODY, S.S.** : Effect of aging on the fluorescence lifetime of chloroplasts. - Z. Naturforsch. *37C* : 881 - 883, 1982.

*48845 - **BRODY, S.S., PORTER, G., TREDWELL, C.J., BARBER, J.** : Picosecond energy transfer in *Anacystis nidulans*. - Photobiochem. Photobiophys. *2* : 11 - 14, 1981.

*48846 - **BRODY, S.S., TREADWELL, C., BARBER, J.** : Picosecond energy transfer in *Porphyridium cruentum* and *Anacystis nidulans*. - Biophys. J. *34* : 439 - 449, 1981.

48847 - BROUERS, M., COLLARD, F., JEANFILS, J., JEANSON, A. : Immobilization and stabilization of green and blue green algae in crosslinked serumalbumin glutaraldehyde and in polyurethane matrices. - In : HALL, D.O., PALZ, W. (ed.) : Photochemical, Photoelectrochemical and Photobiological Processes. Vol.1. Pp. 134 - 139. D.Reidel Publ.Co., Dordrecht - Boston - London 1982.

*48848 - BROVCHENKO, M.I. : Sostav sakharov v apoplaste v svyazi s zagruzkoĭ okonchaniĭ provodyashcheĭ sistemy lista sakharnoĭ svekly. [Composition of sugars in the apoplast in relation to the load on the inlets of conductive system of sugar beet leaves.] - In : Sovremennye Problemy Fiziologii i Biokhimii Sakharnoĭ Svekly. Pp. 61 - 66. Naukova Dumka, Kiev 1981. [Photosynthate translocation; in R.]

48849 - BROWN, J.S. : Chlorophylls of higher plants and algae. - In : ZABORSKY, O.R. (ed.) : CRC Handbook of Biosolar Resources. Vol.I/1. Pp. 63 - 68. CRC Press, Boca Raton 1982.

48850 - BROWN, J.S. : Spectral analyses of antenna chlorophyll a in barley. - Carnegie Inst. Washington Year Book 81 : 34 - 37, 1982.

48851 - BROWN, J.S., ANDERSON, J.M., GRIMME, L.H. : Antenna chlorophyll a complexes in mutant and developing barley. - Photosynthesis Res. 3 : 279 - 291, 1982.

*48852 - BROWN, J.S., SCHOCH, S. : Spectral analyses of chlorophyll-protein complexes. - Carnegie Inst. Washington Year Book 80 : 15 - 16, 1981.

48853 - BROWN, J.S., SCHOCH, S. : Comparison of chlorophyll a spectra in wild-type and mutant barley chloroplasts grown under day or intermitent light. - Photosynthesis Res. 3 : 19 - 30, 1982.

48854 - BROWN, L.M. : Photosynthetic and grown responses to salinity in a marine isolate of *Nannochloris bacillaris (Chlorophyceae)*. - J. Phycol. 18 : 483 - 488, 1982.

48855 - BROWN, L.M., McLACHLAN, J. : Atypical carotenoids for the *Rhodophyceae* in in the genus *Gracilaria (Gigartinales)*. - Phycologia 21 : 9 - 16, 1982.

48856 - BROWN, P.H., OUTLAW, W.H.,Jr. : Effect of fusicoccin on dark $^{14}CO_2$ fixation by *Vicia faba* guard cell protoplasts. - Plant Physiol. 70 : 1700 - 1703, 1982.

48857 - BROWN, R.H. : Response of terrestrial plants to light quality, light intensity, temperature, CO_2, and O_2. - In : ZABORSKY, O.R. (ed.) : CRC Handbook of Biosolar Resources. Vol.I/2. Pp. 185 - 212. CRC Press, Boca Raton 1982.

48858 - BROWN, S.B., HOLROYD, J.A., VERNON, D.I., TROXLER, R.F., SMITH, K.M. : The effect of N-methylprotoporphyrin IX on the synthesis of photosynthetic pigments in *Cyanidium caldarium*. Further evidence for the role of haem in the biosynthesis of plant bilins. - Biochem. J. 208 : 487 - 491, 1982.

48859 - BRUCE, B.D., FULLER, R.C., BLANKENSHIP, R.E. : Primary photochemistry in the facultatively aerobic green photosynthetic bacterium *Chloroflexus aurantiacus*. - Proc. nat. Acad. Sci. USA 79 : 6532 - 6536, 1982.

48860 - BRUIST, M.F., HAMMES, G.G. : Mechanism for catalysis and regulation of adenosine 5'-triphosphate hydrolysis by chloroplast coupling factor 1. - Biochemistry 21 : 3370 - 3377, 1982.

48861 - BRULFERT, J., GUERRIER, D., QUEIROZ, O. : Photoperiodism and Crassulacean acid metabolism. II. Relations between leaf aging and photoperiod in Crassulacean acid metabolism induction. - Planta 154 : 332 - 338, 1982.

48862 - BRULFERT, J., MÜLLER, D., KLUGE, M., QUEIROZ, O. : Photoperiodism and Crassulacean acid metabolism. I. Immunological and kinetic evidences for different patterns of phosphoenolpyruvate carboxylase isoforms in photoperiodically inducible and non-inducible Crassulacean acid metabolism plants. - Planta 154 : 326 - 331, 1982.

48863 - BRULFERT, J., QUEIROZ, O. : Photoperiodisms and Crassulacean acid metabolism. III. Different characteristics of the photoperiod-sensitive and non-sensitive isoforms of phosphoenolpyruvate carboxylase and Crassulacean acid metabolism operation. - Planta 154 : 339 - 343, 1982.

48864 – BRULFERT, J., VIDAL, J., GADAL, P., QUEIROZ, O. : Daily rhythm of phosphoenolpyruvate carboxylase in Crassulacean acid metabolism plants. Immunological evidence for the absence of a rhythm in protein synthesis. – Planta
156 : 92 – 94, 1982.

48865 – BRUNE, D.C., GONZÁLES, I. : Measurements of photosynthetic sulfide oxidation by *Chlorobium* using a sulfide ion selective electrode. – Plant Cell
Physiol. *23* : 1323 – 1328, 1982.

*48866 – BRUUN, J.-E., GRÖNLUND, L. : A comparison of methods for estimating phyto-
plankton daily primary production and production capacity off Tvärminne,
south coast of Finland, in 1979. – Meri *9* : 107 – 115, 1981/2.

48867 – BRYANT, D.A. : Phycoerythrocyanin and phycoerythrin: properties and occurrence in cyanobacteria. – J. gen. Microbiol. *128* : 835 – 844, 1982.

*48868 – BUBICZ, M. : Zawartość składników pokarmowych w różnych odmianach kapusty
pastewnej *Brassica oleracea* L. var. *acephala* DC. [Contents of nutrients in
different cultivars of *Brassica oleracea* L. var. *acephala* DC.] – Ann. Univ.
Mariae Curie-Skłodowska, Sect.E *31* : 265 – 279, 1976. [Car; in Pol., ab :
E, R.]

48869 – BUCHANAN, B.B. : Carbon dioxide assimilation in photosynthetic bacteria. –
In : ZABORSKY, O.R. (ed.) : CRC Handbook of Biosolar Resources. Vol.I/1.
Pp. 175 – 180. CRC Press, Boca Raton 1982.

*48870 – BUCHANAN, B.B., HUTCHESON, S.W., MAGYAROSY, A.C., MONTALBINI, P. : Photo-
synthesis in healthy and diseased plants. – In : AYRES, P.G. (ed.) :
Effects of Disease on the Physiology of the Growing Plant. Pp. 13 – 28.
Cambridge University Press, Cambridge 1981.

48871 – BUCHHOLZ, K., GOOD, R.E. : Density, age structure, biomass and net annual
aboveground productivity of dwarfed *Pinus rigida* MILL. from the New Jersey
Pine Barren Plains. – Bull. Torrey bot. Club *109* : 24 – 34, 1982.

48872 – BUCKS, R.R., BOXER, S.G. : Synthesis and spectroscopic properties of a novel cofacial chlorophyll-based dimer. – J. amer. chem. Soc. *104* : 340 – 343,
1982.

48873 – BUCKS, R.R., NETZEL, T.L., FUJITA, I., BOXER, S.G. : Picosecond spectroscopic study of chlorophyll-based models for the primary photochemistry
of photosynthesis. – J. phys. Chem. *86* : 1947 – 1955, 1982.

48874 – BUDZIKIEWICZ, H. : Mass spectra of carotenoids – labelling studies. – In :
BRITTON, G., GOODWIN, T.W. (ed.) : Carotenoid Chemistry and Biochemistry.
Pp. 155 – 165. Pergamon Press, Oxford – New York – Toronto – Sydney – Pa-
ris – Frankfurt 1982.

48875 – BUENO, A., ATKINS, R.E. : Growth analysis of grain sorghum hybrids. – Iowa
State J. Res. *56* : 367 – 381, 1982.

48876 – BUETOW, D.E. : Isolation of chloroplast polyribosomes from *Euglena graci-
lis*. – In : EDELMAN, M., HALLICK, R.B., CHUA, N.-H. (ed.) : Methods in
Chloroplast Molecular Biology. Pp. 637 – 644. Elsevier Biomedical Press,
Amsterdam – New York – Oxford 1982.

48877 – BUETOW, D.E. : Molecular biology of chloroplasts. – In : GOVINDJEE (ed.) :
Photosynthesis. Vol.II. Pp. 43 – 88. Academic Press, New York – London –
– Paris – San Diego – San Francisco – São Paulo – Sydney – Tokyo – Toronto
1982.

48878 – BUETOW, D.E., GILBERT, C.W. : Polypeptide composition of thylakoid membranes: Two-dimensional gel analysis during development of *Euglena* chloroplasts. – In : AKOYUNOGLOU, G., EVANGELOPOULOS, A.E., GEORGATSOS, J.,
PALAIOLOGOS, G., TRAKATELLIS, A., TSIGANOS, C.P. (ed.) : Cell Function
and Differentiation, Part B. Biogenesis of Energy Transducing Membranes
and Membrane and Protein Energetics. Pp. 139 – 148. Alan R. Liss, New
York 1982.

48879 - BUGROVSKIĬ, V.V., DUDIN, E.B., MELLINA, E.G., TSEL'NIKER, Yu.L. : Modelirovanie produktsionnykh protsessov v chistykh drevostoyakh. [Modelling of
production process in pure forest stands.] - Zh. obshch. Biol. *43* : 480 -
- 488, 1982. [In R, ab : E.]

48880 - BUJTÁS, K., SZIGETI, Z. : Egy újabb herbicidcsalád: a 3,5-dihalogén-4-hi-
droxi-benzonitrilek élettani hatásai és hatásmechanizmusa. [Physiological
effects and mechanism of action of 3,5-dihalogeno-4-hydroxy-benzonitriles.]
- Növénytermelés *31* : 465 - 471, 1982. [Ps, Chl; in Hung., ab : E.]

48881 - BULTS, G., HORWITZ, B.A., MALKIN, S., CAHEN, D. : Photoacoustic measurements
of photosynthetic activities in whole leaves. Photochemistry and gas exchange. - Biochim. biophys. Acta *679* : 452 - 465, 1982.

48882 - BULTS, G., NORDAL, P.-E., KANSTAD, S.O. : *In vivo* studies of gross photosynthesis in attached leaves by means of photothermal radiometry. - Biochim.
biophys. Acta *682* : 234 - 237, 1982.

48883 - BUNCE, J.A. : Low humidity effects on photosynthesis in single leaves of
C_4 plants. - Oecologia *54* : 233 - 235, 1982.

48884 - BUNCE, J.A. : Effect of water stress on photosynthesis in relation to diurnal accumulation of carbohydrates in source leaves. - Can. J. Bot. *60* :
195 - 200, 1982.

48885 - BUNCE, J.A. : Photosynthesis at ambient and elevated humidity over a growing season in soybean. - Photosynthesis Res. *3* : 307 - 311, 1982.

48886 - BUNT, J.S. : Primary productivity of marine benthic algae and macrophytes
in Australia. - In : ZABORSKY, O.R. (ed.) : CRC Handbook of Biosolar Resources. Vol.I/2. Pp. 467 - 470. CRC Press, Boca Raton 1982.

*48887 - BURCEA, M., HURDUC, N. : Influenţa unor caractere morfofiziologice asupra
procesului de formare a recoltei de boabe la grîul de toamna. [The influence of some morpho-physiological characters upon grain yield formation
in winter wheat.] - Probl. Genet. teor. apl. *13* : 121 - 143, 1981. [Growth
analysis; in Roum., ab : E.]

*48888 - BURCZYK, J., HESSE, M. : The ultrastructure of the outer cell wall-layer
of *Chlorella* mutants with and without sporopollenin. - Plant Syst. Evol.
138 : 121 - 137, 1981. [Car.]

48889 - BURGHOFFER, C., COSTES, C. : Localisation des cytochromes chloroplastiques
sur gel de polyacrylamide après électrophorèse en présence de chloral. -
Compt. rend. Acad. Sci. Paris, Sér. III *295* : 275 - 278, 1982.

48890 - BURKARD, G., CANADAY, J., CROUSE, E.J., GLOECKLER, R., GORDON, K., GUILLE-
MAUT, P., KEITH, G., KUNTZ, M., MUMUMBILA, M., OSORIO, M.L., STEINMETZ, A.,
WEIL, J.H. : Protein synthesis in chloroplasts: Sequence studies and gene
mapping studies on chloroplast transfer RNAs. - Ciênc. biol. (Portugal)
7 : 27 - 37, 1982.

48891 - BURKARD, G., STEINMETZ, A., KELLER, M., MUMUMBILA, M., CROUSE, E., WEIL,
J.-H. : Resolution of chloroplast tRNAs by two-dimensional gel electropho-
resis. - In : EDELMAN, M., HALLICK, R.B., CHUA, N.H. (ed.) : Methods in
Chloroplast Molecular Biology. Pp. 347 - 357. Elsevier Biomedical Press,
Amsterdam - New York - Oxford 1982.

48892 - BURKE, J.J., WILSON, R.F., SWAFFORD, J.R. : Characterization of chloroplast
isolated from triazine-susceptible and triazine resistant biotypes of *Brassica campestris* L. - Plant Physiol. *70* : 24 - 29, 1982.

48893 - BURKERT, D. : Vergleichende Untersuchungen zum Wachstumsverlauf an Sorten
von *Festuca arundinacea* SCHREB. und *Lolium perenne* L. während der frühen
vegetativen Entwicklung. - Arch. Acker- Pflanzenb. Bodenk. *26* : 673 - 677,
1982. [Growth analysis.]

48894 - BURKEY, K.O., GROSS, E.L. : Chemical modification of spinach plastocyanin:
Separation and characterization of four different forms. - Biochemistry
21 : 5886 - 5890, 1982.

B48895 - BURTON, W.G. : Post-Harvest Physiology of Food Crops. - Longman, London -
- New York 1982. [Ps.]

48896 - BUSCHMANN, C., GRUMBACH, K.H. : Herbicides which inhibit electron transport or produce chlorosis and their effect on chloroplast development in
radish seedlings II. Pigment excitation, chlorophyll fluorescence and pigment-protein complexes. - Z. Naturforsch. $37C$: 632 - 641, 1982.

48897 - BUSCHMANN, C., LICHTENTHALER, H.K. : The effect of cytokinins on growth
and pigment accumulation of radish seedlings (*Raphanus sativus* L.) grown
in the dark and at different light quanta fluence rates. - Photochem. Photobiol. *35* : 217 - 221, 1982.

48898 - BUSER-SUTER, C., WIEMKEN, A., MATILE, P. : A malic acid permease in isolated vacuoles of a Crassulacean acid metabolism plant. - Plant Physiol. *69* :
456 - 459, 1982.

48899 - BUSHWAY, R.J., WILSON, A.M. : Determination of α- and β-carotene in fruit
and vegetables by high performance liquid chromatography. - Can. Inst.
Food Sci. Technol. J. *15* : 165 - 169, 1982.

*48900 - BUTLER, W.F., JOHNSTON, D.C., SHORE, H.B., FREDKIN, D.R., OKAMURA, M.Y.,
FEHER, G. : The electronic structure of Fe^{2+} in reaction centers from *Rho-
dopseudomonas sphaeroides*. I. Static magnetization measurements. - Biophys.
J. *32* : 967 - 992, 1980.

48901 - BYKOV, O.D., ZELENSKIĬ, M.I. : Fotosintez i produktivnost' sel'skokhozyaĭstvennykh kul'tur. [Photosynthesis and productivity of agricultural crops.]
- Sel'skokhoz. Biol. *17* : 14 - 27, 1982. [In R, ab : E.]

48902 - BYKOV, O.D., ZELENSKIĬ, M.I. : O vozmozhnosti selektsionnogo uluchsheniya
fotosinteticheskikh priznakov sel'skokhozyaĭstvennykh rasteniĭ. [Possibili-
ty of breeding improvement of photosynthetic parameters of agricultural
plants.] - In : Fiziologiya Fotosinteza. Pp. 294 - 310. Nauka, Moskva 1982.
[In R.]

48903 - BYSTROVA, M.I., SAFRONOVA, I.A., KRASNOVSKIĬ, A.A. : Izuchenie molekulyarnoĭ
organizatsii agregirovannykh form protokhlorofilla v tverdykh plenkakh.
[Molecular arrangement of chlorophyll aggregated forms in solid films.] -
Mol. Biol. (Moskva) *16* : 291 - 301, 1982. [In R, ab : E.]

48904 - BYSTRYKH, E.E., NIKOLAEVA, E.K. : Vliyanie kolosa na fotosinteticheskuyu
aktivnost' verkhushechnogo lista pshenitsy. [Ear influence on photosynthetic activity of wheat flag-leaf.] - Sel'skokhoz. Biol. *17* : 488 - 494,
1982. [In R, ab : E.]

48905 - CAFISO, D.S., HUBBELL, W.L., QUINTANILHA, A. : Spin-label probes of light-
-induced electrical potentials in rhodopsin and bacteriorhodopsin. - In :
COLOWICK, S.P., KAPLAN, N.O. (ed.) : Methods in Enzymology. Vol.88. Pp.
682 - 696. Academic Press, New York - London - Paris - San Diego - San
Francisco - São Paulo - Sydney - Tokyo - Toronto 1982.

48906 - CAHEN, D., BULTS, G., CAPLAN, S.R., GARTY, H., MALKIN, S. : Photoacoustic
methods applied to biological systems. - In : HÉLÈNE, C., CHARLIER, M.,
MONTENAY-GARESTIER, T., LAUSTRIAT, G. (ed.) : Trends in Photobiology. Pp.
21 - 32. Plenum Press, New York - London 1982. [Chloroplast, bacteriorhodopsin.]

*48907 - CALDWELL, M.M. : Plant response to solar ultraviolet radiation. - In :
LANGE, O.L., NOBEL, P.S., OSMOND, C.B., ZIEGLER, H. (ed.) : Physiological
Plant Ecology I. Pp. 169 - 197. Springer-Verlag, Berlin - Heidelberg - New
York 1981. [Ps.]

48908 - CALLAGHAN, T.V., SCOTT, R., LAWSON, G.J. : Biofuel production from natural
vegetation in Great Britain. - In : GRASSI, G., PALZ, W. (ed.): Energy
from Biomass. Vol.3. Pp. 30 - 36. D.Reidel Publ.Co., Dordrecht - Boston -
London 1982. [Dry-matter accumulation.]

48909 - CALVIN, M. : Bioconversion of solar energy. - In : HÉLÈNE, C., CHARLIER, M., MONTENAY-GARESTIER, T., LAUSTRIAT, G.(ed.) : Trends in Photobiology. Pp. 645 - 659. Plenum Press, New York - London 1982.

48910 - CALVIN, M., NEMETHY, E.K., REDENBAUGH, K., OTVOS, J.W. : Plants as a direct source of fuel. - Experientia *38* : 18 - 22, 1982.

48911 - CAMACHO RUBIO, F., MARTINEZ SANCHO, E. : Desarrollo de un modelo para explicar la influencia de la intensidad de iluminacion en el crecimiento de *Chlorella pyrenoidosa*. [Model to explain the effect of irradiance on growth of *Chlorella pyrenoidosa*.] - An. Quím. Ser.A *78* : 376 - 380, 1982. [Chl;in Span., ab : E.]

48912 - CAMACHO RUBIO, F., PADIAL VICO, A., MARTINEZ SANCHO, E. : Intensidad media de iluminacion en cultivos de *Chlorella pyrenoidosa*. [Mean irradiance in cultures of *Chlorella pyrenoidosa*.] - An. Quím. Ser. A *78* : 371 - 375, 1982. [In Span., ab : E.]

48913 - CAMARA, B., BARDAT, F., MONÉGER, R. : Sites of biosynthesis of carotenoids in *Capsicum* chromoplasts. - Europe. J. Biochem. *127* : 255 - 258, 1982.

48914 - CAMARA, B., BARDAT, F., SEYE, A., D'HARLINGUE, A., MONÉGER, R. : Terpenoid metabolism in plastids. Localization of α-tocopherol synthesis in *Capsicum* chromoplasts. - Plant Physiol. *70* : 1562 - 1563, 1982. [Chloroplast.]

48915 - CAMERON, A.C., YANG, S.F. : A simple method for the determination of resistance to gas diffusion in plant organs. - Plant Physiol. *70* : 21 - 23, 1982. [Ethane efflux method.]

48916 - CAMM, E.L., GREEN, B.R. : The effects of cations and trypsin on extraction of chlorophyll-protein complexes by octyl glucoside. - Arch. Biochem. Biophys. *214* : 563 - 572, 1982.

48917 - CAMMAERTS, D., JACOBS, M. : Support for the evolution of fraction 1 protein in the genus *Lycopersicon* by gene mutation. - Z. Pflanzenphysiol. *106* : 251 - 256, 1982.

48918 - CAMP, P.J., HUBER, S.C., BURKE, J.J., MORELAND, D.E. : Biochemical changes that occur during senescence of wheat leaves. I.Basis for the reduction of photosynthesis. - Plant Physiol. *70* : 1641 - 1646, 1982.

48919 - CAMPBELL, W.H., BLACK, C.C. : Cellular aspects of C_4 leaf metabolism. - In : CREASY, L.L., HRAZDINA, G. (ed.) : Cellular and Subcellular Localization in Plant Metabolism. Pp. 223 - 248. Plenum Press, New York - London 1982.

48920 - CANAANI, O., CAHEN, D., MALKIN, S. : Photosynthetic chromatic transitions and Emerson enhancement effects in intact leaves studied by photoacoustic. - FEBS Lett. *150* : 142 - 146, 1982.

48921 - CANAANI, O., CAHEN, D., MALKIN, S. : Use of photoacoustic methods in probing development of the photosynthetic apparatus in greening leaves. - In : AKOYUNOGLOU, G., EVANGELOPOULOS, A.E., GEORGATSOS, J., PALAIOLOGOS, G., TRAKATELLIS, A., TSIGANOS, C.P. (ed.) : Cell Function and Differentiation, Part B. Biogenesis of Energy Transducing Membranes and Membrane and Protein Energetics. Pp. 299 - 308. Alan R.Liss, New York 1982.

48922 - CANNISTRARO, S., JORI, G., VAN DE VORST, A. : Quantum yield of electron transfer and of singlet oxygen production by porphyrins: an ESR study. - Photobiochem. Photobiophys. *3* : 353 - 363, 1982.

48923 - CANTOR, M.H., RAHAT, M. : Regulation of respiration and photosynthesis in *Hydra viridis* and in its separate cosymbionts: effect of nutrients. - Physiol. Zool. *55* : 281 - 288, 1982.

48924 - CAPEL, M., BOURQUE, D.P. : Fractionation and comparative analysis of chloroplast ribosomal proteins by two-dimensional gel electrophoresis. - In : EDELMAN, M., HALLICK, R.B., CHUA, N.H. (ed.) : Methods in Chloroplast Molecular Biology. Pp. 1029 - 1043. Elsevier Biomedical Press, Amsterdam - New York - Oxford 1982.

48925 - CAPEL, M.S., BOURQUE, D.P. : Characterization of *Nicotiana tabacum* chloroplast and cytoplasmatic ribosomal proteins. - J. biol. Chem. *257* : 7746 - 7755, 1982.

48926 - CAPORN, S.J.M., LUDWIG, L.J., FLOWERS, T.J. : Potassium deficiency and photosynthesis in tomato. - In : SCAIFE, A. (ed.) : Plant Nutrition. Proc. Ninth Int. Plant Nutr. Colloq. Pp. 78 - 83. Commonwealth Agricultural Bureaux, Farnham Royal 1982.

*48927 - CARBON, B.A., BARTLE, G.A., MURRAY, A.M. : Leaf area index of some eucalypt forests in south-west Australia. - Aust. Forest Res. *9* : 323 - 326, 1979.

48928 - CARDE, J.-P., JOYARD, J., DOUCE, R. : Electron microscopic studies of envelope membranes from spinach plastids. - Biol. Cell *44* : 315 - 324, 1982.

*48929 - CARIAS, J.R., MOURICOUT, M., JULIEN, R. : Chloroplastic methionyl-tRNA synthetase from wheat. - Biochem. biophys. Res. Commun. *98* : 735 - 742, 1981.

48930 - CARLSON, R.W., BAZZAZ, F.A. : Photosynthetic and growth response to fumigation with SO_2 at elevated CO_2 for C_3 and C_4 plants. - Oecologia *54* : 50 - - 54, 1982.

48931 - CARMELI, C., GUTMAN, M. : Rapid light-induced surface charge changes in bacteriorhodopsin. - FEBS Lett. *141* : 88 - 92, 1982.

48932 - CARPENTER, J.H. : Oxygen in aquatic environments. - In : ZABORSKY, O.R. (ed.) : CRC Handbook of Biosolar Resources. Vol.I/2. Pp. 531 - 536. CRC Press, Boca Raton 1982. [Ps.]

*48933 - CARRILLO, N., ARANA, J.L., VALLEJOS, R.H. : An essential carboxyl group at the nucleotide binding site of ferredoxin-NADP+ oxidoreductase. - J. biol. Chem. *256* : 6823 - 6828, 1981.

*48934 - CARRILLO, N., LUCERO, H.A., VALLEJOS, R.H. : Light modulation of chloroplast membrane-bound ferredoxin-NADP+ oxidoreductase. - J. biol. Chem. *256* : 1058 - 1059, 1981.

48935 - CARRILLO, N., VALLEJOS, R.H. : Interaction of ferredoxin-NADP oxidoreductase with the thylakoid membrane. - Plant Physiol. *69* : 210 - 213, 1982.

*48936 - CARTER, J.N., TRAVELLER, D.J. : Effect of time and amount of nitrogen uptake on sugarbeet growth and yield. - Agron. J. *73* : 665 - 671, 1981. [Photosynthates.]

48937 - CASADEVALL, E. : Renewable hydrocarbon production by cultivation of the green alga *Botryococcus braunii*. Investigation of the factors affecting hydrocarbon production. - In : GARSSI, G., PALZ, W. (ed.) : Energy from Biomass. Vol.3. Pp. 142 - 149. D. Reidel Publ. Co., Dordrecht - Boston - - London 1982.

*48938 - CASADIO, R., VENTUROLI, G., MELANDRI, B.A. : Light-induced transmembrane potential difference in chromatophores of photosynthetic bacteria: measurements with an ion-selective minielectrode. - Photobiochem. Photobiophys. *2* : 245 - 253, 1981.

*48939 - CASADORO, G., RASCIO, N. : Morphogenesis of membrane-bound bodies in belladonna (*Atropa belladonna* L.) plastids. - J. Ultrastructure Res. *61* : 186 - - 192, 1977. [Chloroplast.]

48940 - CASTEL, J.R., FERERES, E. : Responses of young almond trees to two drought periods in the field. - J. hort. Sci. *57* : 175 - 187, 1982. [Ps.]

48941 - CASTELFRANCO, P.A., CHERESKIN, B.M. : Biosynthesis of chlorophyll *a*. - In : SCHIFF, J.A., LYMAN, H. (ed.) : On the Origins of Chloroplasts. Pp. 199 - - 218. Elsevier/North-Holland, New York - Amsterdam - Oxford 1982.

48942 - CASTILLO, F., CÁRDENAS, J. : Nitrate reduction by photosynthetic purple bacteria. - Photosynthesis Res. *3* : 3 - 18, 1982.

48943 - CASTORINIS, A., AKOYUNOGLOU, G., ARGYROUDI-AKOYUNOGLOU, J.H. : Correlation between the organization of the pigment-protein complexes and the Chl *a* fluorescence yield of chloroplasts, during development in *Phaseolus vulgaris*. - Photobiochem. Photobiophys. *4* : 283 - 291, 1982.

48944 - CATHEY, H.M., CAMPBELL, L.E. : Plant response to light quality and quantity. - In : CHRISTIANSEN, M.N., LEWIS, C.F. (ed.) : Breeding Plants for Less Favorable Environments. Pp. 213 - 257. John Wiley & Sons, New York - Chichester - Brisbane - Toronto - Singapore 1982.

48945 - ČATSKÝ, J., TICHÁ, I. : Photosynthetic characteristics during ontogenesis of leaves. 6. Intracellular conductance and its components. - Photosynthetica *16* : 253 - 284, 1982.

*B48946 - CAUSTON, D.R., VENUS, J.C. : The Biometry of Plant Growth. - Edward Arnold, London 1981. [Growth analysis.]

48947 - CAWTHON, D.L., MORRIS, J.R. : Relationship of seed number and maturity to berry development, fruit maturation, hormonal changes, and uneven ripening of "Concord" (*Vitis labrusca* L.) grapes. - J. amer. Soc. hort. Sci. *107* : 1097 - 1104, 1982. [Photosynthates.]

*48948 - CENTER, T.D., SPENCER, N.R. : The phenology and growth of water hyacinth (*Eichhornia crassipes* (MART.) SOLMS) in a eutrophic north-central Florida lake. - Aquat. Bot. *10* : 1 - 32, 1981.

*48949 - CERFF, R. : Glyceraldehyde-3-phosphate dehydrogenase (NADP) from *Sinapis alba:* steady state kinetics. - Phytochemistry *17* : 2061 - 2067, 1978.

*48950 - CERFF, R. : Quaternary structure of higher plant glyceraldehyde-3-phosphate dehydrogenases. - Europe. J. Biochem. *94* : 243 - 247, 1979.

48951 - CERFF, R. : Evolutionary divergence of chloroplast and cytosolic glyceraldehyde-3-phosphate dehydrogenases from angiosperms. - Europe. J. Biochem. *126* : 513 - 515, 1982.

48952 - CERFF, R. : Separation and purification of NAD- and NADP-linked glyceraldehyde-3-phosphate dehydrogenases from higher plants. - In : EDELMAN,M.,HALLICK,R.B.,CHUA,N.-H.(ed.):Methods in Chloroplast Molecular Biology.Pp. 683 - -694. Elsevier Biomedical Press, Amsterdam - New York - Oxford 1982.

*48953 - CERFF, R., CHAMBERS, S.E. : Subunit structure of higher plant glyceraldehyde-3-phosphate dehydrogenases (EC 1.2.1.12 and EC 1.2.1.13). - J. biol. Chem. *254* : 6094 - 6098, 1979.

48954 - CERFF, R., KLOPPSTECH, K. : Structural diversity and differential light control of mRNAs coding for angiosperm glyceraldehyde-3-phosphate dehydrogenases. - Proc. nat. Acad. Sci. USA *79* : 7624 - 7628, 1982.

48955 - CERIONE, R.A., HAMMES, G.G. : Structural mapping of nucleotide binding sites on chloroplast coupling factor. - Biochemistry *21* : 745 - 752, 1982.

48956 - CEROVIĆ, Z.G., KALEZIĆ, R., PLESNIČAR, M. : The role of photophosphorylation in SO_2 and SO_3^{2-} inhibition of photosynthesis in isolated chloroplasts. - Planta *156* : 249 - 254, 1982.

48957 - CEULEMANS, R., GABRIELS, R., IMPENS, I. : Influence of fertilization level on gas exchange, structural, and growth characteristics of azalea. - HortScience *17* : 43 - 44, 1982.

48958 - CEULEMANS, R., IMPENS, I. : ECOPASS - a multivariate model used as an index of growth performance of poplar clones. - Forest Sci. *28* : 862 - 867, 1982. [Ps, Chl, resistances.]

48959 - CEULEMANS, R., IMPENS, I. : L'étude des échanges photosynthétiques de CO_2 comme mesure de la croissance et de la production de différentes plantes agricoles, ornamentales et horticoles. - Rev. Agr. *35* : 2631 - 2653, 1982.

48960 - CEULEMANS, R., VAN ASSCHE, F.,IMPENS, I., CLIJSTERS, H. : Ribulose-1,5- -bisphosphate carboxylase activity, chlorophyll and protein concentrations in different *Populus* clones. - Biol. Plant. *24* : 57 - 62, 1982.

48961 - CHAIN, R.K. : Evidence for a reductant-dependent oxidation of chloroplast cytochrome *b*-563. - FEBS Lett. *143* : 273 - 278, 1982.

48962 - CHAIT, B.T., FIELD, F.H. : ^{252}Cf fission fragment ionization mass spectrometry of chlorophyll *a*. - J. amer. chem. Soc. *104* : 5519 - 5521, 1982.

48963 - CHAKRAVARTY, A.S. : On the mechanism of photosynthesis. - Specul. Sci. Technol. *5* : 31 - 41, 1982.

48964 - CHAL"KOVA, M., GEORGIEVA, R. : Karotenoiden s"stav na plodovete na khibridi mezhdu *Lycopersicon esculentum* MILL. i nyakoi divi vidove ot roda *Lycopersicon* MILL. I.F_1 na *L.esculentum L.chmielewskii* i *L.esculentum L.parviflorum*. [Carotenoid composition of fruits in hybrids between *Lycopersicon esculentum* MILL. and some wild-growing species of the genus *Lycopersicon* MILL. I. *Lycopersicon esculentum* × *L.chmielewskii* F_1 and *L.esculentum* × *L.parviflorum* F_1.] - Genet. Selekts. *15* : 101 - 108, 1982. [In Bulg., ab : E, R.]

*48965 - CHALY, N., POSSINGHAM, J.V. : Structure of constricted proplastids in meristematic plant tissues. - Biol. Cell *41* : 203 - 210, 1981.

48966 - CHAMBERLIN, R.J., WILSON, G.L. : Development of yield in two grain-sorghum hybrids. I.Dry weight and carbon-14 studies. - Aust. J. agr. Res. *33* : 1009 - 1018, 1982.

48967 - CHAMONT, S., SOTTA, B., MIGINIAC, É. : Influence de traitments thermiques appliqués au niveau des racines sur la croissance, la floraison et l'équilibre hydrique de *Chenopodium polyspermum* L. - Physiol. vég. *20* : 1 - 10, 1982. [Stomata.]

48968 - CHAMOROVSKY, S.K., CAMMACK, R. : Direct determination of the midpoint potential of the acceptor X in chloroplast photosystem I by electrochemical reduction and ESR spectroscopy. - Photobiochem. Photobiophys. *4* : 195 - - 200, 1982.

48969 - CHAMOROVSKY, S.K., CAMMACK, R. : Effect of temperature on the photoreduction of centres A and B in photosystem I, and the kinetics of recombination. - Biochim. biophys. Acta *679* : 146 - 155, 1982.

48970 - CHAMPIGNY, M.-L. : III.-Relations entre photosynthèse et nutrition azotée minérale. - Compt. rend. Séances Acad. Agr. Fr. *68* : 883 - 892, 1982.

48971 - CHANG, W.Y.B. : Primary productivity and nutrients in the sediment retention basin of Lake Monroe. - Hydrobiologia *87* : 193 - 200, 1982.

48972 - CHANG, W.Y.B., ROSSMANN, R. : The influence of phytoplankton composition on the relative effectiveness of grinding and sonification for chlorophyll extraction. - Hydrobiologia *88* : 245 - 249, 1982.

*48973 - CHAPIN, F.S.,III, TIESZEN, L.L., LEWIS, M.S., MILLER, P.C., McCOWN, B.H. : Control of tundra plant allocation patterns and growth. - In : BROWN, J., MILLER, P.C., TIESZEN, L.L., BUNNELL, F.L. (ed.) : An Arctic Ecosystem: The Coastal Tundra at Barrow, Alaska. Pp. 140 - 185. Dowden, Hutchinson & Ross, Stroudsburg 1980.

*48974 - CHARRIERE-LADREIX, Y., DOUCE, R., JOYARD, J. : Characterization of *o*-methyltransferase activities associated with spinach chloroplast fractions. - FEBS Lett. *133* : 55 - 58, 1981.

48975 - CHATSKI, I., PLESKANKA, Ya., POSPISHILOVA, Ya., SOLAROVA, Ya., TIKHA, I. : Regulirovanie diffuzionnoĭ provodimosti ust'its biologicheskimi i ėkologicheskimi faktorami. [Control of epidermal diffusive conductance by biological and ecological factors.] - Bot. Zh. *67* : 455 - 461, 1982. [In R, ab : E.]

48976 - CHATURVEDI, R., HAUGSTAD, M.K., NILSEN, S. : The relationship between photosynthetic electron transport and photorespiratory $^{14}CO_2$ release after DCMU treatment in the duckweed, *Lemna gibba*. - Physiol. Plant. *56* : 23 - 27, 1982.

48977 - CHAUDHURI, U.N., KANEMASU, E.T. : Effect of water gradient on sorghum growth, water relations and yield. - Can. J. Plant. Sci. *62* : 599 - 607, 1982.

48978 - CHAUVAT, F., CORRE, B., HERDMAN, M., JOSET-ESPARDELLIER, F. : Energetic and metabolic requirements for the germination of akinetes of the cyanobacterium *Nostoc* PCC 7524. - Arch. Microbiol. *133*: 44 - 49, 1982. [Ps.]

*48979 - CHEBAN, A.I. : Vliyanie urovnya azotnogo pitaniya na aktivnost' fermentov pervichnoĭ assimilyatsii i vosstanovleniya uglekisloty v list'yakh yachmenya. [Effect of nitrogen nutrition level on the activity of enzymes of pri-

mary assimilation and reduction of carbon dioxide in barley leaves.] - Tr.
vsesoyuz. nauchno-issled. Inst. Udobr. Agropochvoved. *60* : 32 - 40, 1981.
[In R.]

*48980 - CHEBAN, A.I., YAKUSHINA, T.F. : Deĭstvie urovnya azotnogo pitaniya na foto-
khimicheskuyu aktivnost' khloroplastov raznykh sortov pshenitsy. [Effect
of the level of nitrogen nutrition on the photochemical activity of chloroplasts of various wheat cultivars.] - Tr. vsesoyuz. nauchno-issled. Inst.
Udobr. Agropochvoved. *60* : 40 - 46, 1981. [In R.]

*48981 - CHEN, S.S.C. : Microalgae as single-cell protein: Carbon dioxide solar
and geothermal energies-algae-fish system. - Nat. Sci. Counc. Mon. *7* :
1029 - 1035, 1979. [In Chin., ab : E.]

48982 - CHERESKIN, B.M., CASTELFRANCO, P.A. : Effects of iron and oxygen on chlorophyll biosynthesis. II. Observations on the biosynthetic pathways in isolated etiochloroplasts. - Plant Physiol. *69* : 112 - 116, 1982.

48983 - CHERESKIN, B.M., WONG, Y.-S., CASTEFRANCO, P.A. : *In vitro* synthesis of the
chlorophyll isocyclic ring. Transformation of magnesium-protoporphyrin IX
and magnesium-protoporphyrin IX monomethyl ester into magnesium-2,4-divinyl
pheoporphyrin A_5. - Plant Physiol. *70* : 987 - 993, 1982.

48984 - CHERNAVSKAYA, N.M.,, VASIL'EVA, L.Yu., VASIL'EV, Yu.M., NAPALKINA, O.V.: Rol'
margantsa v rabote fotosistemy II. [Role of manganese in photosystem II.] -
In : Svoĭstva Veshchestv i Stroenie Molekul. Pp. 62 - 68. Kalinin. gos.
Univ., Kalinin 1982. [In R.]

48985 - CHERRY, R.J. : Measurement of rotational diffusion of membrane proteins
using optical probes. - In : HÉLÈNE, C., CHARLIER, M., MONTENAY-GARESTIER,
T., LAUSTRIAT, G. (ed.) : Trends in Photobiology. Pp. 43 - 50. Plenum Press,
New York - London 1982. [Bacteriorhodopsin.]

48986 - CHERRY, R.J. : Transient dichroism of bacteriorhodopsin. - In : COLOWICK,
S.P., KAPLAN, N.O. (ed.) : Methods in Enzymology. Vol.88. Pp. 248 - 254.
Academic Press, New York - London - Paris - San Diego - San Francisco -
- São Paulo - Sydney - Tokyo - Toronto 1982.

48987 - CHERRY, R.J., GODFREY, R.E., PETERS, R. : Mobility of bacteriorhodopsin
in lipid vesicles. - Biochem. Soc. Trans. *10* : 342 - 343, 1982.

48988 - CHETVERIKOVA, N.I., ZHEMCHUGOVA, V.P. : Otnositel'naya rol' list'ev i plo-
dov kak fotosinteziruyushchikh organov v metabolizme sozrevayushchikh
semyan gorokha. [Relative role of leaves and fruits as photosynthesizing
organs in metabolism of ripening·pea seeds.] - Fiziol. Biokhim. kul't. Rast.
14 : 63 - 69, 1982. [In R, ab : E.]

*48989 - CHIFU, E., TOMOAIA-COTIŞEL, M. : Carotene and protein films at the oil/
/water interface. - Stud. Univ. Babeş-Bolyai, Ser. Chem. *26* (2) : 3 - 8,
1981.

48990 - CHIFU, E., TOMOAIA-COTIŞEL, M. : Filme mixte de pigmenţi carotinoidici şi
lipide la interfaţa aer/apă. [Mixed films of carotenoid pigments and lipids at the air/water interface.] - Rev. Chim. *33* : 125 - 131, 1982. [In
Roum., ab : E, R.]

48991 - CHIHARA, M. : Phylogenic relationships among algal phyla. - In : ZABORSKY,
O.R. (ed.) : CRC Handbook of Biosolar Resources. Vol.I/1. Pp. 565 - 567.
CRC Press, Boca Raton 1982. [Ps.]

48992 - CHIKOV, V.I., YARGUNOV, V.G., FEDOSEEVA, É.Z., CHEMIKOSOVA, S.V. : Vliyanie
sootnosheniya mezhdu proizvodstvom i potrebleniem assimilyatov na funktsionirovanie fotosinteticheskogo apparata rasteniĭ. [Influence of the relationship between production and consumption of photosynthates on the operation and consumption of photosynthates on the operation of photosynthetic
apparatus in plants.] - Fiziol. Rast. *29* : 1141 - 1146, 1982. [In R, ab :
E.]

48993 - CHIRAC, C., CASADEVALL, E., LARGEAU, C., METZGER, P. : Influence de la sou-
che et des Bactéries associées sur la productivité en hydrocarbures de
l'Algue *Botryococcus braunii*. - Compt. rend. Acad. Sci. Paris, Sér. III
295 : 671 - 674, 1982.

48994 - CHOTHIA, C., LESK, A.M. : Evolution of proteins formed by β-sheets. I. Plastocyanin and azurin. - J. mol. Biol. *160* : 309 - 323, 1982.

*48995 - CHOUDHURY, N.K., MAHAPATRA, P.K. : Senescence of *Cucurbita maxima* cotyledons
and the effect of gibberellic acid and potassium nitrate on the excised cotyledons. - J. Indian bot. Soc. *56* : 275 - 277, 1977. [Chl.]

48996 - CHRETIENNOT-DINET, M.-J. : Production primaire en baie de Concarneau. Relations algues-bactéries et filtration différentielle. - J. Plankton Res. *4* :
463 - 479, 1982.

48997 - CHRISTELLER, J.T., HARTMAN, F.C. : Inactivation of *Rhodospirillum rubrum*
ribulose bisphosphate carboxylase/oxygenase by the affinity label 2-N-chlo-
roamino-2-deoxypentitol 1,5-bisphosphate. - FEBS Lett. *142* : 162 - 166,
1982.

B48998 - CHRISTIANSEN, M.N., LEWIS, C.F.(ed.): Breeding Plants for Less Favorable Environments. - Wiley-Interscience Publication,John Wiley & Sons, New York -
- Chichester - Brisbane - Toronto - Singapore 1982. [Ps.]

48999 - CHRISTIE, E.K., DETLING, J.K. : Analysis of interference between C_3 and C_4
grasses in relation to temperature and soil nitrogen supply. - Ecology *63* :
1277 - 1284, 1982.

49000 - CHRISTY, A.L., PORTER, C.A. : Canopy photosynthesis and yield in soybean.
- In : GOVINDJEE (ed.) : Photosynthesis. Vol.II. Pp. 499 - 511. Academic
Press, New York - London - Paris - San Diego - San Francisco - São Paulo -
- Sydney - Tokyo - Toronto 1982.

*49001 - CHRŐST, R.J. : The composition and bacterial utilization of DOC released
by phytoplankton. - Kieler Meeresforsch. *1981* (Sonderh.5) : 325 - 332,
1981. [Chl.]

49002 - CHU Chih-ching, SUN Ching-san, LI Shou-quan : [Ultrastructural study of
proplastid ontogeny in tobacco mesophyll cells *in vitro*.] - Acta bot. sin.
24 : 199 - 203, 1982. [In Chin., ab : E.]

49003 - CHUA, N.-H., BARTLETT, S.G., WEISS, M. : Preparation and characterization
of antibodies to chloroplast proteins. - In : EDELMAN, M., HALLICK, R.B.,
CHUA, N.-H. (ed.) : Methods in Chloroplast Molecular Biology. Pp. 1063 -
- 1080. Elsevier Biomedical Press, Amsterdam - New York - Oxford 1982.

49004 - CHUECA, A., BARŐN, M., LŐPEZ-GORGÉ, J. : Acción *in vitro* e *in vivo* de los
biscarbamatos sobre la actividad fotosintética del cloroplasto. [*In vitro*
and *in vivo* action of bis-carbamates on photosynthetic activity of chloroplasts.] - Rev. esp. Fisiol. *38* (Supl.) : 315 - 320, 1982. [In Span., ab :
E.]

*49005 - CHUMAK, L.N. : Pogloshchenie fotosinteticheski aktivnoĭ radiatsii v zavi-
simosti ot arkhitektoniki sorta i gustoty stoyaniya risa. [Absorption of
photosynthetically active radiation depending on cultivar architectonics
and density of rice stand.] - Byull. nauch.-tekh. Inform. vsesoyuz. nauch.-
-issled. Inst. Risa *31* : 18 - 19, 1981. [In R, ab : E.]

*49006 - CHUNAEV, A.S., LADYGIN, V.G., GAVRILENKO, T.A., KRĔLA, L.P., KORNYUSHENKO,
G.A. : Nasledovanie priznaka "otsutstvie khlorofilla *b*" i izmenchivost'
svetosobirayushchego kompleksa v meĭotipicheskom potomstve mutanta C-48
Chlamydomonas reinhardii. [Inheritance of the "chlorophyll *b*-deficiency"
character and variability of the light-harvesting complex in the meiotic
progeny of *Chlamydomonas reinhardii* mutant C-48.] - Genetika *17* : 2013 -
- 2024, 1981. [In R, ab : E.]

49007 - CHUNAEV, A.S., MIRNAYA, O.N., GAEVSKIĬ, N.A. : Izmenchivost' sootnosheniya
khlorofill *a*/khlorofill *b* u *Chlamydomonas reinhardii*. [Variability in the
ratio of chlorophyll *a*/chlorophyll *b* in *Chlamydomonas reinhardii*.] - Vest.
leningrad. Univ. *1982*[9(Biol.2)] : 98 - 102, 1982. [In R, ab : E.]

49008 - CHUNAEV, A.S., MIRNAYA, O.N., GAEVSKIĬ, N.A. : Kachestvennaya otsenka priznaka "otsutstvie khlorofilla *b*" u *Chlamydomonas reinhardii*. [Qualitative
detection of the "chlorophyll *b*-deficiency" character in *Chlamydomonas
reinhardii*.] - Genetika *18* : 1906 - 1909, 1982. [In R, ab : E.]

49009 - CHUNG, B. : Growth analysis of poppies (*Papaver somniferum* L.). - Aust. J. agr. Res. *33* : 233 - 242, 1982.

✶49010 - CHUNG, H.H., TRLICA, M.J. : ^{14}C-distribution and utilization in blue grama as affected by temperature, water potential and defoliation regimes. - Oecologia *47* : 190 - 195, 1980.

49011 - CLAPP, R.E. : Loop currents in chlorophyll *a*. - Theor. chim. Acta *61* : 105 - 133, 1982.

B49012 - CLARK, W.C. (ed.) : Carbon Dioxide Review 1982. - Clarendon Press, Oxford; Oxford University Press, New York 1982.

49013 - CLARKE, I.E., BRAMLEY, P.M., SANDMANN, G., BÖGER, P. : Herbicide action on carotenogenesis in a photosynthetic cell-free system. - In : WINTERMANS, J.F.G.M., KUIPER, P.J.C. (ed.) : Biochemistry and Metabolism of Plant Lipids. Pp. 549 - 554. Elsevier Biomedical Press, Amsterdam 1982.

49014 - CLARKE, I.E., SANDMANN, G., BRAMLEY, P.M., BÖGER, P. : Carotene biosynthesis with isolated photosynthetic membranes. - FEBS Lett. *140* : 203 - 206, 1982.

49015 - CLARKE, J.M., McCAIG, T.N. : Leaf diffusion resistance, surface temperature, osmotic potential and $^{14}CO_2$-assimilation capability as indicators of drought intensity in rape. - Can. J. Plant Sci. *62* : 785 - 789, 1982.

49016 - CLARKE, R.H. : The chlorophyll triplet state and the structure of chlorophyll aggregates. - In : FONG, F.K. (ed.) : Light Reaction Path of Photosynthesis. Pp. 196 - 233. Springer-Verlag, Berlin - Heidelberg - New York 1982.

49017 - CLARKE, R.H., HOTCHANDANI, S., JAGANNATHAN, S.P., LEBLANC, R.M. : The effect of coordinating ligands on the triplet state of chlorophyll. - Photochem. Photobiol. *36* : 575 - 579, 1982.

✶49018 - CLAYTON, R.K., CARL, P., MAAZ, G. : Discussions to I. Primary reactions of photoreception and comparison with photosynthesis. - Biophys. Struct. Mech. *3* : 107 - 116, 1977.

✶49019 - CLOSS, G.L., SITZMANN, E.V. : Measurement of degenerate radical ion-neutral molecule electron exchange by microsecond time-resolved CIDNP. Determination of relative hyperfine coupling constants of radical cations of chlorophyll and derivatives. - J. amer. chem. Soc. *103* : 3217 - 3219, 1981.

49020 - CLOUGH, B.F., ANDREWS, T.J., COWAN, I.R. : Physiological processes in mangroves. - In : CLOUGH, B.F. (ed.) : Mangrove Ecosystems in Australia. Structure, Function and Management. Pp. 193 - 210.Australian National University Press, Canberra 1982. [Ps.]

49021 - CLOUGH, B.F., ATTIWILL, P.M. : Primary productivity of mangroves. - In : CLOUGH, B.F. (ed.) : Mangrove Ecosystems in Australia. Structure, Function and Managemant. Pp. 213 - 222.Australian National University Press, Canberra 1982.

49022 - COCHRANE, M.P., DUFFUS, C.M. : Opportunities for the regulation of grain development. - In : HAWKINS, A.F., JEFFCOAT, B. (ed.) : Opportunities for Manipulation of Cereal Productivity. Monograph 7. Pp. 167 - 178. Brit. Plant Growth Regulator Group, Wantage 1982. [Ps, photosynthates.]

49023 - COCQUEMPOT, M.F., AGUIRRE, R., LISSOLO, T., MONSAN, P., HATCHIKIAN, E.C., THOMAS, D. : Co-immobilization effect on H_2 production by a chloroplast membranes-hydrogenase system. - Biotechnol. Lett. *4* : 313 - 318, 1982.

49024 - COCQUEMPOT, M.F., LARRETA GARDE, V., THOMASSET, B., LISSOLO, T., BARBOTIN, J.N. : Stabilization of biological photosystems: continuous reactor use for hydrogen production through biophotolysis of water. - In : HALL, D.O., PALZ, W. (ed.) : Photochemical, Photoelectrochemical and Photobiological Processes. Pp. 203 - 207. D.Reidel Publ.Co., Dordrecht - Boston - London 1982.

49025 - CODDINGTON, J.M., JOHNS, S.R., WILLING, R.I., KENRICK, J.R., BISHOP, D.G. :
Preparation and comparison of model bilayer systems from chloroplast thylakoid membrane lipids for ^{13}C-NMR studies. - J. biochem. biophys. Meth.
6 : 351 - 356, 1982.

49026 - COGDELL, R.J. : The electron transport components in photosynthetic bacteria. - In : BARBER, J. (ed.) : Electron Transport and Photophosphorylation.
Pp. 177 - 196. Elsevier Biomedical Press, Amsterdam - New York - Oxford
1982.

49027 - COGDELL, R.J., LINDSAY, J.G., VALENTINE, J., DURANT, I. : A further characterisation of the B890 light-harvesting pigment-protein complex from *Rhodospirillum rubrum* strain S1. - FEBS Lett. 150 : 151 - 154, 1982.

49028 - COGDELL, R.J., VALENTINE, J., LINDSAY, J.G., SCHMIDT, K. : The structure
of the bacterial photosynthetic unit. - Biochem. Soc. Trans. 10 : 334 -
- 335, 1982.

49029 - COHEN, C.J., CHILCOTE, D.O., FRAKES, R.V. : Gas exchange and leaf area
characteristics of four tall fescue selections differing in forage yield.
- Crop Sci. 22 : 709 - 711, 1982.

49030 - COHEN, R.R.H., KELLY, M.G., CHURCH, M.R. : The effect of CO_2 on the relationship of photosynthetic rate to light intensity in laboratory phytoplankton cultures. - Arch. Hydrobiol. 94 : 326 - 340, 1982.

49031 - COHEN, S.S. : On the endosymbiotic origins of chloroplasts: Still another
approach to the problem. - In : SCHIFF, J.A., LYMAN, H. (ed.) : On the Origins of Chloroplasts. Pp. 93 - 106. Elsevier/North-Holland, New York - Amsterdam - Oxford 1982.

49032 - COHEN, S.S., MARCU, D.E., BALINT, R.F. : Light-dependent fixation of polyamines into chloroplasts of Chinese cabbage. - FEBS Lett. 114 : 93 - 97,
1982.

*49033 - COLBOW, K. : Energy transfer in photosynthesis. - In : COLBOW, K. (ed.) :
The Physics of Biological Membranes. Pp. 430 - 444. Simon Fraser University,
Burnaby 1975.

49034 - COLEMAN, J.R., SEEMANN, J.R., BERRY, J.A. : RuBP carboxylase in carboxyso-
mes of blue-green algae. - Carnegie Inst. Washington Year Book 81 : 83 - 87,
1982.

49035 - COLIJN, C.M., KOOL, A.J., NIJKAMP, H.J.J. : Protein synthesis in *Petunia
hybrida* chloroplasts isolated from leaves and cell cultures. - Planta 155 :
37 - 44, 1982.

49036 - COLLINS, C.D., BOYLEN, C.W. : Ecological consequences of long-term exposure
of *Anabaena variabilis (Cyanophyceae)* to shifts in environmental factors.
- Appl. environm. Microbiol. 44 : 141 - 148, 1982.

49037 - COLLINS, C.D., BOYLEN, C.W. : Physiological responses of *Anabaena variabilis
(Cyanophyceae)* to instantaneous exposure to various combinations of light
intensity and temperature. - J. Phycol. 18 : 206 - 211, 1982. [Ps.]

49038 - COMINS, H.N., FARQUHAR, G.D. : Stomatal regulation and water economy in
Crassulacean acid metabolism plants: an optimization model. - J. theor.
Biol. 99 : 263 - 284, 1982. [Ps.]

49039 - CONNOLLY, J.S., JANZEN, A.F., SAMUEL, E.B. : Fluorescence lifetimes of chlorophyll *a*: solvent, concentration and oxygen dependence. - Photochem. Photobiol. 36 : 559 - 563, 1982.

49040 - CONNOLLY, J.S., SAMUEL, E.B., JANZEN, A.F. : Effects of solvent on the fluorescence properties of bacteriochlorophyll *a*. - Photochem. Photobiol. 36 :
565 - 574, 1982.

49041 - CONOVER, C.A., POOLE, R.T. : Fluoride induced chlorosis and necrosis of
Dracaena fragrans "Massangeana". - J. amer. Soc. hort. Sci. 107 : 136 -
- 139, 1982.

49042 - CONSTABLE, G.A., RAWSON, H.M. : Distribution of ^{14}C label from cotton leaves: consequences of changed water and nitrogen status. - Aust. J. Plant Physiol. *9* : 735 - 747, 1982.

49043 - COOMBS, J. : Carbon metabolism. - In : COOMBS, J., HALL, D.O.(ed.) : Techniques in Bioproductivity and Photosynthesis. Pp. 81 - 90. Pergamon Press, Oxford - New York - Toronto - Sydney - Paris - Frankfurt 1982.

49044 - COOMBS, J. : Isolation of enzymes. - In : COOMBS, J., HALL, D.O. (ed.) : Techniques in Bioproductivity and Photosynthesis. Pp. 142 - 152. Pergamon Press, Oxford - New York - Toronto - Sydney - Paris - Frankfurt 1982. [Ps enzymes.]

B49045 - COOMBS, J., HALL, D.O. (ed.) : Techniques in Bioproductivity and Photosynthesis. - Pergamon Press, Oxford - New York - Toronto - Sydney - Paris - - Frankfurt 1982.

49046 - COOPER, A. : Tryptophan reactivity. - In : COLOWICK, S.P., KAPLAN, N.O. (ed.) : Methods in Enzymology. Vol.81. Pp. 285 - 288. Academic Press, New York - London - Paris - San Diego - San Francisco - São Paulo - Sydney - Tokyo - Toronto 1982. [Bacteriorhodopsin.]

49047 - COOPER, A. : Calorimetric measurements of light-induced processes. - In : COLOWICK, S.P., KAPLAN, N.O. (ed.) : Methods in Enzymology. Vol.88. Pp. 667 - 673. Academic Press, New York - London - Paris - San Diego - San Francisco - São Paulo - Sydney - Tokyo - Toronto 1982.

49048 - CORK, D.J. : Acid waste gas bioconversion - an alternative to the Claus desulfurization process. - Dev. ind. Microbiol.*23* : 379 - 387, 1982. [Ps.]

49049 - CORK, D.J., MA, S. : Acid-gas bioconversion favors sulfur production. - Biotechnol. Bioeng. Symp. *12* : 285 - 290, 1982. [Ps.]

49050 - CORNIC, G., WOO, K.C., OSMOND, C.B. : Photoinhibition of CO_2-dependent O_2 evolution by intact chloroplasts isolated from spinach leaves. - Plant Physiol. *70* : 1310 - 1315, 1982.

*49051 - CORNILLON, P., DAUPLE, P. : Influence of irrigation rhythm and water supply on growth, water status and yield of egg-plant (*Solanum melongena* L.). - Plant Soil *59* : 365 - 379, 1981. [Stomatal resistance.]

*49052 - CORREIA, O.C.A., CATARINO, F.M. : Effect of temperature and light on gas exchanges of *Ceratonia siliqua*. - Portug. Acta biol., Ser.A *16* : 141 - 150, 1980.

49053 - CORTIJO, M., ALONSO, A., GOMEZ-FERNANDEZ, J.C., CHAPMAN, D. : Intrinsic protein-lipid interactions. Infrared spectroscopic studies of gramicidin A, bacteriorhodopsin and Ca^{2+}-ATPase in biomembranes and reconstituted systems. - J. mol. Biol. *157* : 597 - 618, 1982.

49054 - COSPER, E. : Effects of diurnal fluctuations in light intensity on the efficiency of growth of *Skeletonema costatum* (GREV.) CLEVE *(Bacillariophyceae)* in a cyclostat. - J. exp. mar. Biol. Ecol. *65* : 229 - 239, 1982. [Chl.]

49055 - COSPER, E. : Effects of variations in light intensity on the efficiency of growth of *Skeletonema costatum (Bacillariophyceae)* in a cyclostat. - J. Phycol. *18* : 360 - 368, 1982. [Chl.]

49056 - COSPER, E. : Influence of light intensity on diel variations in rates of growth, respiration and organic release of a marine diatom: comparison of diurnally constant and fluctuating light. - J. Plankton Res. *4* : 705 - 724, 1982.

49057 - COSTA, B., GULIK-KRZYWICKI, T., REISS-HUSSON, F., RIVAS, E. : Fusion entre chromatophores extraits de *Rhodopseudomonas spheroides* et liposomes. - Compt. rend. Acad. Sci. Paris, Sér. III *295* : 517 - 522, 1982.

49058 - COSTA, S., COSTANZO, E., GRILLO, N., RUBBINO, A. : Source of entropy decrease in biosystems. - Nuovo Cimento *1D* (1) : 1 - 8, 1982. [Ps.]

49059 - COSTE, B., NIVAL, P., MINAS, H.J. : Analyse des relations entre les conditions hydrologiques, les sels nutritifs et la chlorophylle des eaux super-
ficielles d'une zone d'upwelling (Côtes de Mauritanie, mars-avril 1974). -
Rapp. Proc.-verb. Réun. Cons. int. Explor. Mer *180* : 108 - 113, 1982.

49060 - COSTES, C. : Le rendement quantique maximal de l'assimilation photosynthétique du bioxyde de carbon: ses composantes et ses variations dans les cultures et dans les végétations non cultivées. - Compt. rend. Acad. agr. Fr.
68 : 847 - 858, 1982.

49061 - COTTON, N.P.J., JACKSON, J.B. : The kinetics of carotenoid absorption
changes in intact cells of photosynthetic bacteria. - Biochim. biophys.
Acta *679* : 138 - 145, 1982.

49062 - COTTON, T.M., VAN DUYNE, R.P. : Resonance Raman scattering from *Rhodopseu-*
domonas sphaeroides reaction centers adsorbed on a silver electrode. -
FEBS Lett. *147* : 81 - 84, 1982.

49063 - COUDRET, A., FÉRARD, G., LASCÈVE, G. : Choc osmotique et mouvements d'eau
chez *Plantago*. - Physiol. vég. *20* : 711 - 720, 1982. [Ps.]

49064 - COUDRET, A., FERRON, F. : Action d'un choc osmotique sur le métabolisme
photosynthétique d'un *Triticum* et d'une *Aegilops*. - Photosynthetica *16* :
217 - 225, 1982.

49065 - COUDRET, A., FERRON, F. : Régulation et dé-régulation du métabolisme photosynthétique en réponse aux variations du milieu chez des végétaux de type
C_3. - Agronomie *2*: 429 - 436, 1982.

49066 - COUGHLAN, S.J., HEBER, U. : The role of glycinebetaine in the protection
of spinach thylakoids against freezing stress. - Planta *156* : 62 - 69, 1982.

49067 - COUGHLAN, S.J., WYN JONES, R.G. : Glycinebetaine biosynthesis and its control in detached secondary leaves of spinach. - Planta *154* : 6 - 17, 1982.

49068 - COURNIER, S., CROUZIS, J.-P., RAMBIER, M., PARIS-PIREYRE, N. : Relations
entre la fixation de Ca^{2+}, l'empilement des thylakoïdes et le caractère
calcicole ou calcifuge chez deux espèces de Vigne. - Physiol. vég. *20* :
423 - 432, 1982.

49069 - COUSENS, R. : The effect of exposure to wave action on the morphology and
pigmentation of *Ascophyllum nodosum* L. Le Jolis in south-eastern Canada. -
Bot. mar. *25* : 191 - 195, 1982.

49070 - COVENEY, M.F. : Bacterial uptake of photosynthetic carbon from freshwater
phytoplankton. - Oikos *38* : 8 - 20, 1982.

49071 - COWAN, I.R. : Regulation of water use in relation to carbon gain in higher
plants. - In : LANGE, O.L., NOBEL, P.S., OSMOND, C.B., ZIEGLER, H. (ed.) :
Physiological Plant Ecology II. Water Relations and Carbon Assimilation.
Pp. 589 - 613. Springer-Verlag, Berlin - Heidelberg - New York 1982. [Ps.]

49072 - COWAN, I.R., RAVEN, J.A., HARTUNG, W., FARQUHAR, G.D. : A possible role for
abscisic acid in coupling stomatal conductance and photosynthetic carbon
metabolism in leaves. - Aust. J.Plant Physiol. *9* : 489 - 498, 1982.

*49073 - COX, R.P., ANDERSSON, B. : Lateral and transverse organization of cytochro-
mes in the chloroplast thylakoid membranes. - Biochem. biophys. Res. Commun.
103 : 1336 - 1342, 1981.

49074 - COX, R.P., OLSEN, L.F. : The organisation of the electron transport chain
in the thylakoid membrane. - In : BARBER, J. (ed.) : Electron Transport
and Photophosphorylation. Pp. 49 - 79. Elsevier Biomedical Press, Amsterdam - New York - Oxford 1982.

49075 - COXSON, D.S., HARRIS, G.P., KERSHAW, K.A. : Physiological-environmental
interactions in lichens. XV. Contrasting gas exchange patterns between a
lichenized and non-lichenized terrestrial *Nostoc* cyanophyte. - New Phytol.
92 : 561 - 572, 1982.

49076 - COYNE, P.I., BINGHAM, G.E. : Variation in photosynthesis and stomatal conductance in an ozone-stressed ponderosa pine stand: light response. - Forest Sci. *28* : 257 - 273, 1982.

49077 - COYNE, P.I., BRADFORD, J.A., DEWALD, C.L. : Leaf water relations and gas exchange in relation to forage production in four asiatic bluestems. - Crop Sci. *22* : 1036 - 1040, 1982.

49078 - CRAMER, W.A., CROFTS, A.R. : Electron and proton transport. - In : GOVIND-JEE (ed.) : Photosynthesis. Vol.1. Pp. 387 - 467. Academic Press, New York - - London - Paris - San Diego - San Francisco - São Paulo - Sydney - Tokyo - - Toronto 1982.

49079 - CRAWFORD, M.S., WANG, W., JENSEN, K.G. : Identification of the primary lesion in a protoporphyrin accumulating mutant of *Chlamydomonas reinhardtii*. - Mol. gen. Genet. *188* : 1 - 6, 1982.

49080 - CRÉACH, E., STEWART, C.R. : Effects of aminoacetonitrile on net photosynthesis, ribulose-1,5-bisphosphate levels, and glycolate pathway intermediates. - Plant Physiol. *70* : 1444 - 1448, 1982.

B49081 - CREASY, L.L., HRAZDINA, G. (ed.) : Cellular and Subcellular Localization in Plant Metabolism. Recent Advances in Phytochemistry. Volume 16. - Plenum Press, New York - London 1982. [Ps.]

49082 - CRESPI, H.L. : The isolation of deuterated bacteriorhodopsin from fully deuterated *Halobacterium halobium*. - In : COLOWICK, S.P., KAPLAN, N.O.(ed.): Methods in Enzymology. Vol.88. Pp. 3 - 5. Academic Press, New York - - London - Paris - San Diego - San Francisco - São Paulo - Sydney - Tokyo - - Toronto 1982.

✲49083 - CRESSWELL, E.G., GRIME, J.P. : Induction of a light requirement during seed development and its ecological consequences. - Nature *291* : 583 - 585, 1981. [Chl.]

49084 - CRITCHLEY, C. : Stimulation of photosynthetic electron transport in a salt--tolerant plant by high chloride concentrations. - Nature *298* : 483 - 485, 1982.

49085 - CRITCHLEY, C., BAIANU, I.C., GOVINDJEE, GUTOWSKY, H.S. : The role of chloride in O_2 evolution by thylakoids from salt-tolerant higher plants. - Biochim. biophys. Acta *682* : 436 - 445, 1982.

49086 - CROATTO, U. : Energy from macroalgae of the Venice lagoon. - In : GARSSI,G., PALZ, W. (ed.) : Energy from Biomass. Vol.3. Pp. 114 - 119. D.Reidel Publ. Co., Dordrecht - Boston - London 1982.

49087 - CROMER, R.N., WILLIAMS, E.R. : Biomass and nutrient accumulation in a planted *E. globulus* (LABILL.) fertilizer trial. - Aust. J. Bot. *30* : 265 - 278, 1982.

49088 - CROSBIE, T.M., PEARCE, R.B. : Effects of recurrent phenotypic selection for high and low photosynthesis on agronomic traits in two maize populations. - Crop Sci. *22* : 809 - 813, 1982.

49089 - CROUCH, R.K. : Spin labeling of bacteriorhodopsin. - In : COLOWICK, S.P., KAPLAN, N.O. (ed.) : Methods in Enzymology. Vol.88. Pp. 175 - 177. Academic Press, New York - London - Paris - San Diego - San Francisco - São Paulo - - Sydney - Tokyo - Toronto 1982.

✲49090 - CROUCH, R.K., EBREY, T.G., GOVINDJEE, R. : A bacteriorhodopsin analogue containing the retinal nitroxide free radical. - J. amer. chem. Soc. *103* : 7364 - 7366, 1981.

49091 - CROUGHAN, T.P., RAINS, D.W. : Terrestrial halophytes: habitats, productivity, and uses. - In : ZABORSKY, O.R. (ed.) : CRC Handbook of Biosolar Resources. Vol.I/2. Pp. 245 - 255. CRC Press, Boca Raton 1982.

49092 - CROWDER, M.S., PRINCE, R.C., BEARDEN, A. : Orientation of membrane-bound cytochromes in chloroplasts, detected by low-temperature EPR spectroscopy. - FEBS Lett. *144* : 204 - 208, 1982.

49093 - CSEKE, C., NISHIZAWA, A.N., BUCHANAN, B.B. : Modulation of chloroplast phosphofructokinase by NADPH - a mechanism for linking light to the regulation of glycolysis. - Plant Physiol. *70* : 658 - 661, 1982.

49094 - **CUELLO, J., SABATER, B.** : Control of some enzymes of nitrogen metabolism during senescence of detached barley (*Hordeum vulgare* L.) leaves. - Plant Cell Physiol. *23* : 561 - 565, 1982. [Chl.]

*49095 - **CUENDET, P., GRÄTZEL, M.** : Photoproduction of H_2 from isolated chloroplasts through ultrafine Pt catalysts and different viologen relays. - Photobiochem. Photobiophys. *2* : 93 - 103, 1981.

49096 - **CUENDET, P., GRÄTZEL, M.** : Artificial photosynthetic systems. - Experientia *38* : 223 - 227, 1982.

49097 - **CUENDET, P., GRÄTZEL, M.** : New photosystem I electron acceptors: improvement of hydrogen photoproduction by chloroplasts. - Photochem. Photobiol. *36* : 203 - 210, 1982.

49098 - **CULLEN, J.J.** : The deep chlorophyll maximum: comparing vertical profiles of chlorophyll *a*. - Can. J. Fish. aquat. Sci. *39* : 791 - 803, 1982.

49099 - **CULLEN, J.J., REID, F.M.H., STEWART, E.** : Phytoplankton in the surface and chlorophyll maximum off southern California in August, 1978. - J. Plankton Res. *4* : 665 - 694, 1982.

49100 - **CUNNINGHAME, M.E., HILLMAN, J.R., BOWES, B.G.** : Ultrastructural changes in mesophyll cells of *Larix decidua* × *kaempferi* during leaf maturation and senescence. - Flora *172* : 161 - 172, 1982. [Chloroplast.]

49101 - **CURE, J.D., PATTERSON, R.P., RAPER, C.D.,Jr., JACKSON, W.A.** : Assimilate distribution in soybeans as affected by photoperiod during seed development. - Crop Sci. *22* : 1245 - 1250, 1982.

*49102 - **CURRAN, P.J.** : Multispectral remote sensing for estimating biomass and productivity. - In : SMITH, H. (ed.) : Plants and the Daylight Spectrum. Pp. 65 - 99. Academic Press, London - New York 1981.

49103 - **CUTLER, A.J., CONN, E.E.** : The synthesis, storage and degradation of plant natural products: Cyanogenic glycosides as an example. - In : CREASY, L.L., HRAZDINA, G. (ed.) : Cellular and Subcellular Localization in Plant Metabolism. Pp. 249 - 271. Plenum Press, New York - London 1982. [Chl, Ps enzymes.]

49104 - **CZARNECKI, J.J., ABBOTT, M.S., SELMAN, B.R.** : Photoaffinity labeling with 2-azidoadenosine diphosphate of a tight nucleotide binding site on chlorolast coupling factor 1. - Proc. nat. Acad. Sci. USA *79* : 7744 - 7748, 1982.

49105 - **CZARNOWSKI, M., PLESIŃSKI, S., SIERKA, E.** : Wpływ sztucznych źródeł światła na fotosyntezę pomidorów i ogórków szklarniowych. [The effect of artificial light sources on photosynthesis of greenhouse tomatoes and cucumbers.] - Zesz. nauk. Akad. rol. Krakow. *171* (Ogrodnictwo 9) : 149 - 159, 1982. [In Pol., ab : E, R.]

49106 - **CZECZUGA, B.** : Badania barwników fikobiliproteinowych u glonów. [Investigations on the phycobiliprotein pigments of algae.] - Wiad. bot. *26* : 171 - -186, 1982. [In Pol.]

49107 - **CZÉGÉ, J., DÉR, A., ZIMÁNYI, L., KESZTHELYI, L.** : Restriction of motion of protein side chains during the photocycle of bacteriorhodopsin. - Proc. nat. Acad. Sci. USA *79* : 7273 - 7277, 1982.

49108 - **CZYGAN, F.-C.** : Primäre und sekundäre Carotinoide in chlorokokkalen Algen. - Arch. Hydrobiol. Suppl. *60* (4, Algol. Stud. 29) : 470 - 488, 1982.

49109 - **DACEY, J.W.H., KLUG, M.J.** : Tracer studies of gas circulation in *Nuphar* : $^{18}O_2$ and $^{14}CO_2$ transport. - Physiol. Plant. *56* : 361 - 366, 1982.

49110 - **DACEY, J.W.H., KLUG, M.J.** : Ventilation by floating leaves in *Nuphar*. - Amer. J. Bot. *69* : 999 - 1003, 1982. [Ps.]

49111 - **DAHLMELM, H., FICKER, K.** : Untersuchungen zur subzellulären Lokalisation proteolytischen Enzymes in *Pisum sativum* L. II. Proteolytische Aktivität in Chloroplasten. - Biochem. Physiol. Pflanzen *177* : 167 - 175, 1982.

49112 - DAHNIYA, M.T., OPUTA, C.O., HAHN, S.K. : Investigating source-sink relations in cassava by reciprocal grafts. - Exp. Agr. *18* : 399 - 402, 1982. [Photosynthates.]

49113 - DALE, J.E. : Some effects of temperature and irradiance on growth of the first four leaves of wheat, *Triticum aestivum*. - Ann. Bot. *50* : 851 - 858, 1982.

B49114 - DALE, J.E. : The Growth of Leaves. - Edward Arnold, London 1982.

49115 - DALE, R.F., SCHEERINGA, K.L., HODGES, H.F., HOUSLEY, T.L. : Effect of leaf area, incident radiation, and moisture stress on reflectance on near infrared radiation from a corn canopy. - Agron. J. *74* : 67 - 73, 1982.

*49116 - DALEY, R.J., CARMACK, E.C., GRAY, C.B.J., PHARO, C.H., JASPER, S., WIEGAND, R.C. : The effects of upstream impoundments on the limnology of Kootenay Lake, B.C. - Nat. Water Res. Inst., Inland Waters Direct., Sci. Ser. (Vancouver) *117* : 1 - 98, 1981. [Chl.]

*49117 - DALLINGER, R.F., FARQUHARSON, S., WOODRUFF, W.H., RODGERS, M.A.J. : Vibrational spectroscopy of the electronically excited state. 4. Nanosecond and picosecond time-resolved resonance Raman spectroscopy of carotenoid excited states. - J. amer. chem. Soc. *103* : 7433 - 7440, 1981.

49118 - DAMISCH, W. : Stoffzuwachs und Jahreswitterung bei Weizen. - In : HOFFMANN, P., HIEKE, B.(ed.) : Photosynthese: Regulation und Evolution. (Colloquia Pflanzenphysiologie Nr.5.) Pp. 191 - 201. Humboldt-Universität, Berlin 1982. [Ps.]

49119 - DANAILOV, Zh. : Za vr"zkata na kombinativnata sposobnost s nyakoi morfologichni i fiziologichni pokazateli pri linii domati. [Relation between combining ability and some morphological and physiological characteristics of tomato lines.] - Genet. Selek. (Sofiya) *15* : 449 - 455, 1982. [Ps; in Bulg., ab : E, R.]

49120 - DANDONNEAU, Y. : A method for the rapid determination of chlorophyll plus phaeopigments in samples collected by merchant ships. - Deep-Sea Res., Part A *29* : 647 - 654, 1982.

49121 - DANIELL, H., REBEIZ, C.A. : Chloroplast culture VIII. A new effect of kinetin in enhancing the synthesis and accumulation of protochlorophyllide *in vitro*. - Biochem. biophys. Res. Commun. *104* : 837 - 843, 1982.

49122 - DANIELL, H., REBEIZ, C.A. : Chloroplast culture IX. Chlorophyll(ide) *a* biosynthesis *in vitro* at rates higher than *in vivo*. - Biochem. biophys. Res. Commun. *106* : 466 - 470, 1982.

49123 - DANIELL, H., SAROJINI, G., KULANDAIVELU, G. : Is direct spectrophotometric determination of chlorophyll in pigment extracts of tissues under different physiological conditions valid? - Biochem. biophys. Res. Commun. *105* : 698 - 704, 1982.

49124 - DANIELL, H., SAROJINI, G., WU, S.-M. : Study of proton translocation in chloroplasts - a new approach. - Biochem. biophys. Res. Commun. *107* : 1191 - - 1197, 1982. [Ps.]

*49125 - DANNOWSKI, M., SCHÄFER, W., KÜNKEL, K. : Standortspezifische Aussagen zur Energienutzung von Winterweizen im Freiland. - Arch. Acker- Pflanzenbau Bodenk. *25* : 601 - 610, 1981. [Ps.]

*49126 - DAPAAH, S.K., PHILPOTTS, L.E., MACK, A.R., PEET, F.G. : Effects of soil and seasonal time on accuracy of crop identification in southwestern Ontario from 1974 Landsat-1 imagery. - Can. J. Plant Sci. *57* : 577 - 590, 1977.

49127 - DaPRA, E., SNOZZI, M., BACHOFEN, R. : Distribution of phosphatidylethanolamine in the lipid bilayer of chromatophores of the photosynthetic bacterium *Rhodospirillum rubrum*. - Arch. Microbiol. *133* : 23 - 27, 1982.

49128 - DARLEY, W.M. : General characteristics of phytoplankton: diatoms. - In : ZABORSKY, O.R. (ed.) : CRC Handbook of Biosolar Resources. Vol.I/2. Pp. 33 - - 36. CRC Press, Boca Raton 1982. [Car.]

49129 - DARMENCY, H., GASQUEZ, J. : Differential temperature-dependence of the Hill activity of isolated chloroplasts from triazine resistant and susceptible biotypes of *Polygonum lapathifolium* L. - Plant Sci. Lett. *24* : 39 - 44, 1982.

*49130 - DARWENT, J.R., KALYANASUNDARAM, K., PORTER, G. : Model systems for photosynthesis. VII. Chlorophyll *a* photosensitized reduction of methyl viologen by hydroquinones. - Proc. roy. Soc. London A *373* : 179 - 187, 1980.

*49131 - DAS, V.S.R., RAJENDRUDU, G. : A simple photorespiratory ratio for the delimitation of C_4 from the C_3 plants. - Proc. Indian Acad. Sci., Sec. B *84* : 148 - 153, 1976.

*49132 - DAS, V.S.R., RAJENDRUDU, G. : The photosynthetic efficiency of flag leaf in relation to structural features in some crop plants. - Indian J. Plant Physiol. *20* : 123 - 128, 1977.

49133 - DASGUPTA, S.R., RYAN, M.D. : The electron-transfer kinetics of spinach ferredoxin with strong reductants. - Biochim. biophys. Acta *680*: 242 - 249, 1982.

*49134 - DATZ, G., DÖHLER, G. : Light-dependent changes in the lipid and fatty acid composition of phycocyanin-free photosynthetic lamellae of *Synechococcus*. - Z. Naturforsch. *36 C* : 856 - 862, 1981. [Chl, Bll.]

49135 - DAUN, J.K. : The relationship between rapeseed chlorophyll, rapeseed oil chlorophyll and percentage green seeds. - J. amer. Oil Chem. Soc. *59* : 15 - - 18, 1982.

49136 - DAVE, Y.S., RAO, K.S. : Plastid ultrastructure in the cambium of teak (*Tectona grandis* L.f.). - Ann. Bot. *49* : 425 - 427, 1982.

*49137 - DAVIDSON, E., COGDELL, R.J. : The polypeptide composition of the B850 light--harvesting pigment-protein complex from *Rhodopseudomonas sphaeroides*, R26.1. - FEBS Lett. *132* : 81 - 84, 1981.

49138 - DAVIDSON, V.L., KNAFF, D.B. : The electrochemical proton gradient in the photosynthetic purple sulphur bacterium *Chromatium vinosum*. - Photochem. Photobiol. *36* : 551 - 558, 1982.

*49139 - DAVIDYUK, L.P., VSHIVKOVA, G.F. : Sravnitel'noe izuchenie karotinoidov v list'yakh belo- i zheltomyasykh sortov persika. [Comparative study of carotenoids in leaves of white- and yellow-fleshed peach cultivars.] - Tr. nikitsk. bot. Sada *83* [AKIMOV, Yu.A. (ed.) : Biologicheski Aktivnye Veshchestva Plodovykh, Pryanoaromaticheskikh i Dekorativnykh Rasteniĭ]:103 - 110, 1981. [In R, ab : E.]

49140 - DAVIES, A.G., SLEEP, J.A., HARBOUR, D.S. : Germanic acid inhibition of carbon fixation in natural phytoplankton assemblages. - Limnol. Oceanogr. *27* : 357 - 361, 1982.

49141 - DAVIES, B.H. : Carotenoids of algae and higher plants. - In : ZABORSKY, O.R. (ed.) : CRC Handbook of Biosolar Resources. Vol.I/1. Pp. 73 - 81. CRC Press, Boca Raton 1982.

49142 - DAVIES, F.S., JOHNSON, C.R. : Water stress, growth, and critical water potentials of rabbiteye blueberry (*Vaccinium ashei* READE). - J. amer. Soc. hort. Sci. *107* : 6 - 8, 1982. [Stomatal resistance.]

*49143 - DAVIES, G.E., BERG, S.P. : Melittin inhibition and uncoupling of spinach thylakoids. - Arch. Biochem. Biophys. *211* : 297 - 304, 1981. [Ps.]

49144 - DAVIES, I. : Developmental characteristics of grass varieties in relation to their herbage production. 6. Spring and summer growth of Sabrina hybrid ryegrass and RvP Italian ryegrass as influenced by the date of the initial cut in spring and the length of the succeeding growth period. - J. agr. Sci. *98* : 47 - 64, 1982. [Dry-matter accumulation.]

49145 - DAVIES, W.J., RODRIGUEZ, J.L., FISCUS, E.L. : Stomatal behaviour and water movement through roots of wheat plants treated with abscisic acid. - Plant Cell Environm. *5* : 485 - 493, 1982. [Stomatal resistance.]

49146 - **DAVIS, P.G., SIEBURTH, J.McN.** : Differentiation of phototrophic and heterotrophic nanoplankton populations in marine waters by epifluorescence microscopy. - Ann. Inst. océanogr. (Paris) *58* : 249 - 259, 1982. [Chl.]

49147 - **DAVISON, A.** : The effects of fluorides on plant growth and forage quality. -
In : UNSWORTH, M.H., ORMROD, D.P. (ed.) : Effects of Gaseous Air Pollution
in Agriculture and Horticulture. Pp. 267 - 291. Butterworth Scientific,
London - Boston - Sydney - Wellington - Durban - Toronto 1982. [Resistan-
ces.]

49148 - **DAWES, C.J.** : General characteristics of red macroalgae: *Eucheuma.* - In :
ZABORSKY, O.R. (ed.) : CRC Handbook of Biosolar Resources. Vol.I/2. Pp. 55 -
- 62. CRC Press, Boca Raton 1982. [Ps.]

49149 - **DAWES, C.J.** : Primary productivity of macroalgae in Florida, the Carribean,
and the South Atlantic. - In : ZABORSKY, O.R. (ed.) : CRC Handbook of Biosolar Resources. Vol.I/2. Pp. 441 - 445. CRC Press, Boca Raton 1982.

49150 - **DAY, W., PARKINSON, K.J.** : Application to wheat and barley of two leaf photosynthesis models for C_3 plants. - Plant Cell Environm. *5* : 501 - 507,
1982.

*49151 - **DAYHOFF, M.O., SCHWARTZ, R.M.** : Evidence on the origin of eukaryotic mitochondria from protein and nucleic acid sequences. - Ann. New York Acad.
Sci. *361* : 92 - 104, 1981. [Chloroplast.]

49152 - **DAZA, L.M., DONAIRE, J.P.** : Lipid biosynthesis by chloroplasts from olive
tree leaves. - Physiol. Plant. *54* : 207 - 212, 1982.

49153 - **DEAN, C., LEECH, R.M.** : Genome expression during normal leaf development.
1. Cellular and chloroplast numbers and DNA, RNA, and protein levels in
tissues of different ages within a seven-day-old wheat leaf. - Plant Phy-
iol. *69* : 904 - 910, 1982. [RuBPC.]

49154 - **DEAN, C., LEECH, R.M.** : Genome expression during normal leaf development.
2. Direct correlation between ribulose bisphosphate carboxylase content
and nuclear ploidy in a polyploid series of wheat. - Plant Physiol. *70* :
1605 - 1608, 1982.

49155 - **DEAN, T.J., PALLARDY, S.G., COX, G.S.** : Photosynthetic responses of black
walnut (*Junglans nigra*) to shading. - Can. J. Forest Res. *12* : 725 - 730,
1982.

49156 - **DEBNATH, R., MUKHERJI, S.** : Barium effects in *Phaseolus aureus, Cephalandra
indica, Canna indica, Beta vulgaris, Triticum aestivum* and *Lactuca sativa.* -
Biol. Plant. *24* : 423 - 429, 1982. [Chl, Car.]

49157 - **DEBUS, R.J., VALKIRS, G.E., OKAMURA, M.Y., FEHER, G.** : Localization of the
secondary quinone-binding site in reaction centers from *Rhodopseudomonas
sphaeroides* R-26 by antibody inhibition of electron transfer. - Biochim.
biophys. Acta *682* : 500 - 503, 1982.

49158 - **DEEVA, N.M., MIKHAĬLOV, V.V., REZUNKOVA, N.A.** : Sezonnye izmeneniya indeksa
listovoĭ poverkhnosti osnovnykh dominantov driadovo-osokovo-mokhovoĭ melko-
bugorkovoĭ tundry (Zapadnyĭ Pamir). [Seasonal changes in leaf area index of
the main dominants of the *Dryas*-sedge-moss hummock tundra (West Pamir).]-
Bot. Zh. *67* : 224 - 228, 1982. [In R.]

*49159 - **DEGENS, E.T., KEMPE, S.** : Istoriya CO_2. [History of CO_2.] - In : Kompleks-
nyĭ Global'nyĭ Monitoring Zagryazneniya Okruzhayushcheĭ Prirodnoĭ Sredy.
Pp. 50 - 61. Leningrad 1980. [In R.]

49160 - **DE HAAN, H., DE BOER, T., KRAMER, H.A., VOERMAN, J.** : Applicability to
light absorbance as a measure of organic carbon in humic lake water. - Water
Res. *16* : 1047 - 1050, 1982.

49161 - **DEI, M.** : A two-fold action of benzyladenine on chlorophyll formation in
etiolated cucumber cotyledons. - Physiol. Plant. *56* : 407 - 414, 1982.

49162 - **DeJONG, T.M.** : Leaf nitrogen content and CO_2 assimilation capacity in peach.
- J. amer. Soc. hort. Sci. *107* : 955 - 959, 1982.

49163 - DeJONG, T.M., DRAKE, B.G., PEARCY, R.W. : Gas exchange responses of Chesa-
peake Bay tidal marsh species under field and laboratory conditions. - Oecologia *52* : 5 - 11, 1982.

49164 - DeJONG, T.M., PHILLIPS, D.A. : Water stress effects on nitrogen assimilation
and growth of *Trifolium subterraneum* L. using dinitrogen or ammonium nitrate. - Plant Physiol. *69* : 416 - 420, 1982. [Ps.]

49165 - DE KLERK-KIEBERT, Y.M., KNEPPERS, T.J.A., BAKKER, P.A.H.M., SCHALK, H.H. :
Comparison of dry matter synthesis and photosynthetic characteristics of
chlorophyllous and non-chlorophyllous cell suspension cultures of soybean
(*Glycine max.*L.). - Z. Pflanzenphysiol. *105* : 445 - 456, 1982.

49166 - DE KOK, J. : Photoreactions of the chromophore cations of denatured species of C-phycocyanin and allophycocyanin. - Photochem. Photobiol. *35* : 849 -
- 851, 1982.

*49167 - DEKOV, I., PETROVA, L., K"DREV, T. : Morfometrichni i kolichestveni izsled-
vaniya na khloroplasti v tsarevichni rasteniya, otglezhdani pri magneziev
i voden nedostig. [Morphological and quantitative study of chloroplasts in
maize, grown under unsufficient magnesium and water supply.] - In : Fiziologiya na Rasteniyata. Vol.5. Pp. 261 - 264. Sofiya 1980. [In Bulg.]

49168 - DE LA CRUZ, A.A. : Effects of oil on phytoplankton metabolism in natural
and experimental estuarine ponds. - Mar. environm. Res. *7* : 257 - 263,
1982. [Chl.]

*49169 - DeLANGE, R.J., WILLIAMS, L.C., GLAZER, A.N. : The amino acid sequence of
the β subunit of allophycocyanin. - J. biol. Chem. *256* : 9558 - 9566,
1981.

49170 - DELEHANTY, J., FARMERIE, W.G., CHANG, S., BARNETT, W.E. : The evolution of
chloroplasts as determined by transfer RNA sequence analysis. - In : SCHIFF,
J.A., LYMAN, H. (ed.) : On the Origins of Chloroplasts. Pp. 307 - 313. Else-
vier/North-Holland, New York - Amsterdam - Oxford 1982.

49171 - DELEPELAIRE, P., CHUA, N.-H. : Isolation, purification, and characterization
of chlorophyll-protein complexes by polyacrylamide gel electrophoresis at
low temperatures. - In : EDELMAN, M., HALLICK, R.B., CHUA, N.-H. (ed.) :
Methods in Chloroplast Molecular Biology. Pp. 835 - 843. Elsevier Biomedical Press, Amsterdam - New York - Oxford 1982.

49172 - DELRIEU, M.-J. : Evidence for unequal misses in oxygen flash yield sequence in photosynthesis. - Z. Naturforsch. *38C* : 247 - 258, 1982.

49173 - DELSOIR, J.P., DELTOUR, B., LEDENT, J.-F. : Effet de divers insecticides
sur la croissance juvenile et la photosynthèse du maïs. - Med. Fac. Land-
bouww. Rijksuniv. Gent *47* : 211 - 217, 1982.

49174 - DE LUCA D'ORO, G.M., TRIPPI, V.S. : Regulación de clorofilas y proteínas
solubles por cinetina y cicloheximida, en condiciones de luz y oscuridad,
durante la senescencia foliar en *Phaseolus vulgaris* L. [Chlorophylls and soluble proteins regulation by kinetin and cycloheximide, under light and
darkness, during leaf aging of *Phaseolus vulgaris* L.] - Phyton *42* : 73 -
- 78, 1982. [In Span., ab : E.]

49175 - DE LUCA REBELLO, A., MOREIRA, I.N.S. : The influence of various seawater
components on the buffer capacity for CO_2. - Mar. Chem. *11* : 33 - 41, 1982.

49176 - DEMERS, S., LEGENDRE, L. : Water column stability and photosynthetic capacity of estuarine phytoplankton: Long-term relationships. - Mar. Ecol., Progr.
Ser. *7* : 337 - 340, 1982.

49177 - DEMETER, S. : Binary oscillation of the thermoluminescence of chloroplasts
preilluminated by flashes prior to inhibitor addition. - FEBS Lett. *144* :
97 - 100, 1982.

49178 - DEMIDENKO, A.A., PETROV, É.G. : O mekhanizme singlet-tripletnykh perekho-
dov v vosstanovlennykh reaktsionnykh tsentrakh fotosinteziruyushchikh bakteriĭ. [Mechanisms of singlet-triplet transitions in reduced centres of
photosynthesizing bacteria.] - Mol. Biol. (Moskva) *16* : 1203 - 1210, 1982.
[In R, ab : E.]

49179 - DEMIDENKO, A.A., PETROV, É.G. : Vliyanie magnitnogo polya na kinetiku perenosa êlektrona mezhdu dimerom khlorofilla i feofitinom v fotosintetiches-kikh reaktsionnykh tsentrakh. [Influence of magnetic fields on kinetics of electron transfer between chlorophyll dimer and pheophytin in photosynthetic reaction centres.] - Eesti NSV Tead. Akad. Toim., Füüs., Mat. *31* : 219 - 223, 1982. [In R, ab : E, Est.]

49180 - DEN BLANKEN, H.J., HOFF, A.J. : High-resolution optical absorption-difference spectra of the triplet state of the primary donor in isolated reaction centers of the photosynthetic bacteria *Rhodopseudomonas sphaeroides* R-26 and *Rhodopseudomonas viridis* measured with optically detected magnetic resonance at 1.2 K. - Biochim. biophys. Acta *681* : 365 - 374, 1982.

49181 - DEN BLANKEN, H.J., VAN DER ZWET, G.P., HOFF, A.J. : Study of the long-wavelength fluorescence band at 920 nm of isolated reaction centers of the photosynthetic bacterium *Rhodopseudomonas sphaeroides* R-26 with fluorescence--detected magnetic resonance in zero field. - Biochim. biophys. Acta *681* : 375 - 382, 1982.

49182 - DEN BLANKEN, H.J., VAN DER ZWET, G.P., HOFF, A.J. : Electron spin resonance in zero field of the photoinduced triplet state in isolated reaction centers of *Rhodopseudomonas sphaeroides* R-26 detected by the singlet ground state absorbance (ADMR). - Chem. Phys. Lett. *85* : 335 - 338, 1982.

49183 - DENCHER, N.A., HEYN, M.P. : Preparation and properties of monomeric bacterio-rhodopsin. - In : COLOWICK, S.P., KAPLAN, N.O. (ed.) : Methods in Enzymology. Vol.88. Pp. 5 - 10. Academic Press, New York - London - Paris - San Diego - San Francisco - São Paulo - Sydney - Tokyo - Toronto 1982.

49184 - DENCHER, N.A., HILDEBRAND, E. : Photobehavior of *Halobacterium halobium*. - In : COLOWICK, S.P., KAPLAN, N.O. (ed.) : Methods in Enzymology. Vol.88. Pp. 420 - 426. Academic Press, New York - London - Paris - San Diego - San Francisco - São Paulo - Sydney - Tokyo - Toronto 1982.

49185 - DENCHER, N.A., RAFFERTY, C.N., SPERLING, W. : Photochemistry and isomer determination of 13-*cis*-and *trans*-bacteriorhodopsin. - In : COLOWICK, S.P., KAPLAN, N.O. (ed.) : Methods in Enzymology. Vol.88. Pp. 167 - 174. Academic Press, New York - London - Paris - San Diego - San Francisco - São Paulo - Sydney - Tokyo - Toronto 1982.

49186 - DENHOLM, J.V., CONNOR, D.J. : Potential photosynthesis in trellis-type orchard canopies. - Aust. J. Plant Physiol. *9* : 629 - 640, 1982.

49187 - DENNIS, W.D., WOLEDGE, J. : Photosynthesis by white clover leaves in mixed clover/ryegrass swards. - Ann. Bot. *49* : 627 - 635, 1982.

49188 - DENNISON, W.C., ALBERTE, R.S. : Photosynthetic responses of *Zostera marina* L. (eelgrass) to *in situ* manipulations of light intensity. - Oecologia *55* : 137 - 144, 1982.

49189 - deNOYELLES, F., KETTLE, W.D., SINN, D.E. : The responses of plankton communities in experimental ponds to atrazine, the most heavily used pesticide in the United States. - Ecology *63* : 1285 - 1293, 1982. [Ps.]

49190 - DE OLIVEIRA, S.A., BLANCO, S.A., ENGLEMAN, E.M. : Influéncia do boro nos parámetros morfológicos e fisiológicos de crescimento do feijoeiro. [Effect of boron on the morphological and physiological growth parameters of bean.] - Pesq. agropec. bras. *17* : 683 - 688, 1982. [Growth analysis; in Port., ab : E.]

49191 - DESAI, T.S., TATAKE, V.G., SANE, P.V. : A slow component od delayed light emission as a function of temperature mimics glow peaks in photosynthetic membranes. Evidence for identity. - Biochim. biophys. Acta *681* : 383 - 387, 1982.

49192 - DESAI, T.S., TATAKE, V.G., SANE, P.V. : High temperature peak on glow curve of the photosynthetic membrane. - Photosynthetica *16* : 129 - 133, 1982.

49193 - DESHPANDE, R., NIMBALKAR, J.D. : Effect od salt-stress on translocation of photosynthates in pigeon pea. - Plant Soil *65* : 129 - 132, 1982.

49194 - DESJARDINS, R.L., BRACH, E.J., ALVO, P., SCHUEPP, P.H. : Aircraft monitoring of surface carbon dioxide exchange. - Science *216* : 733 - 735, 1982.

*49195 - DEVASSY, V.P., BHATTATHIRI, P.M.A. : Phytoplankton ecology of the Cochin
backwater. - Indian J. mar. Sci. *3* : 46 - 50, 1974. [Primary production.]

*49196 - DEVASSY, V.P., BHATTATHIRI, P.M.A. : Distribution of phytoplankton & chlo-
rophyll *a* around Little Andaman Island. - Indian J. mar. Sci. *10* : 248 -
- 252, 1981.

*49197 - DEVASSY, V.P., BHATTATHIRI, P.M.A., QASIM, S.Z. : *Trichodesmium* phenomenon.
- Indian J. mar. Sci. *7* : 168 - 186, 1978. [Chl.]

*49198 - DEVASSY, V.P., BHATTATHIRI, P.M.A., QASIM, S.Z. : Succession of organisms
following *Trichodesmium* phenomenon. - Indian J. mar. Sci. *8* : 89 - 93, 1979.
[Chl.]

49199 - DE VECCHI, L. : Structure and development of plastids in hypsophylls of
Euphorbia pulcherima and *Bouganwillea* sp. - Cytologia *47* : 1 - 9, 1982.

49200 - DEVLIN, R.M., KARCZMARCZYK, S.J., ZBIEC, I.I., SARAS, C.N. : Influence of
norflurazon on the light activation of oxyfluorfen. - Proc. annu. Meet.
northeast. Weed Sci. Soc. *36* : 81 - 85, 1982. [Car.]

49201 - DE WILTON, A., HALEY, L.V., KONINGSTEIN, J.A. : Detection limitations of
photomultiplier tubes in pulsed-laser emission spectroscopy: time-resolved
fluorescence spectra of chlorophyll solutions. - Can. J. Chem. *60* : 2198 -
- 2206, 1982.

49202 - DHALIWAL, A.S., MALIK, C.P., SINGH, M.B. : Pattern of non-photosynthetic
CO_2 fixation in stigmatic & stylar tissues of *Brassica campestris* following
incompatible pollination. - Indian J. exp. Biol. *20* : 462 - 464, 1982.

*49203 - DIAMANTOGLOU, S., MELETIOU-CHRISTOU, M.S. : Changes of storage lipids, fat-
ty acids and carbohydrates in vegetative parts of Mediterranean evergreen
sclerophylls during one year. - In : MARGARIS, N.S., MOONEY, H.A. (ed.) :
Components of Productivity of Mediterranean-Climate Regions. Basic and
Applied Aspects. Pp. 121 - 127. Dr.W.Junk Publ., The Hague - Boston - London
1981. [Photosynthates.]

*49204 - DIAMANTOGLOU, S., MITRAKOS, K. : Leaf longevity in Mediterranean evergreen
sclerophylls. - In : MARGARIS, N.S., MOONEY, H.A. (ed.) : Components of
Productivity of Mediterranean-Climate Regions. Basic and Applied Aspects.
Pp. 17 - 19. Dr.W.Junk Publ., The Hague - Boston - London 1981. [Leaf life
span.]

49205 - DIAZ, F.M., KOHASHI, J.S. : Distribucion de materia seca en el frijol (*Phaseolus vulgaris* L.) bajo condiciones de campo. [Distribution of aboveground
dry matter of bean (*Phaseolus vulgaris* L.) under field conditions.] - Turrialba *32* : 19 - 27, 1982. [In Span., ab : E.]

49206 - DICKSON, R.E., NELSON, E.A. : Fixation and distribution of ^{14}C in *Populus
deltoides* during dormancy induction. - Physiol. Plant. *54* : 393 - 401,
1982.

49207 - DICKSON, R.E., SHIVE, J.B.,Jr. : $^{14}CO_2$ fixation, translocation, and carbon
metabolism in rapidly expanding leaves of *Populus deltoides*. - Ann. Bot.
50 : 37 - 47, 1982.

49208 - DIEPENBROCK, W., GEISLER, G. : Einfluß der Stickstoffernährung auf den Gehalt an Chlorophyll und Galaktolipiden alter und junger Rapsblätter. - Z.
Pflanzenernähr. Bodenk. *145* : 2 - 9, 1982.

49209 - DIERSTEIN, R., DREWS, G. : Membrane differentiation and assembly of the
pigment-protein complexes of the photosynthetic bacterium *Rhodopseudomonas
capsulata*. - In : AKOYUNOGLOU, G., EVANGELOPOULOS, A.E., GEORGATSOS, J.,
PALAIOLOGOS, G., TRAKATELLIS, A., TSIGANOS, C.P. (ed.) : Cell Function and
Differentiation, Part B. Biogenesis of Energy Transducing Membranes and
Membrane and Protein Energetics. Pp. 247 - 256. Alan R. Liss, New York
1982.

49210 - DIJAK, M., ORMROD, D.P. : Some physiological and anatomical characteristics associated with differential ozone sensitivity among pea cultivars. - Environm. exp. Bot. *22* : 395 - 402, 1982. [Stomatal resistance.]

49211 - DILLEY, R.A. : Quinones. - In : ZABORSKY, O.R. (ed.) : CRC Handbook of Biosolar Resources. Vol.I/1. Pp. 167 - 171. CRC Press, Boca Raton 1982.

49212 - DILLEY, R.A., PROCHASKA, L.J., BAKER, G.M., TANDY, N.E., MILLNER, P.A. : Proton-membrane interactions in chloroplast bioenergetics. - In : SLAYMAN, C.L. (ed.) : Electrogenic Ion Pumps. (Curr. Topics Membr. Transport. Vol. 16.) Pp. 345 - 369. Academic Press, New York 1982.

*49213 - DILWORTH, M.F., GANTT, E. : Phycobilisome-thylakoid topography on photosynthetically active vesicles of *Porphyridium cruentum*. - Plant Physiol. *67* : 608 - 612, 1981.

49214 - DI MARCO, G., GREGO, S., TRICOLI, D. : Ribulose bisphosphate carboxylase-oxygenase in field grown wheat in two contrasting season. - Agrochimica *26* : 146 - 156, 1982.

*49215 - DIVATE, M.R., PANDEY, R.M. : Salt tolerance in grapes. III. Effects of salinity on chlorophyll, photosynthesis and respiration. - Indian J. Plant Physiol. *24* : 74 - 79, 1981.

*49216 - DOBRINSKIĬ, L.N., MALAFEEV, Yu.M., KRYAZHIMSKIĬ, F.V. : Materialy po sutochnoĭ dinamike uglekislotnogo balansa lugovogo fitotsenoza. [Diurnal dynamics of carbon dioxide balance in meadow phytocenoses.] - In : Ėkologicheskie Issledovaniya v Lesnykh i Lugovykh Biogeotsenozakh Ravninnogo Zaural'ya. Pp. 5 - 9. Inst. Ėkol. Rast. Zhivot. Ural'. Nauch. Tsentr Akad. Nauk SSSR, Sverdlovsk 1978. [In R.]

49217 - DODD, J.L., LAUENROTH, W.K., HEITSCHMIDT, R.K. : Effect of controlled SO_2 exposure·on net primary production and plant biomass dynamics. - J. Range Manage. *35* : 572 - 579, 1982.

49218 - DODGE, A.D. : The role of light and oxygen in the action of photosynthetic inhibitor herbicides. - In : MORELAND, D.E., St.JOHN, J.B., HESS, F.D. (ed.) : Biochemical Responses Induced by Herbicides. Pp. 57 - 77. Amer. Chem. Soc., Washington 1982.

49219 - DOGBO, O., CAMARA, B., MONÉGER, R. : Étude de l'activité chlorophylle synthétase dans les chloroplastes et chromoplastes isolés de fruit de Poivron (*Capsicum annuum* L.). - Compt. rend. Acad. Sci. Paris,Sér. III *295* : 477 - - 480, 1982.

49220 - DÖHLER, G. : CO_2 fixation in *Anabaena cylindrica*. - Z. Naturforsch. *37 C* : 213 - 217, 1982.

49221 - DÖHLER, G. : Effect of UV-B radiation on the marine diatom *Bellerochea yucatanensis*. - In : BAUER, H., CALDWELL, M.M., TEVINI, M., WORREST, R.C. (ed.) : Biological Effects of UV-B Radiation. Pp. 211 - 215. Gesellsch. Strahlen- Umweltforsch., München 1982.

49222 - DÖHLER, G. : Metabolismus von ^{14}C-Glycin, ^{14}C-Serin und ^{14}C-Homoserin in *Synechococcus*. - Biochem. Physiol. Pflanzen *177* : 244 - 250, 1982. [Ps.]

*49223 - DÖHLER, G., BARCKHAUSEN, R., RUPPEL, M. : Ultrastructure of differently pigmented *Synechococcus* cells. - Z. Naturforsch. *36 C* : 907 - 909, 1981.

49224 - DÖHLER, G., LECLERC, J.-C. : Photosynthetic adaptation in *Synechococcus* cells. - Z. Naturforsch. *37 C* : 1075 - 1080, 1982.

*49225 - DÖHLER, G., ROßLENBROICH, H.-J. : Photosynthetic assimilation of ^{15}N-ammonia and ^{15}N-nitrate in the marine diatoms *Bellerochea yucatensis* (VON STOSCH) and *Skeletonema costatum*. - Z. Naturforsch. *36C* : 834 - 839, 1981.

49226 - DOI, M., TAKAMIYA, K.-I., NISHIMURA, M. : Isolation and purification of membrane-bound cytochrome *b*-560 from photosynthetic bacterium *Chromatium vinosum*. - Photosynthesis Res. *3* : 131 - 139, 1982.

49227 - DOI, M., TAKAMIYA, K.-I., NISHIMURA, M. : Localization of membrane-bound cytochromes of photosynthetic bacterium *Chromatium vinosum*. - Photosynthesis Res. *3* : 357 - 361, 1982.

49228 - DOLEY, D. : Photosynthetic productivity of forest canopies in relation to solar radiation and nitrogen cycling. - Aust. Forest Res. *12* : 245 - 261, 1982.

49229 - DONG, C.-N., ARTECA, R.N. : Changes in photosynthetic rates and growth following root treatments of tomato plants with phytohormones. - Photosynthesis Res. *3* : 45 - 52, 1982.

49230 - DONKIN, M.E., TAFFS, J., MARTIN, E.S. : A study of the *in-vitro* regulation of phosphoenolpyruvate carboxylase from the epidermis of *Commelina communis* by malate and glucose-6-phosphate. - Planta *155* : 416 - 422, 1982.

49231 - DORAISWAMY, P.C., THOMPSON, D.R. : A crop moisture stress index for large areas and its application in the prediction of spring wheat phenology. - Agr. Meteorol. *27* : 1 - 15, 1982. [Growth analysis.]

49232 - DORNE, A.-J., BLOCK, M.A., JOYARD, J., DOUCE, R. : Studies on the localization of enzymes involved in galactolipid metabolism in chloroplast envelope membranes. - In : WINTERMANS, J.F.G.M., KUIPER, P.J.C. (ed.) : Biochemistry and Metabolism of Plant Lipids. Pp. 153 - 164. Elsevier Biomedical Press B.V., Amsterdam 1982.

49233 - DORNE, A.J., BLOCK, M.A., JOYARD, J., DOUCE, R. : The galactolipid:galactolipid galactosyltransferase is located on the outer surface of the outer membrane of the chloroplast envelope. - FEBS Lett. *145* : 30 - 34, 1982.

49234 - DORNE, A.-J., CARDE, J.-P., JOYARD, J., BÖRNER, T., DOUCE, R. : Polar lipid composition of a plastid ribosome-deficient barley mutant. - Plant Physiol. *69* : 1467 - 1470, 1982.

49235 - DÖRNEMANN, D., SENGER, H. : Physical and chemical properties of chlorophyll RCI extracted from photosystem I of spinach leaves and from green algae. - Photochem. Photobiol. *35* : 821 - 826, 1982.

*49236 - DOROSHEK, A.S. : O dinamike bioêlektricheskikh perekhodnykh protsessov list'ev rasteniĭ. [Dynamics of light-induced bioelectric transition processes in leaves.] - In : Svetozavisimaya Bioêlektricheskaya Aktivnost' List'ev Rasteniĭ. Pp. 84 - 91, III - IV. Ural'. Gos. Univ., Sverdlovsk 1980. [Ps; in R, ab : E.]

*49237 - DOROSHEK, A.S., KANDAUROVA, G.S. : Vliyanie prostranstvenno-neodnorodnogo magnitnogo polya na svetozavisimuyu bioêlektricheskuyu aktivnost' list'ev rasteniĭ. [Influence of spatial-heterogeneous magnetic field on light-dependent bioelectric activity of plant leaves.] - In : Svetozavisimaya Bioêlektricheskaya Aktivnost' List'ev Rasteniĭ. Pp. 105 - 107, 119. Ural'. Gos. Univ., Sverdlovsk 1977. [Ps; in R, ab : E.]

*49238 - DOROSHEK, A.S., KORONA, O.A. : O initsiiruyushchikh mekhanizmakh svetozavisimoĭ bioêlektricheskoĭ reaktsii rasteniĭ. [Primary mechanisms of light-dependent electroresponses of plants.] - In : Svetozavisimaya Bioêlektricheskaya Aktivnost' List'ev Rasteniĭ. Pp. 92 - 96, IV. Ural'. Gos. Univ., Sverdlovsk 1980. [Ps; in R, ab : E.]

*49239 - DOROSHENKO, T.N., PETIBSKAYA, V.S. : Osobennosti fotosinteza i dykhaniya sortov risa intensivnogo tipa. [Photosynthesis and respiration peculiarities of rice cultivars of the intensive type.] - Byull. nauch.-tekh. Inf. vsesoyuz. nauch.-issled. Inst. Risa *28* : 38 - 42, 1980. [In R, ab : E.]

49240 - DOS SANTOS, C.P., HALL, D.O. : Thylakoid polypeptides of light and dark aged chloroplasts. - Plant Physiol. *70* : 795 - 802, 1982.

49241 - DOSSOU-YOVO, S., PRIOUL, J.-L., DEMARLY, Y. : Croissance et photosynthèse comparées de phénovariants de riz.-Effet de l'ombrage. - Agronomie *2* : 493 - - 502, 1982.

49242 - DOUCE, R., JOYARD, J. : Purification of the chloroplast envelope. - In : EDELMAN, M., HALLICK, R.B., CHUA, N.-H.(ed.) : Methods in Chloroplast Molecular Biology. Pp. 239 - 256. Elsevier Biomedical Press, Amsterdam - New York - Oxford 1982.

49243 - DOUGLAS, T.J., VILLALOBOS, V.M., THOMPSON, M.R., THORPE, T.A. : Lipid and pigment changes during shoot initiation in cultured explants of *Pinus radiata*. - Physiol. Plant. *55* : 470 - 477, 1982.

49244 - DOUILLARD, R., BERGERON, E., SCALBERT, A. : Quelques caractéristiques des lipoxygénases solubles chloroplastiques et non chloroplastiques des feuilles de Pois. - Physiol. vég. *20* : 377 - 384, 1982. [Chl.]

49245 - DOUILLARD, R., BURGHOFFER, C., COSTES, C. : Structure excitonique de complexes hydroéthanoliques de la lutéine et de la zéaxanthine. - Physiol. vég. *20* : 123 - 136, 1982.

*49246 - DOUKAS, A.G., PANDE, A., SUZUKI, T., CALLENDER, R.H., HONIG, B., OTTOLENGHI, M. : On the mechanism of hydrogen-deuterium exchange in bacteriorhodopsin. - Biophys. J. *33* : 275 - 280, 1981.

49247 - DOWNER, N.W., ENGLANDER, J.J. : Tritium-hydrogen exchange kinetics. - In : COLOWICK, S.P., KAPLAN, N.O. (ed.): Methods in Enzymology. Vol.88. Pp. 673 - - 676. Academic Press, New York - London - Paris - San Diego - San Francisco - São Paulo - Sydney - Tokyo - Toronto 1982. [Bacteriorhodopsin.]

49248 - DOWNTON, W.J.S., BERRY, J.A. : Chlorophyll fluorescence at high temperature. - Biochim. biophys. Acta *679* : 474 - 478, 1982.

*49249 - DOWNTON, W.J.S., BJÖRKMAN, O., PIKE, C.S. : Consequences of increased atmospheric concentrations of carbon dioxide for growth and photosynthesis of higher plants. - In : PEARMAN, G.I. (ed.) : Carbon Dioxide and Climate: Australian Research. Pp. 143 - 151. Australian Academy of Science, Canberra 1980.

*49250 - DRACHEV, L.A., FROLOV, V.N., KAULEN, A.D., LIBERMAN, E.A., OSTROUMOV, S.A., PLAKUNOVA, V.G., SEMENOV, A.Yu., SKULACHEV, V.P. : Reconstitution of biological molecular generators of electric current. Bacteriorhodopsin. - J. biol. Chem. *251* : 7059 - 7065, 1976.

49251 - DRACHEV, L.A., KAULEN, A.D., SKULACHEV, V.P., VOYTSITSKY, V.M. : Bacteriorhodopsin-mediated photoelectric responses in lipid/water systems. - J. Membr. Biol. *65* : 1 - 12, 1982.

*49252 - DRASKOVITS, R.M. : Photosynthetic activity of species in a beechwood II. Spring-summer aspects. - Ann. Univ. Sci. budapest. Rolando Eötvös nom., Sect. biol. *22-23* : 57 - 64, 1980-81.

*49253 - DRASKOVITS, R.M., ÁBRÁNYI, A. : Effect of the illumination in different types of forests. - Ann. Univ. Sci. budapest. Rolando Eötvös nom., Sect. biol. *22-23* : 65 - 70, 1980-81.

*49254 - DRAXLER, G. : Gaswechselmessungen an *Spirogyra*-Arten am natürlichen Standort. - Verh. zool.-bot. Ges. (Wien) *116-117* : 83 - 98, 1978.

49255 - DRECHSLER, Z., NEUMANN, J. : Inhibition of oxygen evolution in chloroplasts by ferricyanide. - Plant Physiol. *70* : 840 - 843, 1982.

49256 - DREWS, G. : Composition and development of the bacterial photosynthetic apparatus. - In : KAPLAN, N.O., ROBINSON, A. (ed.) : From Cyclotrons to Cytochromes. Pp. 355 - 366. Academic Press, New York 1982.

49257 - DRING, M.J., BROWN, F.A. : Photosynthesis of intertidal brown algae during and after periods of emersion: a renewed search for physiological causes of zonation. - Mar. Ecol. *8* : 301 - 308, 1982.

49258 - DRING, M.J., JEWSON, D.H. : What does ^{14}C uptake by phytoplankton really measure? A theoretical modelling approach. - Proc. roy. Soc. London B *214* : 351 - 368, 1982.

49259 - DRON, M., RAHIRE, M., ROCHAIX, J.-D. : Sequence of the chloroplast DNA region of *Chlamydomonas reinhardii* containing the gene of the large subunit of ribulose bisphosphate carboxylase and parts of its flanking genes. - J. mol. Biol. *162*: 775 - 793, 1982.

49260 - DRUCKMANN, S., OTTOLENGHI, M., PANDE, A., PANDE, J., CALLENDER, R.H. : Acid-
-base equilibrium of the Schiff base in bacteriorhodopsin. - Biochemistry
21 : 4953 - 4959, 1982.

49261 - DRUMM-HERREL, H., MOHR, H. : Effect of blue/UV light on anthocyanin synthesis in tomato seedlings in the absence of bulk carotenoids. - Photochem.
Photobiol. *36* : 229 - 233, 1982.

49262 - DU, S.-H., FANG, S.C. : Uptake of elemental mercury vapor by C_3 and C_4 species. - Environm. exp. Bot. *22* : 437 - 443, 1982. [Resistances.]

49263 - DUBERTRET, G., LEFORT-TRAN, M. : Chloroplast molecular structure with par-
ticular reference to thylakoids and envelopes. - In : BUETOW, D.E. (ed.) :
The Biology of *Euglena*. Vol.3. Physiology. Pp. 253 - 312. Academic Press,
New York 1982.

49264 - DUBROVSKIĬ, V.T., BALASHOV, S.P., SINESHCHEKOV, O.A., CHEKULAEVA, L.N.,
LITVIN, F.F. : Fotoindutsirovannye izmeneniya kvantovykh vykhodov foto-
khimicheskogo tsikla prevrashcheniĭ bakteriorodopsina i transmembrannogo
perenosa protonov v kletkakh *Halobacterium halobium*. [Light-induced changes
in quantum yields of the photochemical cycle of bacteriorhodopsin conversion
and transmembrane proton transfer in *Halobacterium halobium* cells.] - Biokhimiya *47* : 1230 - 1240, 1982. [In R, ab : E.]

49265 - DUCRUET, J.M., DE PRADO, R. : Comparison of inhibitory activity of amides
derivatives in triazine-resistant and -susceptible chloroplasts from *Chenopodium album* and *Brassica campestris*. - Pestic. Biochem. Physiol. *18* : 253 -
- 261, 1982.

49266 - DUERR, E.O., KUMAZAWA, S., MITSUI, A. : Nitrogen fixation in free-living
algae. - In : ZABORSKY, O.R. (ed.) : CRC Handbook of Biosolar Resources.
Vol.I/1. Pp. 251 - 264. CRC Press, Boca Raton 1982. [Ps.]

49267 - DUERR, E.O., MITSUI, A. : Salinity preference and tolerance of aquatic photosynthetic organisms. - In : ZABORSKY, O.R. (ed.) : CRC Handbook of Biosolar Resources. Vol.I/2. Pp. 223 - 244. CRC Press, Boca Raton 1982.

49268 - DUFOUR, P. : Modèles semi-empiriques de la production phytoplanktonique
en milieu lagunaire tropical (Côte-d'Ivoire). - Acta oecol.,Oecol.gen. *3* :
223 - 239, 1982.

49269 - DUFOUR, P., DURAND, J.-R. : La production végétale des lagunes de Côte
d'Ivoire. - Rev. Hydrobiol. trop. *15* : 209 - 230, 1982. [Chl.]

49270 - DUGGAN, J.X., MELLER, E., GASSMAN, M.L. : Catabolism of 5-aminolevulinic
acid to CO_2 by etiolated barley leaves. - Plant Physiol. *69* : 19 - 22,
1982.

49271 - DUGGAN, J.X., REBEIZ, C.A. : Chloroplast biogenesis. 37. Induction of chlorophyllide *a* (E459 F675) accumulation in higher plants. - Plant Sci. Lett.
24 : 27 - 37, 1982.

49272 - DUGGAN, J.X., REBEIZ, C.A. : Chloroplast biogenesis. 38. Quantitative detection of a chlorophyllide *b* pool in higher plants. - Biochim. biophys.
Acta *679* : 248 - 260, 1982.

49273 - DUGGAN, J.X., REBEIZ, C.A. : Chloroplast biogenesis. 42. Conversion of divinyl chlorophyllide *a* to monovinyl chlorophyllide *a in vivo* and *in vitro*. -
Plant Sci. Lett. *27* : 137 - 145, 1982.

49274 - DUJARDIN, E. : Extinction of the *in-vivo* low-temperature fluorescence of
chlorophyll *a* by long-wavelength-absorbing quenchers formed from protochlorophyllide. - In: AOKOYUNOGLOU, G.,EVANGELOPOULOS,A.E.,GEORGATSOS,J., PALAIOLO-
GOS, G., TRAKATELLIS, A., TSIGANOS, C.P. (ed.) : Cell Function and Differentiation, Part B. Biogenesis of Energy Transducing Membranes and Membrane
and Protein Energetics. Pp. 43 - 52. Alan R. Liss, New York 1982.

49275 - DUJARDIN, E., DELCAMBE, P., LACROSSE, C., SIRONVAL, C. : Biomass produced
by fresh water algae in luke-warm water. - In : GARSSI, G., PALZ, W. (ed.):
Energy from Biomass. Vol.3. Pp. 135 - 141. D. Reidel Publ.Co., Dordrecht -
- Boston - London 1982.

49276 - DUKE, S.O., WICKLIFF, J.L., VAUGHN, K.C., PAUL, R.N. : Tentoxin does not cause chlorosis in greening mung bean leaves by inhibiting photophosphorylation. - Physiol. Plant. *56* : 387 - 398, 1982.

49277 - DUNCAN, M.J., HARRISON, P.J. : Comparison of solvents for extracting chlorophylls from marine macrophytes. - Bot. mar. *25* : 445 - 447, 1982.

49278 - DUNGEY, N.O., DAVIES, D.D. : Protein turnover in the attached leaves of non-stress and stressed barley seedlings. - Planta *154* : 435 - 440, 1982. [Chl.]

☆49279 - DUNHAM, K.R., SELMAN, B.R. : Interactions of inorganic phosphate with spinach coupling factor 1. Effects on ATPase and ADP binding activities. - J. biol. Chem. *256* : 10044 - 10049, 1981.

49280 - DURAND-CHASTEL, H. : General characteristics of blue-green algae (*Cyanobacteria*) : *Spirulina*. - In : ZABORSKY, O.R. (ed.) : CRC Handbook of Biosolar Resources. Vol.I/2. Pp. 19 - 23. CRC Press, Boca Raton 1982. [Chl, Car.]

49281 - DÜRING, H., LOVEYS, B.R. : Diurnal changes in water relations and abscisic acid in field grown *Vitis vinifera* cvs. I. Leaf water potential components and leaf conductance under humid temperate and semiarid conditions. - Vitis *21* : 223 - 232, 1982.

49282 - DURYEA, M.L., LAVENDER, D.P. : Water relations, growth, and survival of root-wrenched Douglas-fir seedlings. - Can. J. Forest Res. *12* : 545 - 555, 1982.

49283 - DUSTAN, P. : Depth-dependent photoadaptation by zooxanthellae of the reef coral *Montastrea annularis*. - Mar. Biol. *68* : 253 - 264, 1982. [Ps.]

49284 - DUTHION, C. : Effets d'une courte période d'excès d'eau sur la croissance et la production du maïs. - Agronomie *2* : 125 - 132, 1982.

☆49285 - DUTTA, P.K., DALLINGER, R., SPIRO, T.G. : Resonance CARS (coherent anti--Stokes Raman scattering) line shapes via Frank-Condon scattering: Cytochromes *c* and β-carotene. - J. chem. Phys. *73* : 3580 - 3585, 1980.

49286 - DUTTON, P.L., GUNNER, M.R., PRINCE, R.C. : Systematic modification of electron transfer kinetics in a biological protein: Replacement of the primary ubiquinone of photochemical reaction centers with other quinones. - In : HÉLÈNE, C., CHARLIER, M., MONTENAY-GARESTIER, T., LAUSTRIAT, G. (ed.) : Trends in Photobiology. Pp. 561 - 570. Plenum Press, New York - London 1982.

49287 - DUVIGNEAUD, P. : Solar energy bioconversion at the ecosystem level. - In : HÉLÈNE, C., CHARLIER, M., MONTENAY-GARESTIER, T., LAUSTRIAT, G. (ed.) : Trends in Photobiology. Pp. 597 - 617. Plenum Press, New York - London 1982.

☆49288 - DVIHALLY, Zs.T. : Über die Primärproduktion des oberen Donauabschnittes in Ungarn. - Ann. Univ. Sci. budapest. Rolando Eötvös nom., Sect. biol. *22-23* : 29 - 33, 1980-81.

49289 - DWARTE, D.M., VESK, M. : Cytochemical localization of biliproteins with silicotungstic acid. - J. Microsc. *126* : 197 - 200, 1982.

49290 - DWARTE, D.M., VESK, M. : Freeze-fracture thylakoid ultrastructure of representative members of "chlorophyll *c*" algae. - Micron *13* : 325 - 326, 1982.

49291 - DYKENS, J.A., SHICK, J.M. : Oxygen production by endosymbiotic algae controls superoxide dismutase activity in their animal host. - Nature *297* : 579 - 580, 1982.

49292 - DZHANUMOV, D.A., BOCHAROV, E.A., KLIMOV, S.V. : Vliyanie temperatury na soderzhanie khloroplastnykh lipidov v pervom liste prorostkov ozimoĭ pshenitsy. [Effect of temperature on the content of chloroplast lipids in the first leaf of winter wheat seedlings.] - Fiziol. Rast. *29* : 1067 - 1074, 1982. [In R, ab : E.]

49293 - **DZIĘCIOŁ, U., MICHALCZUK, L., ANTOSZEWSKI, R.** : Respiratory losses of photo-synthates in the process of translocation. - Acta Physiol. Plant. *4* : 129 - - 138, 1982.

49294 - **EBBESEN, T.W., DELGADO, O., VALLA, A., GIRAUD, M., SAITO, Y., TACHIBANA, H., WADA, A.** : Energy transfer from sodium N-alkyl carbazole sulfonate to zinc tetraphenylporphyrin in micellar solutions. - Photochem. Photobiol. *35* : 665 - 669, 1982. [Ps model.]

49295 - **EBREY, T.G.** : Primary events in bacteriorhodopsin. - In : **ALFANO, R.R.** (ed.): Biological Events Probed by Ultrafast Laser Spectroscopy. Pp. 271 - 280. Academic Press, New York - London - Paris - San Diego - San Francisco - - São Paulo - Tokyo - Toronto 1982.

49296 - **EBREY, T.G.** : Synthetic pigments of rhodopsin and bacteriorhodopsin. - In : **COLOWICK, S.P., KAPLAN, N.O.** (ed.) : Methods in Enzymology. Vol.88. Pp. 516 - - 521. Academic Press, New York - London - Paris - San Diego - San Francis-co - São Paulo - Sydney - Tokyo - Toronto 1982.

49297 - **ECKARDT, F.E., HEERFORDT, L., JØRGENSEN, H.M., VAAG, P.** : Photosynthetic production in Greenland as related to climate, plant cover and grazing pressure. - Photosynthetica *16* : 71 - 100, 1982.

49298 - **EDEN, G., FUCHS, G.** : Total synthesis of acetyl coenzyme A involved in auto-trophic CO_2 fixation in *Acetobacterium woodii*. - Arch. Microbiol. *133* : 66 - - 74, 1982.

49299 - **EDWARDS, G.E., KU, M.S.B., HATCH, M.D.** : Photorespiration in *Panicum milio-ides*, a species with reduced photorespiration. - Plant Cell Physiol. *23* : 1185 - 1195, 1982.

49300 - **EDWARDS, W.D., HEAD, J.D., ZERNER, M.C.** : On the electronic excited states of model chlorophyll. - J. amer. chem. Soc. *104* : 5833 - 5834, 1982.

49301 - **EHWALD, R., JOVTCHEV-SCHACHNER, G.** : Einbau von 2-Desoxy-D-Glucose in Stär-ke - Schlußfolgerungen über den Kohlenhydrattransport zwischen Plastiden und Cytoplasma. - In : **HOFFMANN, P., HIEKE, B.** (ed.) : Photosynthese : Re-gulation und Evolution. (Colloquia Pflanzenphysiologie Nr.5.) Pp. 83 - 86. Humboldt-Universität, Berlin 1982.

49302 - **EICKMEIER, W.G.** : Protein synthesis and photosynthetic recovery in the re-surrection plant, *Selaginella lepidophylla*. - Plant Physiol. *69* : 135 - 138, 1982.

49303 - **EIGENBERG, K.E., CROASMUN, W.R., CHAN, S.I.** : Chlorophyll *a* in bilayer mem-branes. I. The thermal phase diagram with distearoylphosphatidylcholine. - Biochim. biophys. Acta *679* : 353 - 360, 1982.

49304 - **EIGENBERG, K.E., CROASMUN, W.R., CHAN, S.I.** : Chlorophyll *a* in bilayer mem-branes. II. Interaction with distearoylphosphatidylcholine by NMR. - Bio-chim. biophys. Acta *679* : 361 - 368, 1982.

49305 - **EISENBERGER, P., OKAMURA, M.Y., FEHER, G.** : The electronic structure of Fe^{2+} in reaction centers from *Rhodopseudomonas sphaeroides*. II. Extended X-ray fine structure studies. - Biophys. J. *37* : 523 - 538, 1982.

49306 - **EISENSTEIN, L.** : Effect of viscosity on the photocycle of bacteriorhodopsin. - In : **COLOWICK, S.P., KAPLAN, N.O.** (ed.) : Methods in Enzymology. Vol.88. Pp. 297 - 305. Academic Press, New York - London - Paris - San Diego - San Francisco - São Paulo - Sydney - Tokyo - Toronto 1982.

49307 - **ELFIMOV, E.I., VOZNYAK, V.M., PROKHORENKO, I.R.** : Vliyanie magnitnogo polya na fotofizicheskie protsessy v svetosobirayushchem pigmentnom apparate purpurnykh fotosinteziruyushchikh bakteriĭ. [Effect of magnetic field on photophysical processes in light-gathering pigment apparatus of purple pho-tosynthetic bacteria.] - Dokl. Akad. Nauk SSSR *264* : 248 - 252, 1982. [In R.]

49308 - **ELLER, B.M.** : Die Strahlungsabsorption von *Argyroderma pearsonii* (N.E.BROWN) SCHW. in der Vegetations- und Ruheperiode. - Ber. deut. bot. Ges. *95* : 333 - - 340, 1982.

49309 - **ELLIOTT, D.C.** : Levels of membrane components regulated by cytokinins and by temperature and aging pretreatments of cotyledons. - Plant Sci. Lett. *26* : 311 - 323, 1982. [Car.]

49310 - **ELLIS, B.A., KUMMEROW, J.** : Temperature effect on growth rates of *Eriophorum vaginatum* roots. - Oecologia *54* : 136 - 137, 1982. [Ps.]

49311 - **ELNER, J.K., WILDISH, D.J., JOHNSTON, D.W.** : Carbon-14 assimilation by algal communities of oligotrophic ponds treated with formulated aminocarb. - Arch. environm. Contam. Toxicol. *11* : 675 - 679, 1982.

49312 - **ELSAHOOKIE, M.M., ELDABAS, E.E.** : One leaf dimension to estimate leaf area in sunflowers. - Z. Acker- Pflanzenbau *151* : 199 - 204, 1982.

49313 - **EL-SAYED, M.A.** : Time-resolved chromophore resonance Raman and protein fluorescence of the intermediates of the proton pump photocycle of bacteriorhodopsin. - In : COLOWICK, S.P., KAPLAN, N.O. (ed.) : Methods in Enzymology. Vol.88. Pp. 617 - 625. Academic Press, New York - London - Paris - San Diego - San Francisco - São Paulo - Sydney - Tokyo - Toronto 1982.

49314 - **EL-SAYED, M.A.** : Time-resolved chromophore resonance Raman and protein fluorescence of intermediated in some photobiological changes. - In : HÉLÈNE, C., CHARLIER, M., MONTENAY-GARESTIER, T., LAUSTRIAT, G. (ed.) : Trends in Photobiology. Pp. 1 - 10. Plenum Publ. Corp., New York 1982. [Bacteriorhodopsin.]

49315 - **EL-SAYED, M.A., HSIEH, C.-L., NICOL, M.** : Resonance Raman spectra of picosecond transients: Application to bacteriorhodopsin. - In : EISENTHAL, K.B., HOCHSTRASSER, R.M., KAISER, W., LAUBEREAU, A. (ed.) : Picosecond Phenomena III. Pp. 302 - 306. Springer-Verlag, Berlin - Göttingen - New York 1982.

*49316 - **EL-SAYED, S.Z., BIGGS, D.C., STOCKWELL, D., WARNER, R., MEYER, M.** : Biogeography and metabolism of phytoplankton and zooplankton in the Ross Sea, Antarctica. - Antarctic J. *13* (4) : 131 - 133, 1978. [Chl.]

49317 - **EL-SAYED, S.Z., WEBER, L.H.** : Spatial and temporal variations in phytoplankton biomass and primary productivity in the Southwest Atlantic and the Scotia Sea. - Polar Biol. *1* : 83 - 90, 1982.

49318 - **ELSTNER, E.F.** : Oxygen activation and oxygen toxicity. - Annu. Rev. Plant Physiol. *33* : 73 - 96, 1982. [Ps.]

*49319 - **ELSTNER, E.F., OSSWALD, W.** : Chlorophyll photobleaching and ethane production in dichlorophenyldimethylurea-(DCMU) or paraquat-treated *Euglena gracilis* cells. - Z. Naturforsch. *35 C* : 129 - 135, 1980.

49320 - **EMEL'YANOV, L.G.** : Fiziologicheskie i produktsionnye poteri u rasteniĭ na torfyanoĭ pochve pri teplovom stresse. [Physiological and productivity losses in plants on peat soil under thermal stress.] - Dokl. Akad. Nauk belorus. SSR *26* : 853 - 856, 864, 1982. [In R, ab : E.]

49321 - **ENDO, H., HOSOYA, H., KOYAMA, T., ICHIOKA, M.** : Isolation of 10-hydroxypheophorbide *a* as a photosensitizing pigment from alcohol-treated *Chlorella* cells. - Agr. biol. Chem. *46* : 2183 - 2193, 1982.

49322 - **ENDRESS, A.G., SUAREZ, S.J., TAYLOR, O.C.** : Photosynthetic and respiratory consequences of hydrogen chloride gas exposures of *Phaseolus vulgaris* L. and *Spinacea oleracea* L. - Environm. Pollut. Ser. A *29* : 13 - 26, 1982.

49323 - **ENGELMAN, D.M.** : An implication of the structure of bacteriorhodopsin. Globular membrane proteins are stabilized by polar interactions. - Biophys. J. *37* : 187 - 188, 1982.

49324 - **ENGELMAN, D.M., GOLDMAN, A., STEITZ, T.A.** : The identification of helical segments in the polypeptide chain of bacteriorhodopsin. - In : COLOWICK, S.P., KAPLAN, N.O. (ed.) : Methods in Enzymology. Vol.88. Pp. 81 - 88. Academic Press, New York - London - Paris - San Diego - San Francisco - - São Paulo - Sydney - Tokyo - Toronto 1982.

49325 - ENGLERT, G. : N.M.R. of carotenoids. - In : BRITTON, G., GOODWIN, T.W. (ed.): Carotenoids Chemistry and Biochemistry. Pp. 107 - 134. Pergamon Press, Oxford - New York - Toronto - Sydney - Paris - Frankurt 1982.

49326 - ENOS, W.T., ALFICH, R.A., HESKETH, J.D., WOOLEY, J.T. : Interactions among leaf photosynthetic rates, flowering and pod set in soybeans. - Photosynthesis Res. *3* : 273 - 278, 1982.

*49327 - ERDÉLYI, G., FRIDVALSZKY, L., FÁRI, M., GRACZA, P., BISZTRAY, Gy. : The photosynthetic rate of three evergreen species in connection with the reversible metamorphosis of plastids. - In : Proc. 19th Hung. Annu. Meeting Biochem. Pp. 235 - 236. Budapest 1979.

49328 - ERIKSEN, A.B., NILSEN, S. : The effect of deep placement and surface application of nitrogen fertilizers at different light intensities on growth and yield of wetland rice. - Plant Soil *68* : 341 - 351, 1982. [Dry-matter accumulation.]

49329 - ERIKSEN, F.I., WHITNEY, A.S. : Growth and N fixation of some tropical forage legumes as influenced by solar radiation regimes. - Agron. J. *74* : 703 - 709, 1982.

49330 - ERNST, D.E.W., SCHULZE, E. : Chlorophyll determination in the field by fluorometry. - Arch. Hydrobiol. Beih. (Ergebn. Limnol.) *16* : 55 - 61, 1982.

49331 - ESCHBACH, J.M., MASSIMINO, D., MENDOZA, A.M.R. : Effet d'une carence en chlore sur la germination, la croissance et la photosynthèse du cocotier. - Oléagineux *37* : 115 - 125, 1982.

49332 - ESCHRICH, W., BURCHARDT, R. : Reactivation of phloem export in mature maize leaves after a dark period. - Planta *155* : 444 - 448, 1982. [Photosynthates.]

49333 - ESKINS, K., KWOLEK, W.F., HARRIS, L. : The accumulation of accessory pigments as a function of chlorophyll a. A comparison of development and genetic control. - Physiol. Plant. *54* : 409 - 413, 1982.

49334 - ESPIE, G.S., COLMAN, B. : Photosynthesis and inorganic carbon transport in isolated *Asparagus* mesophyll cells. - Plant Physiol. *70* : 649 - 654, 1982.

*49335 - ESPINOZA, W., DA SILVA, E.M., DE SOUZA, O.C. : Irrigação de trigo em solo de cerrado. [Response of wheat to irrigation in a "cerrado" soil.] - Pesq. agropec. bras. *15* : 107 - 115, 1980. [Growth analysis; in Port., ab : E.]

49336 - ESTABLIER, R., LUBIÁN, L.M. : Composición de pigmentos en *Nannochloris maculata* BUTCHER y *N. oculata* DROOP (CCAP, 251/6). Implicaciones de tipo taxonómico. [Pigment composition of *Nannochloris maculata* BUTCHER and *N. oculata* DROOP (CCAP, 251/6). Taxonomic implications.] - Invest. pesq. *46* : 451 - 457, 1982. [In Span., ab : E.]

49337 - ESTEP, M.L.F. : Stable isotopic composition of algae and bacteria that inhabit hydrothermal environments in Yellowstone National Park. - Carnegie Inst. Washington Year Book *81* : 402 - 410, 1982. [$\delta^{13}C$.]

*49338 - ESTRADA, M. : Biomasse et production phytoplanctonique dans la Mediterranée occidentale, au debut de l'automne. - Rapp. Commun. int. Mer méditer. *27* (7) : 65 - 66, 1981.

49339 - ESTRADA, M. : Phytoplankton of the Western Mediterranean at the beginning of autumn. - Int. Rev. ges. Hydrobiol. *67* : 517 - 532, 1982. [Chl.]

49340 - EUGSTER, C.H. : New carotenoid structures and stereochemistry. - In : BRITTON, G., GOODWIN, T.W. (ed.) : Carotenoid Chemistry and Biochemistry. Pp. 1 - 26. Pergamon Press, Oxford - New York - Toronto - Sydney - Paris - Frankfurt 1982.

49341 - EVANS, L.S. : Biological effects of acidity in precipitation on vegetation: a review. - Environm. exp. Bot. *22* : 155 - 169, 1982. [Stomatal resistance.]

49342 - EVANS, M.C.W., ATKINSON, Y.E., NUGENT, J.H.A. : The mechanism of electron transfer in photosynthetic reaction centres and oxygen evolution by plants. - In : HALL, D.O., PALZ, W. (ed.) : Photochemical, Photoelectrochemical and Photobiological Processes. Vol.1. Pp. 125 - 128. D. Reidel Publ. Co., Dordrecht - Boston - London 1982.

49343 - EVANS, M.C.W., DINER, B.A., NUGENT, J.H.A. : Characteristics of the photosystem reaction centre. I. Electron acceptors. - Biochim. biophys. Acta
682 : 97 - 105, 1982.

49344 - EVANS, M.C.W., NUGENT, J.H.A., TILLING, L.A., ATKINSON, Y.E. : Direct determination of the oxidation reduction potential of the iron-quinone electron
acceptor (Q) in photosystem II in *Chlamydomonas reinhardtii*. - FEBS Lett.
145 : 176 - 178, 1982.

49345 - EWEL, J., BENEDICT, F., BERISH, C., BROWN, B., GLIESSMAN, S., AMADOR, M.,
BERMÚDEZ, R., MARTÍNEZ, A., MIRANDA, R., PRICE, N. : Leaf area, light
transmission, roots and leaf damage in nine tropical plant communities. -
Agro-Ecosystems *7* : 305 - 326, 1982.

49346 - EZE, J.M.O., DUMBROFF, E.B. : A comparison of the Bradford and Lowry methods
for the analysis of protein in chlorophyllous tissue. - Can. J. Bot. *60* :
1046 - 1049, 1982. [Chl.]

49347 - FÁBRY, A., VAŠÁK, J., MIKOLÁŠ, J. : Produkční procesy při tvorbě asimilač-
ního povrchu u ozimé řepky. [Production processes in assimilation surface
formation in winter rape.] - In : Sborník Vědecké Konference s Mezinárodní
Účastí k 30. Výročí Založení Agronomické Fakulty Vysoké Školy Zemědělské
v Praze. Pp. 375 - 388. Vysoká Škola Zemědělská, Praha 1982. [In Czech,
ab : E, G, R.]

49348 - FADEEVA, L.M., NIZOVSKAYA, N.V., ZHAKSYBEKOVA, K.E., KONONENKO, A.A.,
KRENDELEVA, T.E. : Ėlektronnyĭ transport i sopryazhennye protsessy v subkhloroplastnykh chastitsakh, obogashchennykh fotosistemoĭ I, v prisutstvii
tetrametilparafenilendiaminov. [Electron transport and conjugated processes
in subchloroplast particles enriched by photosystem I in the presence of
tetramethyl-p-phenylene diamine.] - Biol. Nauki *1982* (5) : 30 - 36, 1982.
[In R.]

*49349 - FAENSEN-THIEBES, A., OVERDIECK, D. : Wirkungen des Cadmiums auf Veränderun-
gen von CO_2-Gaswechsel und Transpiration bei *Phaseolus vulgaris* L. nach
Ozon-Begasung. - Verhandl. Ges. Ökol. *9* : 277 - 281, 1981.

*49350 - FAGANELI, J., FANUKO, N., MALEJ, A., STEGNAR, P., VUKOVIČ, A. : Primary
production in the Gulf of Triest (North Adriatic). - Rapp. Commun. int.
Mer méditer. *27* (7) : 69 - 71, 1981. [Chl.]

49351 - FAGANELI, J., FANUKO, N., STEGNAR, P., VUKOVIČ, A. : Raziskovanja primarne
pelaške bioprodukcije v Tržaškem zalivu. [Studies on primary pelagic bio-
production in the Gulf of Trieste (North Adriatic).] - Acta adriat. *23* :
53 - 60, 1982. [In Croat., ab : E.]

49352 - FAHL, J.I., MACHADO, E.C., PEREIRA, A.R., ARRUDA, H.V., LORENZI, J.O. :
Características fisiológicas de três cultivares de mandioca. [Physiological
characteristics of three cassava cultivars.] - Pesq. agropec. bras. *17* :
399 - 405, 1982. [Growth analysis; in Port., ab : E.]

49353 - FAHR, A., BAMBERG, E. : Photocurrents of dark-adapted bacteriorhodopsin on
black lipid membranes. - FEBS Lett. *140* : 251 - 253, 1982.

*49354 - FAHR, A., LÄUGER, P., BAMBERG, E. : Photocurrent kinetics of purple-membra-
ne sheets bound to planar bilayer membranes. - J. Membrane Biol. *60* : 51 -
- 62, 1981.

49355 - FAILS, B.S., LEWIS, A.J., BARDEN, J.A. : Anatomy and morphology of sun- and
shade-grown *Ficus benjamina*. - J. amer. Soc. hort. Sci. *107* : 754 - 757,
1982. [Growth analysis.]

49356 - FAILS, B.S., LEWIS, A.J., BARDEN, J.A. : Net photosynthesis and transpiration of sun- and shade-grown *Ficus benjamina* leaves. - J. amer. Soc. hort.
Sci. *107* : 758 - 761, 1982.

49357 - FAILS, B.S., LEWIS, A.J., BARDEN, J.A. : Light acclimatization potential
of *Ficus benjamina*. - J. amer. Soc. hort. Sci. *107* : 762 - 766, 1982. [Ps.]

49358 - FAIN, S.R., MURRAY, S.N. : Effects of light and temperature on net photosynthesis and dark respiration of gametophytes and embryonic sporophytes
of *Macrocystis pyrifera*. - J. Phycol. *18* : 92 - 98, 1982.

49359 - FAJER, J., FUJITA, I., DAVIS, M.S., FORMAN, A., HANSON, L.K., SMITH, K.M.:
Photosynthetic energy transduction. Spectral and redox characteristics of
chlorophyll radicals *in vitro* and *in vivo*. - In : KADISH, K.M. (ed.) :
Electrochemical and Spectrochemical Studies of Biological Redox Components.
Pp. 489 - 513. Americal Chemical Society, Washington 1982.

*49360 - FALER, N., VUKADINOVIĆ, V. : Uloga makro- i mikroelemenata u fiziološko-
-biokemijskim procesima šećerne repe. [The role of macro- and microelements
in physiological and biochemical processes in sugar beet.] - Posebna Izd.
srpska Akad. Nauka Umet. *538* , Od. prirod.-mat. Nauka *54* [BELIĆ, J. (ed.) :
Fiziologija Šećerne Repe] : 73 - 95, 1981. [Ps; in Serb., ab : E.]

49361 - FALTYNOWICZ, M., LECHOWICZ, W., POSKUTA, J. : Abscisic acid as a factor
in regulation of photosynthetic carbon metabolism of pea seedlings. - Acta
Soc. Bot. Pol. *51* : 229 - 240, 1982.

49362 - FAN Cihui : [Effect of light intensity on photosynthetic character in rice
plants.] - Yunnan Zhiwu Yanjiu [Acta bot. yunnanica] *4* : 207 - 210, 1982.
[In Chin.]

49363 - FANJUL, L., JONES, H.G. : Rapid stomatal responses to humidity. - Planta
154 : 135 - 138, 1982. [Stomatal resistance.]

49364 - FARMER, D.M., TAKAHASHI, M. : Effects of vertical mixing on photosynthetic
responses. - Jap. J. Limnol. *43* : 173 - 181, 1982.

49365 - FARQUHAR, G.D., BALL, M.C., CAEMMERER, S. von, ROKSANDIC, Z. : Effect of
salinity and humidity on $\delta^{13}C$ value of halophytes - evidence for diffusion-
al isotope fractionation determined by the ratio of intercellular/atmosphe-
ric partial pressure of CO_2 under different environmental conditions. -
Oecologia *52* : 121 - 124, 1982.

49366 - FARQUHAR, G.D., CAEMMERER, S. von : Modelling of photosynthetic response
to environmental conditions. - In : LANGE, O.L., NOBEL, P.S., OSMOND, C.B.,
ZIEGLER, H. (ed.) : Physiological Plant Ecology II. Water Relations and Carbon Assimilation. Pp. 549 - 587. Springer-Verlag, Berlin - Heidelberg -
New York 1982.

*49367 - FARQUHAR, G.D., FIRTH, P.M., WETSELAAR, R., WEIR, B. : On the gaseous ex-
change of ammonia between leaves and the environment: determination of the
ammonia compensation point. - Plant Physiol. *66* : 710 - 714, 1980. [Ps,
photorespiration.]

49368 - FARQUHAR, G.D., O'LEARY, M.H., BERRY, J.A. : On the relationship between
carbon isotope discrimination and the intercellular carbon dioxide concentration in leaves. - Aust. J. Plant Physiol. *9* : 121 - 137, 1982.

49369 - FARQUHAR, G.D., SHARKEY, T.D. : Stomatal conductance and photosynthesis. -
Annu. Rev. Plant Physiol. *33* : 317 - 345, 1982.

49370 - FARVER, O., SHAHAK, Y., PECHT, I. : Electron uptake and delivery sites on
plastocyanin in its reactions with the photosynthetic electron transport
system. - Biochemistry *21* : 1885 - 1890, 1982.

49371 - FAUST, M.A., NORRIS, K.H. : Rapid *in vivo* spectrophotometric analysis of
chlorophyll pigments in intact phytoplankton cultures. - Brit. Phycol. J.
17 : 351 - 361, 1982.

49372 - FAVALI, M.A., FERRARIO, M., BARBIERI, N. : Subcellular distribution of
potassium antimonate precipitates in plant tissues. 1 *Zea mays* leaves. -
Cytobios *35* : 19 - 28, 1982. [Chloroplast.]

49373 - FAWLEY, M., GRIMME, L.H., BROWN, J.S. : Spectral analysis of detergent-
-solubilized photosynthetic membranes. - Carnegie Inst. Washington Year
Book *81* : 38 - 40, 1982.

49374 - FAWLEY, M.W., GRIMME, L.H., BROWN, J.S. : Spectral characterization of pigment-protein complexes from the diatom *Phaeodactylum tricornutum.* - Carnegie Inst. Washington Year Book *81* : 41 - 42, 1982.

*49375 - FAZYLOVA, S.F. : O fotosinteticheskoĭ produktivnosti *Ceratoides ewersmanniana* (STSCHEGL. ex LOSINSK.) BOTSCH. et IKONN. (teresken), vyrashchennogo v usloviyakh razlichnogo vodnogo rezhima pochvy. [Photosynthetic productivity of *Ceratoides ewersmanniana* (STSCHEGL. ex LOSINSK.) BOTSCH. et IKONN. (teresken) grown under various soil moisture regime.] - Uzb. biol. Zh. *1980*(5) : 21 - 24, 1980. [In R.]

49376 - FEDINA, I.S. : Effect of nitrate and ammonium nitrogen on photosynthetic intensity depending on bicarbonate concentration and accompanying ion. - Dokl. bolg. Akad. Nauk *35* : 505 - 508, 1982.

49377 - FEDTKE, C. : Modes of herbicide action as determined with *Chlamydomonas reinhardii* and Coulter counting. - In : MORELAND, D.E., St.JOHN, J.B., HESS, F.D. (ed.) : Biochemical Responses Induced by Herbicides. Pp. 231 - - 250. American Chemical Society, Washington 1982. [Ps, Chl.]

49378 - FEHR, W.R. : Control of iron-deficiency chlorosis in soybeans by plant breeding. - J.Plant Nutr. *5* : 611 - 621, 1982.

49379 - FEICK, R.G., FITZPATRICK, M., FULLER, R.C. : Isolation and characterization of cytoplasmic membranes and chlorosomes from the green bacterium *Chloroflexus aurantiacus.* - J. Bacteriol. *150* : 905 - 915, 1982.

49380 - FEIERABEND, J. : Inhibition of chloroplast ribosome formation by heat in higher plants. - In : EDELMAN, M., HALLICK, R.B., CHUA, N.-H. (ed.) : Methods in Chloroplast Molecular Biology. Pp. 671 - 680. Elsevier Biomedical Press, Amsterdam - New York - Oxford 1982.

49381 - FEIERABEND, J., WINKELHÜSENER, T. : Nature of photooxidative events in leaves treated with chlorosis-inducing herbicides. - Plant Physiol. *70* : 1277 - 1282, 1982.

49382 - FEKETE, G., TUBA, Z. : Photosynthetic activity in the stages of sandy succession. - Acta bot. Acad. Sci. hung. *28* : 291 - 296, 1982.

49383 - FELDMAN, R.I., SIGMAN, D.S. : The synthesis of enzyme-bound ATP by soluble chloroplast coupling factor 1. - J. biol. Chem. *257* : 1676 - 1683, 1982.

49384 - FELKER, P., CLARK, P.R., OSBORN, J.F., CANNELL, G.H. : Biomass estimation in a young stand of mesquite (*Prosopis* spp.), ironwood (*Olneya tesota*), palo verde (*Cercidium floridium,* and *Parkinsonia aculeata*), and leucaena (*Leucaena leucocephala*). - J. Range Manage. *35* : 87 - 89, 1982.

*49385 - FELLER, M.C. : Biomass and nutrient distribution in two eucalypt forest ecosystems. - Aust. J. Ecol. *5* : 309 - 333, 1980.

*49386 - FELTON, R.H. : Primary redox reactions of metalloporphyrins. - In : DOLPHIN, D. (ed.) : Porphyrins. Vol.5. Pp. 53 - 125. Academic Press, New York 1978. [Chl.]

49387 - FENTON, R., MANSFIELD, T.A., JARVIS, R.G. : Evaluation of the possibilities for modifying stomatal movement. - In : McLAREN, J.S. (ed.) : Chemical Manipulation of Crop Growth and Development. Pp. 19 - 37. Butterworth Scientific, London - Boston - Durban - Singapore - Sydney - Toronto - Wellington 1982. [Ps.]

49388 - FERET, P.P. : Effect of moisture stress on the growth of *Pinus ponderosa* DOUGL. ex. LAWS seedlings in relation to their field performance. - Plant Soil *69* : 177 - 186, 1982. [Growth analysis.]

49389 - FERREE, D.C., PALMER, J.W. : Effect of spur defoliation and ringing during bloom on fruiting, fruit mineral level, and net photosynthesis of 'Golden Delicious' apple. - J. amer. Soc. hort. Sci. *107* : 1182 - 1186, 1982.

49390 - FERRON, F., COUDRET, A., L'HARDY-HALOS, M.-Th. : Chlorure de sodium et voie du glycolate chez l'*Aglaothamnion chadefaudii* L'HARDY-HALOS. - Photosynthetica *16* : 43 - 48, 1982. [Ps.]

49391 - FERRON, F., COUDERT, A., ZINSOU, C., COSTES, C. : Action de l'énergie lumineuse incidente sur le métabolisme photosynthétique de la christophine
(*Sechium edule* SWARTZ). - Agronomie 2 : 621 - 627, 1982.

49392 - FERTE, N., MEUNIER, J.-C., RICARD, J., BUC, J., SAUVE, P. : Molecular properties and thioredoxin-mediated activation of spinach chloroplastic NADP-
-malate dehydrogenase. - FEBS Lett. *146* : 133 - 138, 1982.

*49393 - FETISOVA, Z.G., BORISOV, A.Yu. : Picosecond time scale of heterogeneous
excitation energy transfer from accessory light-harvesting bacterioviridin
antenna to main bacteriochlorophyll *a* antenna in photoactive pigment-protein complexes obtained from *Chlorobium limicola,* a green bacterium. -
FEBS Lett. *114* : 323 - 326, 1980.

*49394 - FETISOVA, Z.G., HARCHENKO, S.G. : New complex method of picosecond-time
scale measurement of fluorescence lifetime with phase fluorometer and its
application to determination of excitation transfer rates in photosynthetic
pigment antenna. - In : Proceedings. IV International Seminar on Energy
Transfer in Condensed Matter. Pp. 132 - 134. Prague 1981.

49395 - FEUCHT, D., SCHMITZ, M., HÖFNER, W. : Veränderungen der Blattspreiten und
des Chlorophyllgehaltes der Fahnenblätter von Winterweizen durch Wachstumsregulatoren. - Z. Pflanzenern. Bodenk. *145* : 288 - 295, 1982.

49396 - FEUILLADE, J., FEUILLADE, M., JOLIVET, E. : Photosynthetic metabolism in
the cyanophyta *Oscillatoria rubescens* D.C. II. Carbon metabolism under
nitrogen starvation. - Arch. Microbiol. *131* : 107 - 111, 1982.

49397 - FIEDLER, E., SOLL, J., SCHULTZ, G. : The formation of homogentisate in the
biosynthesis of tocopherol and plastoquinone in spinach chloroplasts. -
Planta *155* : 511 - 515, 1982.

49398 - FIELD, C., BERRY, J.A., MOONEY, H.A. : A portable system for measuring
carbon dioxide and water vapour exchange of leaves. - Plant Cell Environm.
5 : 179 - 186, 1982.

49399 - FIELD, C., CHIARIELLO, N., WILLIAMS, W.E. : Determinants of leaf temperature
in California *Mimulus* species at different altitudes. - Oecologia *55* : 414 -
- 420, 1982. [Ps.]

49400 - FIELD, S.D., EFFLER, S.W. : Photosynthesis-light mathematical formulations.
- J. environm. Eng. Div. *108* (EE 1) : 199 - 203, 1982.

*49401 - FILATOV, G.V., SUPONINA, S.L. : Ob osobennost'yakh funktsionirovaniya fer-
mentativnykh protsessov pri fotosinteze. [Peculiarities of functioning of
enzymatic processes in photosynthesis.] - In : Biologiya, Selektsiya i
Semenovodstvo Zernovykh Kul'tur. Pp. 90 - 96. Nauch.-issled. Inst. Sel'skogo Khozyaĭstva TsChP Im. V.V.Dokuchaeva, Kamennaya Step' 1981. [In R.]

49402 - FILATOV, G.V., SUPONINA, S.L. : Osobennosti fotosinteticheskogo apparata
geterozisnykh gibridov kukuruzy. [Properties of the photosynthetic appara-
tus of heterotic maize hybrids.] - In : Selektsiya Polevykh i Kormovykh
Kul'tur v Tsental'no-Chernozemnoĭ Zone. Pp. 16 - 21. Nauch.-issled. Inst.
Sel'skogo Khozyaĭstva TsChP Im. A.A. Dokuchaeva, Kamennaya Step' 1982. [In
R.]

49403 - FILBIN, G.J., HOUGH, R.A. : *In situ* photosynthesis and respiration in the
littoral sedge, *Scirpus acutus* MUHL. - J. Freshwater Ecol. *1* : 443 - 450,
1982.

49404 - FILIPPOVA, L.A., MAMUSHINA, N.S., ZALENSKIĬ, O.V. : O funktsionirovanii
osnovnykh ètapov temnogo dykhaniya vo vremya fotosinteza. [Metabolic changes of dark respiration in the course of photosynthesis.] - Bot. Zh. *67* :
1169 - 1178, 1982. [In R, ab : E.]

49405 - FILIPPOVICH, I.I., NOZDRINA, V.N., SYCHEV, A.R. : Osobennosti strukturnoĭ
i prostranstvennoĭ organizatsii poliribosom khloroplastov i ikh uchastie
v sborke tilakoidov gran. [Features of the structural and spatial organization of polyribosomes of chloroplasts and their participation in assembly
of thylakoid grana.] - In : Sborka Predbiologicheskikh i Biologicheskikh
Struktur. Pp. 262 - 266, 342. Nauka, Moskva 1982. [In R.]

*49406 - FIRL, J., FROMMEYER, D., ELSTNER, E.F. : Isolation and identification of an oxygen reducing factor (ORF) from isolated spinach chloroplast lamellae. - Z. Naturforsch. *36 C* : 284 - 294, 1981.

*49407 - FISCHER, K., METZNER, H. : Bicarbonate effects on photosynthetic electron transport. I. Concentration dependence and influence on manganese reincorporation. - Photobiochem. Photobiophys. *2* : 133 - 140, 1981.

*49408 - FISCHER, U. : A high potential iron sulfur protein of the purple sulfur bacterium *Thiocapsa roseopersicina*. - Z. Naturforsch. *35 C* : 150 - 153, 1980.

*49409 - FISCHER, U., TRÜPER, H.G. : Cytochrome *c*-550 of *Thiocapsa roseopersicina*: Properties and reduction by sulfide. - FEMS Lett. *1* : 87 - 90, 1977.

*49410 - FISCHER, U., TRÜPER, H.G. : Some properties of cytochrome *c'* and other hemoproteins of *Thiocapsa roseopersicina*. - Curr. Microbiol. *3* : 41 - 44, 1979.

*49411 - FISH, L.E., JAGENDORF, A.T. : Light-induced increase in the number and activity of ribosomes bound to pea chloroplast thylakoids *in vivo*. - Plant Physiol. *69* : 814 - 825, 1982.

49412 - FISH, L.E., JAGENDORF, A.T. : High rates of protein synthesis by isolated chloroplasts. - Plant Physiol. *70* : 1107 - 1114, 1982.

49413 - FISHER, D.G., EVERT, R.F. : Studies on the leaf of *Amaranthus retroflexus* *(Amaranthaceae)* : chloroplast polymorphism. - Bot. Gaz. *143* : 146 - 155, 1982.

49414 - FISHER, K.A. : Preparation of planar membrane monolayers for spectroscopy and electron microscopy. - In : COLOWICK, S.P., KAPLAN, N.O. (ed.) : Methods in Enzymology. Vol.88. Pp. 230 - 235. Academic Press, New York - London - - Paris - San Diego - San Francisco - São Paulo - Sydney - Tokyo - Toronto 1982. [Bacteriorhodopsin.]

49415 - FISHER, M.J., CHARLES-EDWARDS, D.A. : A physiological approach to the analysis of crop growth data. 3. The effects of repeated short term soil water deficits on the growth of spaced plants of the legume, *Macroptilium atropurpureum* cv. Siratro. - Ann. Bot. *49* : 341 - 346, 1982. [Growth analysis.]

49416 - FISHER, T.R., CARLSON, P.R., BARBER, R.T. : Carbon and nitrogen primary productivity in three North Carolina estuaries. - Estuar. coast. Shelf Sci. *15* : 621 - 644, 1982.

49417 - FITT, W.K., PARDY, R.L., LITTLER, M.M. : Photosynthesis, respiration, and contribution to community productivity of the symbiotic sea anemone *Anthopleura elegantissima* (BRANDT, 1835). - J. exp. mar. Biol. Ecol. *61* : 213 - - 232, 1982.

49418 - FITZWATER, S.E., KNAUER, G.A., MARTIN, J.H. : Metal contamination and its effect on primary production measurements. - Limnol. Oceanogr. *27* : 544 - - 551, 1982.

B49419 - Fiziologiya Fotosinteza. [Physiology of Photosynthesis.] - Nauka, Moskva 1982. [In R.]

49420 - FLEISCHMAN, D., PERKINS, S. : P985$^+$-Q$_B$-recombination luminescence from *Rhodopseudomonas viridis*. - Annu. Rep. C.F.Kettering Res. Lab. *1982* : 39 - 40,1982.

49421 - FLETCHER, R.A., KALLIDUMBIL, V., BHARDWAJ, S.N. : Effects of fusicoccin on fresh weight and chlorophyll levels in cucumber cotyledons. - Plant Cell Physiol. *23* : 717 - 719, 1982.

B49422 - FLETCHER, W.W., KIRKWOOD, R.C. : Herbicides and Plant Growth Regulators. - Granada, London - Toronto - Sydney - New York 1982. [Ps.]

49423 - FLOENER, L., BOTHE, H. : Metabolic activities in *Cyanophora paradoxa* and its cyanelles. II. Photosynthesis and respiration. - Planta *156* : 78 - 83, 1982.

49424 - FLORET, C., PONTANIER, R., RAMBAL, S. : Measurement and modelling of primary production and water use in a south Tunisian steppe. - J. Arid Environm. *5* : 77 - 90, 1982.

49425 - FLOS, J. : Producción primaria, clorofila *a* y visibilidad del disco de Secchi en el golfo de Vizcaya. [Primary production, chlorophyll *a* and Secchi disc visibility in the Bay of Biscay.] - Invest. pesq. *46* : 215 - 230, 1982. [In Span., ab : E.]

49426 - FLÜGEL, M., GROSS, J. : Farbstoff- und Plastidenveränderungen während der Fruchtreife bei Zuckermelonen, *Cucumis melo* cv. Galia. - Angew. Bot. *56* : 393 - 406, 1982.

49427 - FLÜGGE, U.-I. : Biogenesis of the chloroplast phosphate translocator. - FEBS Lett. *140* : 273 - 276, 1982.

49428 - FLÜGGE, U.I., STITT, M., FREISL, M., HELDT, H.W. : On the participation of phosphoribulokinase in the light regulation of CO_2 fixation. - Plant Physiol. *69* : 263 - 267, 1982.

49429 - FOLEY, T., BEALE, S.I. : δ-aminolevulinic acid formation from γ,δ-dioxovaleric acid in extracts of *Euglena gracilis*. - Plant Physiol. *70* : 1495 - - 1502, 1982.

49430 - FOLEY, T., DZELZKALNS, V., BEALE, S.I. : δ-aminolevulinic acid synthase of *Euglena gracilis* : Regulation of activity. - Plant Physiol. *70* : 219 - 226, 1982.

49431 - FOLTÝN, J., VLASÁK, M., ŠVORCOVÁ, A. : Prostorová struktura rostlin s různým počtem klasů v zapojeném porostu pšenice. [Spatial structure of plants with different ear number in closed wheat stand.] - Rost. Výroba (Praha) *28* : 773 - 782, 1982. [In Czech, ab : E, G, R.]

49432 - FONDY, B.R., GEIGER, D.R. : Diurnal pattern of translocation and carbohydrate metabolism in source leaves of *Beta vulgaris* L. - Plant Physiol. *70* : 671 - 676, 1982. [Ps.]

B49433 - FONG, F.K. (ed.) : Light Reaction Path of Photosynthesis. - Springer-Verlag, Berlin - Heidelberg - New York 1982.

49434 - FONG, F.K. : Free energy change for quantum storage in photosynthesis. - In: FONG, F.K. (ed.) : Light Reaction Path of Photosynthesis. Pp. 1 - 6. Springer-Verlag, Berlin - Heidelberg - New York 1982.

49435 - FONG, F.K. : Light path of carbon reduction in photosynthesis. - In : FONG, F.K. (ed.) : Light Reaction Path of Photosynthesis. Pp. 277 - 321. Springer-Verlag, Berlin - Heidelberg - New York 1982.

49436 - FONG, F.K., KUSUNOKI, M., GALLOWAY, L., MATTHEWS, T.G., LYTLE, F.E., HOFF, A.J., BRINKMAN, F.A. : Elementary reconstitution of the water splitting light reaction in photosynthesis. 1. Time-resolved fluorescence and electron spin resonance studies of chlorophyll *a* dihydrate photoreaction with water in nonpolar solutions. - J. amer. chem. Soc. *104* : 2759 - 2767, 1982.

49437 - FORD, R.C., CHAPMAN, D.J., BARBER, J., PEDERSEN, J.Z., COX, R.P. : Fluorescence polarization and spin-label studies of the fluidity of stromal and granal chloroplast membranes. - Biochim. biophys. Acta *681* : 145 - 151, 1982.

49438 - FORD, W.E., TOLLIN, G. : Direct observation of electron transfer across a lipid bilayer: pulsed laser photolysis of an asymmetric vesicle system containing chlorophyll, methyl viologen and EDTA. - Photochem. Photobiol. *36* . 009 - 819, 1982.

49439 - FORD, W.E., TOLLIN, G. : Chlorophyll photosensitized electron transfer in phospholipid vesicle bilayers: inside *vs* outside asymmetry. - Photochem. Photobiol. *36* : 647 - 655, 1982.

49440 - FORK, D.C., ÖQUIST, G., HOCH, G.E. : Fluorescence emission from photosystem I at room temperature in the red alga *Porphyra perforata*. - Plant Sci. Lett. *24* : 249 - 254, 1982.

49441 - FORK, D.C., SATOH, K. : Light-induced changes of energy distribution between the two photosystems in the extreme thermophilic blue-green alga *Synechococcus lividus*. - Carnegie Inst. Washington Year Book *81* : 45 - 49, 1982.

49442 - FORK, D.C., SATOH, K. : State transitions in the green alga *Scenedesmus* and in a mutant lacking chlorophyll *b*. - Carnegie Inst. Washington Year Book *81* : 54 - 58, 1982.

49443 - FORSETH, I.N., EHLERINGER, J.R. : Ecophysiology of two solar-tracking desert winter annuals. I. Photosynthetic acclimation to growth temperature. - Aust. J. Plant Physiol. *9* : 321 - 332, 1982.

49444 - FORSETH, I.N., EHLERINGER, J.R. : Ecophysiology of two solar tracking desert winter annuals. II. Leaf movements, water relations and microclimate. - Oecologia *54* : 41 - 49, 1982. [Stomatal resistance.]

*49445 - FORSSKÅHL, M., SUNDBERG, A. : Abundance, biomass, species composition of phyto- and zooplankton and their interrelations at the entrance to the Gulf of Finland in 1979. - Meri *9* : 43 - 55, 1981/2. [Primary productivity.]

49446 - FOSTER, J.G., EDWARDS, G.E., WINTER, K. : Changes in levels of phosphoenol-pyruvate carboxylase with induction of Crassulacean acid metabolism in *Mesembryanthemum crystallinum* L. - Plant Cell Physiol. *23* : 585 - 594, 1982.

49447 - FOY, R.H., GIBSON, C.E. : Photosynthetic characteristics of planktonic blue-green algae: the response of twenty strains grown under high and low light. - Brit. Phycol. J. *17* : 169 - 182, 1982.

49448 - FOY, R.H., GIBSON, C.E. : Photosynthetic characteristics of planktonic blue-green algae: changes in photosynthetic capacity and pigmentation of *Oscillatoria redekei* VAN GOOR under high and low light. - Brit. Phycol. J. *17* : 183 - 193, 1982.

49449 - FOYER, C., WALKER, D., LATZKO, E. : The regulation of cytoplasmic fructose 1,6-bisphosphatase in relation to the control of carbon flow to sucrose. - Z. Pflanzenphysiol. *107* : 457 - 465, 1982.

49450 - FOYER, C., WALKER, D., SPENCER, C., MANN, B. : Observations on the phosphate status and intracellular pH of intact cells, protoplasts and chloroplasts from photosynthetic tissue using phosphorus-31 nuclear magnetic resonance. - Biochem. J. *202* : 429 - 434, 1982.

49451 - FRACKOWIAK, D., BAUMAN, D., STILLMAN, M.J. : Circular dichroism and magnetic circular dichroism spectra of chlorophylls in nematic liquid crystals. I. Electric and weak magnetic field effects on the dichroism spectra. - Biochim. biophys. Acta *681* : 273 - 285, 1982.

49452 - FRĄCKOWIAK, D., PIEŃKOWSKA, H., SZURKOWSKI, J. : Excitation energy transfer between phycoerythrin, phycocyanin and chlorophyllin in polyvinyl alcohol films. - Photosynthetica *16* : 496 - 508, 1982.

49453 - FRADKIN, L.I., DOMANSKIĬ, V.P., SAMOĬLENKO, A.G., SHLYK, A.A. : Pigmentnye sistemy fotosinteza v techenie pervogo chasa deětiolyatsii list'ev yachmenya. [Pigment systems of photosynthesis in the first hour of barley leaves deetiolation.] - Dokl. Akad. Nauk SSSR *264* : 732 - 736, 1982. [In R.]

49454 - FRADO, T.M., STERN, A.I. : Photosynthesis and chloroplast development in primary leaves of *Phaseolus vulgaris* illuminated with continuous far-red light. - Z. Pflanzenphysiol. *105* : 255 - 265, 1982.

49455 - FRANCEY, R.J., FARQUHAR, G.D. : An explanation of $^{13}C/^{12}C$ variations in tree rings. - Nature *297* : 28 - 31, 1982.

49456 - FRANK, H.A., MACHNICKI, J., FELBER, M. : Carotenoid triplet states in photosynthetic bacteria. - Photochem. Photobiol. *35* : 713 - 718, 1982.

49457 - FRANK, H.A., McGANN, W.J., MACKNICKI, J., FELBER, M. : Magnetic field effects on the fluorescence of two reaction centerless mutants of *Rhodopseudomonas capsulata*. - Biochem. biophys. Res. Commun. *106* : 1310 - 1317, 1982.

49458 - FRANKE, W., LOOSEN, R. : Beiträge zur Biologie der Nutzpflanzen. 1. Über den grünen Farbstoff der Keimblätter der Echten Pistazie oder Grünmandel (*Pistacia vera* L.). - Angew. Bot. *56* : 307 - 313, 1982.

49459 - FRANKEL, R.D., FORSYTH, J.M. : Application of nanosecond X-ray diffraction techniques to bacteriorhodopsin. - In : COLOWICK, S.P., KAPLAN, N.O. (ed.): Methods in Enzymology. Vol.88. Pp. 276 - 281. Academic Press, New York - - London - Paris - San Diego - San Francisco - São Paulo - Sydney - Tokyo - - Toronto 1982.

49460 - FRASCH, W.D., SELMAN, B.R. : Mechanism of phosphorylation catalyzed by chloroplast coupling factor 1. Stereochemistry. - Biochemistry *21* : 3636 - 3643, 1982.

49461 - FREEDMAN, B., DUINKER, P.N., MORASH, R., PRAGER, U. : A comparison of measurements of the standing crops of biomass and nutrients in a conifer stand in Nova Scotia. - Can. J. Forest Res. *12* : 499 - 502, 1982.

49462 - FREEMAN, B.M., KLIEWER, W.M., STERN, P. : Influence of windbreaks and climatic region on diurnal fluctuation of leaf water potential, stomatal conductance, and leaf temperature of grapevine. - Amer. J. Enol. Viticult. *33* : 233 - 236, 1982.

49463 - FREEMAN, T.P., DUYSEN, M.E., OLSON, N.H., WILLIAMS, N.D. : Electron transport and chloroplast ultrastructure of a chlorophyll-deficient mutant of wheat. - Photosynthesis Res. *3* : 179 - 189, 1982.

49464 - FREĬBERG, A.M., TIMPMANN, K.Ė., TAMKIVI, R.P., AVARMAA, R.A. : Izuchenie pikosekundnoĭ kinetiki izlucheniya fragmentov khloroplastov s pomoshch'yu sinkhronno nakachivaemogo lazera na krasitele i spektrokhronografa. [Investigation of picosecond fluorescence fluorescence kinetics of chloroplast fragments by synchronously-pumped dye-laser and spectrochronograph.] - Eesti NSV Tead. Akad. Toim., Füüs., Mat. *31* : 200 - 207, 1982. [In R, ab : E, Est.]

49465 - FRENCH, C.S. : Dye photooxidation by the chlorophyll-protein complex of *Nostoc*. - Carnegie Inst. Washington Year Book *81* : 43 - 45, 1982.

49466 - FREYE, E. : Abhängigkeit der CO_2-Assimilation bei *Vicia faba* L. (var.*minor* cv. Fribo) von Beleuchtungsstärke und Kohlendioxidgehalt der Luft. - In : UNGER, K., SCHUH, J. (ed.) : Umwelt-Stress. Pp. 237 - 240. Martin-Luther- -Universität, Halle 1982.

49467 - FREYE, E., MERBACH, W. : Beziehungen zwischen Kohlenstoffhaushalt und symbiotischer N_2-Fixierung bei Ackerbohnen (*Vicia faba* L.var. *minor*). - In : HOFFMANN, P., HIEKE, B. (ed.) : Photosynthese: Regulation und Evolution. (Colloquia Pflanzenphysiologie Nr.5.) Pp. 255 - 259. Humboldt-Universität, Berlin 1982.

49468 - FRICKE, H., VARESCHI, E. : A scleractinian coral (*Plerogyra sinuosa*) with "photosynthetic organs". - Mar. Ecol.-Progr. Ser. *7* : 273 - 278, 1982.

*49469 - FRIEDRICH, J.W., HUFFAKER, R.C. : The relationship between ribulose bisphosphate carboxylase concentration and photosynthesis. - In : LYONS, J.M., VALENTINE, R.C., PHILLIPS, D.A., RAINS, D.W., HUFFAKER, R.C. (ed.) : Genetic Engineering of Symbiotic Nitrogen Fixation and Conservation of Fixed Nitrogen. Pp. 305 - 312. Plenum Press, New York - London 1981.

49470 - FRIES, L. : Vanadium an essential element for some marine macroalgae. - Planta *154* : 393 - 396, 1982. [Chl.]

49471 - FRISCHKNECHT, P.M., ELLER, B.M., BAUMANN, T.W. : Purine alkaloid formation and CO_2 gas exchange in dependence of development and of environmental factors in leaves of *Coffea arabica* L. - Planta *156* : 295 - 301, 1982.

49472 - FRITZ-SHERIDAN, R.P. : Impact of the herbicide Magnacide-H (2-Propenal) on algae. - Bull. environm. Contam. Toxicol. *28* : 245 - 249, 1982. [Ps.]

49473 - FROLOV, A.K., GORYSHINA, T.K. : Osobennosti fotosinteticheskogo apparata nekotorykh drevesnykh porod v gorodskikh usloviyakh. [Peculiarities of the photosynthetic apparatus of trees in the urban environment.] - Bot. Zh. *67* : 599 - 609, 1982. [In R, ab : E.]

49474 - FRY, B., LUTES, R., NORTHAM, M., PARKER, P.L., OGDEN, J. : A $^{13}C/^{12}C$ comparison of food webs in Caribbean seagrass meadows and coral reefs. - Aquat. Bot. *14* : 389 - 398, 1982.

49475 - FU, C.F., HEW, C.S. : Crassulacean acid metabolism in orchids under water stress. - Bot. Gaz. *143* : 294 - 297, 1982.

49476 - FUESLER, T.P., HANAMOTO, C.M., CASTELFRANCO, P.A. : Separation of Mg-protoporphyrin IX and Mg-protoporphyrin IX monomethyl ester synthesized by developing cucumber etioplasts. - Plant Physiol. *69* : 421 - 423, 1982.

49477 - FUHRER, J., KAUR-SAWHNEY, R., SHIH, L.-M., GALSTON, A.W. : Effects of exogenous 1,3-diaminopropane and spermidine on senescence of oat leaves. II. Inhibition of ethylene biosynthesis and possible mode of action. - Plant Physiol. *70* : 1597 - 1600, 1982. [Chl.]

49478 - FUHRMAN, M.H., KOUKKARI, W.L. : Characteristics of the circadian rhythm in diffusive resistance of *Abutilon theophrasti* leaves in humid and dry environments. - Chronobiologia *9* : 21 - 32, 1982. [Stomatal resistance.]

49479 - FUJIIE, A. : Ecological studies on the population of the pear leaf miner, *Bucculatrix pyrivorella* KUROKO *(Lepidoptera, Lyonetiidae)* VI. Effects of injury by the pear leaf miner on leaf fall and photosynthesis of the pear tree. - Appl. Entomol. Zool. *17* : 188 - 193, 1982.

49480 - FUJITA, I., FAJER, J., CHANG, C.-K., WANG, C.-B., BERGKAMP, M.A., NETZEL, T.L. : Solvent and structural effects on picosecond electron transfer reactions in diporphyrin models of the photosystem II reaction center of green plants. - J. phys. Chem. *86* : 3754 - 3759, 1982.

49481 - FUJITA, I., NETZEL, T.L., CHANG, C.K., WANG, C.-B. : Picosecond photochemistry of a cofacial diporphyrin containing iron(III) and zinc(II) : Mimicking electron transfer between cytochrome *c* and the primary electron donor in reaction centers of photosynthetic bacteria. - Proc. nat. Acad. Sci. USA *79* : 413 - 417, 1982.

49482 - FUJITA, Y., MURATA, N. : Photosynthetic pigments. - Recent Progr. nat. Sci. Jap. *7* : 3 - 7, 1982.

*49483 - FUKAZAWA, N., ISHIMARU, T., TAKAHASHI, M., FUJITA, Y. : A mechanism of "red tide" formation. I. Growth rate estimate by DCMU-induced fluorescence increase. - Mar. Ecol.-Progr. Ser. *3* : 217 - 222, 1980.

*49484 - FUKSHANSKY, L., KAZARINOVA, N. : Extension of the Kubelka-Munk theory of light propagation in intensely scattering materials to fluorescent media. - J. opt. Soc. Amer. *70* : 1101 - 1111, 1980. [Chl.]

*49485 - FUKUCHI, M. : Chlorophyll-a content in the surface water along the course of the FUJI to and from Antarctica in 1976-1977. - Antarctic Res. *60* : 57 - - 69, 1977.

*49486 - FUKUCHI, M. : Phytoplankton chlorophyll stocks in the Antarctic Ocean. - J. oceanogr. Soc. Jap. *36* : 73 - 84, 1980.

*49487 - FUKUCHI, M., SASAKI, H. : Phytoplankton and zooplankton standing stocks and downward flux of particulate material around fast ice edge of Lützow-Holm Bay, Antarctica. - Mem. nat. Inst. polar Res., Ser.E *34* : 13 - 36, 1981.

49488 - FUKUCHI, M., TAMURA, S. : Chlorophyll a distribution in the Indian sector of the Antarctic Ocean in 1978-1979. - Antarctic Record *74* : 143 - 162, 1982.

49489 - FUKUDA, K., TOYAMA, S. : Electron microscope studies on the morphogenesis of plastids XI. Ultrastructural changes of the chloroplasts in tomato leaves treated with ethylene and kinetin. - Cytologia (Tokyo) *47* : 725 - 736, 1982.

49490 - FUKUYAMA, M., SHIMAMURA, M., USHIYAMA, M., OIKAWA, M. : A new portable solarimeter of visible wavelength and it's application to estimate standing crop on pasture. - Bull. nat. Grassl. Res. Inst. *21* : 79 - 87, 1982.

49491 - FURBANK, R.T., BADGER, M.R. : Photosynthetic oxygen exchange in attached leaves of C_4 monocotyledons. - Aust. J. Plant Physiol. *9* : 553 - 558, 1982.

49492 - FURBANK, R.T., BADGER, M.R., OSMOND, C.B. : Photosynthetic oxygen exchange in isolated cells and chloroplasts of C_3 plants. - Plant Physiol. *70* : 927 - - 931, 1982.

49493 - FURUNO, T., SAITO, H. : [Seasonal variations of litter falls and primary consumption by herbivorous insects in *Chamaecyparis obtusa* plantations in Owase, Mie Prefecture and Kamikitayama, Nara Prefecture.] - J. Jap. Forest Soc. *64* : 177 - 186, 1982. [Primary production; in Jap., ab : E.]

49494 - FURUYA, K. : Measurement of phytoplankton standing stock using an image analyzer system. - Bull. Plankton Soc. Jap. *29* : 131 - 132, 1982.

49495 - FURYAEV, E.A., BELYANIN, V.N., TERSKOV, I.A. : Izmenenie opticheskikh kharakteristik odinochnykh kletok khlorelly v tsikle razvitiya. [Changes in optical characteristics of single cells of *Chlorella* in the developmental cell cycle.] - Fiziol. Rast. *29* : 728 - 736, 1982. [In R, ab : E.]

*49496 - GÁBORČÍK, N. : Príspevok k štúdiu produkčného procesu trvalého trávneho porastu v rozličných podmienkach dusíkatej výživy. [Production process of perennial grass stand in different conditions of nitrogen nutrition.] - Ved. Práce výsk. Ústavu Lúk Pasienok Banskej Bystrici *14* : 37 - 49, 1979. [In Slovak, ab : E, R.]

*49497 - GÁBORČÍK, N. : Výkon čistej fotosyntézy trvalého trávneho porastu v rôznych podmienkach minerálnej výživy. [Net assimilation rate of permanent grassland stand under different mineral nutrition.] - In : Ekológia Trávneho Porastu. Pp. 142 - 152. Českoslov. Ved.-Tech. Spol.,Pobočka Výsk. Ústave Lúk Pasienkov, Banská Bystrica 1980. [In Slovak, ab : E.]

*49498 - GÁBORČÍK, N. : Obsah chlorofylu a dusíka v listoch niektorých tráv mierneho pásma. [Content of chlorophyll and nitrogen in leaves of some temperate grasses.] - Ved. Práce výsk. Ústavu Lúk Pasienkov Banskej Bystrici *15* : 151 - 163, 1981. [In Slovak, ab : E, R.]

*49499 - GÁBORČÍK, N. : Množstvo chlorofylu v listoch reznačky laločnatej (*Dactylis glomerata* L.) rastúcej v kontrastných podmienkach dusíkatej výživy. [Chlorophyll content in leaves of orchard grass (*Dactylis glomerata* L.) growing in contrast conditions of nitrogen nutrition.] - Ved. Práce výsk. Ústavu Lúk Pasienok Banskej Bystrici *16* : 15 - 26, 1981. [In Slovak, ab : E, R.]

49500 - GÁBORČÍK, N. : Analýza produkčného procesu asociácie *Anthoxantho-Agrostietum*. [Analysis of production process of the association *Anthoxantho-Agrostietum*.] - In : ŠPÁNIK, F. (ed.) : Vplyv Prostredia na Fytoprodukciu. Pp. 59 - 67. Slovenská Bioklimatologická Spoločnosť SAV, Bratislava 1982. In [Growth analysis; in Slovak.]

*49501 - GÁBORČÍK, N., GÁBORČÍK, Š. : Ciele a fyziologické základy šľachtenia tráv. [Aims and physiological bases of grass breeding.] - Stud. Inform. ÚVTIZ (Praha), Rostl. Výroba *1981*(2) : 1 - 64, 1981. [Ps; In Slovak, ab : E, R.]

49502 - GACHKOVSKIĬ, V.F. : O lyuminestsentsii khlorofilla v list'yakh rasteniĭ. [Chlorophyll luminescence in plant leaves.] - Dokl. Akad. Nauk SSSR *262* : 1265 - 1268, 1982. [In R.]

49503 - GALLAGHER, J.C. : Physiological variation and electrophoretic banding patterns of genetically different seasonal populations of *Skeletonema costatum* (*Bacillariophyceae*). - J. Phycol. *18* : 148 - 162, 1982.[Ps, Chl.]

49504 - GALLEGOS, C.L., PLATT, T. : Phytoplankton production and water motion in surface mixed layers. - Deep-Sea Res. Pt.A *29* : 65 - 76, 1982.

49505 - GALLING, G. : Isolation of chloroplast mutants in *Chlorella*. - In : EDELMAN, M., HALLICK, R.B., CHUA, N.-H. (ed.) : Methods in Chloroplast Molecular Biology. Pp. 65 - 72. Elsevier Biomedical Press, Amsterdam - New York - Oxford 1982.

49506 - GALLOWAY, R.E., METS, L. : Non-Mendelian inheritance of 3-(3,4-dichlorophe-
nyl)-1,1-dimethylurea-resistant thylakoid membrane properties in *Chlamydomonas*. - Plant Physiol. *70* : 1673 - 1677, 1982.

49507 - GALMICHE, J.M., GIRAULT, G. : Synthesis of ATP induced in pea chloroplasts
by single turnover flashes. - FEBS Lett. *146* : 123 - 128, 1982.

49508 - GANAGO, I.B., KLIMOV, V.V., GANAGO, A.O., SHUVALOV, V.A., EROKHIN, Y.E. :
Linear dikhroizm and orientation of pheophytin, the intermediary electron
acceptor in photosystem II reaction centers. - FEBS Lett. *140* : 127 - 130,
1982.

49509 - GANAGO, I.B., KLIMOV, V.V., GANAGO, A.O., SHUVALOV, V.A., EROKHIN, Yu.E. :
Lineǐnyǐ dichroizm izmeneniǐ pogloshcheniya pri fotovosstanovlenii feofitina v orientirovannykh preparatakh fotosistemy II. [Linear dichroism of absorption changes during pheophytin photoreduction in oriented preparations
of photosystem II.] - Dokl. Akad. Nauk SSSR *263* : 479 - 483, 1982. [In R.]

49510 - GANCHEV, S.P., ǏOTSOVA-BAURENSKA, N.M., MESHINEV, T.A., BOYADZHIǏSKI, M.N.:
Vliyanie na svetlinniya rezhim v"rkhu rastezha i razvitieto na *Potentilla
fruticosa* L. v mestnostta Beglika (Zapadni Rodopi). [The effect of the
light conditions on the growth and development of *Potentilla fruticosa* L.
in Beglika locality (Western Rhodopes).] - Ekologiya (Sofiya) *10* : 3 - 12,
1982. [Chl; in Bulg., ab : E, R.]

49511 - GANGADHARAN, C., NAYAK, S.K. : Screening for variation in response to high
O_2 levels among rice genotypes and its significance. - Cereal Res. Commun.
10 : 87 - 93, 1982. [Ps.]

49512 - GANTT, E. : Rapporteur's summary: Plastids and their precursors. - In :
SCHIFF, J.A., LYMAN, H. (ed.) : On the Origins of Chloroplasts. Pp. 107 -
- 113. Elsevier/North-Holland, New York - Amsterdam - Oxford 1982.

49513 - GAO (KAO) Yu-zhu, ZHONG Wang : [On the relationship between photorespiration
and photosynthesis II. Effects of environmental factors on photosynthesis
and its relation to photorespiration.] - Acta Phytophysiol. sin. *8* : 373 -
- 384, 1982. [In Chin., ab : E.]

49514 - GAPONENKO, V.I., NIKOLAEVA, G.N., SHEVCHUK, S.N., KUPERMAN, N.I., DOVNAR,
V.S. : Obnovlenie khlorofilla kak kriteriǐ fotosinteza i urozhaya rasteniǐ
rzhi i yachmenya. [Chlorophyll regeneration as criterion of photosynthesis
and yield of rye and barley plants.] - Dokl. Akad. Nauk belorus. SSR *26*(1):
74 - 77, 95 - 96, 1982. [In R, ab : E.]

49515 - GAPONENKO, V.I., SHEVCHUK, S.N., BALEVA, E.F., ZHEBRAKOVA, I.V. : Deǐstvie
sveta raznoǐ dliny volny na obnovlenie khlorofilla i fotosintez u prorostkov kukuruzy. [Effect of various wavelengths of light on chlorophyll turnover and photosynthesis in maize seedlings.] - Fiziol. Rast. *29* : 713 -
- 719, 1982. [In R, ab : E.]

49516 - GARDEMANN, A., STITT, M., HELDT, H.W. : Regulation of spinach ribulose-5-
-phosphate kinase by 3-phosphoglycerate. - FEBS Lett. *137* : 213 - 216, 1982.

49517 - GARG, B.K., KATHJU, S., VYAS, S.P., LAHIRI, A.N. : Influence of soil fertility on the growth and metabolism of wheat under salt stress. - Biol. Plant.
24 : 290 - 295, 1982. [Ps, Chl.]

49518 - GARGAS, M., GARGAS, E. : Growth physiological conditions of marine microalgae in the topmost 10 cm of the sediment. - Vatten *38* : 189 - 198, 1982.
[Chl.]

49519 - GARGAS, M., GARGAS, E. : Influence of temperature and oxygen conditions on
phytomicrobenthos stored for longer periods in darkness. - Vatten *38* :
306 - 316, 1982. [Chl.]

49520 - GARRARD, L.A., VAN, T.K. : General characteristics of freshwater vascular
plants. - In : ZABORSKY, O.R. (ed.) : CRC Handbook of Biosolar Resources.
Vol.I/2. Pp. 75 - 85. CRC Press, Boca Raton 1982. [Ps.]

49521 - GARRITY, D.P., WATTS, D.G., SULLIVAN, C.Y., GILLEY, J.R. : Moisture deficits
and grain sorghum performance: Evapotranspiration-yield relationships. -
Agron. J. *74* : 815 - 820, 1982. [Growth analysis.]

*49522 - GÄRTNER, W., OESTERHELT, D., TOWNER, P., HOPF, H., ERNST, L. : 13-(trifluo-romethyl)retinal forms an active and far-red-shifted chromophore in bacte-riorhodopsin. - J. amer. chem. Soc. *103* : 7642 - 7643, 1981.

49523 - GASCHO, G.J., SHIH, S.F. : Sugarcane. - In : TEARE, I.D., PEET, M.M. (ed.) : Crop-Water Relations. Pp. 445 - 480. John Wiley & Sons, New York - Chiches-ter - Brisbane - Toronto - Singapore 1982. [Ps.]

49524 - GASQUEZ, J., BARRALIS, G., AIGLE, N. : Distribution et extension de la ré-sistance chloroplastique aux triazines chez les adventices annuelles en France. - Agronomie *2* : 119 - 124, 1982.

49525 - GASSMANN, G., GILLBRICHT, M. : Correlations between phytoplankton, organic detritus and carbon in North Sea waters during the Fladenground Experiment (FLEX '76). - Helgol. Meeresuntersuch. *35* : 253 - 262, 1982. [Primary pro-duction.]

49526 - GAST, P., MUSHLIN, R.A., HOFF, A.J. : Nonuniform transfer of electron spin polarization in reaction centers of the photosynthetic bacterium *Rhodopseu-domonas sphaeroides*. - J. phys. Chem. *86* : 2886 - 2891, 1982.

*B49527 - GATES, D.M. : Biophysical Ecology. - Springer-Verlag, New York - Heidelberg - - Berlin 1980. [Ps.]

49528 - GAUDILLÈRE, J.-P. : La photorespiration, son coût énergétique. - Compt.rend. Acad. agr. Fr. *68* : 872 - 882, 1982.

49529 - GAUSMAN, H.W. : Visible light reflectance, transmittance, and absorptance of differently pigmented cotton leaves. - Remote Sens. Environm. *13* : 233 - - 238, 1982.

49530 - GAVALAS, N.A., CARAVATAS, S., MANETAS, Y. : Factors affecting a fast and reversible inactivation of photosynthetic phosphoenolpyruvate carboxylase. - Photosynthetica *16* : 49 - 58, 1982.

49531 - GEACINTOV, N.E., BRETON, J. : Application and pulsed lasers to the study of energy transfer and fluorescence phenomena in photosynthetic systems. - In : GARETZ, B.A., LOMBARDI, J.R. (ed.) : Advances in Laser Spectroscopy. Pp. 213 - 237. Heyden & Sons, Philadelphia 1982.

49532 - GEACINTOV, N.E., BRETON, J. : Exciton annihilation and other nonlinear high-intensity excitation effects. - In : ALFANO, R.R. (ed.) : Biological Events Probed by Ultrafast Laser Spectroscopy. Pp. 157 - 191. Academic Press, New York - London - Paris - San Diego - San Francisco - São Paulo - - Sydney - Tokyo - Toronto 1982.

49533 - GEACINTOV, N.E., BRETON, J. : Laser studies of primary processes in photo-synthesis. - In : HÉLÈNE, C., CHARLIER, M., MONTENAY-GARESTIER, T., LAUS-TRIAT, G. (ed.) : Trends in Photobiology. Pp. 549 - 559. Plenum Press, New York - London 1982.

49534 - GEIGER, D.R., GIAQUINTA, R.T. : Translocation of photosynthate. - In : GOVINDJEE (ed.) : Photosynthesis. Vol.II. Pp. 345 - 386. Academic Press, New York - London - Paris - San Diego - San Francisco - São Paulo - Sydney - - Tokyo - Toronto 1982.

49535 - GELLER, G.N., SMITH, W.K. : Influence of leaf size, orientation, and ar-rangement on temperature and transpiration in three high-elevation, large--leafed herbs. - Oecologia *53* : 227 - 234, 1982. [Stomatal resistance.]

49536 - GENT, M.P.N. : Effect of defoliation and depodding on long distance trans-location and yield in y-shaped soybean plants. - Crop Sci. *22* : 245 - 250, 1982.

49537 - GENT, M.P.N. : Effect of defoliation and depodding on $^{14}CO_2$-assimilation and photosynthate distribution in Y-shaped soybean plants. - Crop Sci. *22* : 860 - 867, 1982.

49538 - GEORGIEVA, M., K"DREV, T. : Izsledvane na promenite na membrannite lipidi v khloroplastite na tsarevitsata pod vliyanie na magneziev nedostig. [In-vestigation of changes in membrane lipids of maize chloroplasts under the influence of magnesium deficit.] - Fiziol. Rast.(Sofiya) *8*(3): 12 - 17,1982. [In Bulg., ab : E, R.]

49539 - GEPSTEIN, S. : Light-induced H⁺ secretion and the relation to senescence of oat leaves. - Plant Physiol. *70* : 1120 - 1124, 1982. [Chl.]

49540 - GERBER, G.E., KHORANA, H.G. : Primary structure of bacteriorhodopsin: Sequencing methods for membrane proteins. - In : COLOWICK, S.P., KAPLAN, N.O. (ed.) : Methods in Enzymology. Vol.88. Pp. 56 - 74. Academic Press, New York - London - Paris - San Diego - San Francisco - São Paulo - Sydney - - Tokyo - Toronto 1982.

49541 - GERDAY, C., MICHEL-WOLWERTZ, M.-P., BROUERS, M. : Some properties of purified fractions from bean etioplast membranes. - In : AKOYUNOGLOU, G., EVANGELOPOULOS, A.E., GEORGATSOS, J., PALAIOLOGOS, G., TRAKATELLIS, A., TSIGANOS, C.P. (ed.) : Cell Function and Differentiation, Part B. Biogenesis of Energy Transducing Membranes and Membrane and Protein Energetics. Pp. 25 - 32. Alan R. Liss Inc., New York 1982.

49542 - GERDES, H.-H., BEHRENDS, W., KINDL, H. : Biosynthesis of a microbody matrix enzyme in greening cotyledons. Glycollate oxidase synthesized *in vivo* and *in vitro*. - Planta *156* : 572 - 578, 1982.

49543 - GEROLA, P.D., GARLASCHI, F.M., FORTI, G., JENNINGS, R.C. : Effects of cations on the adhesion between membranes vesicles obtained by digitonin fractionation of spinach chloroplasts. - Biochim. biophys. Acta *679* : 101 - 109, 1982.

49544 - GERSHONI, J.M., SHOCHAT, S., MALKIN, S., OHAD, I. : Functional organization of the chlorophyll-containing complexes of *Chlamydomonas reinhardi*. A study of their formation and interconnection with reaction centers in the greening process of the y-1 mutant. - Plant Physiol. *70* : 637 - 644, 1982.

49545 - GERSTL, S.A.W., ZARDECKI, A. : Effects of aerosols on photosynthesis. - Nature *300* : 436 - 437, 1982.

49546 - GERWICK, B.C. : Responses of terrestrial plants to mineral nutrients. - In : ZABORSKY, O.R. (ed.) : CRC Handbook of Biosolar Resources. Vol.I/2. Pp. 213 - 222. CRC Press, Boca Raton 1982. [Ps.]

49547 - GEST, H. : The comparative biochemistry of photochemistry: Milestones in a conceptual zigzag. - In : KAPLAN, N.O., ROBINSON, A. (ed.) : From Cyclotrons to Cytochromes. Pp. 305 - 321. Academic Press, New York - London 1982.

49548 - GHANOTAKIS, D.F., YERKES, C.T., BABCOCK, G.T. : The role of reagents accelerating the deactivation reactions of water-splitting enzyme system Y (ADRY reagents) in destabilizing high-potential oxidizing equivalents generated in chloroplast photosystem II. - Biochim. biophys. Acta *682* : 21 - 31, 1982.

49549 - GHOLZ, H.L. : Environmental limits on aboveground net primary production, leaf area, and biomass in vegetation zones of the Pacific Northwest. - Ecology *63* : 469 - 481, 1982.

49550 - GHOLZ, H.L., FISHER, R.F. : Organic matter production and distribution in slash pine (*Pinus elliottii*) plantations. - Ecology *63* : 1827 - 1839, 1982.

49551 - GIERSCH, C. : Photophosphorylation by chloroplasts: effects of low concentrations of ammonia and methylamine. - Z. Naturforsch. *37 C* : 242 - 250, 1982.

49552 - GIESKES, W.W.C., KRAAY, G.W. : Effect of enclosure in large plastic bags on diurnal change in oxygen concentration in tropical ocean water. - Mar. Biol. *70* : 99 - 104, 1982.

49553 - GIETL, C., HOCK, B. : Organelle-bound malate dehydrogenase isoenzymes are synthesized as higher molecular weight precursors. - Plant Physiol. *70* : 483 - 487, 1982.

49554 - GIFFORD, D.J., COSSINS, E.A. : Relationships between glycollate and formate metabolism in greening barley leaves. - Phytochemistry *21* : 1485 - 1490, 1982.

49555 - **GIFFORD, R.M.** : Global photosynthesis in relation to our food and energy
needs. - In : GOVINDJEE (ed.) : Photosynthesis. Vol.II. Pp. 459 - 495.
Academic Press, New York - London - Paris - San Diego - San Francisco -
- São Paulo - Sydney - Tokyo - Toronto 1982.

49556 - **GIFFORD, R.M., JENKINS, C.L.D.** : Prospects of applying knowledge of photosynthesis toward improving crop production. - In : GOVINDJEE (ed.) : Photosynthesis. Vol.II. Pp. 419 - 457. Academic Press, New York - London -
- Paris - San Diego - San Francisco - São Paulo - Sydney - Tokyo - Toronto
1982.

*49557 - **GILBERT, C.W., BUETOW, D.E.** : Gel electrophoresis of chloroplast polypeptides: comparison of one-dimensional and two-dimensional gel analyses of
chloroplast polypeptides from *Euglena gracilis*. - Plant Physiol. *67* : 623 -
- 628, 1981.

49558 - **GILLANDERS, B., TAYLOR, J.A., MACKENDER, R.O.** : UDP-galactosyl-1,2-diacyl-
glycerol galactosyltransferase activity in developing oat leaf plastids. -
In : WINTERMANS, J.F.G.M., KUIPER, P.J.C. (ed.) : Biochemistry and Metabolism of Plant Lipids. Pp. 191 - 196. Elsevier Biomedical Press, Amsterdam
1982.

49559 - **GILLER, Yu.E., ASOEVA, L.M.** : O deĭstvii spetsificheskikh ingibitorov biosinteza RNK i belka na protsessy migratsii ėnergii v khloroplastakh gorokha. [Effect of specific inhibitors of RNA and protein biosynthesis on
energy migration processes in pea chloroplasts.] - Fiziol. Rast. *29* : 387 -
- 392, 1982. [In R, ab : E.]

49560 - **GILLETTE, D.A.** : Decomposition of annual patterns of atmospheric carbon
dioxide concentrations: a preliminary interpretation of one year of data
at 13 globally distributed locations. - Atmos. Environm. *16* : 2537 - 2542,
1982.

B49561 - **GIL MARTINEZ, F., IRIARTE AMBEL, J., JIMENEZ PARRONDO, M.S.** : La Fotosinte-
sis C₄.(Revision del Sindrome Kranz). [C_4 Photosynthesis. (Revision of the
Kranz Syndrome).] - Coleccion maior 1. Univ. de la Laguna, La Laguna 1982.
[In Span.]

49562 - **GILMOUR, D.J., HIPKINS, M.F., BONEY, A.D.** : The effects of salt stress on
the primary processes of photosynthesis in *Dunaliella tertiolecta*. - Plant
Sci. Lett. *26* : 325 - 330, 1982.

49563 - **GIMENEZ-GALLEGO, G., RAMÍREZ-PONCE, M.P., LAUZURICA, P., RAMÍREZ, J.M.** :
Photooxidase system of *Rhodospirillum rubrum* III. The role of rhodoquinone
and ubiquinone in the activity of preparations of chromatophores and photoreaction centers. - Europe. J. Biochem. *121* : 343 - 347, 1982.

49564 - **GIMÉNEZ-GALLEGO, G., SUANZES, P., RAMÍREZ, J.M.** : Functional bacterial photoreaction centres with only one type of protein. - FEBS Lett. *149* : 59 -
- 62, 1982.

49565 - **GINGRICH, J.C., BLAHA, L.K., GLAZER, A.N.** : Rod substructure in cyanobacterial phycobilisomes: analysis of *Synechocystis* 6701 mutants low in phycoerythrin. - J. Cell Biol. *92* : 261 - 268, 1982.

49566 - **GINGRICH, J.C., WILLIAMS, R.C., GLAZER, A.N.** : Rod substructure in cyanobacterial phycobilisomes: phycoerythrin assembly in *Synechocystis* 6701
phycobilisomes. - J. Cell Biol. *95* : 170 - 178, 1982.

49567 - **GINS, V.K., TIKHONOV, A.N., MUKHIN, E.N., RUUGE, Ė.K.** : Osobennosti funktsionirovaniya dvukh molekulyarnykh form ferredoksina gorokha v tsepi
ėlektronnogo transporta khloroplastov. [Two molecular forms of pea ferredoxin in the electron transport chain of chloroplasts.] - Biokhimiya *47* :
1859 - 1866, 1982. [In R, ab : E.]

*49568 - **GINZBURG, C.** : Metabolic changes in *Gladiolus* cormels during the break of
dormancy: the role of dark CO_2 fixation. - Plant Physiol. *68* : 1105 - 1109,
1981.

49569 - GIRAULT, G., GALMICHE, J.-M., LEMAIRE, C., STULZAFT, O. : Binding and exchange of nucleotides on the chloroplast coupling factor CF_1. The role of magnesium. - Europe. J. Biochem. *128* : 405 - 411, 1982.

49570 - GIRS, G.I., ZUBAREVA, O.N., ELAGIN, I.N. : Vliyanie nizkikh temperatur pochvy na soderzhanie pigmentov v khvoe sosny. [Influence of low soil temperature on pigment content in *Pinus silvestris* L. needles.] - Lesovedenie *1982* (3) : 53 - 60, 1982. [In R, ab : E.]

49571 - GIURGEVICH, J.R., DUNN, E.L. : Seasonal patterns of daily net photosynthesis, transpiration and net primary productivity of *Juncus roemerianus* and *Spartina alterniflora* in a Georgia salt marsh. - Oecologia *52* : 404 - 410, 1982.

*49572 - GLÄTTLI, R., ELLER, B.M., WANNER, H. : Temperaturabhängigkeit der Dunkelatmung und der Nettophotosynthese bei *Coffea arabica* L. - Ber. schweiz. bot. Ges. *90* : 189 - 193, 1980.

49573 - GLAZER, A.N. : Phycobilisomes: structure and dynamics. - Annu. Rev. Microbiol. *36* : 173 - 198, 1982.

49574 - GLEADOW, R.M., ROWAN, K.S. : Invasion by *Pittosporum undulatum* of the forests of central Victoria. III. Effects of temperature and light on growth and drought resistance. - Aust. J. Bot. *30* : 347 - 357, 1982. [Growth analysis.]

49575 - GLEIXNER, G., KARG, V., KIS, P. : Rapid preparation of pure chlorophyll *a*. - Experientia *38* : 303 - 304, 1982.

49576 - GLICK, H.L., BELL, W.C., SHAYKEWICH, C.F., LACROIX, L.J., BRACH, E.J. : Field spectral reflectance measurements of small grain crops. - Can. J. Plant Sci. *62* : 71 - 79, 1982.

49577 - GLICK, R.E., ZILINSKAS, B.A. : Role of the colorless polypeptides in phycobilisome reconstitution from separated phycobiliproteins. - Plant Physiol. *69* : 991 - 997, 1982.

49578 - GLOOSCHENKO, W.A., HARPER, N.S. : Net aerial primary production of a James Bay, Ontario, salt marsh. - Can. J. Bot. *60* : 1060 - 1067, 1982.

49579 - GLOSER, J. : Proměnlivý metabolismus sukulentních rostlin. [Changeable metabolism in succulents.] - Vesmír (Praha) *61* (1) : 10 - 14, 1982. [Ps; in Czech.]

49580 - GLYAUBERTENE, V.F. : Biologicheskaya i biokhimicheskaya kharakteristika perspektivnykh silosnykh rasteniĭ. (13. Soderzhanie karotina i askorbinovoĭ kisloty v nadzemnoĭ chasti sorta Veĭrikha, okopnika shershavogo, sil'fii pronzennolistnoĭ i maral'ego kornya. [Biological and biochemical characteristics of perspective silo plants. (13. Amount of carotene and ascorbic acid in overground part of *Polygonum Weyrichii, Symphytum asperum* LER., *Rhaponticum carthamoides* (WILLD.) ILJIN and *Silphium perfoliatum* L.] - Liet. TSR Mokslу Akad. darbai, Ser. C *1982* (1) : 81 - 87, 1982. [In R, ab : E, Lithu.]

49581 - GNAUCK, A. : Strukturelle und funktionelle Änderungen in aquatischen Ökosystemen. - In : UNGER, K., SCHUH, J. (ed.) : Umwelt-Stress. Pp. 335 - 344. Martin-Luther-Universität, Halle 1982.

49582 - GODDE, D., TREBST, A. : A NADH dehydrogenase bound to the photosynthetic membrane of hydrogen adapted *Chlamydomonas reinhardii*. - In : HALL, D.O., PALZ, W. (ed.) : Photochemical, Photoelectrochemical and Photobiological Processes. Pp. 213 - 217. D.Reidel Publ.Co., Dordrecht - Boston - London 1982.

49583 - GODIK, V.I., KOTOVA, E.A., BORISOV, A.Yu. : Nanosecond recombination luminescence of purple bacteria. The lifetime temperature dependence in *Rhodospirillum rubrum* chromatophores. - Photobiochem. Photobiophys. *4* : 219 - - 226, 1982.

49584 - GOEYENS, L., POST, E., DEHAIRS, F., VANDENHOUDT, A., BAEYENS, W. : The use of high pressure liquid chromatography with fluorimetric detection for chlorophyll A determination in natural extracts of chloropigment and their degradation products. - Int. J. environm. anal. Chem. *12* : 51 - 63, 1982.

*49585 - GOGEL, G., LEWIS, A. : Effect of iodination on the pK of Schiff base deprotonation and M_{412} production in purple membrane. - Biochem. biophys. Res. Commun. *103* : 175 - 181, 1981. [Bacteriörhodopsin.]

49586 - GOLBECK, J.H., WARDEN, J.T. : Electron spin resonance studies of the bound iron-sulfur centers in Photosystem I. Photoreduction of center A occurs in the absence of center B. - Biochim. biophys. Acta *681* : 77 - 84, 1982.

49587 - GOL'D, V.M., BELONOG, N.P., MOGIL'NAYA, O.A., GAEVSKIĬ, N.A., GRIGOR'EV, Yu.S., MORGUN, V.N. : Osobennosti organizatsii fotosinteticheskogo apparata tsvetkov semeĭstva orkhidnykh. [Peculiarities of the organization of photosynthetic apparatus in flowers of orchids.] - Fiziol. Rast. *29* : 1109 - 1113, 1982. [In R, ab : E.]

49588 - GOLDBERG, D.E. : Comparison of factors determining growth rates of deciduous vs. broad-leaf evergreen trees. - Amer. Midland Naturalist *108* : 133 - 143, 1982.

49589 - GOL'DFEL'D, M.G. : Paramagnitnye tsentry èlektrontransportnoĭ tsepi fotosinteza vysshikh rasteniĭ. [Paramagnetic centres of electron transport chain of higher plant photosynthesis.] - Biofizika *27* : 954 - 965, 1982. [In R, ab : E.]

49590 - GOL'DFEL'D, M.G., DMITROVSKIĬ, L.G., BLYUMENFEL'D, L.A. : Vliyanie redoks--agentov i khelatorov na svoĭstva ATPazy khloroplastov. [The effect of redox and chelating reagents on the properties of chloroplast ATPase.] - Mol. Biol. (Moskva) *16* : 183 - 189, 1982. [In R, ab : E.]

*49591 - GOLDMAN, J.C., TAYLOR, C.D., GLIBERT, P.M. : Non-linear time-course uptake of carbon and ammonium by marine phytoplankton. - Mar. Ecol. *6* : 137 - 148, 1981.

49592 - GOLIK, K.N. : Osobennosti produktsionnogo protsessa sortov yarovoĭ pshenitsy raznoĭ urozhaĭnosti. [Peculiarities of productivity process in spring wheat cultivars with different yields.] - Sel'skokhoz. Biol. *17* : 641 - - 646, 1982. [In R, ab : E.]

49593 - GOLLER, M., HAMPP, R., ZIEGLER, H. : Regulation of the cytosolic adenylate ratio as determined by rapid fractionation of mesophyll protoplasts of oat. Effect of electron transfer inhibitors and uncouplers. - Planta *156*: 255 - - 263, 1982.

*49594 - GOL'TSEV, V.N., TODOROV, S.I., VENEDIKTOV, P.S. : Izsledvane vr"zkata mezhdu intenziteta na zabavenata fluorestsentsiya na khloroplasti ot grakh i parametri na funktsioniraneto na fotosistema II. [Investigations of the relationship between the delayed fluorescence intensity and the photosystem II functional parameters in pea chloroplasts.] - In : Fiziologiya na Rasteniyata. Vol.5. Pp. 627 - 631. Inst. Fiziol. Rast. "M. Popov", Sofiya 1980. [In Bulg., ab : E.]

49595 - GOLVANO, M.P., FELIPE, M.R., CINTAS, A.M. : Influence of nitrogen sources on chloroplast development in wheat seedlings. - Physiol. Plant. *56* : 353 - - 360, 1982.

49596 - GOMEZ, I., DEL CAMPO, F.F., RAMIREZ, J.M. : The antenna system of *Rhodospirillum rubrum:* Derivative analysis of the major near infrared absorption band of chromatophores. - FEBS Lett. *141* : 185 - 188, 1982.

49597 - GOMEZ, I., PICOREL, R., RAMIREZ, J.M., PEREZ, R., DEL CAMPO, F.F. : Reversible oxidation of antenna bacteriochlorophyll in two photoreaction centerless mutants of *Rhodospirillum rubrum.* - Photochem. Photobiol. *35* : 399 - 403, 1982.

49598 - GOODWIN, J.B., GARAGORRY, F.L., ESPINOSA, W., SANS, L.M., YOUNGDAHL, L.J.: Modelling soil-water-plant relationships in the Cerrado soils of Brazil: The case of maize (*Zea mays* L.). - Agr. Systems *8* : 115 - 127, 1982. [Growth analysis.]

49599 - GOPAL, B., SHARMA, K.P. : Studies of wetlands in India with emphasis on structure, primary production and management. - Aquat. Bot. *12* : 81 - 91, 1982. [Chl.]

49600 - GORBACH, N.V., KHODASEVICH, É.V., MEL'NIKOVA, L.M., KOBZAR', N.N. : Izmenenie kolichestvennykh kharakteristik fonda khlorofillov v lishaĭnikakh pod vozdeĭstviem zagryazneniya atmosfernogo vozdukha SO_2. [Changes in quantitative characteristics of chlorophyll pools in lichens in SO_2-contaminated air.] - Dokl. Akad. Nauk belorus. SSR *26* : 850 - 852, 864, 1982. [In R, ab : E.]

49601 - GORDON, A.J., HESKETH, J.D., PETERS, D.B. : Soybean leaf photosynthesis in relation to maturity classification and stage of growth. - Photosynthesis Res. *3* : 81 - 93, 1982.

49602 - GORDON, A.J., RYLE, G.J.A., MITCHELL, D.F., POWELL, C.E. : The dynamics of carbon supply from leaves of barley plants grown in long or short days. - J. exp. Bot. *33* : 241 - 250, 1982.

49603 - GORDON, H.R. : Penetration of radiant energy into the aquatic environment. - In : ZABORSKY, O.R. (ed.) : CRC Handbook of Biosolar Resources. Vol.I/2. Pp. 503 - 511. CRC Press, Boca Raton 1982.

49604 - GORDON, H.R., CLARK, D.K., BROWN, J.W., BROWN, O.B., EVANS, R.H. : Satellite measurement of the phytoplankton pigment concentration in the surface waters of a warm core Gulf Stream ring. - J. mar. Res. *40* : 491 - 502, 1982.

49605 - GORDON, T.R., DUNIWAY, J.M. : Effects of powdery mildew infection on the efficiency of CO_2 fixation and light utilization by sugar beet leaves. - Plant Physiol. *69* : 139 - 142, 1982.

49606 - GORDON, T.R., DUNIWAY, J.M. : Photosynthesis in powdery mildewed sugar beet leaves. - Phytopathology *72* : 718 - 723, 1982.

49607 - GORDON, T.R., DUNIWAY, J.M. : Stomatal behavior and water relations in sugar beet leaves infected by *Erysiphe polygoni*. - Phytopathology *72* : 723 - - 726, 1982.

49608 - GÖRÖG, K., MUSCHINEK, Gy., MUSTÁRDY, L.A., FALUDI-DÁNIEL, Á. : Comparative studies of safeners for the prevention of EPTC injury in maize. - Weed Res. *22* : 27 - 33, 1982. [Ps.]

49609 - GORYSHINA, T.K., KISELEVA, T.M. : Ob ékologicheskoĭ labil'nosti fotosinteticheskogo apparata lista u nekotorykh kustarnichkov i trav elovogo lesa. II. Pigmenty lista. [Ecological lability of the photosynthetic apparatus in some micro-shrubs and herbs of the spruce forest. II. Leaf pigments.] - Vestn. leningrad. Univ. *1982* (3) : 31 - 37, 1982. [In R, ab : E.]

49610 - GORYSHINA, T.K., SËKE ZOYA, A. : Sravnitel'noe issledovanie sezonnoĭ dinamiki struktury assimilyatsionnogo apparata lista u rannevesennogo éfemeroida v dubovykh lesakh tsentral'noĭ i vostochnoĭ Evropy. [A comparative study of seasonal dynamics of leaf assimilatory-apparatus structure in a spring ephemeroid in oak forests of central and east Europe.] - Lesovedenie *1982* (1) : 61 - 67, 1982. [In R, ab : E.]

46911 - GORYSHINA, T.K., ZABOTINA, L.N., KISELEVA, T.M., PRUZHINA, E.G. : Osobennosti fotosinteticheskogo apparata lista i ego sezonnoĭ dinamiki u trav i kustarnichkov elovogo lesa. [Peculiarities of leaf photosynthetic apparatus and its seasonal dynamics in some herbs and micro-shrubs of the spruce forest.] - Vestnik leningrad. Univ. *1982* [15(Biol.3)] : 21 - 28, 1982. [In R, ab : E.]

49612 - GOSIEWSKI, W., NILWIK, H.J.M., BIERHUIZEN, J.F. : The influence of temperature on photosynthesis of different tomato genotypes. - Sci. Hort. *16* : 109 - 115, 1982.

49613 - GOSSE, G., CHARTIER, M., VARLET-GRANCHER, C., BONHOMME, R. : Interception du rayonnement utile à la photosynthèse chez la luzerne: Variations et modélisation. - Agronomie *2* : 583 - 588, 1982.

49614 - **GOSTIMSKIĬ, S.A., KARVOVSKAYA, E.A., SINESHCHEKOV, V.A., BELYAEVA, O.B.** : Ėlektronno-mikroskopicheskie i spektral'nye svoĭstva zheltykh letal'nykh mutantov gorokha. [Electron microscopic and spectral properties of yellow lethal mutants of pea.] - Genetika *18* : 124 - 132, 1982. [Chl; in R, ab : E.]

49615 - **GOUNARIS, K., SEN, A., QUINN, P.J.** : Polyunsaturated fatty acids may have a structural role in the chloroplast membrane. - Biochem. Soc. Trans. *10* : 408, 1982.

49616 - **GOUNOT, M.** : Écosystème prairial.1.Analyse et modélisation de l'écosystème prairial. - Acta oecol. - Oecol. gen. *3* : 7 - 28, 1982. [Production, carbon flux.]

49617 - **GOUNOT, M., YU, O., N'KANDZA, J.** : Écosystème prairial. 3.2. Insertion de la morphogenèse dans les modèles de productivité primaire. - Acta oecol. - Oecol. gen. *3* : 53 - 74, 1982.

49618 - **GOURDON, F., PLANCHON, C.** : Responses of photosynthesis to irradiance and temperature in soybean, *Glycine max* (L.) MERR. - Photosynthesis Res. *3* : 31 - 43, 1982.

49619 - **GOVINDARAJAN, A.G., POOVAIAH, B.W.** : Effect of root zone carbon dioxide enrichment on ethylene inhibition of carbon assimilation in potato plants. - Physiol. Plant. *55* : 465 - 469, 1982.

B49620 - **GOVINDJEE (ed.)** : Photosynthesis. Volume I. Energy Conversion by Plants and Bacteria. - Academic Press, New York - London - Paris - San Diego - San Francisco - São Paulo - Sydney - Tokyo - Toronto 1982.

B49621 - **GOVINDJEE (ed.)** : Photosynthesis. Volume II. Development, Carbon Metabolism, and Plant Productivity. - Academic Press, New York - London - Paris - - San Diego - San Francisco - São Paulo - Sydney - Tokyo - Toronto 1982.

49622 - **GOVINDJEE, WHITMARSH, J.** : Introduction to photosynthesis: Energy conversion by plants and bacteria. - In : GOVINDJEE (ed.) : Photosynthesis. Vol.I. Pp. 1 - 16. Academic Press, New York - London - Paris - San Diego - San Francisco - São Paulo - Sydney - Tokyo - Toronto 1982.

49623 - **GOWEN, R.J., TETT, P., WOOD, J.B.** : The problem of degradation products in the estimation of chlorophyll by fluorescence. - Arch. Hydrobiol. Beih. (Ergebn. Limnol.) *16* : 101 - 106, 1982.

49624 - **GRAAN, T., ORT, D.R.** : Photophosphorylation associated with synchronous turnovers of the electron-transport carriers in chloroplasts. - Biochim. biophys. Acta *682* : 395 - 403, 1982.

49625 - **GRÄBER, P.** : Phosphorylation in chloroplasts: ATP synthesis driven by $\Delta\psi$ and by ΔpH of artificial or light-generated origin. - In : SLAYMAN, C.L. (ed.) : Electrogenic Ion Pumps. Pp. 215 - 245. Academic Press, New York 1982.

49626 - **GRÄBER, P., RÖGNER, M., BUCHWALD, H.-E., SAMORAY, D., HAUSKA, G.** : Field-driven ATP synthesis by the chloroplast coupling factor complex reconstituted into liposomes. - FEBS Lett. *145* : 35 - 40, 1982.

49627 - **GRACE, J., PITCAIRN, C.E.R., RUSSELL, G., DIXON, M.** : The effects of shaking on the growth and water relations of *Festuca arundinacea* SCHREB. - Ann. Bot. *49* : 207 - 215, 1982. [Stomatal resistance.]

49628 - **GRAHAM, D.** : Carbonic anhydrases (carbonate dehydratases) from plants. - In : ZABORSKY, O.R. (ed.) : CRC Handbook of Biosolar Resources. Vol.I/1. Pp. 215 - 229. CRC Press, Boca Raton 1982.

49629 - **GRALL, J.R., LE CORRE, P., TRÉGUER, P.** : Short-term variability of primary production in coastal upwelling off Morocco. - Rapp. P.-v. Réun. Cons. int. Mer *180*: 221 - 227, 1982.

49630 - **GRANBERG, K., HARJULA, H.** : Nutrient dependence of phytoplankton production in brown-water lakes with special reference to Lake Päijänne. - Hydrobiologia *86* : 129 - 132, 1982. [Chl.]

49631 - GRÄTZEL, M. : Artificial photosynthesis, energy- and light-driven electron transfer in organized molecular assemblies and colloidal semiconductors. - Biochim. biophys. Acta *683* : 221 - 244, 1982.

*49632 - GRAY, J.C., HOOPER, E.A., PERHAM, R.N. : Subunit stoichiometry of tobacco ribulose 1,5-bisphosphate carboxylase. - FEBS Lett. *114* : 237 - 239, 1980.

49633 - GRECHKIN, A.N., TARCHEVSKIĬ, I.A. : Ob uchastii polyarnykh lipidov v obrazovanii polienovykh zhirnykh kislot v zeleneyushchikh prorostkakh pshenitsy. [Involvement of polar lipids in polyenoic fatty acids formation in greening wheat seedlings.] - Biokhimiya *47* : 1007 - 1014, 1982. [Chloroplast; in R, ab : E.]

49634 - GREEN, B.R. : Protein synthesis by isolated *Acetabularia* chloroplasts. Synthesis of the two minor chlorophyll *a* complexes *in vitro*. - Europe. J. Biochem. *128* : 543 - 546, 1982.

49635 - GREEN, B.R., CAMM, E.L. : The nature of the light-harvesting complex as defined by sodium dodecyl sulfate polyacrylamide gel electrophoresis. - Biochim. biophys. Acta *681* : 256 - 262, 1982.

49636 - GREEN, B.R., CAMM, E.L., VAN HOUTEN, J. : The chlorophyll-protein complexes of *Acetabularia*. A novel chlorophyll *a/b* complex which forms oligomers. - Biochim. biophys. Acta *681* : 248 - 255, 1982.

49637 - GREEN, J.M., WILLIAMS, G.J. III : The subdominant status of *Echinocereus viridiflorus* and *Mammillaria vivipara* in the shortgrass prairie: The role of temperature and water effects on gas exchange. - Oecologia *52* : 43 - 48, 1982.

49638 - GREEN, T.G.A., SNELGAR, W.P. : A comparison of photosynthesis in two thalloid liverworts. - Oecologia *54* : 275 - 280, 1982.

49639 - GREEN, T.G.A., SNELGAR, W.P. : Carbon dioxide exchange in lichens: Relationship between the diffusive resistance of carbon dioxide and water vapour. - Lichenologist *14* : 255 - 260, 1982.

49640 - GREENBAUM, E. : Photosynthetic hydrogen and oxygen production: kinetic studies. - Science *215* : 291 - 293, 1982.

49641 - GREENING, M.T., BUTTERFIELD, F.J., HARRIS, N. : Chloroplast ultrastructure during senescence and regreening of flax cotyledons. - New Phytol. *92* : 279 - 285, 1982.

49642 - GREGORY, R.P.F., BORBÉLY, G., DEMETER, S., FALUDI-DÁNIEL, Á. : Chiroptical properties of chlorophyll-protein complexes separated on Deriphat/polyacrylamide gel. - Biochem. J. *202* : 25 - 29, 1982.

49643 - GRESSEL, J. : Triazine herbicide interaction with a 32 000 M_r thylakoid protein - alternative possibilities. - Plant Sci. Lett. *25* : 99 - 106, 1982.

49644 - GRIESS, H. : Eine neue Leuchte für die Pflanzenbestrahlung und die rationelle Anordnung mehrerer Leuchten zu Lichtfeldern. - Arch. Züchtungsforsch. *12* : 249 - 256, 1982.

49645 - GRIFFIN, J.L., WATSON, V.H. : Production and quality of four bermudagrasses as influenced by rainfall patterns. - Agron. J. *74* : 1044 - 1047, 1982.

49646 - GRIFFITH, M., BROWN, G.N., HUNER, N.P.A. : Structural changes in thylakoid proteins during cold acclimation and freezing of winter rye (*Secale cereale* L. cv. Puma). - Plant Physiol. *70* : 418 - 423, 1982.

49647 - GRIFFITHS, R.P., CALDWELL, B.A., MORITA, R.Y. : Seasonal changes in microbial heterotrophic activity in subarctic marine waters as related to phytoplankton primary productivity. - Mar. Biol. *71* : 121 - 127, 1982.

49648 - GRIFFITHS, W.T., BEER, N.S. : Site of synthesis of NADPH. Protochlorophyllide oxidoreductase in rye (*Secale cereale*). - Plant Physiol. *70* : 1014 - - 1018, 1982.

49649 - GRIGOR'EV, Yu.S., MORGUN, V.N., GAEVSKIĬ, N.A. : Issledovanie svetoindutsi-
ruemykh izmeneniĭ millisekundnogo poslesvecheniya khloroplastov gorokha.
[Light-induced changes of ms-delayed light emission in pea chloroplasts.] -
Biofizika 27 : 973 - 976, 1982. [In R, ab : E.]

49650 - GRIMME, L.H., FAWLEY, M., BROWN, J.S. : Spectral properties of degreened
cells of Chlorella fusca. - Carnegie Inst. Washington Year Book 81 : 37 -
- 38, 1982.

49651 - GRIMSTAD, S.O. : Lampetyper og plantebestråling 2. Virkning av lampetype
og strålingsfluktstetthet på vekst og utvikling av salat (Lactuca sativa L.)
dyrket i veksthus under ulike naturlige lysforhold. [Light sources and
plant irradiation 2. Effect of light source and irradiance on growth and
development of lettuce (Lactuca sativa L.) grown in greenhouse under different natural light conditions.] - Meld. norg. Landbrukshøgskole 61 (2) :
1 - 24, 1982. [In Norw., ab : E.]

49652 - GRIMSTAD, S.O. : Lampetyper og plantebestråling 3. Virking av lampetype
og strålingsfluktstetthet på innhold av klorofyll, L-askorbinsyre og glu-
kose i salat (Lactuca sativa L.) dyrket i veksthus under ulike naturlige
lysforhold. [Light sources and plant irradiation 3.Effect of light source
and irradiance on the content of chlorophyll, L-ascorbic acid and glucose
in lettuce (Lactuca sativa L.) grown in greenhouse under different natural
conditions.] - Meld. norg. Landbrukshøgskole 61 (3) : 1 - 25, 1982. [In
Norw., ab : E.]

49653 - GRODZINSKI, B., BOESEL, I., HORTON, R.F. : Ethylene release from leaves
of Xanthium strumarium L. and Zea mays L. - J. exp. Bot. 33 : 344 - 354,
1982. [Ps.]

*49654 - GROMA, G.I., STRUŽINSKÝ, R., KARVALY, B.E. : A model system for bacterio-
rhodopsin chromophore. - Acta biochim. biophys. Acad. Sci. hung. 16 : 211 -
- 217, 1981.

49655 - GROSS, J. : Changes of chlorophylls and carotenoids in developing strawberry fruits (Fragaria ananassa) cv. Tenira. - Gartenbauwissenschaft 47 :
142 - 144, 1982.

49656 - GROSS, J. : Pigment changes in the pericarp of the chinese gooseberry or
kiwi fruit (Actinidia chinensis) cv. Bruno during ripening. - Gartenbauwissenschaft 47 : 162 - 167, 1982.

49657 - GROSS, J. : Chlorophyll and carotenoid pigments in Ribes fruits. - Sci.
Hort. 18 : 131 - 136, 1982/83.

49658 - GROSS, J. : Photosynthetic dynamics in varying light environments: a model
and its application to whole leaf carbon gain. - Ecology 63 : 84 - 93,
1982.

49659 - GROSSMAN, A.R., BARTLETT, S.G., SCHMIDT, G.W., MULLET, J.E., CHUA, N.-H.:
Optimal conditions for post-translational uptake of proteins by isolated
chloroplasts. In vitro synthesis and transport of plastocyanin, ferredoxin-
-NADP$^+$ oxidoreductase, and fructose-1,6-bisphosphatase. - J. biol. Chem.
257 : 1558 - 1563, 1982.

49660 - GROSSMANN, K., JUNG, J. : Zur methodischen Erfassung pflanzlicher Senes-
zenzvorgänge. - Z. Acker- Pflanzenbau 151 : 149 - 165, 1982. [Chl.]

49661 - GROUZIS, J.-P., RAMBIER, M., GRIGNON, C. : The stacking of the thylakoids
of two leguminosae. Differential responses to H$^+$ and divalent cations. -
Biochim. biophys. Acta 679 : 131 - 137, 1982.

49662 - GRUMBACH, K.H. : Herbicides which inhibit electron transport or produce
chlorosis and their effect on chloroplast development in radish seedlings.
I. Chlorophyll a fluorescence transients and photosystem II activity. -
Z. Naturforsch. 37 C : 268 - 275, 1982.

49663 - GRUMBACH, K.H. : Herbicides which inhibit electron transport or produce
chlorosis and their effect on chloroplast development in radish seedlings.
III. Plastid pigment and quinone composition. - Z. Naturforsch. 37 C : 642 -
- 650, 1982.

49664 - GRUMBACH, K.H., LICHTENTHALER, H.K. : Chloroplast pigments and their biosynthesis in relation to light intensity. - Photochem. Photobiol. *35* :
209 - 212, 1982.

49665 - GRUMBACH, K.H., MUNGENAST, P., RITZ, J. : Biosynthesis and degradation
of chlorophylls in relation to the developmental stages of a plastid. - In :
WINTERMANS, J.F.G.M., KUIPER, P.J.C. (ed.) : Biochemistry and Metabolism
of Plant Lipids. Pp. 559 - 564. Elsevier Biomedical Press, Amsterdam 1982.

☆49666 - GUAN Chunyun, WANG Guohuai, ZHAO Juntian : [The preliminary investigation
on heterosis and early prediction in heterosis selection of hybrids of rapeseed (*Brassica napus*).] - Acta genet. sin. *7* : 55 - 63, 1980. [Ps, Chl;
in Chin., ab : E.]

49667 - GUDIN, C. : Solar biotechnology: Microalgae production in tubular photo-
reactors for energy and chemicals. - In : GRASSI, G., PALZ, W. (ed.) :
Energy from Biomass. Vol.3. Pp. 131 - 134. D.Reidel Publ. Co., Dordrecht -
- Boston - London 1982.

49668 - GUEHL, J.-M. : Potentiel de photosynthèse hivernale du Douglas (*Pseudotsuga menziesii* MIRB.) en relation avec le régime thermique. - Ann. Sci. forest. *39* : 239 - 258, 1982.

49669 - GUERRERO, M.G., RAMOS, J.L., LOSADA, M. : Photosynthetic production of
ammonia. - Experientia *38* : 53 - 58, 1982.

49670 - GUËT, C., TRÉMOLIÈRES, A., LECHARNY, A. : The effect of monochromatic light
on *trans*-hexadecenoic acid and chlorophyll accumulation in etiolated leaves of *Vigna sinensis* L. - Photochem. Photobiol. *35* : 283 - 284, 1982.

49671 - GUIKEMA, J., SHERMAN, L. : Protein composition and architecture on the
photosynthetic membranes from the cyanobacterium, *Anacystis nidulans* R2. -
Biochim. biophys. Acta *681* : 440 - 450, 1982.

49672 - GUILLON, P., CHERBUIN, A., MOUTOT, F., COUSIN, R., JOLIVET, E. : Effet de
la mutation afila sur les caractéristiques photosynthétiques du Pois (*Pisum sativum* L.). - Compt. rend. Acad. Sci. Paris, Sér. III *294* : 231 - 234,
1982.

49673 - GULLIKSEN, O.M., HUSHOVD, O.T., TEXMON, I., NORDBY, Ø. : Changes in respiration, photosynthesis and protein composition during induced synchronous
formation of gametes and zoospores in *Ulva mutabilis* FØYN. - Planta *156* :
33 - 40, 1982.

49674 - GULOTTY, R.J., FLEMING, G.R., ALBERTE, R.S. : Low-intensity picosecond
fluorescence kinetics and excitation dynamics in barley chloroplasts. -
Biochim. biophys. Acta *682* : 322 - 331, 1982.

49675 - GULYA, T.J., BANTTARI, E.E. : Apical chlorosis of sunflower caused by
Pseudomonas syringae pv. *tagetis*. - Plant Dis. *66* : 598 - 600, 1982.

49676 - GULYAEV, B.A., TETEN'KIN, V.L., MATORIN, D.N. : Polyarizatsiya bystroĭ
i zamedlennoĭ fluorestsentsii fotosistemy II. [Polarization of prompt and
delayed fluorescence of photosystem II.] - Biofizika *27* : 42 - 48, 1982.
[In R, ab : E.]

49677 - GULYAEV, B.I., KIRIZIĬ, D.A., MILOV, M.A. : Nakoplenie sukhogo veshchestva
i gazoobmen rasteniĭ svekly v usloviyakh razlichnykh intensivnosti i pro-
dolzhitel'nosti oblucheniya. [Accumulation of dry matter and gas exchange
in beet plants under various irradiance and irradiation time.] - Fiziol.
Biokhim. kul't. Rast. *14* : 523 - 528, 1982. [In R, ab : E.]

49678 - GUNASEELAN, T., KRISHNASWAMI, R., RAO, M.R.K. : Heterosis and photosynthetic rate in inter-racial hybrids of *Gossypium hirsutum* L. - Coton Fibres
trop. *37* : 277 - 278, 1982.

49679 - GUTELMACHER, B.L., PETROVA, N.A. : Production of individual species of algae and its role in the productivity of phytoplankton in Ladoga Lake. -
Int. Rev. ges. Hydrobiol. *67* : 613 - 624, 1982.

49680 - GUTTERIDGE, S., PARRY, M.A.J., SCHMIDT, C.N.G. : The reactions between active and inactive forms of wheat ribulosebisphosphate carboxylase and ef-
fectors. - Europe. J. Biochem. *126* : 597 - 602, 1982.

49681 - GYSI, J.R., CHAPMAN, D.J. : Phycobilins and phycobiliproteins of algae. -
In : ZABORSKY, O.R. (ed.) : CRC Handbook of Biosolar Resources. Vol.I/1. Pp.
83 - 102. CRC Press, Boca Raton 1982.

49682 - GYURJÁN, I., NAGY, A.H., ERDÖS, G., PALESS, Gy., KERESZTES, Á., KOVÁCS, P.,
SZIGETI, Z. : Photosynthetic functions and thylakoid membrane polypeptide
composition in light-sensitive mutants of *Chlamydomonas reinhardii*. - Photosynthesis Res. *3* : 255 - 271, 1982.

49683 - HACHTEL, W. : Biosynthesis and assembly of thylakoid membrane proteins in
isolated chloroplasts from *Vicia faba* L.: The P700-chlorophyll *a*-protein. -
Z. Pflanzenphysiol. *107* : 383 - 394, 1982.

49684 - HÄDER, D.-P. : Coupling of photomovement and photosynthesis in desmids. -
Cell Motility *2* : 73 - 82, 1982.

49685 - HÄDER, D.-P., POFF, K.L. : Spectrophotometric measurements of plastoquinone photoreduction in the blue-green alga, *Phormidium uncinatum*. - Arch.
Microbiol. *131* : 347 - 350, 1982.

*49686 - HADLEY, N.F., SZAREK, S.R. : Productivity of desert ecosystems.- BioScience
31 : 747 - 753, 1981.

49687 - HAEHNEL, W. : On the functional organization of electron transport from
plastoquinone to photosystem I. - Biochim. biophys. Acta *682* : 245 - 257,
1982.

49688 - HAEHNEL, W., NAIRN, J.A., REISBERG, P., SAUER, K. : Picosecond fluorescence kinetics and energy transfer in chloroplasts and algae. - Biochim. biophys. Acta *680* : 161 - 173, 1982.

49689 - HALL, A.E. : Mathematical models of plant water loss and plant water relat-
ions. - In : LANGE, O.L., NOBEL, P.S., OSMOND, C.B., ZIEGLER, H. (ed.) :
Physiological Plant Ecology II. Water Relations and Carbon Assimilation.
Pp. 231 - 261. Springer-Verlag, Berlin - Heidelberg - New York 1982.
[Ps.]

49690 - HALL, D.O. : Solar energy through biology: fuels from biomass. - Experientia *38* : 3 - 10, 1982.

B49691 - HALL, D.O., BARNARD, G.W., MOSS, P.A. : Biomass for Energy in the Developing
Countries. - Pergamon Press, Oxford - New York - Toronto - Sydney - Paris -
Frankfurt 1982.

*49692 - HALL, D.O., CAMMACK, R., RAO, K.K. : Chemie und Biologie der Eisen-Schwefel-
-Proteine. - Chem. unserer Zeit *11* (6) : 165 - 176, 1977. [Ps.]

*49693 - HALL, D.O., CHARTIER, P. : Biomass in Europe. - In : GLENN, B.H., FRANTA,
G.E. (ed.) : Proceedings AS-ISES 1981 Annual Meeting. Pp. 196 - 200. Publ.
AS-ISES, Newark 1981.

49694 - HALL, D.O., COOMBS, J. : Biomass facts and figures. - In : COOMBS, J.,
HALL, D.O. (ed.) : Techniques in Bioproductivity and Photosynthesis. Pp.
159 - 166. Pergamon Press, Oxford - New York - Toronto - Sydney - Paris -
- Frankfurt 1982.

49695 - HALL, D.O., GISBY, P.E., RAO, K.K. : Biophotolysis of water for H_2 production using immobilized and synthetic catalysts. - In : HÉLÈNE, C.,
CHARLIER, M., MONTENAY-GARESTIER, T., LAUSTRIAT, G. (ed.) : Trends in Photobiology. Pp. 587 - 595. Plenum Press, New York - London 1982.

49696 - HALL, H.K., McWHA, J.A. : Abscisic acid and wheat leaf senescence: the effect of pre-treating the intact plant. - Z. Pflanzenphysiol. *106* : 371 -
- 373, 1982. [Chl.]

49697 - HALL, R.G., LARSON, K.L. : Water stress of alfalfa during stress and recovery. - Can. J. Plant Sci. *62* : 639 - 647, 1982. [Stomatal resistance.]

49698 - HALLENBECK, P.C., JOUANNEAU, Y., VIGNAIS, P.M. : Purification and molecular properties of a soluble ferredoxin from *Rhodopseudomonas capsulata*. -
Biochim. biophys. Acta *681* : 168 - 176, 1982.

*49699 - HALLENBECK, P.C., KOCHIAN, L.V., BENEMANN, J.R. : Hydrogen evolution cata-
lyzed by hydrogenase in cultures of cyanobacteria. - Z. Naturforsch. *36 C* :
87 - 92, 1981.

49700 - HÄLLGREN, J.-E. : Field photosynthesis; monitoring with $^{14}CO_2$. - In :
COOMBS, J., HALL, D.O. (ed.) : Techniques in Bioproductivity and Photosyn-
thesis. Pp. 36 - 44. Pergamon Press, Oxford - New York - Toronto - Sydney -
- Paris - Frankfurt 1982.

49701 - HÄLLGREN, J.-E., GEZELIUS, K. : Effects of SO_2 on photosynthesis and ribulose bisphosphate carboxylase in pine tree seedlings. - Physiol. Plant.
54 : 153 - 161, 1982.

49702 - HÄLLGREN, J.-E., SUNDBOM, E., STRAND, M. : Photosynthetic responses to low
temperature in *Betula pubescens* and *Betula tortuosa*. - Physiol. Plant. *54* :
275 - 282, 1982.

49703 - HALLICK, R.B., CHELM, B.K., OROZCO, E.M.,Jr., RUSHLOW, K.E., GRAY, P.W. :
Organization and expression of the chloroplast genome of *Euglena gracilis*.
- In : SCHIFF, J.A., LYMAN, H. (ed.) : On the Origins of Chloroplasts. Pp.
297 - 306. Elsevier/North-Holland, New York - Amsterdam - Oxford 1982.

49704 - HALTERLEIN, A.J. : Bean. - In : TEARE, I.D., PEET, M.M. (ed.) : Crop-Water
Relations. Pp. 157 - 186. John Wiley & Sons, New York - Chichester - Brisbane - Toronto - Singapore 1982. [Ps.]

49705 - HALVA, E., HRABĚ, F. : Příspěvek ke studiu analýzy růstu nadzemní biomasy
bobu obecného (*Faba vulgaris* MOENCH.) na zelenou píci v závlahových pod-
mínkách. [Growth analysis of aerial biomass of irrigated horse bean (*Faba
vulgaris* MOENCH.) grown for green fodder.] - Acta Univ. Agr., Fac. agron.
(Brno) *A28* (3/4) : 239 - 255, 1980. [In Czech, ab : E, G, R.]

49706 - HAMANAKA, T., HIRAKI, K., MITSUI, T. : X-ray diffraction studies of purple
membranes reconstituted from brown membrane. - In : COLOWICK, S.P., KAPLAN,
N.O. (ed.) : Methods in Enzymology. Vol.88. Pp. 268 - 271. Academic Press,
New York - London - Paris - San Diego - San Francisco - São Paulo - Sydney -
- Tokyo - Toronto 1982. [Bacteriorhodopsin.]

49707 - HAMBLIN, A.P., TENNANT, D., COCHRANE, H. : Tillage and the growth of a
wheat crop in a loamy sand. - Aust. J. agr. Res. *33* : 887 - 897, 1982.
[Growth analysis.]

49708 - HAMPE, T., MARSCHNER, H. : Effect of sodium on morphology, water relations
and net photosynthesis of sugar beet leaves. - Z. Pflanzenphysiol. *108* :
151 - 162, 1982.

49709 - HAMPP, R., GOLLER, M., ZIEGLER, H. : Adenylate levels, energy charge, and
phosphorylation potential during dark-light and light-dark transition in
chloroplasts, mitochondria, and cytosol of mesophyll protoplasts from *Avena sativa* L. - Plant Physiol. *69* : 448 - 455, 1982.

49710 - HAMPP, R., OUTLAW, W.H.,Jr., TARCZYNSKI, M.C. : Profile of basic carbon
pathways in guard cells and other leaf cells of *Vicia faba* L. - Plant Physiol. *70* : 1582 - 1585, 1982.

49711 - HAMZE, M., NIMAH, M. : Iron content during lime-induced chlorosis with
two citrus rootstocks. - J. Plant Nutr. *5* : 797 - 804, 1982.

49712 - HAND, D.W. : CO_2 enrichment, the benefits and problems. - Sci. Hort. *33* :
14 - 43, 1982. [Ps.]

49713 - HAND, J.M., YOUNG, E., VASCONCELOS, A.C. : Leaf water potential, stomatal
resistance, and photosynthetic response to water stress in peach seedlings.
- Plant Physiol. *69* : 1051 - 1054, 1982.

49714 - HANGARTER, R.P., GOOD, N.E. : Energy thresholds for ATP synthesis in chloroplasts. - Biochim. biophys. Acta *681* : 397 - 404, 1982.

49715 - HANKS, R.J., RASMUSSEN, V.P. : Predicting crop production as related to plant water stress. - Adv. Agron. *35* : 193 - 215, 1982.

49716 - HANSEN, P. : Assimilation and carbohydrate utilization in apple. - In : Proceedings 21st International Horticultural Congress. Volume I. Pp. 257 - - 268. International Society for Horticultural Science, Wageningen 1982. [Growth analysis.]

49717 - HANSON, A.D., HITZ, W.D. : Metabolic responses of mesophytes to plant water deficits. - Annu. Rev. Plant Physiol. *33* : 163 - 203, 1982. [Ps.]

49718 - HANSON, J.D. : Effect of light, temperature and water stress on net photosynthesis in two populations of honey mesquite. - J. Range Manage. *35* : 455 - 459, 1982.

49719 - HANSON, W.D., WEST, D.R. : Source-sink relationships in soybeans. 1. Effects of source manipulation during vegetative growth on dry matter distribution. - Crop Sci. *22* : 372 - 376, 1982.

49720 - HANSSON, Ö., ANDRÉASSON, L.-E. : EPR-detectable magnetically interacting manganese ions in the photosynthetic oxygen-evolving system after continuous illumination. - Biochim. biophys. Acta *679* : 261 - 268, 1982.

49721 - HARASHIMA, K., NAKAGAWA, M., MURATA, N. : Photochemical activities of bacteriochlorophyll in aerobically grown cells of aerobic heterotrophs, *Erythrobacter* species (OCh 114) and *Erythrobacter longus* (OCh 101). - Plant Cell Physiol. *23* : 185 - 193, 1982.

49722 - HARAUX, F., KOUCHKOVSKY, Y. de : Further investigation of the lateral and transversal proton currents at the thylakoid membrane level by hydrogen- -deuterium exchange. - Biochim. biophys. Acta *679* : 235 - 247, 1982.

*49723 - HARAZONO, Y., YABUKI, K. : [Studies on the effect of leaf-boundary layer resistance on the matter production of crop. (1) Effects of wind direction to leaf and angle of attack of air to leaf on the boundary layer resistance of sweet potato leaf.] - J. agr. Meteorol. *37* : 103 - 110, 1981. [In Jap., ab : E.]

49724 - HARAZONO, Y., YABUKI, K. : [Studies on the effect of leaf-boundary layer resistance on the matter production of crop. (2) Distributions of the photosynthetic rate caused by the local difference in the boundary layer structures of individual leaves.] - J. agr. Meteorol. *38* : 231 - 238, 1982. [In Jap., ab : E.]

*49725 - HARBRON, S., FOYER, C., WALKER, D. : The purification and properties of sucrose-phosphate synthetase from spinach leaves: The involvement of this enzyme and fructose bisphosphatase in the regulation of sucrose biosynthesis. - Arch. Biochem. Biophys. *212* : 237 - 246, 1981. [Photosynthates.]

49726 - HARDER, H.J., CARLSON, R.E., SHAW, R.H. : Leaf photosynthetic response to foliar fertilizer applied to corn plants during grain fill. - Agron. J. *74* : 759 - 761, 1982.

49727 - HARDER, H.J., CARLSON, R.E., SHAW, R.H. : Photosynthesis in corn in relationship to limited soil water. - Iowa State J. Res. *57* : 21 - 31, 1982.

49728 - HARDING, L.W.,Jr., PRÉZELIN, B.B., SWEENEY, B.M., COX, J.L. : Diel oscillations of the photosynthesis-irradiance (P-I) relationship in natural assemblages of phytoplankton. - Mar. Biol. *67* : 167 - 178, 1982.

49729 - HARDING, L.W.,Jr., PRÉZELIN, B.B., SWEENEY, B.M., COX, J.L. : Primary production as influences ., diel periodicity of phytoplankton. - Mar. Biol. *67* : 179 - 186, 1982.

49730 - HARI, P., KELLOMÄKI, S., MÄKELÄ, A., ILONEN, P., KANNINEN, M., KORPILAHTI, E., NYGRÉN, M. : Metsikön varhaiskehityksen dynamiikka. [Dynamics of early development of tree stand.] - Acta forest. fenn. *177* : 1 - 42, 1982. [Ps; in Finn., ab : E.]

49731 - HARIVANDI, M.A., BUTLER, J.D. : Factor associated with iron chlorosis of Kentucky bluegrass cultivars. - J. Plant Nutr. *5* : 569 - 573, 1982.

✷49732 - **HARRIS, W., FORDE, B.J., HARDACRE, A.K.** : Temperature and cutting effects
          on the growth and competitive interaction of ryegrass and paspalum. 1. Dry
          matter production, tiller numbers, and light interception. - New Zeal. J.
          agr. Res. *24* : 299 - 307, 1981.

✷49733 - **HARRIS, W., FORDE, B.J., HARDACRE, A.K.** : Temperature and cutting effects
          on the growth and competitive interaction of ryegrass and paspalum II.
          Interspecific competition. - New Zeal. J. agr. Res. *24* : 309 - 320, 1981.

  49734 - **HARRISON, P.A., BLACK, C.C.** : Two-dimensional electrophoretic mapping of
          proteins of bundle sheath and mesophyll cells of the C₄ grass *Digitaria
          sanguinalis* (L.) SCOP. (crabgrass). - Plant Physiol. *70* : 1359 - 1366,
          1982.

  49735 - **HARRISON, W.G., PLATT, T., IRWIN, B.** : Primary production and nutrient
          assimilation by natural phytoplankton populations of the eastern Canadian
          Arctic. - Can. J. Fisheries aquat. Sci. *39* : 335 - 345, 1982.

✷49736 - **HARTMANN, H.D., FORCHE, E.** : Bestimmungen der "sink"-Intensität von Seiten-
          sprossen mittels radioaktiver Substanzen. - Gartenbauwissenschaft *45* : 4 -
          - 6, 1980.

  49737 - **HARTMANN, T.** : Ammonium assimilation and nitrogen partitioning. - Progr.
          Bot. *44* : 154 - 164, 1982. [Photorespiration.]

  49738 - **HARU, K., NAITO, K., SUZUKI, H.** : Differential effects of benzyladenine
          and potassium on DNA, RNA, protein and chlorophyll contents and on expan-
          sion growth of detached cucumber cotyledons in the dark and light. - Phy-
          siol. Plant. *55* : 247 - 254, 1982.

✷49739 - **HARVEY, G.W., KEISTER, D.L.** : Energy-linked reactions in photosynthetic
          bacteria: P$_i$ ⇌ HOH oxygen exchange catalyzed by the membrane-bound inorga-
          nic pyrophosphatase of *Rhodospirillum rubrum*. - Arch. Biochem. Biophys.
          208 : 426 - 430, 1981.

  49740 - **HARWOOD, J.L., JONES, A.V.H.M., THOMAS, H.** : Leaf senescence in a non-yel-
          lowing mutant of *Festuca pratensis*. III. Total acyl lipids of leaf tissue
          during senescence. - Planta *156* : 152 - 157, 1982.

  49741 - **HASE, E., OH-HAMA, T., TSUJI, H.** : Development of photosynthetic apparatus.
          - Recent Progr. nat. Sci. Jap. *7* : 47 - 51, 1982.

  49742 - **HASE, T., MATSUBARA, H., HUTBER, G.N., ROGERS, L.J.** : Amino acid sequences
          of *Nostoc* strain MAC ferredoxins I and II. - J. Biochem. (Tokyo) *92* : 1347 -
          - 1355, 1982.

  49743 - **HASE, T., YAMANASHI, H., MATSUBARA, H.** : Purification and amino acid se-
          quence of a fern (*Gleichenia japonica*) ferredoxin. - J. Biochem. (Tokyo)
          91 : 341 - 346, 1982.

  49744 - **HASEGAWA, Y.** : General characteristics of brown macroalgae: *Laminaria*. -
          In : **ZABORSKY, O.R.** (ed.) : CRC Handbook of Biosolar Resources. Vol.I/2.
          Pp. 63 - 68. CRC Press, Boca Raton 1982. [Ps.]

  49745 - **HASHIMOTO, H., MURAKAMI,S.** : Chloroplast replication and loss of chloroplast
          DNA induced by nalidixic acid in *Euglena gracilis*. - Cell Struct. Funct.
          7 : 111 - 120, 1982.

  49746 - **HASLER, M., RUFFNER, H.P., RAST, D.M.** : High-yield isolation of grape leaf
          protoplasts as an instrument in physiological research. - Experientia *38* :
          564 - 565, 1982. [Ps.]

  49747 - **HÄSLER, R.** : Net photosynthesis and transpiration of *Pinus montana* on east
          and north facing slopes at alpine timberline. - Oecologia *54* : 14 - 22,
          1982.

  49748 - **HASS, W.** : Die Verteilung essentieller Fettsäuren und Lipide zwischen plas-
          tidischem und nichtplastidischem Kompartiment der Zelle in Primärblättern
          ausgewählter Weizenevolutionsformen. - In : **HOFFMANN, P., HIEKE, B.** (ed.):
          Photosynthese: Regulation und Evolution. (Colloquia Pflanzenphysiologie
          Nr.5.) Pp. 87 - 92. Humboldt-Universität, Berlin 1982.

B49749 - **HASSAL, K.A.** : The Chemistry of Pesticides. Their Metabolism, Mode of Action and Uses in Crop Protection. - Verlag Chemie, Weinheim - Deerfield Beach - Basel 1982. [Ps, Chl, Car.]

49750 - **HASUMI, H.** : Analyses of optical absorption and circular dichroism spectra of spinach ferredoxin at alkaline pH. - J. Biochem. (Tokyo) *92* : 1049 - - 1057, 1982.

49751 - **HASUMI, H., NAKAMURA, S., KOGA, K., YOSHIZUMI, H., PARCELLS, J.H., KIMURA, T.** : Further physicochemical studies on the complex formation between iron-sulfur proteins and flavoproteins from spinach chloroplast and beef adrenal cortex electron-transfer systems. - J. Biochem. (Tokyo) *91* : 135 - 141, 1982.

49752 - **HATA, M., ABE, S., HATA, M.** : Occurrence of peridinin-chlorophyll *a*-protein complex in red tide dinoflagellate *Prorocentrum micans.* - Bull. jap. Soc. sci. Fish. *48*: 459 - 461, 1982.

49753 - **HATA, M., HATA, M.** : Isolation and properties of peridinin-chlorophyll *a*-protein complex from the brick-red-colored oyster, *Crassostrea gigas.* - Comp. Biochem. Physiol. *72 B* : 631 - 635, 1982.

49754 - **HATA, M., HATA, M., NAKAMURA, K., FUJIWARA, H.** : [Brick-red coloration of oyster *Crassostrea gigas.*] - Bull. jap. Soc. sci. Fish. *48* : 975 - 979, 1982. [Chl, in Jap., ab : E.]

49755 - **HATCH, M.D.** : Photosynthetic carbon dioxide assimilation via the C_4 pathway. - In : **ZABORSKY, O.R.** (ed.) : CRC Handbook of Biosolar Resources. Vol. I/1. Pp. 185 - 189. CRC Press, Boca Raton 1982.

49756 - **HATCH, M.D.** : C_4 acid decarboxylases. - In : **ZABORSKY, O.R.** (ed.) : CRC Handbook of Biosolar Resources. Vol. I/1. Pp. 211 - 213. CRC Press, Boca Raton 1982.

49757 - **HATCH, M.D.** : Properties and regulation of adenylate kinase from *Zea mays* leaf operating in C_4 pathway photosynthesis. - Aust. J. Plant Physiol. *9* : 287 - 296, 1982.

49758 - **HATCH, M.D., TSUZUKI, M., EDWARDS, G.E.** : Determination of NAD malic enzyme in leaves of C_4 plants. Effects of malate dehydrogenase and other factors. - Plant Physiol. *69* : 483 - 491, 1982.

49759 - **HATTERSLEY, P.W.** : $\delta^{13}C$ values of C_4 types in grasses. - Aust. J. Plant Physiol. *9* : 139 - 154, 1982.

49760 - **HATTERSLEY, P.W., WATSON, L., JOHNSTON, C.R.** : Remarkable leaf anatomical variations in *Neurachne* and allies (*Poaceae*) in relation to C_3 and C_4 photosynthesis. - Bot. J. linnean Soc. *84* : 265 - 272, 1982.

49761 - **HATZIOS, K.K., HOWE, C.M.** : Influence of the herbicides hexazinone and chlorsulfuron on the metabolism of isolated soybean leaf cells. - Pestic. Biochem. Physiol. *17* : 207 - 214, 1982.

49762 - **HAUPT, W.** : Light-mediated movement of chloroplasts. - Annu. Rev. Plant Physiol. *33* : 205 - 233, 1982.

49763 - **HAUPT, W.** : Physiology of movement. - Progr. Bct. *44* : 221 - 230, 1982. [Chloroplast.]

49764 - **HAUSKA, G., GABELLINI, N., HURT, E., KRINNER, M., LOCKAU, W.** : Cytochrome *b/c* complexes with polyprenylquinol: cytochrome *c* oxidoreductase activity from *Anabaena variabilis* and *Rhodopseudomonas sphaeroides* GA: comparison of preparations from chloroplasts and mitochondria. - Biochem. Soc. Trans. *10* : 340 - 341, 1982.

49765 - **HAWKER, J.S., SMITH, G.M.** : Salt tolerance and regulation of enzymes of starch synthesis in cassava (*Manihot esculenta* CRANTZ). - Aust. J. Plant Physiol. *9* : 509 - 518, 1982. [Ps.]

49766 - **HAWKINS, A.F.** : Light interception, photosynthesis and crop productivity. - Outlook Agr. *11* : 104 - 113, 1982.

B49767 - HAWKINS, A.F., JEFFCOAT, B. (ed.) : Opportunities for Manipulation of Cereal Productivity (Monograph No.7). - British Plant Growth Regulator Group, Wantage 1982. [Ps.]

49768 - HAWKINS, C.M., LEWIS, J.B. : Benthic primary production on a fringing coral reef in Barbados, West Indies. - Aquat. Bot. *12* : 355 - 363, 1982.

49769 - HAWORTH, P., ARNTZEN, C.J., TAPIE, P., BRETON, J. : Orientation of pigments in the thylakoid membrane and in the isolated chlorophyll-protein complexes of higher plants. I. Determination of optimal conditions for linear dichroism measurement. - Biochim. biophys. Acta *679* : 428 - 435, 1982.

49770 - HAWORTH, P., KYLE, D.J., ARNTZEN, C.J. : A demonstration of the physiological role of membrane phosphorylation in chloroplasts, using the bipartite and tripartite models of photosynthesis. - Biochim. biophys. Acta *680* : 343 - 351, 1982.

49771 - HAWORTH, P., KYLE, D.J., HORTON, P., ARNTZEN, C.J. : Chloroplast membrane protein phosphorylation. - Photochem. Photobiol. *36* : 743 - 748, 1982.

49772 - HAWORTH, P., TAPIE, P., ARNTZEN, C.J., BRETON, J. : Orientation of pigments in the thylakoid membrane and in the isolated chlorophyll-protein complexes of higher plants. II. Linear dichroism spectra of isolated pigment-protein complexes oriented in polyacrylamide gels at 300 and 100 K. - Biochim. biophys. Acta *682* : 152 - 159, 1982.

49773 - HAWORTH, P., TAPIE, P., ARNTZEN, C.J., BRETON, J. : Orientation of pigments in the thylakoid membrane and in the isolated chlorophyll-protein complexes of higher plants. IV. The 100 K linear dichroism spectra of thylakoids from wild-type and chlorophyll *b*-less barley thylakoids. - Biochim. biophys. Acta *682* : 504 - 506, 1982.

49774 - HAYASHI, H., MIYAO, M., MORITA, S. : Absorption and fluorescence spectra of light-harvesting bacteriochlorophyll-protein complexes from *Rhodopseudomonas palustris* in the near-infrared region. - J. Biochem. (Tokyo) *91* : 1017 - 1027, 1982.

49775 - HAYASHI,H., NAKANO, M., MORITA, S. : Comparative studies of protein properties and bacteriochlorophyll contents of bacteriochlorophyll-protein complexes from spectrally different types of *Rhodopseudomonas palustris*. - J. Biochem. (Tokyo) *92* : 1805 - 1811, 1982.

49776 - HAYASHI, H., NOZAWA, T., HATANO, M., MORITA, S. : Circular dichroism of bacteriochlorophyll *a* in light-harvesting bacteriochlorophyll-protein complexes from *Rhodopseudomonas palustris*. - J. Biochem. (Tokyo) *91* : 1029 - - 1038, 1982.

49777 - HAYWARD, T.L., VENRICK, E.L. : Relation between surface chlorophyll, integrated chlorophyll and integrated primary production. - Mar. Biol. *69* : 247 - 252, 1982.

49778 - HAZEMOTO, N., KAMO, N., KONDO, M., KOBATAKE, Y. : The quenching effect of blue light on halorhodopsin. - Biochim. biophys. Acta *682* : 67 - 74, 1982.

49779 - HEARNSHAW, G.F., PROCTOR, M.C.F. : The effect of temperature on the survival of dry bryophytes. - New Phytol. *90* : 221 - 228, 1982. [Chl.]

49780 - HEATH, R.L., FREDERICK, P.E., CHIMIKLIS, P.E. : Ozone inhibition of photosynthesis in *Chlorella sorokiniana*. - Plant Physiol. *69* : 229 - 233, 1982.

49781 - HEATHCOTE, P., WARDEN, J.T. : Detection of chemically-induced dynamic electron polarisation (CIDEP) in whole cells and membrane fractions of *Chlorobium limicola* f. *thiosulphatophilum*. - FEBS Lett. *140* : 277 - 281, 1982.

49782 - HEBER, U., TAKAHAMA, U., NEIMANIS, S., SHIMIZU-TAKAHAMA, M. : Transport as the basis of the Kok effect. Levels of some photosynthetic intermediates and activation of light-regulated enzymes during photosynthesis of chloroplasts and green leaf protoplasts. - Biochim. biophys. Acta *679* : 287 - 299, 1982.

*49783 - HEGDE, D.M. : Dry matter and diosgenin production and tuber development
         in medicinal yam in relation to planting material and nitrogen fertilizat-
         ion. - Indian J. Agron. *26* : 289 - 296, 1981. [Growth analysis.]

 49784 - HEGSETH, E.N. : Chemical and species composition of the phytoplankton during
         the first spring bloom in Trondheimsfjorden, 1975. - Sarsia *67* : 131 - 141,
         1982. [Chl.]

 49785 - HEICHEL, G.H. : Alfalfa. - In : TEARE, I.D., PEET, M.M. (ed.) : Crop-Water
         Relations. Pp. 127 - 156. John Wiley & Sons, New York - Chichester - Bris-
         bane - Toronto - Singapore 1982. [Ps.]

 49786 - HEINONEN, P. : On the annual variation of phytoplankton biomass in Finnish
         inland waters. - Hydrobiologia *86* : 29 - 31, 1982.

 49787 - HEINZE, B., WARTENBERG, A. : Differently oriented chlorophylls in *Mesotae-
         nium caldariorum* detected by microphotometrical dichroism measurements *in
         vivo*. - Biochim. biophys. Acta *681* : 212 - 219, 1982.

 49788 - HELDER, R.J., VAN HARMELEN, M. : Carbon assimilation pattern in the sub-
         merged leaves of the aquatic angiosperm: *Vallisneria spiralis* L. - Acta
         bot. neerl. *31* : 281 - 295, 1982.

 49789 - HELLINGWERF, K.J., DE VRIJ, W., KONINGS, W.N. : Wavelength dependence of
         energy transduction in *Rhodopseudomonas sphaeroides:* action spectrum of
         growth. - J. Bacteriol. *151* : 534 - 541, 1982.

 49790 - HENDERSON, R., JUBB, J.S., ROSSMANN, M.G. : A contracted form of the tri-
         gonal purple membrane of *Halobacterium halobium*. - J. mol. Biol. *154* :
         501 - 514, 1982.

 49791 - HENDREN, R.W. : Rapporteur's summary: Origin and evolution of chloroplast
         metabolism. - In : SCHIFF, J.A., LYMAN, H. (ed.) : On the Origins of Chlo-
         roplasts. Pp. 219 - 225. Elsevier/North-Holland, New York - Amsterdam -
         - Oxford 1982.

 49792 - HENDRIX, J.E. : Sugar translocation in two members of the *Cucurbitaceae*.
         - Plant Sci. Lett. *25* : 1 - 7, 1982.

 49793 - HENNINGSEN, K.W., STUMMANN, B.M. : Use of mutants in the study of chloro-
         plast biogenesis. - In : PARTHIER, B., BOULTER, D. (ed.) : Nucleic Acids
         and Proteins in Plants II. Structure, Biochemistry and Physiology of Nucle-
         ic Acids. Pp. 597 - 644. Springer-Verlag, Berlin - Heidelberg - New York
         1982.

 49794 - HENRY, L.E.A., STRASSER, R.J., SIEGENTHALER, P.-A. : Alteration in the
         acyl lipid composition of thylakoids induced by aging and its effect on thy-
         lakoid structure. - Plant Physiol. *69* : 531 - 536, 1982.

 49795 - HENRY, R., TUNDISI, J.G. : Efeitos de enriquecimento artificial por nitra-
         to de fosfato no crescimento da comunidade fitoplanctônica da represa do
         lobo ("Broa", Brotas - Itirapina, SP). [Effects of artificial enrichment
         with nitrate and phosphate on the growth of planktonic community in the
         Lobo Reservoir ("Broa", Brotas, Itirapina, SP).] - Ciência Cultura *34* :
         518 - 524, 1982. [Chl; in Port., ab : E.]

 49796 - HENRY, R., TUNDISI, J.G. : Évidence of limitation by molybdenum and nitro-
         gen on the growth of the phytoplankton community of the Lobo Reservoir
         (São Paulo, Brazil). - Rev. Hydrobiol. trop. *15* : 201 - 208, 1982. [Chl.]

 49797 - HENSON, I.E. : Abscisic acid and water relations of rice (*Oryza sativa* L.):
         Sequential responses to water stress in the leaf. - Ann. Bot. *50* : 9 - 24,
         1982. [Stomatal resistance.]

 49798 - HENSON, I.E., ALAGARSWAMY, G., BIDINGER, F.R., MAHALAKSHMI, V. : Stomatal
         responses of pearl millet (*Pennisetum americanum* [L.] LEEKE) to leaf water
         status and environmental factors in the field. - Plant Cell Environm. *5* :
         65 - 74, 1982. [Stomatal resistance.]

 49799 - HERMAN, A.W. : Spatial and temporal variability of chlorophyll distribut-
         ions and geostrophic current estimates on the Peru Shelf at 9S. - J. mar.
         Res. *40* : 185 - 207, 1982.

*49800 - HERN, S.C., LAMBOU, V.W., WILLIAMS, L.R., TAYLOR, W.D. : Modifications of
         model predicting trophic state of lakes. Adjustment of models to account
         for the biological manifestations of nutrients. - US environm. Protect. Agen-
         cy Rep. *EPA-600/3/-81-001* : I - IX, 1 - 38, 1981. [Chl.]

49801 - HERNDON, C.S., NORTON, I.L., HARTMAN, F.C. : Reexamination of the binding
        site for pyridoxal 5'-phosphate in ribulosebisphosphate carboxylase/oxyge-
        nase from *Rhodospirillum rubrum*. - Biochemistry *21* : 1380 - 1385, 1982.

49802 - HERVE, S., HEINONEN, P. : Some factors affecting the determination of chlo-
        rophyll *a* in algal samples. - Ann. bot. fenn. *19* : 211 - 217, 1982.

49803 - HERZOG, H. : Relation of source and sink during grain filling period in
        wheat and some aspects of its regulation. - Physiol. Plant. *56* : 155 - 160,
        1982.

49804 - HERZOG, H., GEISLER, G. : Influence of ear size, leaf area and cytokinin
        applications on the flag leaf development in wheat. - Z. Acker- Pflanzen-
        bau *151* : 128 - 136, 1982. [Chl.]

49805 - HERZOG, H., STAMP, P. : Chlorophyll content and RuBP carboxylase activity
        in assimilating organs in relation to kernel growth of "gigas", semidwarf
        and normal spring wheats. - Z. Pflanzenzücht. *88* : 127 - 136, 1982.

49806 - HESKETH, J.D., WOOLLEY, J.T., PETERS, D.B. : Predicting photosynthesis. -
        In : GOVINDJEE (ed.) : Photosynthesis. Vol.II. Pp. 387 - 418. Academic
        Press, New York - London - Paris - San Diego - San Francisco - São Paulo -
        - Sydney - Tokyo - Toronto 1982.

49807 - HESLA, B.I., TIESZEN, L.L., IMBAMBA, S.K. : A systematic survey of C_3 and
        C_4 photosynthesis in the *Cyperaceae* of Kenya, East Africa. - Photosynthe-
        tica *16* : 196 - 205, 1982.

49808 - HESS, B., KUSCHMITZ, D., ENGELHARD, M. : Bacteriorhodopsin. - In : MARTONOSI,
        A.N. (ed.) : Membranes and Transport. Vol.2. Pp. 309 - 318. Plenum Publ.
        Corp., New York 1982.

49809 - HESSE, N., LENZ, F. : Einfluss der Wasserversorgung auf Transpirations- und
        Netto-Photosyntheseraten bei Stangenbohnen (*Phaseolus vulgaris* L. var. *vul-
        garis*). - Gartenbauwissenschaft *47* : 145 - 152, 1982.

49810 - HESSE, N., LENZ, F. : Einfluss der Wasserversorgung auf den Wasserverbrauch
        und auf das Wachstum von Stangenbohnen (*Phaseolus vulgaris* var. *vulgaris*).
        - Gartenbauwissenschaft *47* : 259 - 264, 1982. [Growth analysis.]

49811 - HETHERINGTON, N.B., HILTON, J. : Some causes of bias in the measurement of
        dissolved oxygen using certain modifications of the Winkler method. - Ana-
        lyst *107* : 110 - 113, 1982.

49812 - HETHERINGTON, S.E., HALLAM, N.D., SMILLIE, R.M. : Ultrastructural and com-
        positional changes in chloroplast thylakoids of leaves of *Borya nitida* du-
        ring humidity-sensitive degreening. - Aust. J. Plant Physiol. *9* : 601 - 609,
        1982.

49813 - HETHERINGTON, S.E., SMILLIE, R.M. : Humidity-sensitive degreening and re-
        greening of leaves of *Borya nitida* LABILL. as followed by changes in chlo-
        rophyll fluorescence. - Aust. J. Plant Physiol. *9* : 587 - 599, 1982.

49814 - HETHERINGTON, S.E., SMILLIE, R.M. : Tolerance of *Borya nitida,* a poikilo-
        hydrous angiosperm, to heat, cold and high-light stress in the hydrated
        state. - Planta *155* : 76 - 81, 1982. [Chl.]

49815 - HETHERINGTON, S.E., SMILLIE, R.M., HALLAM, N.D. : *In vivo* changes in chloro-
        plast thylakoid membrane activity during viable and non-viable dehydration
        of a drought-tolerant plant, *Borya nitida*. - Aust. J. Plant Physiol. *9* :
        611 - 621, 1982.

49816 - HEUER, B., HANSEN, M.J., ANDERSON, L.E. : Light modulation of phosphofruc-
        tokinase in pea leaf chloroplasts. - Plant Physiol. *69* : 1404 - 1406, 1982.

49817 - **HEUER, B., PLAUT, Z.** : Activity and properties of ribulose-1,5-bisphospha-
te carboxylase of sugarbeet plants grown under saline conditions. - Physiol.
Plant. *54* : 505 - 509, 1982.

49818 - **HEWITT, H.G., GARROD, J.F., COPPING, L.G., GREENWOOD, D.** : The effect of
BTS 44584, a ternary sulphonium growth retardant, on net photosynthesis
and yield in soyabeans. - In : **McLAREN, J.S.** (ed.) : Chemical Manipulation
of Crop Growth and Development. Pp. 221 - 235. Butterworth Scientific,
London - Boston - Durban - Singapore - Sydney - Toronto - Wellington 1982.

49819 - **HEYN, M.P., DENCHER, N.A.** : Reconstitution of monomeric bacteriorhodopsin
into phospholipid vesicles. - In : **COLOWICK, S.P., KAPLAN, N.O.** (ed.) :
Methods in Enzymology. Vol.88. Pp. 31 - 35. Academic Press, New York - London - Paris - San Diego - San Francisco - São Paulo - Sydney - Tokyo -
- Toronto 1982.

49820 - **HIEKE, B.** : Photosynthetischer Elektronentransport und Biomassebildung bei
ausgewählten Evolutionsformen des Weizens. - Wiss. Z. Humboldt-Univ.Berlin,
math.-nat. Reihe *31* : 91 - 119, 1982.

49821 - **HIGUCHI, T.** : Gaseous CO_2 transport through the aerenchyma and intercellular
spaces in relation to the uptake of CO_2 by rice roots. - Soil Sci. Plant
Nutr. *28* : 491 - 497, 1982.

49822 - **HILDEBRAND, D.F., HYMOWITZ, T.** : Carotene and chlorophyll bleaching by soybeans with and without seed lipoxygenase-1. - J. agr. Food Chem. *30* : 705 -
- 708, 1982.

*49823 - **HILL, B.H.** : Distribution and production of *Justicia americana* in the New
River, Virginia. - Castanea *46* : 162 - 169, 1981.

49824 - **HILL, B.H., WEBSTER, J.R.** : Periphyton production in an Appalachian river. -
Hydrobiologia *97* : 275 - 280, 1982.

49825 - **HILL, R.** : Cytochromes and redox systems in photosynthesis. - In : **KAPLAN,
N.O., ROBINSON, A.** (ed.) : From Cyclotrons to Cytochromes. Pp. 299 - 303.
Academic Press, New York 1982.

49826 - **HILLERDAL-HAGSTRÖMER, K., MATTSON-DJOS, E., HELLKVIST, J.** : Field studies
of water relations and photosynthesis in Scots pine. II. Influence of irrigation and fertilization on needle water potential of young pine trees. -
Physiol. Plant. *54* : 295 - 301, 1982.

49827 - **HINCHIGERI, S.B., RICHARDS, W.R.** : The reaction mechanism of *S*-adenosyl-L-
-methionone: magnesium protoporphyrin methyltransferase from *Euglena gracilis.* - Photosynthetica *16* : 554 - 560, 1982.

49828 - **HIND, G.** : An experiment to test quantitative techniques. - In : **COOMBS, J.,
HALL, D.O.** (ed.) : Techniques in Bioproductivity and Photosynthesis. Pp.
91 - 93. Pergamon Press, Oxford - New York - Toronto - Sydney - Paris -
- Frankfurt 1982. [Cytochromes.]

49829 - **HIND, G.** : Photosynthetic energy conversion. - In : **COOMBS,J., HALL, D.O.**
(ed.) : Techniques in Bioproductivity and Photosynthesis. Pp. 112 - 117.
Pergamon Press, Oxford - New York - Toronto - Sydney - Paris - Frankfurt
1982.

49830 - **HIRAI, A.** : Isoelectrofocusing of non-carboxylated fraction I protein from
green callus. - Plant Sci. Lett. *25* : 37 - 41, 1982.

49831 - **HIRAI, G., SUZUKI, A., TAKAHASHI, M., YAMAUCHI, A., UENO, E.** : [Studies on
the effects of relative humidity of the atmosphere upon the growth and
physiology of rice plant. I. Relations of relative humidity of the atmosphere to leaf-emergence rate and morphology of plant at the seedling stage.]
- Jap. J. Crop Sci. *51* : 301 - 309, 1982. [In Jap., ab : E.]

49832 - **HIREL, B., VIDAL, J., GADAL, P.** : Evidence for a cytosolic-dependent light
induction of chloroplastic glutamine synthetase during greening of etiolated rice leaves. - Planta *155* : 17 - 23, 1982.

49833 - **HIROKAWA, T., HATA, M., TAKEDA, H.** : Correlation between the starch level
and the rate of starch synthesis during the developmental cycle of *Chlorella ellipsoidea.* - Plant Cell Physiol. *23* : 813 - 820, 1982. [Ps.]

49834 - HIROTA, O., TAKEDA, T. : [Studies on utilization of solar radiation by crop stands. IV. Estimated absorptivities of solar energy in each leaf layer of rice and soybean canopies.] - Jap. J. Crop Sci. *51* : 151 - 158, 1982. [In Jap., ab : E.]

49835 - HIRSCH, J., NEEF, E., FINK, F. : The yield of chlorophyll *a* fluorescence as a means to test the various deexcitation mechanisms in the antenna system of green plants. - Biochim. biophys. Acta *681* : 15 - 20, 1982.

49836 - HITZ, W.D., LADYMAN, J.A.R., HANSON, A.D. : Betaine synthesis and accumulation in barley during field water-stress. - Crop Sci. *22* : 47 - 54, 1982. [Chl.]

49837 - HIYAMA, T., SAKURAI, H., IKEGAMI, I. : Photosystem I : Reaction center and photochemistry. - Recent Progr. nat. Sci. Jap. *7* : 15 - 18, 1982.

49838 - HLADÍK, J., PANČOŠKA, P., SOFROVÁ, D. : The influence of carotenoids on the conformation of chlorophyll-protein complexes isolated from the cyanobacterium *Plectonema boryanum*. Absorption and circular dichroism study. - Biochim. biophys. Acta *681* : 263 - 272, 1982.

49839 - HO, K.K., KROGMANN, D.W. : Photosynthesis. - In : CARR, N.G., WHITTON, B.A. (ed.) : The Biology of Cyanobacteria. Pp. 191 - 214. Blackwell Scientific Publications, Oxford, University of California Press, Berkeley 1982.

49840 - HO, L.C., BAKER, D.A. : Regulation of loading and unloading in long distance transport systems. - Physiol. Plant. *56* : 225 - 230, 1982.

*49841 - HO, T.-F., McINTOSH, A.R., BOLTON, J.R. : Intramolecular photochemical electron transfer in a linked porphyrin-quinone molecule as a model for the primary step of photosynthesis. - Nature *286* : 254 - 256, 1980.

49842 - HO, Y.-K., WANG, J.H. : Effect of pyridine homologues on proton flux through the $CF_0 \cdot CF_1$ complex and photophosphorylation in chloroplasts. - J. Bioenerg. Biomembr. *14* : 97 - 113, 1982.

49843 - HOAGLAND, R.E., DUKE, S.O. : Biochemical effects of glyphosate [*N*-(phosphonomethyl)glycine]. - In : MORELAND, D.E., ST.JOHN, J.B., HESS, F.D. (ed.) : Biochemical Responses Induced by Herbicides. Pp. 175 - 205. Amer. Chem. Soc., Washington 1982. [Ps, Chl.]

49844 - HOCKER, H.W. Jr. : Effects of thinning on biomass growth in young *Populus tremuloides* plots. - Can. J. Forest Res. *12* : 731 - 737, 1982. [Growth analysis.]

*49845 - HODANJOVA, D. : Radijacioni režim, struktura useva i fotosinteza šećerne repe. [Radiation regime, canopy structure and photosynthesis of sugar beet.] - Posebna Izd. srpska Akad. Nauka Umet. *538* , Od. prirod.-mat. Nauka *54* [BELIČ, J. (ed.) : Fiziologija Šećerne Repe] : 39 - 56, 1981. [In Serb., ab : E.]

49846 - HODDINOTT, J., HALL, L.M. : The responses of photosynthesis and translocation rates to changes in the ζ ratio of light. - Can. J. Bot. *60* : 1285 - - 1291, 1982.

49847 - HODGSON, A.S. : The effects of duration, timing and chemical amelioration of short-term waterlogging during furrow irrigation of cotton in a cracking grey clay. - Aust. J. agr. Res. *33* : 1019 - 1028, 1982. [Growth analysis.]

49848 - HOFF, A.J. : ESR and ENDOR of primary reactants in photosynthesis. - Biophys. Struct. Mechan. *8* : 107 - 150, 1982.

49849 - HOFF, A.J. : Photooxidation of the reaction center chlorophylls and structural properties of photosynthetic reaction centers. - In : FONG, F.K. (ed.) : Light Reaction Path of Photosynthesis. Pp. 80 - 151. Springer-Verlag, Berlin - Heidelberg - New York 1982.

49850 - HOFFMAN-FALK, H., MATTOO, A.K., MARDER, J.B., EDELMAN, M., ELLIS, R.J. : General occurrence and structural similarity of the rapidly synthesized, 32,000-dalton protein of the chloroplast membrane. - J. biol. Chem. *257* : 4583 - 4587, 1982.

49851 - HOFFMANN, F., SCHÄFER, W. : Ausnutzung der photosynthetisch aktiven Sonnen-
strahlung durch optimal versorgte Winterweizen- und Zuckerrübenbestände. -
In : HOFFMANN, P., HIEKE, B. (ed.) : Photosynthese: Regulation und Evolution. (Colloquia Pflanzenphysiologie Nr.5.) Pp. 202 - 209. Humboldt-Universität, Berlin 1982.

49852 - HOFFMANN, F., WEIRAUCH, M. : Dynamische Optimierung eines Ertragsbildungs-
modells von Zuckerrüben zur Verwendung bei der operativen Steuerung der
Wasser- und Stickstoffversorgung. - Arch. Acker- Pflanzenbau Bodenk. *26* :
77 - 85, 1982. [Growth analysis.]

B49853 - HOFFMANN, P., HIEKE, B. (ed.) : Photosynthese: Regulation und Evolution.
(Colloquia Pflanzenphysiologie Nr.5.) - Humboldt-Universität, Berlin 1982.

49854 - HOFLACHER, H., BAUER, H. : Light acclimation in leaves of the juvenile and
adult life phases of ivy (*Hedera helix*). - Physiol. Plant. *56* : 177 - 182,
1982. [Ps.]

*49855 - HÖFNER, W., BRÜCKNER, U. : Einfluß von Wachstumsregulatoren auf Ertrag und
Ertragskomponenten bei Getreide. - Kali-Briefe *15* : 277 - 285, 1980. [Ps.]

*49856 - HÖFNER, W., FEUCHT, D., BRÜCKNER, U. : Beeinflussung der Ähren- und Kornent-
wicklung von Sommerweizen durch Wachstumsregulatoren. - Z. Acker- Pflanzenbau *149* : 177 - 182, 1980. [Dry-matter accumulation.]

49857 - HÖFNER, W., KÜHN, H. : Effect of growth regulator combinations on ear de-
velopment, assimilate translocation and yield in cereal crops. - In :
McLAREN, J.S. (ed.) : Chemical Manipulation of Crop Growth and Development.
Pp. 375 - 390. Butterworth Scientific, London - Boston - Durban - Singapore -
- Sydney - Toronto - Wellington 1982.

49858 - HOHMAN, T.C., McNEIL, P.L., MUSCATINE, L. : Phagosome-lysosome fusion inhi-
bited by algal symbionts of *Hydra viridis*. - J. Cell Biol. *94* : 56 - 63,
1982. [Ps.]

49859 - HÖJEBERG, B., LIND, C., KHORANA, H.G. : Reconstitution of bacteriorhodopsin
vesicles with *Halobacterium halobium* lipids. Effects of variations in lipid
composition. - J. biol. Chem. *257* : 1690 - 1694, 1982.

49860 - HOLADAY, A.S., HARRISON, A.T., CHOLLET, R. : Photosynthetic/photorespirato-
ry CO_2 exchange characteristics of the C_3-C_4 intermediate species, *Moricandia arvensis*. - Plant Sci. Lett. *27* : 181 - 189, 1982.

49861 - HOLLOWAY, P.S., VAN VELDHUIZEN, R.M., STUSHNOFF, C., WILDUNG, D.K. : Effects
of light intensity on vegetative growth of lingonberries. - Can. J. Plant
Sci. *62* : 965 - 968, 1982.

49862 - HOLMES, M.G., WAGNER, E. : The influence of chlorophyll on the spectral
control of elongation growth in *Chenopodium rubrum* L. hypocotyls. - Plant
Cell Physiol. *23* : 745 - 750, 1982.

49863 - HOLTUM, J.A.M., WINTER, K. : Activity of enzymes of carbon metabolism during the induction of Crassulacean acid metabolism in *Mesembryanthemum
crystallinum* L. - Planta *155* : 8 - 16, 1982.

49864 - HOLZAPFEL, A., ZERBE, R., WILD, A. : The effect of indole-3-acetic acid
on plants cultivated under different light intensities. - Z. Pflanzenphysiol. *108* : 409 - 417, 1982.[Chl.]

49865 - HOLZWARTH, A.R., WENDLER, J., WEHRMEYER, W. : Picosecond time resolved
energy transfer in isolated phycobilisomes from *Rhodella violacea (Rhodophyceae)*. - Photochem. Photobiol. *36* : 479 - 487, 1982.

49866 - HONIG, B. : Theoretical aspects of photoisomerization in visual pigments
and bacteriorhodopsin. - In : ALFANO, R.R. (ed.) : Biological Events Probed
by Ultrafast Laser Spectroscopy. Pp. 281 - 297. Academic Press, New York -
- London - Paris - San Diego - San Francisco - São Paulo - Sydney - Tokyo -
- Toronto 1982.

49867 - HOOBER, J.K., BEDNARIK, D., KELLER, B.J., MARKS, D.B. : Regulatory aspects of thylakoid membrane formation in *Chlamydomonas reinhardtii* y-1. - In : AKOYUNOGLOU, G., EVANGELOPOULOS, A.E., GEORGATSOS, J., PALAIOLOGOS, G., TRAKATELLIS, A., TSIGANOS, C.P. (ed.) : Cell Function and Differentiation, Part B. Biogenesis of Energy Transducing Membranes and Membrane and Protein Energetics. Pp. 127 - 137. Alan R. Liss Inc., New York 1982.

49868 - HOOBER, J.K., MARKS, D.B., KELLER, B.J., MARGULIES, M.M. : Regulation of accumulation of the major thylakoid peptides in *Chlamydomonas reinhardtii* y-1 at 25 °C and 38 °C. - J. Cell Biol. *95* : 552 - 558, 1982.

49869 - HOPE, A.B., RANSON, D., DIXON, P.G. : Photophosphorylation in chloroplasts with varied proton motive force (PMF): I. The PMF and its onset. - Aust. J. Plant Physiol. *9* : 385 - 397, 1982.

49870 - HOPE, A.B., RANSON, D., DIXON, P.G. : Photophosphorylation in chloroplasts with varied proton motive force (PMF): II. Phosphorylation and the PMF. - Aust. J. Plant Physiol. *9* : 399 - 407, 1982.

49871 - HOPEWELL, W.D., FUKUMOTO, J.M. : Time-resolved protein fluorescence measurements of intermediates in the photocycle of bacteriorhodopsin. - In : COLOWICK, S.P., KAPLAN, N.O. (ed.) : Methods in Enzymology. Vol.88. Pp. 306 - 310. Academic Press, New York - London - Paris - San Diego - San Francisco - São Paulo - Sydney - Tokyo - Toronto 1982.

49872 - HOPKINS, W.G. : Formation of chloroplast pigments in a temperature-sensitive, virescent mutant of maize. - Can. J. Bot. *60* : 737 - 740, 1982.

49873 - HORIO, T. : [Cyclic electron transport system of photosynthetic bacteria and its reorganization.] - In : YOSHIDA, M. (ed.) : Fuoto Baiorojii: Koseiri Gensho No Shoki Katei. Pp. 306 - 322. Kodansha Saientifiku, Tokyo 1982. [In Jap.]

49874 - HORNER, R.R., SCHRADER, G.C. : Relative contributions of ice algae, phytoplankton, and benthic microalgae to primary production in nearshore regions of the Beaufort Sea. - Arctic *35* : 485 - 503, 1982.

*49875 - HORNER, R.R., WELCH, E.B. : Stream periphyton development in relation to current velocity and nutrients. - Can. J. Fish. aquat. Sci. *38* : 449 - 457, 1981.

49876 - HORSTMANN, U. : Phytoplankton productivity in the Baltic Sea, North Sea, and Atlantic Ocean. - In : ZABORSKY, O.R. (ed.) : CRC Handbook of Biosolar Resources. Vol.I/2. Pp. 407 - 416. CRC Press, Boca Raton 1982.

49877 - HORTON, P., BLACK, M.T. : On the nature of the fluorescence decrease due to phosphorylation of chloroplast membrane proteins. - Biochim. biophys. Acta *680* : 22 - 27, 1982.

49878 - HORTON, R.F., WOODROW, L., BOESEL, I., GRODZINSKI, B. : Light, carbon dioxide and ethylene metabolism in photosynthetic tissue. - In : JACKSON, M.B., GROUT, B., MacKENZIE, I.A. (ed.) : Growth Regulators in Plant Senescence. (Monograph 8.) Pp.93 - 101. British Plant Growth Regulator Group, Wantage 1982.

49879 - HORWOOD, J. : Algal production in the west-central North-Sea. - J. Plankton Res. *4* : 103 - 124, 1982. [Chl.]

49880 - HORWOOD, J., NICHOLS, J.H., HARROP, R. : Seasonal changes in net phytoplankton of the west-central North Sea. - J. mar. biol. Assoc. UK *62* : 15 - - 23, 1982. [Chl.]

49881 - HOSKER, R.P.,Jr., LINDBERG, S.E. : Review: Atmospheric deposition and plant assimilation of gases and particles. - Atmos. Environm. *16* : 889 - 910, 1982.

49882 - HOTCHANDANI, S., LEBLANC, R.M., CLARKE, R.H., FRAGATA, M. : Zero field optical detection of magnetic resonance of triplet state of chlorophyll *a* in lipid bilayer vesicles. - Photochem. Photobiol. *36* : 235 - 240, 1982.

49883 - HOUCHINS, J.P., HIND, G. : Pyridine nucleotides and H_2 as electron donors to the respiratory and photosynthetic electron-transfer chains and to nitrogenase in *Anabaena* heterocysts. - Biochim. biophys. Acta *682* : 86 - 96, 1982.

49884 - HOUGEN, C.L., MELLER, E., GASSMAN, M.L. : Magnesium protoporphyrin monoester destruction by extracts of etiolated red kidney bean leaves. - Plant Sci. Lett. *24* : 289 - 294, 1982.

49885 - HOUGHTON, J.D., HONEYBOURNE, C.L., SMITH, K.M., TABBA, H.D., JONES, O.T.G.: The use of *N*-methylprotoporphyrin dimethyl ester to inhibit ferrochelatase in *Rhodopseudomonas sphaeroides* and its effect in promoting biosynthesis of magnesium tetrapyrroles. - Biochem. J. *208* : 479 - 486, 1982.

49886 - HOURSIANGOU-NEUBRUN, D., LÜTTKE, A., ARAPIS, G., PUISEUX-DAO, S., BONOTTO, S. : Apicobasal gradient of chloroplast DNA synthesis and distribution in *Acetabularia*. - In : AKOYUNOGLOU, G., EVANGELOPOULOS, A.E., GEORGATSOS, J., PALAIOLOGOS, G., TRAKATELLIS, A., TSIGANOS, C.P. (ed.) : Cell Function and Differentiation, Part B. Biogenesis of Energy Transducing Membranes and Membranes and Protein Energetics. Pp. 333 - 345. Alan R. Liss Inc., New York 1982.

49887 - HOWE, C.J., AUFFRET, A.D., DOHERTY, A., BOWMAN, C.M., DYER, T.A., GRAY, J.C.: Location and nucleotide sequence of the gene for the proton-translocating subunit of wheat chloroplast ATP synthase. - Proc. nat. Acad. Sci. USA *79* : 6903 - 6907, 1982.

49888 - HOWELL, J.M., VIETH, W.R. : Biophotolytic membranes: simplified kinetic model of photosynthetic electron transport. - J. mol. Catalysis *16* : 245 - - 298, 1982.

*49889 - HOWES, B.L., HOWARTH, R.W., TEAL, J.M., VALIELA, I. : Oxidation-reduction potentials in a salt marsh: Spatial patterns and interactions with primary production. - Limnol. Oceanogr. *26* : 350 - 360, 1981.

49890 - HOWITZ, K.T., McCARTY, R.E. : pH dependence and kinetics of glycolate uptake by intact pea chloroplasts. - Plant Physiol. *70* : 949 - 952, 1982.

49891 - HØYER-HANSEN, G., MØLLER, B.L., HENRY, L.E.A., CASADORO, G. : Thylakoid polypeptide synthesis and assembly in wild-type and mutant barley. - In : AKOYUNOGLOU, G., EVANGELOPOULOS, A.E., GEORGATSOS, J., PALAIOLOGOS, G., TRAKATELLIS, A., TSIGANOS, C.P. (ed.) : Cell Function and Differentiation, Part B. Biogenesis of Energy Transducing Membranes and Membrane and Protein Energetics. Pp. 111 - 125. Alan R. Liss Inc., New York 1982.

49892 - HSU, B.-D., ARNON, D.I. : Inhibition of photosynthetic electron transport in chloroplasts by UHDBT, a synthetic analogue of cellular benzoquinones. - Photobiochem. Photobiophys. *4* : 187 - 193, 1982.

49893 - HUANG, A.H.C. : Metabolism in plant peroxisomes. - In : CREASY, L.L., HRAZDINA, G. (ed.) : Cellular and Subcellular Localization in Plant Metabolism. Pp. 85 - 123. Plenum Press, New York - London 1982.

49894 - HUANG Huici : [Thermodynamic implications of photosynthesis.] - Shengwu Huaxue Yu Shengwu Wuli Jinzhan *46* : 12 - 15, 1981. [In Chin.]

49895 - HUANG, K.-S., LIAO, M.-J., GUPTA, C.M., ROYAL, N., BIEMANN, K., KHORANA, H.G. : The site of attachment of retinal in bacteriorhodopsin. The ε-amino group in Lys-4 is not required for proton translocation. - J. biol. Chem. *257* : 8596 - 8599, 1982.

49896 - HUANG, K.-S., RADHAKRISHNAN, R., BAYLEY, H., KHORANA, H.G. : Orientation of retinal in bacteriorhodopsin as studied by cross-linking using a photosensitive analog of retinal. - J. biol. Chem. *257* : 13616 - 13623, 1982.

49897 - HUANG Zhuo-hui, SHEN Yun-gang, WEI Jia-mian : [Studies on the coupling mechanism of photophosphorylation IV. Studies on the fluorescence and the action site of aureomycin on coupling factor 1.] - Acta Phytophysiol. sin. *8* : 253 - 265, 1982. [In Chin., ab : E.]

49898 - HUBAC, C., GUERRIER, D., VIEIRA DA SILVA, J. : Étude de la transpiration
du Cottonier (*Gossypium hirsutum* L.) en relation avec la photopériode. -
Acta oecol.-Oecol.Plant. *3* : 279 - 289, 1982. [Stomatal resistance.]

*49899 - HUBAC, C., VIEIRA DA SILVA, J. : Indicateurs métaboliques de contraintes
mésologiques. - Physiol. vég. *18* : 45 - 53, 1980. [Photosynthates.]

49900 - HUBER, S.C. : Photosynthetic carbon metabolism in chloroplasts. - In :
CREASY, L.L., HRAZDINA, G. (ed.) : Cellular and Subcellular Localization
in Plant Metabolism. Pp. 151 - 184. Plenum Press, New York - London 1982.

49901 - HUBER, S.C., ISRAEL, D.W. : Biochemical basis for partitioning of photosynthetically fixed carbon between starch and sucrose in soybean (*Glycine
max* MERR.) leaves. - Plant Physiol. *69* : 691 - 696, 1982.

49902 - HUCHZERMEYER, B. : Energy transfer inhibition induced by nitrofen. - Z. Naturforsch. *37 C* : 787 - 792, 1982.

49903 - HUFF, A. : Peroxidase-catalysed oxidation of chlorophyll by hydrogen peroxide. - Phytochemistry *21* : 261 - 265, 1982.

*49904 - HUFF, A., ABDEL-BAR, M.Z., RODNEY, D.R., ROTH, R.L., GARDNER, B.R. : Enhan-
cement of citrus regreening and peel lycopene by trickle irrigation. -
HortScience *16* : 301 - 302, 1981. [Chl, Car.]

49905 - HUMPHRY-BAKER, R., LILIE, J., GRÄTZEL, M. : *In vitro* analogues of photosystem II. Combined flash photolytic and conductometric study of light-
-induced oxygen evolution from water mediated by colloidal RuO_2-TiO_2. -
J. amer. chem. Soc. *104* : 422 - 425, 1982.

49906 - HUNER, N.P.A., CARTER, J.V., WOLD, F. : Effects of reducing agent on the
conformation of the isolated subunits of ribulose bisphosphate carboxylase-
-oxygenase from cold-hardened and unhardened rye. - Z. Pflanzenphysiol.
106 : 69 - 80, 1982.

49907 - HUNER, N.P.A., HAYDEN, D.B. : Changes in the heterogeneity of ribulosebisphosphate carboxylase-oxygenase in winter rye induced by cold hardening. -
Can. J. Biochem. *60* : 897 - 903, 1982.

B49908 - HUNT, R. : Plant Growth Curves. The Functional Approach to Plant Growth
Analysis. - Edward Arnold, London 1982.

49909 - HUNT, R. : Plant growth analysis: Second derivatives and compounded second
derivatives of splined growth curves. - Ann. Bot. *50* : 317 - 328, 1982.

49910 - HUNTER, C.N., PENNOYER, J.D., NIEDERMAN, R.A. : Assembly and structural
organization of pigment-protein complexes in membranes of *Rhodopseudomonas
sphaeroides*. - In : AKOYUNOGLOU, G., EVANGELOPOULOS, A.E., GEORGATSOS, J.,
PALAIOLOGOS, G., TRAKATELLIS, A., TSIGANOS, C.P. (ed.) : Cell Function and
Differentiation, Part B. Biogenesis of Energy Transducing Membranes and
Membrane and Protein Energetics. Pp. 257 - 265. Alan R. Liss Inc., New
York 1982.

*49911 - HUPPATZ, J.L., PHILLIPS, J.N., RATTIGAN, B.M. : Cyanocrylates. Herbicidal
and photosynthetic inhibitory activity. - Agr. biol. Chem. (Tokyo) *45* :
2769 - 2773, 1981.

49912 - HURLEY, J.K., TOLLIN, G. : Photochemical energy conversion in chlorophyll-
-containing lipid bilayer vesicles. - Solar Energy *28* : 187 - 196, 1982.

49913 - HURT, E., HAUSKA, G. : Identification of the polypeptides in the cytochrome b_6/f complex from spinach chloroplasts with redox-center-carrying subunits. - J. Bioenerg. Biomembr. *14* : 405 - 424, 1982.

49914 - HURT, E., HAUSKA, G. : Involvement of plastoquinone bound within the isolated cytochrome b_6/f complex from chloroplasts in oxidant-induced reduction of cytochrome b_6. - Biochim. biophys. Acta *682* : 466 - 473, 1982.

49915 - HURT, E.C., HAUSKA, G., SHAHAK, Y. : Electrogenic proton translocation
by the chloroplast cytochrome b_6/f complex reconstituted into phospholipid
vesicles. - FEBS Lett. *149* : 211 - 216, 1982.

49916 - HÜSEMANN, W. : Photoautotrophic growth of cell suspension cultures from *Chenopodium rubrum* in an airlift fermenter. - Protoplasma *113* : 214 - 220, 1982.

49917 - HUSHOVD, O.T., GULLIKSEN, O.M., NORDBY, Ø. : Isolation of chloroplast membranes and electrophoretic separation of chlorophyll-containing proteins from *Ulva mutabilis* FØYN. - Bot. mar. *25* : 155 - 161, 1982.

49918 - HUTLEY-BULL, P.D., SCHWABE, W.W. : Morphogenesis in the wheat apex as influenced by environment and plant growth regulators. - In : HAWKINS, A.F., JEFFCOAT, B. (ed.) : Opportunities for Manipulation of Cereal Productivity. (Monograph 7.) Pp. 150 - 166. Brit. Plant Growth Regulator Group, Wantage 1982. [Leaf-area formation.]

*49919 - HUTSON, K.G., ROGERS, L.J. : Ferredoxins from two cyanobacteria capable of heterotrophic growth in the dark. - FEMS Microbiol. Lett. *7* : 279 - 284, 1980.

49920 - HYEON, S.-B., NISHIDA, M., OHSAKA, A., KIM, J.-M., SUZUKI, A. : A simple bioassay for chemicals active to the photosynthetic or respiratory systems of plants. - Agr. biol. Chem. (Tokyo) *46* : 811 - 812, 1982.

B49921 - IDSO, S.B. : Carbon Dioxide: Friend or Foe? - IBR Press, Tempe 1982.

49922 - IDSO, S.B., REGINATO, R.J., RADIN, J.W. : Leaf diffusion resistance and photosynthesis in cotton as related to a foliage temperature based plant water stress index. - Agr. Meteorol. *27* : 27 - 34, 1982.

49923 - IGNAT'EV, A.R., SHABAEVA, É.V., POLEVAYA, V.S. : Sposob vyrashchivaniya avtotrofnoĭ kul'tury tkani *Ruta graveolens*. [A technique for cultivating autotrophic *Ruta graveolens* tissue culture.] - Fiziol. Rast. *29* : 181 - - 182, 1982. [Chl; in R.]

49924 - IGNAT'EVSKAYA, M.A., MINEEVA, L.A.: Vliyanie kislotnosti sredy na rost i pigmentnyĭ sostav *Cyanidium caldarium* v zavisimosti ot sposoba sushchestvovaniya. [Effect of pH on growth and pigment composition of *Cyanidium caldarium* in relation to mode of existence.] - Fiziol. Rast. *29* : 586 - 590, 1982. [In R, ab : E.]

49925 - IIDA, T., SHIOZAWA, H., KOBAYASHI, H., MITAMURA, T. : The relationship between the photocurrent of the chloroplast-methylviologen photoelectrochemical system and its derivatives of photosystems I and II. - Agr. biol. Chem. (Tokyo) *46* : 275 - 277, 1982.

49926 - IKAI, K., UEZONO, T., OHTA, H., HAMADA, K., TANAKA, F. : [Effect of alloxydim on growth, chlorophyll content and anthocyanin accumulation of crabgrass (*Digitaria adscendens* HENR.).] - Zasso Kenkyu *27* : 121 - 125, 1982. [In Jap., ab : E.]

49927 - IKE, I.F. : Effects of water deficits on transpiration, photosynthesis and leaf conductance in cassava. - Physiol. Plant. *55* : 411 - 414, 1982.

49928 - IKEGAMI, I. : [Chlorophyll proteins.] - In : YOSHIDA, M. (ed.) : Fuoto Baiorojii: Koiseiri Gensho No Shoki Katei. Pp. 77 - 93. Kodansha Saientifiku, Tokyo 1982. [In Jap.]

49929 - IKEUCHI, M., MURAKAMI, S. : Behavior of the 36,000-dalton protein in the internal membranes of squash etioplasts during greening. - Plant Cell Physiol. *23* : 575 - 583, 1982.

49930 - IKEUCHI, M., MURAKAMI, S. : Measurement and identification of NADPH: protochlorophyllide oxidoreductase solubilized with Triton X-100 from etioplast membranes of squash cotyledons. - Plant Cell Physiol. *23* : 1089 - 1099, 1982.

*49931 - **IL'INA, M.D., BORISOV, A.Yu.** : The fluorescence lifetime and quantum yield
of chlorophyll a in Triton X-100 solutions. - Biochim. biophys. Acta *637* :
540 - 545, 1981.

49932 - **IL'INA, M.D., BORISOV, A.Yu.** : Fotookislenie *P700* v preparatakh fotosistemy 1 s raznym soderzhaniem antennogo khlorofilla *a*. [Photooxidation of
P700 in photosystem 1 preparations with various content of antenna chlorophyll *a*.] - Biokhimiya *47* : 1954 - 1962, 1982. [In R, ab : E.]

49933 - **ILMAVIRTA, V.** : Dynamics of phytoplankton in Finnish lakes. - Hydrobiologia
86 : 11 - 20, 1982.

49934 - **IL'NITSKIĬ, O.A.** : Zavisimost' intensivnosti fotosinteza ot dinamiki vodnogo obmena khrizantemy i alychi pri razlichnoĭ skorosti i stepeni ikh
obezvozhivaniya. [Dependence of photosynthetic rate on the dynamics of
water relations in *Chrysanthemum* and *Prunus divaricata* at different rate
and degree of their dehydration.] - Fiziol. Biokhim. kul't. Rast. *14* : 175 -
- 181, 1982. [In R, ab : E.]

49935 - **IL'YANKOVA, T.I., SEMENOVICH, N.D., SHLYK, A.A.** : Metabolicheskie osobennosti bakteriokhlorofilla pigment-belkovykh kompleksov *Chromatiaceae*. [Metabo-
lic characteristics of bacteriochlorophyll in pigment-protein complexes
from *Chromatiaceae*.] - Dokl. Akad. Nauk belorus. SSR *26* : 1129 - 1132,
1982. [In R, ab : E.]

49936 - **IL'YASHUK, E.M.** : Blok-kamera dlya opredeleniya intensivnosti gazoobmena
nebol'shikh ili raschlenennykh listovykh plastinok. [Block-chamber for
determining gas exchange rate in small or lobated leaf blades.] - Fiziol.
Biokhim. kul't. Rast. *14* : 498 - 502, 1982. [In R, ab : E.]

49937 - **IMADA, O., SAITO, Y.** : [Studies on light sources used for culture of laver
Porphyra.] - Bull. jap. Soc. sci. Fish. *48* : 1517 - 1524, 1982. [Ps; in
Jap., ab : E.]

49938 - **IMAI, K., OGURA, F., MURATA, Y.** : Photosynthesis and respiration of papaya
(*Carica papaya* L.) leaves. - Acta oecol.-Oecol. Plant. *3* : 399 - 407,
1982.

49939 - **IMAIZUMI, M., HIRAOKA, T.** : Cytochemical study of tetrazolium reduction by
isolated *Vicia* chloroplasts under illumination. - Acta histochem. cytochem.
15 : 58 - 67, 1982.

49940 - **IMAIZUMI, M., HIRAOKA, T.** : Cytochemical study of diaminobenzidine oxidation by isolated *Vicia* chloroplasts under illumination. - Acta histochem.
cytochem. *15* : 208 - 222, 1982.

49941 - **IMAKI, T.** : [Effects of light intensity on the crop photosynthesis of mat
rush (*Juncus decipiens* NAKAI).] - Jap. J. Crop Sci. *51* : 65 - 69, 1982.
[In Jap., ab : E.]

49942 - **IMHOFF, J.F., TRÜPER, H.G.** : Taxonomic classification of photosynthetic
bacteria (anoxyphotobacteria, phototrophic bacteria). - In : **ZABORSKY, O.R.**
(ed.) : CRC Handbook of Biosolar Resources. Vol.I/1. Pp. 513 - 522. CRC
Press, Boca Raton 1982. [Ps.]

49943 - **INAMINE, G.S., NIEDERMAN, R.A.** : Development and growth of photosynthetic
membranes of *Rhodospirillum rubrum*. - J. Bacteriol. *150* : 1145 - 1153,
1982.

*49944 - **INOUE, S.** : Model reactions of biochemical carbon dioxide fixations. - In :
INOUE, S., YAMAZAKI, N. (ed.) : Organic and Bio-organic Chemistry of Carbon Dioxide. Pp. 253 - 274. Kodansha Ltd., Tokyo 1981.

49945 - **INOUE, Y., FURUYA, M.** : Absorption spectra changes caused by a laser flash
of red light in etiolated maize leaves. - Plant Sci. Lett. *24* : 267 - 273,
1982.

49946 - **INOUE, Y., SHIBATA, K.** : Thermoluminescence from photosynthetic apparatus.
- In : **GOVINDJEE** (ed.) : Photosynthesis. Vol.1. Pp. 507 - 533. Academic
Press, New York - London - Paris - San Diego - San Francisco - São Paulo -
- Sydney - Tokyo - Toronto 1982.

49947 - INOUE, Y., YAMASHITA, T., SATOH, K. : Photosystem II: Reaction center and
oxygen evolution. - Rec. Progr. nat. Sci. Jap. *7* : 19 - 24, 1982.

49948 - INUBUSHI, K., WADA, H., TAKAI, Y. : [Easily decomposable organic matter in
paddy soils (II) Chlorophyll-type compounds in Apg horizons.] - Nippon Dojo
Hiryogaku Zasshi [Jap. J. Soil Sci. Plant Nutr.]*53* : 277 - 282, 1982. [In
Jap.]

*49949 - ĬORDANOV, I., KAPCHINA, V., IVANOVA, Ĭ. : Gormonal'naya regulyatsiya inten-
sivnosti i produktov fotosinteza u giatsinta (*Hyacinthus orientalis* L.).
[Hormonal regulation of photosynthetic rate and products in hyacinth (*Hya-
cinthus orientalis* L.).] - In : GEORGIEV, G.Kh., BAK"RDZHIEVA, N.T.,
KOLEVA, S.T. (ed.) : Samoregulyatsiya Metabolizma Rasteniĭ. Pp. 280 - 287.
Sofiĭskiĭ Universitet Imeni Klimenta Okhridskogo, Sofia 1981. [In R, ab :
E.]

49950 - IP, S.Y., BRIDGER, J.S., CHIN, C.T., MARTIN, W.R.B., RAPER, W.G.C. :
Algal growth in primary settled sewage. The effects of five key variables.
- Water Res. *16* : 621 - 632, 1982.

49951 - IRELAND, C.R., SCHWABE, W.W. : Studies on the role of photosynthesis in
the photoperiodic induction of flowering in the short-day plants *Kalanchoe
blossfeldiana* POELLNIZ and *Xanthium pensylvanicum* WALLR. I. The requirement
for CO_2 during photoperiodic induction. - J. exp. Bot. *33* : 738 - 747,
1982.

49952 - IRELAND, C.R., SCHWABE, W.W. : Studies on the role of photosynthesis in the
photoperiodic induction of flowering in the short-day plants *Kalanchoe
blossfeldiana* POELLNIZ and *Xanthium pennsylvanicum* WALLR. II. The effect
of chemical inhibitors of photosynthesis. - J. exp. Bot. *33* : 748 - 760,
1982.

49953 - IRIYAMA, K., YOSHIURA, M. : Selection of adsorbents for the separation of
chlorophylls. - J. Liquid Chromatogr. *5* : 2211 - 2216, 1982.

49954 - IROSHNIKOVA, G.A., RAKHIMBERDIEVA, M.G., KARAPETYAN, N.V. : Izuchenie modi-
fitsiruyushchikh pigmentatsiyu mutatsiĭ u shtammov *Chlamydomonas reinhardii*
raznoĭ ploidnosti. Soobshchenie III. Kharakteristika narusheniĭ fotosinteticheskogo apparata pri mutatsiyakh v lokuse ℓts1. [Investigation of pig-
mentation-modifying mutations in *Chlamydomonas reinhardii* strains of different ploidy. III. Characterization of the damages of the photosynthetic
apparatus as a result of mutations in ℓts1 locus.] - Genetika *18* : 1817 -
- 1824, 1982. [In R, ab : E.]

49955 - ISEBRANDS, J.G., NELSON, N.D. : Crown architecture of short-rotation, intensively cultured *Populus* II. Branch morphology and distribution of leaves within the crown of *Populus* "Tristis" as related to biomass production.
- Can. J. Forest Res. *12* : 853 - 864, 1982.

49956 - ISHAG, H.M. : The influence of irrigation frequency on growth and yield
of groundnuts (*Arachis hypogaea* L.) under arid conditions. - J. agr. Sci.
99 : 305 - 310, 1982. [Growth analysis.]

49957 - ISHII, R., SCHMID, G.H. : Studies on $^{18}O_2$-uptake in the light by entire
plants of different tobacco mutants. - Z. Naturforsch. *37 C* : 93 - 101,
1982.

49958 - ISONO, T., KATOH, T. : Cylindrical phycobilisomes from a blue-green alga,
Anabaena variabilis. - Plant Cell Physiol. *23* : 1347 - 1355, 1982.

49959 - ITOH, R., ITOH, S., SUGAWA, M., OISHI, O., TABATA, K., OKADA, M., NISHIMURA,
M., YAKUSHIJI, E. : Isolation of crystalline water-soluble chlorophyll
proteins with different chlorophyll *a* and *b* contents from stems and leaves
of *Lepidium virginicum*. - Plant Cell Physiol. *23* : 557 - 560, 1982.

49960 - ITOH, S. : Movement of ions and change in the electrical potential profile
across the membrane of the *Rhodopseudomonas sphaeroides* chromatophore studied by the absorbance change of carotenoid. - Plant Cell Physiol. *23* :
595 - 605, 1982.

49961 - ITOH, S., MORITA, S. : Decay of membrane potential under phosphorylating
        conditions in chloroplasts with *in vivo* activated ATPase. - Biochim. bio-
        phys. Acta *682* : 413 - 419, 1982.

49962 - IVANISHCHEV, V.V., AKHMEDOV, Yu.D. : O vozmozhnosti obrazovaniya kompleksa
        fermentami karboksiliruyushchey fazy fotosinteza. [Possibility of complex
        formation by enzymes of the carboxylating phase of photosynthesis.] - Dokl.
        Akad. Nauk tadzh. SSR *25* : 302 - 305, 1982. [In R, ab : Tajik.]

49963 - IVANISHCHEV, V.V., AKHMEDOV, Yu.D., NASYROV, Yu.S. : O nekotorykh fiziko-
        -khimicheskikh svoystvakh ribozo-5-fosfat-izomerazy iz list'ev shpinata.
        [Some physico-chemical properties of ribose-5-phosphate isomerase from
        spinach leaves.] - Biokhimiya *47* : 1526 - 1531, 1982. [In R, ab : E.]

*49964 - IVANISHCHEV, V.V., NASYROV, Yu.S. : Ob oligomernoy strukture ribozo-
        -5-fosfat izomerazy.   [Oligomeric structure of ribose-5-phosphate
        isomerase.] - Dokl. Akad. Nauk tadzh. SSR *24* : 751 - 754, 1981. [In R,
        ab : Tajik.]

49965 - IVLEV, A.A., KALOSHIN, A.G., KOROLEVA, M.Ya. : Izotopnye effekty ugleroda
        v uglevodakh i aminokislotakh fotosinteziruyushchikh organizmov. [Carbon
        isotope effects in saccharides and amino acids of photosynthesizing orga-
        nisms.] - Stud. biophys. *88* : 39 - 46, 1982. [In R, ab : E.]

49966 - IVLEV, A.A., KNYAZEV, D.A. : Izotopnye effekty ugleroda i ikh veroyatnye
        metabolicheskie mekhanizmy. [Carbon isotope effects and their probable
        metabolic mechanisms.] - Izv. timiryazev. sel'skokhoz. Akad. *1982* (3) :
        3 - 11, 1982. [In R, ab : E.]

49967 - IVLEV, A.A., KOROLEVA, M.Ya. : Fraktsionirovanie izotopov ugleroda v pro-
        tsessakh assimilyatsii CO_2 i dissimilyatsii organicheskikh veshchestv v
        zhivykh organizmakh. [Fractionation of carbon isotopes in processes of
        CO_2 assimilation and dissimilation of organic substances in living organi-
        sms.] - Stud. biophys. *88* : 47 - 54, 1982. [In R, ab : E.]

49968 - IWAKI, H., KUROIWA, S., TOTSUKA, T., ARUGA, Y., TAKAHASHI, M. : Ecological
        aspects. - Recent Progr. nat. Sci. Jap. *7* : 71 - 77, 1982. [Ps.]

49969 - IWASA, T., TAKEDA, K., TOKUNAGA, F., SCHERRER, P.S., PACKER, L. : Photo-
        reaction of tyrosin-iodinated bacteriorhodopsin at low temperature. - Bio-
        sci. Rep. *2* : 949 - 958, 1982.

49970 - IZAWA, S. : Basic mechanisms of photosynthetic electron transport and ATP
        synthesis in oxygen-evolving organisms. - In : ZABORSKY, O.R. (ed.) : CRC
        Handbook of Biosolar Resources. Vol.I/1. Pp. 107 - 113. CRC Press, Boca
        Raton 1982.

49971 - IZBAVITELEV, S.P., IVANCHENKO, V.M. : Osobennosti fotoindutsirovannogo
        rasseyaniya sveta khloroplastami v polosakh izbiratel'noy dispersii. [Pecu-
        liarities of the photoinduced light scattering by the chloroplasts in the
        selective dispersion bands.] - Dokl. Akad. Nauk belorus. SSR *26* : 454 - 457,
        479, 1982. [In R, ab : E.]

49972 - JABLONSKI, P.P., ANDERSON, J.W. : Light-dependent reduction of hydrogen pe-
        roxide by ruptured pea chloroplasts. - Plant Physiol. *69* : 1407 - 1413,
        1982.

49973 - JABLONSKI, P.P., ANDERSON, J.W. : Light-dependent reduction of selenite
        by sonicated pea chloroplasts. - Phytochemistry *21* : 2179 - 2184, 1982.

49974 - JACKSON, J.B. : Evidence that the ionic conductivity of the cytoplasmic
        membrane of *Rhodopseudomonas capsulata* is dependent upon membrane poten-
        tial. - FEBS Lett. *139* : 139 - 143, 1982.

*49975 - JACKSON, J.B., CLARK, A.J. : Carotenoid absorption band shifts and distri-
        bution of butyltriphenylphosphonium ions as membrane potential indicators
        in intact cells of photosynthetic bacteria.   - In : PALMIERI, F.,
        QUAGLIARIELLO, E., SILIPRANDI, N., SLATER, E.C. (ed.) : Vectorial

Reactions in Electron and Ion Transport in Mitochondria and Bacteria. Pp. 371 - 379. Elsevier/North-Holland Biomedical Press, Amsterdam 1981.

49976 - JACOBSON, J.S. : Ozone and the growth and productivity of agricultural crops. - In : UNSWORTH, M.H., ORMROD, D.P. (ed.) : Effects of Gaseous Pollution in Agriculture and Horticulture. Pp. 293 - 304. Butterworth Scientific, London - Boston - Sydney - Wellington - Durban - Toronto 1982. [Photosynthates.]

49977 - JACQUES, G., DE BILLY, G., PANOUSE, M. : Biomasse et production primaire dans les secteurs antarctique et subantarctique de l'Océan Indien (mars 1980). - Com. nat. fr. Rech. antarct. *53* (Production Pélagique dans le Secteur Antarctique de l'Océan Indien) : 87 - 99, 1982. [Chl.]

49978 - JAGENDORF, A.T. : Isolation of chloroplast coupling factor (CF$_1$) and of its subunits. - In : EDELMAN, M., HALLICK, R.B., CHUA, N.-H. (ed.) : Methods in Chloroplast Molecular Biology. Pp. 881 - 898. Elsevier Biomedical Press, Amsterdam - New York - Oxford 1982.

49979 - JAGER, A. de : Effects of localized supply of H_2PO_4, NO_3, SO_4, Ca and K on the production and distribution of dry matter in young maize plants. - Neth. J. agr. Sci. *30* : 193 - 203, 1982.

49980 - JAGGARD, K.W., LAWRENCE, D.K., BISCOE, P.V. : An understanding of crop physiology in assessing a plant growth regulator on sugar beet. - In : McLAREN, J.S. (ed.) : Chemical Manipulation of Crop Growth and Development. Pp. 139 - 150. Butterworth Scientific, London - Boston - Durban - Singapore - Sydney - Toronto - Wellington 1982. [Chl.]

49981 - JAMES, L., SCHWARTZBACH, S.D. : Differential regulation of phosphoglycolate and phosphoglycerate phosphatases in *Euglena*. - Plant Sci. Lett. *27* : 223 - - 232, 1982. [Photorespiration.]

49982 - JANA, S., CHOUDHURI, M.A. : Changes in the activities of ribulose 1,5-bisphosphate and phosphoenolpyruvate carboxylases in submerged aquatic angiosperms during ageing. - Plant Physiol. *70* : 1125 - 1127, 1982.

49983 - JANA, S., CHOUDHURI, M.A. : Changes occurring during aging and senescence in a submerged aquatic angiosperm *(Potamogeton pectinatus)*. - Physiol. Plant. *55* : 356 - 360, 1982. [Chl.]

49984 - JANATKOVA, H., WILDNER, G.F. : Isolation and characterisation of metribuzin--resistant *Chlamydomonas reinhardii* cells. - Biochim. biophys. Acta *682*: 227 - 233, 1982. [Ps.]

*49985 - JANDOVÁ, B., SLADKÝ, Z. : The importance of cotyledons for the formation of chlorophyll and the character of endogenous regulators of pea plants *(Pisum sativum* L.). - Scripta Fac. Sci. Univ. J.E.Purkyně brunensis, Biol. 1, *9*: 1 - 10, 1979.

49986 - JANERO, D.R., BARRNETT, R. : Thylakoid membrane biogenesis in *Chlamydomonas reinhardtii* 137$^+$. II. Cell-cycle variations in the synthesis and assembly of pigment. - J. Cell Biol. *93* : 411 - 416, 1982. [Chl, Car.]

*49987 - JANIK, J.J., TAYLOR, W.D. : Estimating phytoplankton biomass and productivity. - Environm. Water Qual. oper. Stud. misc. Paper *E-81-2* : 1 - 23, 1981.

*49988 - JANISZOWSKA, W., WIŁKOMIRSKI, B., WOJCIECHOWSKI, Z.A., KASPRZYK, Z. : Studies on primary and secondary products of photosynthesis. - Pol. ecol. Stud. *7* : 433 - 451, 1981.

*49989 - JANJIČ, V. : Fiziološke osnove otpornosti *Setaria glauca* na djelovanje herbicida glifosata i cijanazina. [Physiological bases of the resistance of *Setaria glauca* to the effects of herbicides glyphosate and cyanazine.] - Fragm. herbol. jugosl. *10* (2) : 45 - 50, 1981. [Chl; In Croat., ab : E.]

49990 - JANZEN, D.H. : Ecological distribution of chlorophyllous developing embryos among perennial plants in a tropical deciduous forest. - Biotropica *14* : 232 - 236, 1982.

49991 - **JARRETT, H.W., BROWN, C.J., BLACK, C.C., CORMIER, M.J.** : Evidence that calmodulin is in the chloroplast of peas and serves a regulatory role in photosynthesis. - J. biol. Chem. *257* : 13795 - 13804, 1982.

49992 - **JATIMLIANSKY, J.-R., BISMUTH, E., CHAMPIGNY, M.-L.** : Participation de la photosynthèse dans la feuille étendard et l'épi à l'élaboration des réserves du grain chez les Blés différant par la constitution génomique, le niveau de ploïdie et le degré de sélection. - Compt. rend. Acad. Sci. Paris, Sér. III *295* : 407 - 412, 1982.

49993 - **JATIMLIANSKY, J.R., CHAMPIGNY, M.-L., PRIOUL, J.-L., BISMUTH, É., MOYSE,A.:** Influence du nitrate sur la croissance et la photosynthèse nette du Soja pourvu ou non de nodosités. - Physiol. vég. *20* : 407 - 422, 1982.

49994 - **JÀVOR, B., REQUADT, C., STOECKENIUS, W.** : Box-shaped halophilic bacteria. - J. Bacteriol. *151* : 1532 - 1542, 1982. [Bacteriorhodopsin.]

*49995 - **JAVORKOVÁ, A., GÁBORČÍK, N.** : Výskum fotosyntetickej produktivity trvalého trávneho porastu v podmienkach stredného Slovenska. [Photosynthetic productivity of a permanent grass cover under the conditions of central Slovakia.] - Pol'nohospodárstvo *24* : 786 - 795, 1978. [In Slovak, ab : E, R.]

49996 - **JAYNES, J.M., VERNON, L.P.** : The cyanelle of *Cyanophora paradoxa*: almost a cyanobacterial chloroplast. - Trends biochem. Sci. *7* : 22 - 24, 1982.

49997 - **JEANFILS, J., COCQUEMPOT, M.F., COLLARD, F.** : Low temperature (77 K) emission spectra and photosystem activities of chloroplasts and thylakoids immobilized in a cross-linked albumin matrix and stored at room temperature in daylight. - Photosynthetica *16*: 245 - 250, 1982.

49998 - **JELLINGS, A.J., LEECH, R.M.** : The importance of quantitative anatomy in the interpretation of whole leaf biochemistry in species of *Triticum, Hordeum* and *Avena*. - New Phytol. *92* : 39 - 48, 1982.

49999 - **JENKINS, C.L.D., HATCH, M.D.** : Why the predominance of C_4 plants amongst the world's worst weeds. - CSIRO Div. Plant Ind. Rep. *1981-82* : 12 - 18, 1982.

50000 - **JENKINS, C.L.D., ROGERS, L.J., KERR, M.W.** : Glycollate oxidase inhibition and its effect on photosynthesis and pigment formation in *Hordeum vulgare*. - Phytochemistry *21* : 1849 - 1858, 1982.

50001 - **JENKINS, C.L.D., ROGERS, L.J., KERR, M.W.** : Glycollate oxidase inhibition and its effect on photosynthesis and pigment formation in *Zea mays*. - Phytochemistry *21* : 1859 - 1863, 1982.

50002 - **JENNER, C.F.** : Movement of water and mass transfer into developing grains of wheat. - Aust. J. Plant Physiol. *9* : 69 - 82, 1982. [Photosynthates.]

50003 - **JENNINGS, R.C., GEROLA, P.D., GARLASCHI, F.M., FORTI, G.** : Studies on the kinetics of cation-associated fluorescence changes in chloroplast membranes. - FEBS Lett. *142* : 167 - 170, 1982.

50004 - **JENSEN, K.F.** : An analysis of the growth of silver maple and eastern cottonwood seedlings exposed to ozone. - Can. J. Forest Res. *12* : 420 - 424, 1982. [Growth analysis.]

50005 - **JENSEN, N.-H., NIELSEN, A.B., WILBRANDT, R.** : Chlorophyll *a* sensitized trans-cis photoisomerization of *all-trans*-β-carotene. - J. amer. chem. Soc. *104* : 6117 - 6119, 1982.

50006 - **JOHAL, S., BOURQUE, D.P.** : Preparation of crystalline RuBPCase from higher plants. - In : **EDELMAN, M., HALLICK, R.B., CHUA, N.-H.** (ed.) : Methods in Chloroplast Molecular Biology. Pp. 783 - 792. Elsevier Biomedical Press, Amsterdam - New York - Oxford 1982.

50007 - **JOHNSON, C.R., NELL, T.A., ROSENBAUM, S.E., LAURITIS, J.A.** : Influence of light intensity and drought stress on *Ficus benjamina* L. - J. amer. Soc. hort. Sci. *107* : 252 - 255, 1982. [Ps, Chl.]

50008 - JOHNSON, J.D., FERRELL, W.K. : The relationship of abscisic acid metabolism to stomatal conductance in Douglas-fir during water stress. - Physiol. Plant. *55* : 431 - 437, 1982.

*50009 - JOHNSON, J.H., LEWIS, A., GOGEL, G. : Kinetic resonance Raman spectroscopy of carotenoids: A sensitive kinetic monitor of bacteriorhodopsin mediated membrane potential changes. - Biochem. biophys. Res. Commun. *103* : 182 - - 188, 1981.

50010 - JOHNSTON, D.S., CLARK, A.D., KEMP, C.M., CHAPMAN, D. : An evaluation of the charge-transfer model for the chromophores of the retinal-containing proteins, rhodopsin and bacteriorhodopsin. - Biochim. biophys. Acta *679* : 400 - 403, 1982.

50011 - JOLIVET, E., MOYSE, A. : Les types métaboliques végétaux du point de vue photosynthétique. - Compt. rend. Séances Acad. Agr. Fr. *68* : 859 - 871, 1982.

50012 - JOLLIFFE, P.A., EATON, G.W., LOVETT DOUST, J. : Sequential analysis of plant growth. - New Phytol. *92* : 287 - 296, 1982. [Growth analysis.]

*50013 - JONES, J.B., ROANE, C.W., WOLF, D.D. : The effects of *Septoria nodorum* and *Xanthomonas translucens* f. sp. *undulosa* on photosynthesis and transpiration of wheat flag leaves. - Phytopathology *71* : 1173 - 1177, 1981.

50014 - JONES, J.R., HOYER, M.V. : Sportfish harvest predicted by summer chlorophyll-a concentration in midwestern lakes and reservoirs. - Trans. amer. Fish. Soc. *111* : 176 - 179, 1982.

50015 - JONES, J.W., BARFIELD, C.S., BOOTE, K.J., SMERAGE, G.H., MANGOLD, J. : Photosynthetic recovery of peanuts to defoliation at various growth stages. - Crop Sci. *22* : 741 - 746, 1982.

50016 - JONES, J.W., HAMMOND, L.C., BOOTE, K.J. : Predicting crop yield in response to irrigation practices. - Florida agr. Res. *1* : 20 - 22, 1982.

50017 - JONES, M.B., COLLETT, B., BROWN, S. : Sward growth under cutting and continuous stocking managements: sward canopy structure, tiller density and leaf turnover. - Grass Forage Sci. *37* : 67 - 73, 1982. [Growth analysis.]

50018 - JONES, R.C., ADAMS, M.S. : Seasonal variations in photosynthetic response of algae epiphytic on *Myriophyllum spicatum* L. - Aquat. Bot. *13* : 317 - - 330, 1982.

50019 - JONES, T., MANSFIELD, T.A. : Studies on dry matter partitioning and distribution of ^{14}C-labelled assimilates in plants of *Phleum pratense* exposed to SO_2 pollution. - Environm. Pollut. *28* : 199.- 207, 1982.

*50020 - JORDAN, D., GOVINDJEE : Bicarbonate stimulation of electron flow in thylakoids. - In : Golden Jubilee Commemoration Volume of the National Academy of Sciences (India).Pp.369 - 378.National Academy of Sciences,Calcutta 1980.

50021 - JORDAN, W.R. : Cotton. - In : TEARE, I.D., PEET, M.M. (ed.) : Crop-Water Relations. Pp. 213 - 254. John Wiley & Sons, New York - Chichester - Brisbane - Toronto - Singapore 1982. [Ps.]

50022 - JØRGENSEN, N.O.G. : Heterotrophic assimilation and occurrence of dissolved free amino acids in a shallow estuary. - Mar. Ecol.-Progr. Ser. *8* : 145 - - 159, 1982. [Chl.]

50023 - JOSHI, G.V., BHOSALE, L.J. : Estuarine ecosystems of India. - In : SEN, D. N., RAJPUROHIT, K.S. (ed.) : Contributions to the Ecology of Halophytes. Pp. 21 - 33. Dr.W.Junk Publ., The Hague - Boston - London 1982. [Ps.]

50024 - JOUANNEAU, Y., WILLISON, J.C., COLBEAU, A., HALLENBECK, P.H., RIOLACCI, C., VIGNAIS, P.M. : Enhancement of the photoproduction of H_2 by *Rhodopseudomonas capsulata:* optimization in continuous culture, role of uptake hydrogenase, genetic characterization and economic evaluation. - In : HALL, D.O., PALZ, W. (ed.) : Photochemical, Photoelectrochemical and Photobiological Processes. Vol.1. Pp. 174 - 179. D.Reidel Publ. Co., Dordrecht - Boston - - London 1982.

50025 - JOUY, M. : Energy transfer from protochlorophyllide to the $C_{695-682}$ chlorophyllide at 77 K. - Photosynthetica *16* : 13 - 16, 1982.

50026 - JOUY, M. : Energy transfer from protochlorophyllide to the $C_{695-682}$ chlorophyllide at room temperature. - Photosynthetica *16* : 123 - 128, 1982.

50027 - JOUY, M. : Spectra of photoinduced absorbance changes in the irradiated etiolated *Phaseolus vulgaris* leaves after rapid dark spectral shift. - Photosynthetica *16* : 176 - 179, 1982.

50028 - JOUY, M. : Photoactivity of the $C_{695-682}$ chlorophyllide-protein complex in the irradiated etiolated leaf of *Phaseolus vulgaris*. - Photosynthetica *16* : 180 - 183, 1982.

50029 - JOUY, M. : Effect of age of etiolated leaves of *Phaseolus vulgaris* on the 695 nm fluorescence kinetics during first irradiation. - Photosynthetica *16* : 234 - 238, 1982.

50030 - JOY, K.W. : Assimilation of nitrate, nitrite, and ammonia in vascular plants. - In : ZABORSKY, O.R. (ed.) : CRC Handbook of Biosolar Resources. Vol.I/1. Pp. 237 - 243. CRC Press, Boca Raton 1982. [Ps.]

*50031 - JOYARD, J., CHUSEL, M., DOUCE, R. : Is the chloroplast envelope a site of galactolipid synthesis? Yes! - In : APPELQVIST, L.-Å., LILJENBERG, C. (ed.): Advances in the Biochemistry and Physiology of Plant Lipids. Pp. 181 - 186. Elsevier/North-Holland Biomedical Press, Amsterdam - New York - Shannon 1979.

50032 - JOYARD, J., DOUCE, R. : The chloroplast envelope: structure and biochemical properties. - In : AKOYUNOGLOU, G., EVANGELOPOULOS, A.E., GEORGATSOS, J., PALAIOLOGOS, G., TRAKATELLIS, A., TSIGANOS, C.P. (ed.) : Cell Function and Differentiation, Part B. Biogenesis of Energy Transducing Membranes and Membrane and Protein Energetics. Pp. 77 - 89. Alan R. Liss Inc., New York 1982.

50033 - JOYARD, J., GROSSMAN, A., BARTLETT, S.G., DOUCE, R., CHUA, N.-H. : Characterization of envelope membrane polypeptides from spinach chloroplasts. - J. biol. Chem. *257* : 1095 - 1101, 1982.

50034 - JUDEL, G.K., MENGEL, K. : Effect of shading on nonstructural carbohydrates and their turnover in culms and leaves during the grain filling period of spring wheat. - Crop Sci. *22* : 958 - 962, 1982. [Photosynthates.]

50035 - JUNGE, W., FOERSTER, V., HONG, Y.Q. : Water oxidation with visible light by green plants. The role of protons. - In : HALL, D.O., PALZ, W. (ed.) : Photochemical, Photoelectrochemical and Photobiological Processes. Vol.1. Pp. 140 - 146. D.Reidel Publ.Co., Dordrecht - Boston - London 1982.

50036 - JUNGE, W., JACKSON, J.B. : The development of electrochemical potential gradients across photosynthetic membranes. - In : GOVINDJEE (ed.) : Photosynthesis. Vol.1. Pp. 589 - 646. Academic Press, New York - London - - Paris - San Diego - San Francisco - São Paulo - Sydney - Tokyo - Toronto 1982.

50037 - JURIK, T.W., CHABOT, J.F., CHABOT, B.F. : Effects of light and nutrients on leaf size, CO_2 exchange, and anatomy in wild strawberry (*Fragaria virginiana*). - Plant Physiol. *70* : 1044 - 1048, 1982.

50038 - JURSINIC, P., GOVINDJEE : Effects of hydroxylamine and silicomolybdate on the decay in delayed light emission in the 6-100 µs range after a single 10 ns flash in pea thylakoids. - Photosynthesis Res. *3* : 161 - 177, 1982.

50039 - JURSINIC, P., STEMLER, A. : A second range component of the reoxidation of the primary photosystem II acceptor, Q. Effects of bicarbonate depletion in chloroplasts. - Biochim. biophys. Acta *681* : 419 - 428, 1982.

50040 - JÜTTNER, F., WIEDEMANN, E., WURSTER, K. : Excretion of S- and O-methyl esters and other volatile compounds by *Ochromonas danica*. - Phytochemistry *21* : 2185 - 2188, 1982. [Chl.]

50041 - KABAKI, N., TAJIMA, K. : [Effect of phosphorus deficiency on RuBP carboxylase-oxygenase activity in rice plants.] - Jap. J. Crop Sci. *51* : 332 -
- 337, 1982. [In Jap., ab : E.]

50042 - KABSCH, U. : Veränderungen im Kohlenhydratstoffwechsel in mit *Sphaerotheca
fuliginea* infizierten Gurkenblättern. - Z. Pflanzenkrankh. Pflanzensch.
89 : 113 - 124, 1982.

*50043 - KACPERSKA, A., LEWAK, S., MACIEJEWSKA, U., MALESZEWSKI, S. : Photosynthe-
sis control by environmental factors. - Pol. ecol. Stud. *7* : 377 - 386,
1981.

50044 - KAHN, J.S. : The ε subunit of the chloroplast coupling factor 1 from *Euglena gracilis*. A possible role in controlling ATPase activity. - Plant
Physiol. *70* : 451 - 455, 1982.

50045 - KAIN, J.M., HOLT, T.J. : Biomass from offshore sea areas. - In : GRASSI,
G., PALZ, W. (ed.) : Energy from Biomass. Vol.3. Pp. 120 - 125. D.Reidel
Publ. Co., Dordrecht - Boston - London 1982. [Algae mass cultivation.]

50046 - KAISER, G., MARTINOIA, E., WIEMKEN, A. : Rapid appearance of photosynthetic
products in the vacuoles isolated from barley mesophyll protoplasts by a
new fast method. - Z. Pflanzenphysiol. *107* : 103 - 113, 1982.

*50047 - KAISER, W., RENK, H., SCHULZ, S. : Die Primärproduktion der Ostsee. - Geod.
geophys. Veröffentl., Reihe 4,*33* : 27 - 52, 1981.

*50048 - KAISER, W., SCHULZ, S., KELL, V. : Die Wirkung der Pollution auf das Phy-
toplankton und seine Primärproduktion. - Geod. geophys. Veröffentl., Reihe
4,*33* : 53 - 60, 1981.

50049 - KAISER, W.M. : Correlation between changes in photosynthetic activity and
changes in total protoplast volume in leaf tissue from hygro-, meso- and
xerophytes under osmotic stress. - Planta *154* : 538 - 545, 1982.

50050 - KAITALA, V., HARI, P., VAPAAVUORI, E., SALMINEN, R. : A dynamic model for
photosynthesis. - Ann. Bot. *50* : 385 - 396, 1982.

50051 - KAK, S.N., KAUL, B.L. : Selective radioprotective role of dithiothreitol
in barley *Hordeum vulgare*. - Indian J. exp. Biol. *20* : 776 - 777, 1982.
[Chl.]

*B50052 - KAKHNOVICH, L.V. : Fotosinteticheskiĭ Apparat i Svetovoĭ Rezhim. [Photo-
synthetic Apparatus and Light Regime.] - Izdatel'stvo Beloruss. gosudarstv.
Univ. Imeni V.I. Lenina, Minsk 1980. [In R.]

50053 - KAKUNO, T., YAGI, T., HORIO, T. : Bioconversion of solar energy. - Recent
Progr. nat. Sci. Jap. *7* : 67 - 70, 1982.

50054 - KAKUNO, T., YAMASHITA, J., HORIO, T. : Cytochromes of photosynthetic bacteria. - In : ZABORSKY, O.R. (ed.) : CRC Handbook of Biosolar Resources.
Vol.I/1. Pp. 143 - 146. CRC Press, Boca Raton 1982.

*50055 - KALER, V.L., FRIDLYAND, L.E. : Modelirovanie adaptatsii fotosintichesko-
go apparata rastitel'noĭ kletki k izmenyayushchimsya usloviyam vneshneĭ
sredy. [Modelling of adaptation of photosynthetic apparatus of plant cell
to changing environmental conditions.] - In : Teoreticheskie Osnovy i Koli-
chestvennye Metody Programmirovaniya Urozhaev. Pp. 24 - 38. Leningrad
1979. [In R.]

50056 - KALIR, A., FLOWERS. T.J. : The effect of salts on malate dehydrogenase
from leaves of *Zea mays*. - Phytochemistry *21* : 2189 - 2193, 1982.

*50057 - KALIR, A., POLJAKOFF-MAYBER, A. : Changes in activity of malate dehydro-
genase, catalase, peroxidase and superoxide dismutase in leaves of *Hali-
mone portulacoides* (L.) AELLEN exposed to high sodium chloride concentration. - Ann. Bot. *47* : 75 - 85, 1981. [Chl.]

50058 - KALISKI, O., OTTOLENGHI, M. : Branching pathways in the photocycle of bacteriorhodopsin. - Photochem. Photobiol. *35* : 109 - 115, 1982.

50059 - KAMIŃSKA, Z., MALESZEWSKI, S. : Glycolic acid pathway in photosynthetic
        carbon metabolism of bean leaves at various oxygen concentrations. - Z.
        Pflanzenphysiol. *108* : 201 - 206, 1982.

50060 - KAMIYA, A., MIYACHI, S. : General characteristics of green microalgae:
        Chlorella. - In : ZABORSKY, O.R. (ed.) : CRC Handbook of Biosolar Resources.
        Vol.I/2. Pp. 25 - 32. CRC Press, Boca Raton 1982.

50061 - KAMO, N., RACANELLI, T., PACKER, L. : Simultaneous measurements of proton
        movement and membrane potential changes in the wild-type and mutant *Halo-
        bacterium halobium* vesicles. - In : COLOWICK, S.P., KAPLAN, N.O. (ed.) :
        Methods in Enzymology. Vol.88. Pp. 356 - 360. Academic Press, New York -
        - London - Paris - San Diego - San Francisco - São Paulo - Sydney - Tokyo -
        - Toronto 1982.

50062 - KAN, K., TAMURA, Y., MAKI, T., KOSEKI, M., NAOI, Y. : [Studies on determin-
        ation of degradative substances of chlorophyll and chlorophyllase activity.]
        - Annu. Rep. Tokyo metr. Res. Lab. P.H. *33* : 208 - 213, 1982. [In Jap.]

50063 - KANAZAWA, Y., SATO, A., ORSOLINO, R.S. : Above-ground biomass and the growth
        of giant ipil-ipil [*Leucaena leucocephala* (LAM.) DE WIT] plantations in
        northern Mindanao Island, Philippines. - J. agr. Res. quart. *16*: 209 - 217,
        1982. [Growth analysis.]

*50064 - KANDA, H., FUKUCHI, M. : Surface chlorophyll *a* concentration along the
        course of the FUJI to and from Antarctica in 1977 - 1978. - Antarctic Res.
        66 : 37 - 49, 1979.

50065 - KANDYA, A.K. : Caloric content and energy dynamics in six tropical dry de-
        ciduous forest tree species. - Indian J. Forest. *5* : 192 - 195, 1982.

50066 - KANNAN, S., RAMANI, S. : Zinc-stress response in some sorghum hybrids and
        parent cultivars: significance of pH reduction and recovery from chlorosis.
        - J. Plant Nutr. *5* : 219 - 227, 1982.

50067 - KANNINEN, M., HARI, P., KELLOMÄKI, S. : A dynamic model for above-ground
        and dry matter production in a forest community. - J. appl. Ecol. *19* : 465 -
        - 476, 1982.

50068 - KANTCHEVA, M.R., POPDIMITROVA, N., STOYLOV, S. : Electrophoretic mobility
        of purple membrane from *Halobacterium halobium*. - Stud. biophys. *90* : 125 -
        - 126, 1982.

50069 - KANTCHEVA, M.R., POPDIMITROVA, N.G., STOYLOV, S.P. : Microelectrophoretic
        properties of purple membrane particles. - Dokl. bolg. Akad. Nauk *35* :
        633 - 635, 1982. [Bacteriorhodopsin.]

50070 - KAPLAN, A., ZENVIRTH, D., REINHOLD, L., BERRY, J.A. : Involvement of a pri-
        mary electrogenic pump in the mechanism for HCO_3^- uptake by the cyanobac-
        terium *Anabaena variabilis*. - Plant Physiol. *69* : 978 - 982, 1982.

50071 - KAPLAN, S., ARNTZEN, C.J. : Photosynthetic membrane structure and function.
        - In : GOVINDJEE (ed.) : Photosynthesis. Vol.1. Pp. 65 - 151. Academic
        Press, New York - London - Paris - San Diego - San Francisco - São Paulo -
        - Sydney - Tokyo - Toronto 1982.

50072 - KAPLAN, S., FORNARI, C., CHORY, J., YEN, B. : Chlorophyll-binding proteins:
        strategies and developments for DNA cloning in *Rhodopseudomonas sphaeroides*.
        - In : : HOLLAENDER, A., DeMOSS, R.D., KAPLAN, S., KONISKY, J., SAVAGE, D.,
        WOLFE, R.S. (ed.) : Genetic Engineering of Microorganisms for Chemicals.
        Pp. 245 - 258. Plenum Publishing Co., New York 1982.

50073 - KARADGE, B.A., CHAVAN, P.D. : Salinity and initial products of photosynthe-
        sis in peanut (*Arachis hypogaea* L.). - Biovigyanam *8* : 95 - 96, 1982.

50074 - KARAMANOS, A.J., ELSTON, J., WADSWORTH, R.M. : Water stress and leaf growth
        of field beans (*Vicia faba* L.) in the field: water potentials and laminar
        expansion. - Ann. Bot. *49* : 815 - 826, 1982. [Growth analysis.]

*50075 - KARAPETYAN, N.W., FISCHER, K., METZNER, H. : Bicarbonate effect on fluorescence induction in thylakoids. - Photobiochem. Photobiophys. 2 : 141 - 147, 1981.

50076 - KARIYA, K., MATSUZAKI, A., MACHIDA, H., TSUNODA, K. : [Distribution of chlorophyll content in leaf blade of rice plant.] - Jap. J. Crop Sci. 51 : 134 - - 135, 1982. [In Jap.]

50077 - KARLMAN, S.-G. : The annual flood regime as a regulatory mechanism for phytoplankton production in Kainji lake, Nigeria. - Hydrobiologia 86 : 93 - 97, 1982.

50078 - KARMANOV, V.G., SOLOV'EV, E.V., ODUMANOVA-DUNAEVA, G.A. : Issledovanie vzaimosvyazi fotosinteza i dykhaniya korneĭ u *Perilla ocymoides (Labiatae)*. [Studies on the relationship between photosynthesis and root respiration in *Perilla ocymoides (Labiatae)*.] - Bot. Zh. 67 : 761 - 770, 1982. [In R, ab : E.]

50079 - KARMARKAR, S.M. : Senescence in mangroves. - In : SEN, D.N., RAJPUROHIT, K. S. (ed.) : Contributions to the Ecology of Halophytes. Pp. 173 - 187. Dr. W. Junk Publ., The Hague - Boston - London 1982. [Ps.]

50080 - KARNAUKHOV, V.N., YASHIN, V.A. : Bezkhlorofil'nyĭ rezhim énergoobespecheniya peridineĭ Sredizemnogo morya. [Energy supply in *Peridiniineae* of the Mediterranean Sea under conditions unfavourable for chlorophyll formation.] - Fiziol. Rast. 29 : 168 - 171, 1982. [In R.]

50081 - KARNEEVA, N.V., BALASHOV, S.P., LITVIN, F.F. : Obnaruzhenie slozhnoĭ struktury spektra pogloshcheniya bakteriorodopsina. [Detection of complex structure of the absorption spectrum of bacteriorhodopsin.] - Dokl. Akad. Nauk SSSR 263 : 725 - 729, 1982. [In R.]

50082 - KARPILOVA, I.F., CHUGUNOVA, N.G., BIL', K.Ya., CHERMNYKH, L.N. : Ontogeneticheskie izmeneniya ul'trastruktury khloroplastov, produktov fotosinteza i ikh ottoka iz list'ev ogurtsov pri ponizhennoĭ nochnoĭ temperature. [Ontogenetic changes in chloroplast ultrastructure, photosynthate pattern and transport from the cucumber leaf under low night temperature.] - Fiziol. Rast. 29 : 113 - 120, 1982. [In R, ab : E.]

50083 - KARUNEN, P., SALIN, M. : Seasonal changes in lipids of photosynthetically active and senescent parts of *Sphagnum fuscum*. - Lindbergia 8 : 35 - 44, 1982.

50084 - KARVALY, B., FUKUMOTO, J.M., HOPEWELL, W.D., EL-SAYED, M.A. : Polarized photochemistry on bacteriorhodopsin. Dichroism of the early photochemical intermediate K_{610}. - J. phys. Chem. 86 : 1899 - 1908, 1982.

50085 - KASEMIR, H., MOHR, H. : Coaction of three factors controlling chlorophyll and anthocyanin synthesis. - Planta 156 : 282 - 288, 1982.

50086 - KATAGIRI, S., ISHII, H., MIYAKE, N. : [Studies on the amounts of dry matter and nutrients in *Sasa* communities.] - Jap. J. Ecol. 32 : 527 - 534, 1982. [In Jap., ab : E.]

50087 - KATAOKA, M., HISATOMI, O., TOKUNAGA, F., WASHIOKA, H., TONOSAKI, A. : Electron microscopic observation on phase transition of purple membrane. - J. Biochem. (Tokyo) 92 : 1667 - 1670, 1982.

50088 - KATOH, S. : [Photochemical reaction center - a chlorophyll-protein.] - Kagaku No Ryoiki 36 : 707 - 715, 1982. [In Jap.]

50089 - KATOH, S. : Plastocyanin. - In : ZABORSKY, O.R. (ed.) : CRC Handbook of Biosolar Resources. Vol.I/1. Pp. 161 - 166. CRC Press, Boca Raton 1982.

50090 - KATSUMI, M., KAZAMA, H., YAMADA, J., MATSUMURA, M. : Gibberellin-induced suppression of chloroplast starch formation from exogenous sucrose in isolated epidermis of light-grown cucumber hypocotyls. - Plant Cell Physiol. 23 : 953 - 958, 1982.

50091 - KATSURA, T., LAM, E., PACKER, L., SELTZER, S. : Light dependent modification of bacteriorhodopsin by tetranitromethane. Interaction of a tyrosine and a tryptophan residue with bound retinal. - Biochem. Int. 5 : 445 - 456, 1982.

50092 - KATTAWAR, G.W., VASTANO, J.C. : Exact 1-D solution to the problem of chlorophyll fluorescence from the ocean. - Appl. Opt. 21 : 2489 - 2492, 1982.

50093 - KATZ, J.J., HINDMAN, J.C. : Photoprocesses in chlorophyll model systems.- In : ALFANO, R.R.(ed.):Biological Events Probed by Ultrafast Laser Spectroscopy. Pp. 119 - 155. Academic Press, New York - London - Paris - San Diego - San Francisco - São Paulo - Sydney - Tokyo - Toronto 1982.

50094 - KAUFMAN, L.S., LYMAN, H. : A 600 nm receptor in *Euglena gracilis:* its role in chlorophyll accumulation. - Plant Sci. Lett. 26 : 293 - 299, 1982.

50095 - KAUFMANN, M.R. : Leaf conductance as a function of photosynthetic photon flux density and absolute humidity difference from leaf to air. - Plant Physiol. 69 : 1018 - 1022, 1982.

50096 - KAUFMANN, M.R. : Evaluation of season, temperature, and water stress effects on stomata using a leaf conductance model. - Plant Physiol. 69 : 1023 - 1026, 1982.

50097 - KAUFMANN, M.R. : Leaf conductance during the final season of a senescing aspen branch. - Plant Physiol. 70 : 655 - 657, 1982.

50098 - KAUFMANN, N., REIDL, H.-H., GOLECKI, J.R., GARCIA, A.F., DREWS, G. : Differentiation of the membrane system in cells of *Rhodopseudomonas capsulata* after transition from chemotrophic to phototrophic growth conditions. - Arch. Microbiol. 131 : 313 - 322, 1982.

50099 - KAUR-SAWHNEY, R., SHIH, L.-M., CEGIELSKA, T., GALSTON, A.W. : Inhibition of protease activity by polyamines: Relevance for control of leaf senescence. - FEBS Lett. 145 : 345 - 349, 1982.

50100 - KAUTSKY, L. : Primary production and uptake kinetics of ammonium and phosphate by *Enteromorpha compressa* in an ammonium sulfate industry outlet area. - Aquat. Bot. 12 : 23 - 40, 1982.

50101 - KAWADA, E., KANAZAWA, T. : Transient changes in the energy state of adenylates and the redox state of pyridine nucleotides in *Chlorella* cells induced by environmental changes. - Plant Cell Physiol. 23 : 775 - 783, 1982. [Ps.]

50102 - KAWASHIMA, C., HIRANO, T. : [A simple method for measuring leaf area in rice plants.] - Jap. J. Crop Sci. 51 : 393 - 394, 1982. [In Jap.]

*50103 - KAZANTSEVA, T.I. : Produktivnost' i dinamika nadzemnoĭ fitomassy pustyn'. [Productivity and dynamics of aboveground biomass of desert vegetation.] - Probl. Osvoeniya Pustyn' 1980 (2) : 76 - 84, 1980. [In R, ab : E.]

50104 - KAZENNOVA, N.V., SAMUILOV, V.D. : Exogenous cytochrome *c*-dependent light--induced membrane potential generation in *Rhodospirillum rubrum* chromatophores. - Biochim. biophys. Acta 681 : 512 - 518, 1982.

50105 - KAZLOVA, Zh.I. : Dinamika kol'kastsi pigmentaŭ u rasline yak kryteryĭ praduktsyĭnastsi agratsěnozaŭ. [Dynamics of pigment contents in a plant as a characteristic of productivity of agrocoenoses.] - Vestsi Akad. Navuk belorus. SSR, Ser. biyal. Navuk 1982 (2) : 16 - 19,122, 1982. [In Belorus., ab : E, R.]

50106 - KE, B. : Primary electron donors. - In : ZABORSKY, O.R. (ed.) : CRC Handbook of Biosolar Resources. Vol.I/1. Pp. 125 - 131. CRC Press, Boca Raton 1982.

50107 - KE, B. : Primary electron acceptors. - In : ZABORSKY, O.R. (ed.) : CRC Handbook of Biosolar Resources. Vol.I/1. Pp. 133 - 142. CRC Press, Boca Raton 1982.

50108 - KE, B., DOLAN, E., SHUVALOV, V.A., KLIMOV, V.V. : Early photochemical events in green plant photosynthesis: absorption and EPR spectroscopic studies. - In : ALFANO, R.R. (ed.) : Biological Events Probed by Ultrafast Laser Spec-

troscopy. Pp. 55 - 77. Academic Press, New York - London - Paris - San Diego - San Francisco - São Paulo - Sydney - Tokyo - Toronto 1982.

50109 - KE, B., FANG, Z.-X., LU, R.-Z., CALVERT, H.E., DOLAN, E. : The presence of
phycobilisomes in *Anabaena* heterocysts. - Annu. Rep. C.F. Kettering Res.
Lab. *1982* : 15, 1982.

50110 - KE, B., INOUÉ, H., BABCOCK, G.T., FANG, Z.-X., DOLAN, E. : Optical and EPR
characterizations of oxygen-evolving Photosystem II subchloroplast fragments isolated from the thermophilic blue-green alga *Phormidium laminosum*.
- Biochim. biophys. Acta *682* : 297 - 306, 1982.

50111 - KE, B., INOUE, H., FANG, Z.-X., DOLAN, E. : Chemistry and photochemistry of
photosystem-II reaction centers. - Annu. Rep. C.F. Kettering Res. Lab. *1982* :
37 - 38, 1982.

*50112 - KE, B., KLIMOV, V.V., INOUE, H., DOLAN, E., FANG, Z.-X., SHAW, E.R. : Chemi-
cal nature of the early electron-acceptor molecules in photosystem II; mechanism and kinetics of light-induced electron-transport reactions. - Annu.
Rep. C.F. Kettering Res. Lab. *1981* : 37 - 39, 1981.

*50113 - KE, B., PETERSON, R., DOLAN, E., CALVERT, H., FANG, Z.-X., LU, R.-Z., SHAW,
E. : Characterization of photoactive heterocysts from *Anabaena variabilis*
cells; role of phycobiliproteins in energy transfer to photosystem I in heterocysts. - Annu. Rep. C.F. Kettering Res. Lab. *1981* : 42 - 43, 1981.

50114 - KEEGSTRA, K., CLINE, K. : Evidence that envelope and thylakoid membranes
from pea chloroplasts lack glycoproteins. - Plant Physiol. *70* : 232 - 237,
1982.

50115 - KEELEY, J.E., BOWES, G. : Gas exchange characteristics of the submerged
aquatic Crassulacean acid metabolism plant, *Isoetes howellii*. - Plant Physiol. *70* : 1455 - 1458, 1982.

50116 - KEELEY, J.E., MORTON, B.A. : Distribution of diurnal acid metabolism in submerged aquatic plants outside the genus *Isoetes*. - Photosynthetica *16* :
546 - 553, 1982.

50117 - KÊÊRBERG, O.F., VIĬL', Yu.A. : Sistemy regulyatsii i énergetika vosstanovi-
tel'nogo pentozofosfatnogo tsikla. [Systems of regulation and energetics of·
reductional pentose-phosphate cycle.] - In : Fiziologiya Fotosinteza. Pp.
104 - 118. Nauka, Moskva 1982. [In R.]

50118 - KEISTER, D.L. : Basic transport and ATP synthesis pathways in photosynthetic bacteria. - In : ZABORSKY, O.R. (ed.) : CRC Handbook of Biosolar Resources. Vol.I/1. Pp. 103 - 106. CRC Press, Boca Raton 1982.

*50119 - KELL, D.B., GRIFFITHS, A.M. : Polarographic assay of the binding of certain
"probe" molecules to illuminated bacteriorhodopsin sheets. - Photobiochem.
Photobiophys. *2* : 105 - 110, 1981.

*50120 - KELLAR, P.E., PAERL, H.W. : Physiological adaptations in response to envi-
ronmental stress during an N_2-fixing *Anabaena* bloom. - Appl. environm. Microbiol. *40* : 587 - 595, 1980. [Chl, Car.]

50121 - KELLER, M., RUTTI, B., STUTZ, E. : Analysis of the *Euglena gracilis* chloroplast genome. Fragment *Eco-I* encodes the gene for the M_r 32000-33000 thylakoid protein of photosystem II reaction center. - FEBS Lett. *149* : 133 - 137,
1982.

50122 - KELLER, T. : Pollution effects on forest and woody ornamentals. - In : Proceedings 21st International Horticultural Congress. Vol. II. Pp.1126 - 1136.
International Society for Horticultural Science, Wageningen 1982. [Ps.]

50123 - KELLEY, D.B., GOODIN, J.R., MILLER, D.R. : Biology of *Atriplex*. - In : SEN,
D.N., RAJPUROHIT, K.S. (ed.) : Contributions to the Ecology of Halophytes.
Pp. 79 - 107. Dr.W.Junk Publ., The Hague - Boston - London 1982. [Ps.]

50124 - KELLOMÄKI, S., PUTTONEN, P., TAMMINEN, H., WESTMAN, C.J. : Effect of nitrogen fertilization on photosynthesis and growth in young Scots pines - preliminary results. - Silva fenn. *16* : 363 - 371, 1982.

50125 - KELLY, G. : Light-mediated activation of enzymes in photosynthetic cells. - Trends biochem.Sci. 7 : 81 - 82, 1982.

50126 - KELLY, G.J., LATZKO, E. : Photosynthesis. Carbon metabolism: the profound effects of illumination on the metabolism of photosynthetic cells. - Progr. Bot. 44 : 103 - 131, 1982.

50127 - KELLY, G.J., ZIMMERMANN, G., LATZKO, E. : Fructose-bisphosphatase from spinach leaf chloroplast and cytoplasm. - In : COLOWICK, S.P., KAPLAN, N.O. (ed.) : Methods in Enzymology. Vol.90. Pp. 371 - 378. Academic Press, New York - London - Paris - San Diego - San Francisco - São Paulo - Sydney - - Tokyo - Toronto 1982.

50128 - KEMP, D.R., BLACKLOW, W.M. : The responsiveness to temperature of the extension rates of leaves of wheat growing in the field under different levels of nitrogen fertilizer. - J. exp. Bot. 33 : 29 - 36, 1982. [Growth analysis.]

50129 - KEMP, P.R., GARDETTO, P.E. : Photosynthetic pathway types of evergreen rosette plants (Liliaceae) of the Chichuahuan desert. - Oecologia 55 : 149 - - 156, 1982.

50130 - KENDALL, W.A., PEDERSON, G.A., HILL, R.R.,Jr. : Root size estimates of red clover and alfalfa based on electrical capacitance and root diameter measurements. - Grass Forage Sci. 37 : 253 - 256, 1982.

50131 - KERFIN, W., BÖGER, P. : Light-induced hydrogen evolution by blue-green algae (Cyanobacteria). - Physiol. Plant. 54 : 93 - 98, 1982.

50132 - KERSHAW, K.A., MACFARLANE, J.D. : Physiological-environmental interactions in lichens XIII. Seasonal constancy of nitrogenase activity, net photosynthesis and respiration, in Collema furfuraceum (AM.) DR. - New Phytol. 90 : 723 - 734, 1982.

50133 - KESSELMEIER, J. : Steroidal saponins in etiolated, greening and green leaves and in isolated etioplasts and chloroplasts of Avena sativa. - Protoplasma 112 : 127 - 132, 1982. [Chl, Car.]

50134 - KESZTHELYI, L. : Orientation of purple membranes by electric field. - In : COLOWICK, S.P., KAPLAN, N.O. (ed.) : Methods in Enzymology. Vol.88. Pp. 287 - - 297. Academic Press, New York - London - Paris - San Diego - San Francisco - - São Paulo - Sydney - Tokyo - Toronto 1982.

50135 - KESZTHELYI, L., ORMOS, P., VÁRÓ, G. : Fast components of the electric response signal of bacteriorhodopsin protein. - Acta phys. Acad. Sci. hung. 53 : 143 - 157, 1982.

50136 - KETTLEWELL, P.S., DAVIES, W.P., HOCKING, T.J. : Disease development and senescence of the flag leaf of winter wheat in response to propiconazole. - J. agr. Sci. 99 : 661 - 663, 1982. [Ps, Chl.]

50137 - KEYS, A.J., BIRD, I.F., CORNELIUS, M.J. : Possible use of chemicals for the control of photorespiration. - In : McLAREN, J.S. (ed.) : Chemical Manipulation of Crop Growth and Development. Pp. 39 - 53. Butterworth Scientific, London - Boston - Durban - Singapore - Sydney - Toronto - Wellington 1982.

50138 - KHADASEVICH, Ě.V., GVARDYYAN, V.N., ARNAUTAVA, A.I., MYSHKAVETS, Ya.N., LIS, P.I. : Tsemnayoe nazapashvanne protakhlarafilidu yak pakazchyk abnaŭlennya khlarafilu ŭ khvoĭnykh. [Dark accumulation of protochlorophyllide as an indicator of chlorophyll regeneration in conifers.] - Vestsi Akad. Navuk belorus. SSR, Ser. biyal. Navuk 1982 (3) : 28 - 33, 122, 1982. [In Belorus., ab : E, R.]

50139 - KHALIFA, F.M., MARSHALL, C., DANIELS, K., WITCOMBE, J.R. : Assimilation of $^{14}CO_2$, assimilate translocation and grain yield in three primitive Nepalese varieties and a European cultivar (Senta) of six-row barley (Hordeum vulgare L.). - Ann. Bot. 50 : 49 - 56, 1982.

50140 - KHAN, A.A., MALHOTRA, S.S. : Ribulose bisphosphate carboxylase and glycollate oxidase from jack pine: effects of sulphur dioxide fumigation. - Phytochemistry 21 : 2607 - 2612, 1982.

50141 - KHANNA, S., NICHOLAS, D.J.D. : Utilization of tetrathionate and ^{35}S-labelled thiosulphate by washed cells of *Chlorobium vibrioforme* f. sp. *thiosulfatophilum*. - J. gen. Microbiol. *128* : 1027 - 1034, 1982. [Ps.]

50142 - KHANNA-CHOPRA, R. : Photosynthesis, photosynthetic enzymes and leaf area development in relation to hybrid vigour in *Sorghum vulgare* L. - Photosynthesis Res. *3* : 113 - 122, 1982.

50143 - KHANNA CHOPRA, R., SINHA, S.K. : Photosynthetic rate and photosynthetic carboxylation enzymes during growth and development in *Cicer arietinum* L. cultivars. - Photosynthetica *16* : 509 - 513, 1982.

B50144 - KHARUK, V.I., TERSKOV, I.A. : Vnelistovye Pigmenty Drevesnykh Rasteniĭ. [Extrafoliar Pigments of Woody Plants.] - Nauka, Sibir. Otd., Novosibirsk 1982. [Ps, Chl, Car; in R.]

50145 - KHITRINA, L.V., DRACHEV, L.A., KAULEN, A.D., CHEKULAEVA, L.N. : Ingibirovanie bakteriorodopsina formalinom i lantanom. [Inhibition of bacteriorhodopsin by formalin and lanthane.] - Biokhimiya *47* : 1763 - 1772, 1982. [In R, ab : E.]

50146 - KHIZHNYAKOV, V.V., TEKHVER, I.Yu. : Perenos èlektronnogo vozbuzhdeniya v khode kolebatel'noĭ relaksatsii. [Transfer of electron excitation during vibrational relaxation.] - Eesti NSV ·Tead. Akad. Toim., Füüs., Mat. *31* : 174 - 179, 1982. [Chl; in R, ab : E, Est.]

50147 - KHMARA, L.A., FEDORENKO, Yu.P. : Rol' margantsa v vydelenii O_2 i belkovom sostave fotosinteticheskikh membran. [Role of manganese in O_2 evolution and protein composition of photosynthetic membranes.] - Dokl. Akad. Nauk SSSR *266* : 488 - 491, 1982. [In R.]

50148 - KHRAMTSOV, V.N. : Produktivnost' *Bromopsis variegata (Poaceae)* v fitotsenozakh teberdinskogo zapovednika. [Productivity of *Bromopsis variegata (Poaceae)* in phytocoenoses of the Teberda reservation.] - Bot. Zh. *67* : 951 - - 959, 1982. [In R.]

50149 - KHURANA, S.C., McLAREN, J.S. : The influence of leaf area, light interception, and season on potato growth and yield. - Potato Res. *25* : 329 - 342, 1982. [Growth analysis.]

50150 - KIDD, G.H., DAVIS, M.E. : Maize RNA polymerases and *in vitro* transcription. - In : SHERIDAN, W.F. (ed.) : Maize for Biological Research. Pp. 169 - 176. Plant Mol. Biol. Assoc., Charlottesville 1982. [Chloroplast.]

50151 - KIMURA, K., MASON, T.L., KHORANA, H.G. : Immunological probes for bacteriorhodopsin. Identification of three distinct antigenic sites on the cytoplasmic surface. - J. biol. Chem. *257* : 2859 - 2867, 1982.

50152 - KIMURA, M., FUNAKOSHI, M., SUDO, S., MASUZAWA, T., NAKAMURA, T., MATSUDA, K. : Productivity and mineral cycling in an oak coppice forest 1. Structure and phytomass of the forest. - Bot. Mag. (Tokyo) *95* : 19 - 33, 1982.

50153 - KIMURA, M., FUNAKOSHI, M., SUDO, S., MASUZAWA, T., NAKAMURA, T., MATSUDA, K.: Productivity and mineral cycling in an oak coppice forest 2. Annual net production of the forest. - Bot. Mag. (Tokyo) *95* : 359 - 373, 1982.

50154 - KIMURA, Y., SUZUKI, A., TAKEMATSU, T., KONNAI, M., TAKEUCHI, Y. : (+)-abscisic acid and two compounds showing chlorophyll degradation activity in *Cuscuta pentagona* ENGELM. - Agr. biol. Chem. *46* : 1071 - 1073, 1982.

50155 - KINDL, H. : The biosynthesis of microbodies (peroxisomes, glyoxysomes). - Int. Rev. Cytol. *80* : 193 - 229, 1982.

50156 - KINET, J.M. : Un abaissement de la température et une élévation de la teneur en CO_2 de l'atmosphère réduisent l'avortement des inflorescences de tomates cultivées en conditions d'éclairement hivernal. - Rev. Agr. *35* : 1767 - 1772, 1982. [Photosynthates.]

50157 - KING, G.I., SCHOENBORN, B.P. : Neutron scattering of bacteriorhodopsin. - In : COLOWICK, S.P., KAPLAN, N.O. (ed.) : Methods in Enzymology. Vol.88. Pp. 241 - 248. Academic Press, New York - London - Paris - San Diego - - San Francisco - São Paulo - Sydney - Tokyo - Toronto 1982.

50158 - KING, R.W., PATRICK, J.W. : Control of assimilate movement in wheat is abscisic acid involved? - Z. Pflanzenphysiol. *106* : 375 - 380, 1982.

50159 - KIPE-NOLT, J.A., STEVENS, S.E.,Jr., BRYANT, D.A. : Growth and chromatic
adaptation of *Nostoc* sp. strain MAC and the pigment mutant R-MAC. - Plant
Physiol. *70* : 1549 - 1553, 1982.

*50160 - KIRBY, J.A., GOODIN, D.B., WYDRZYNSKI, T., ROBERTSON, A.S., KLEIN, M.P.:
State of manganese in the photosynthetic apparatus. 2. X-ray absorption
edge studies on manganese in photosynthetic membranes. - J. amer. chem.
Soc. *103* : 5537 - 5542, 1981.

*50161 - KIRBY, J.A., ROBERTSON, A.S., SMITH, J.P., THOMPSON, A.C., COOPER, S.R.,
KLEIN, M.P. : State of manganese in the photosynthetic apparatus. 1. Extended X-ray absorption fine structure studies on chloroplasts and di-µ-
-oxo-bridged dimanganese model compounds. - J. amer. chem. Soc. *103* : 5529 -
- 5537, 1981.

50162 - KIRICHENKO, E.B. : Properties of the plastid apparatus of the leaves and
generative organs of some angiosperms. - In : Proceedings 21st International
Horticultural Congress. Vol.II. Pp. 1116 - 1125. International Society for
Horticultural Science, Wageningen 1982.

50163 - KIRICHENKO, E.B. : Sostav pigmentov generativnykh organov i zernovok zla-
kov. [Pigment content of generative organs and caryopses of cereals.] - Fiziol. Rast. *29* : 325 - 331, 1982. [In R, ab : E.]

50164 - KIRKHAM, M.B. : Orientation of leaves of winter wheat planted in north-
-south or east-west rows. - Agron. J. *74* : 893 - 898, 1982.

50165 - KIRKHAM, M.B., KANEMASU, E.T. : Wheat. - In : TEARE, I.D., PEET, M.M. (ed.):
Crop-Water Relations. Pp. 481 - 521. John Wiley & Sons, New York - Chichester - Brisbane - Toronto - Singapore 1982. [Ps.]

50166 - KIRKMAN, H., COOK, I.H., REID, D.D. : Biomass and growth of *Zostera capri-
corni* ASCHERS. in Port Hacking, N.S.W., Australia. - Aquat. Bot. *12* : 57 -
- 67, 1982. [Dry-matter accumulation.]

50167 - KISHITANI, S., TSUNODA, S. : Leaf thickness and response of leaf photosynthesis to water stress in soybean varieties. - Euphytica *31* : 657 - 664,
1982.

*50168 - KITSNO, L.V., BORISYUK, V.A., MEDVEDEV, A.A., LINNIK, L.I. : Formirovanie
assimilyatsionnogo apparata i produktivnost' sakharnoĭ svekly v usloviyakh
zarazhennosti pochvy sveklovichnoĭ nematodoĭ. [Formation of assimilation
apparatus and yielding capacity of sugar beet plants under soil infection
with sugar beet nematode.] - Dokl. Akad. Nauk ukr. SSR, Ser. B *1980* (3) :
87 - 90, 1980. [In R, ab : E.]

*50169 - KIYOTA, M., YABUKI, K. : [Transient phenomena of CO_2 exchange of cucumber
leaves with changing humidity.] - J. agr. Meteorol. *36* : 275 - 278, 1981.
[In Jap.]

50170 - KIYOTA, M., YABUKI, K. : [Studies on the carbon dioxide environment for
plant growth (VIII). Changes in photosynthetic rates of cucumber leaves
after the treatment of CO_2-enrichment.] - Seibutsu Kankyo Chosetsu [Envi-
ronm. Control Biol.] *20* : 17 - 23, 1982. [In Jap., ab : E.]

*50171 - KJERFVE, B., McKELLAR, H.N.,Jr. : Time series measurements of estuarine
material fluxes. - In : KENNEDY, V.S. (ed.) : Estuarine Perspectives. Pp.
341 - 357. Academic Press, New York 1980. [Chl.]

50172 - KLAUSNER, R.D., BERMAN, M., BLUMENTHAL, R., WEINSTEIN, J.N., CAPLAN, S.R. :
Compartmental analysis of light-induced proton movement in reconstituted
bacteriorhodopsin vesicles. - Biochemistry *21* : 3643 - 3650, 1982.

50173 - KLEIN, O., PORRA, R.J. : The participation of the Shemin and C_5 pathways
in 5-aminolaevulinate and chlorophyll formation in higher plants and fa-
cultative photosynthetic bacteria. - Hoppe-Seyler's Z. physiol. Chem. *363* :
551 - 562, 1982.

50174 - **KLEIN, S.** : Diversity of chloroplast structure. - In : SCHIFF, J.A., LYMAN, H. (ed.) : On the Origins of Chloroplasts. Pp. 35 - 53. Elsevier/North-Holland, New York - Amsterdam - Oxford 1982.

50175 - **KLEINKOPF, G.E.** : Potato. - In : TEARE, I.D., PEET. M.M. (ed.) : Crop-Water Relations. Pp. 287 - 306. John Wiley & Sons, New York - Chichester - Brisbane - Toronto - Singapore 1982. [Ps.]

50176 - **KLEMER, A.R., FEUILLADE, J., FEUILLADE, M.** : Cyanobacterial blooms: Carbon and nitrogen limitation have opposite effects on the buoyancy of *Oscillatoria*. - Science *215* : 1629 - 1631, 1982.

50177 - **KLIMOV, V.V., ALLAKHVERDIEV, S.I., SHUVALOV, V.A., KRASNOVSKIĬ, A.A.** : Deǐstvie obratimoǐ ėkstraktsii margantsa na svetovye reaktsii preparatov fotosistemy II. [Action of reversible extraction of manganese on light reactions of photosystem II preparations.] - Dokl. Akad. Nauk SSSR *263* : 1001 - 1005, 1982. [In R.]

50178 - **KLIMOV, V.V., ALLAKHVERDIEV, S.I., SHUVALOV, V.A., KRASNOVSKY, A.A.** : Effect of extraction and re-addition of manganese in light reactions of photosystem-II preparations. - FEBS Lett. *148* : 307 - 312, 1982.

50179 - **KLIMOV, V.V., KRASNOVSKIĬ, A.A.** : Uchastie feofitina v pervichnykh protsessakh perenosa ėlektrona v reaktsionnykh tsentrakh fotosistemy II. [Pheophytin participation in primary processes of electron transport in the photosystem II reaction centers.] - Biofizika *27* : 179 - 189, 1982. [In R, ab : E.]

50180 - **KLOET, W.A. de** : The primary production of phytoplankton in Lake Vechten. - Hydrobiologia *95* : 37 - 57, 1982.

50181 - **KLOPPSTECH, K., PFISTERER, J., MEYER, G., MÜLLER, M.** : Control of expression of chloroplast membrane proteins in higher plants. - In : AKOYUNOGLOU, G., EVANGELOPOULOS, A.E., GEORGATSOS, J., PALAIOLOGOS, G., TRAKATELLIS, A., TSIGANOS, C.P. (ed.) : Cell Function and Differentiation, Part B. Biogenesis of Energy Transducing Membranes and Membrane and Protein Energetics. Pp. 101 - 110. Alan R. Liss Inc., New York 1982.

50182 - **KLUGE, M.** : Crassulacean Acid Metabolism (CAM). - In : GOVINDJEE (ed.) : Photosynthesis. Vol.II. Pp. 231 - 262. Academic Press, New York - London - - Paris - San Diego - San Francisco - São Paulo - Sydney - Tokyo - Toronto 1982.

50183 - **KNACKER, T., SCHAUB, H.** : Wachstums- und CO_2-Gaswechsel-Charakteristik der C_4-Pflanze *Amaranthus paniculatus* L. bei Anzucht in verschiedenen Sauerstoffkonzentrationen. - Photosynthetica *16* : 206 - 216, 1982.

50184 - **KNAFF, D.B., DAVIDSON, V.L.** : Light-dependent active transport in prokaryotes. - Photochem. Photobiol. *36* : 721 - 724, 1982. [Ps.]

50185 - **KNAUER, G.A., HEBEL, D., CIPRIANO, F.** : Marine snow: major site of primary production in coastal waters. - Nature *300* : 630 - 631, 1982.

50186 - **KNEUSEL, R.E., MERCHANT, S., SELMAN, B.R.** : Properties of the solvent-stimulated ATPase activity of chloroplast coupling factor 1 from *Chlamydomonas reinhardii*. - Biochim. biophys. Acta *681* : 337 - 344, 1982.

50187 - **KNOTH, R.** : Protein crystalloids in ribosome-deficient plastids of *Aeonium domesticum* cv. *variegatum (Crassulaceae)*. - Planta *156* : 528 - 535, 1982. [Chloroplast.]

50188 - **KNOX, P.P., GARAB, G.I.** : The effect of a permanent electric field on thermoluminescence of chloroplasts. - Photochem. Photobiol. *35* : 733 - 736, 1982.

50189 - **KOBAYASHI, T.** : Picosecond laser spectroscopy. - In : Methods of Experimental Physics. Vol.20. Pp. 163 - 195. Academic Press, New York - London 1982. [Ps.]

50190 - **KOBAYASHI, Y., KÖSTER, S., HEBER, U.** : Light scattering,chlorophyll fluorescence and state of the adenylate system in illuminated spinach leaves. - Biochim. biophys. Acta *682* : 44 - 54, 1982.

50191 - KOCH, J.L., OBERLANDER, R.M., TAMAS, I.A., GERMAIN, J.L., AMMONDSON, D.B.S.:
Evidence of singlet oxygen participation in the chlorophyll-sensitized photooxidation of indoleacetic acid. - Plant Physiol. *70* : 414 - 417, 1982.

*50192 - KOCH, K.E., KENNEDY, R.A. : Effects of seasonal changes in the midwest on
Crassulacean Acid Metabolism (CAM) in *Opuntia humifusa* RAF. - Oecologia *45* :
390 - 395, 1980.

50193 - KOCH, K.E., KENNEDY, R.A. : Crassulacean acid metabolism in the succulent
C_4 dicot, *Portulaca oleracea* L under natural environmental conditions. -
Plant Physiol. *69* : 757 - 761, 1982.

50194 - KOCH, K.E., TSUI, C.-L., SCHRADER, L.E., NELSON, O.E. : Source-sink relations in maize mutants with starch-deficient endosperms. - Plant Physiol.
70 : 322 - 325, 1982. [Photosynthates.]

50195 - KÖCK, M., METZGER, U., SCHLEE, D. : Einfluß von Natriumsulfit auf die Photo-
syntheseaktivität von *Trebouxia* spec. und *Euglena gracilis*. - In : HOFFMANN,
P., HIEKE, B. (ed.) : Photosynthese: Regulation und Evolution. (Colloquia
Pflanzenphysiologie Nr.5.) Pp. 168 - 171. Humboldt-Universität, Berlin
1982.

50196 - KOENIG, F., MØLLER, B.L. : Isolation and characterization of cytochrome
b-559 from chloroplasts and etioplasts of barley. - Carlsberg Res. Commun.
47 : 245 - 262, 1982.

*50197 - KOENIG, F., VERNON, L.P. : Which polypeptides are characteristic for photo-
system II? Analysis of active photosystem II particles from the blue-green
alga *Anacystis nidulans*. - Z. Naturforsch. *34 C* : 295 - 304, 1981.

50198 - KOEPP, R. : Wachstum, Photosynthese und Assimilatverteilung in jungen Mais-
pflanzen nach Einwirkung kleiner Dosen ionisierender Strahlung auf das Saat-
gut. - In : UNGER, K., SCHUH, J. (ed.) : Umwelt-Stress. Pp. 289 - 292.
Martin-Luther-Universität, Halle 1982.

50199 - KOEPP, R., SENONER, M., ZACHARCZUK, K., MIR, F. : Induktionskurven der ver-
zögerten Fluoreszenz von Maisblättern (*Zea mays*). - In : HOFFMANN, P., HIEKE,
B. (ed.) : Photosynthese: Regulation und Evolution. (Colloquia Pflanzenphysiologie Nr.5.) Pp. 115 - 117. Humboldt-Universität, Berlin 1982.

50200 - KOHMOTO, K., NISHIMURA, S., OTANI, H. : Action sites for AM-toxins produced
by the apple pathotype of *Alternaria alternata*. - In : ASADA, Y., BUSHNELL,
W.R., OUCHI, S., VANCE, C.P. (ed.) : Plant Infection. The Physiological
and Biochemical Basis. Pp. 253 - 263. Jap. Sci. Soc. Press, Tokyo, Springer-
-Verlag, Berlin - Heidelberg - New York 1982. [Ps, Chl, chloroplast.]

50201 - KOIKE, H., SATOH, K., KATOH, S. : Heat-stabilities of electron transport
related photosystem I in a thermophilic blue-green alga, *Synechococcus* sp. -
Plant Cell Physiol. *23* : 293 - 299, 1982.

50202 - KOIKE, I., FURUYA, K., OTOBE, H., NAKAI, T., MEMOTO, T., HATTORI, A. :
Horizontal distributions of surface chlorophyll *a* and nitrogenous nutrients
near Bering Strait and Unimak Pass. - Deep-Sea Res. A *29* : 149 - 155,
1982.

50203 - KOIKE, I., HATTORI, A., TAKAHASHI, M., GOERING, J.J. : The use of enclosed
experimental ecosystems to study nitrogen dynamics in coastal waters. - In :
GRICE, G.D., REEVE, M.R. (ed.) : Marine Mesocosms. Biological and Chemical
Research in Experimental Ecosystems. Pp. 291 - 303. Springer-Verlag, New
York - Heidelberg - Berlin 1982.

50204 - KOKUBUN, M., WATANABE, K. : Analysis of the yield-determining process of
field-grown soybeans in relation to canopy structure. VI. Characteristics
of grain production in relation to plant types as affected by planting
patterns and planting densities. - Jap. J. Crop Sci. *51* : 51 - 57, 1982.

50205 - KOLESNIKOV, P.A., DANG SUEN N'Y, ZORÉ, S.V. : Glikolatoksidaza i katalaza
v kletochnykh fraktsiyakh assotsiata azolly s azotfiksiruyushchimi tsiano-
bakteriyami. [Glycolate oxidase and catalase in cell fractions of *Azolla*
associated with nitrogen fixing cyanobacteria.] - Dokl. Akad. Nauk SSSR
266 : 1501 - 1504, 1982. [Chloroplast; in R.]

50206 - KOLESNIKOVA, L.I., TRIBEL', M.M., FRANKEVICH, E.L. : Izuchenie razdeleniya
zaryadov v reaktsionnykh tsentrakh fotosistemy II vysshikh rasteniĭ metodom
magnitnoĭ modulyatsii fluorestsentsii. [Charge separation in the reaction
centres of photosystem 2 of higher plants by the method of fluorescence magnetic modulation.] - Biofizika 27 : 565 - 571, 1982. [In R, ab : E.]

50207 - KOLOMEĬCHENKO, V.V. : Fotosinteticheskie osnovy povysheniya produktivnosti
kormovykh kul'tur. [Photosynthetic bases for improving forage crop performance.] - Vest. sel'skokhoz. Nauki 1982 (7) : 38 - 49, 1982. [In R, ab : E.]

50208 - KOLOMEĬCHENKO, V.V. : Gravimetric determination of leaf area in grasses. -
Photosynthetica 16 : 251 - 252, 1982.

*50209 - KOMAROVA, G.E., ZEVERTAĬLO, T.F., ZVEREVA, V.F., KOZHUKHAR', E.K., SOLONENKO,
T.A. : Soderzhanie fotosinteticheskikh pigmentov v list'yakh khlorofil'nykh
mutantov i selektsionnykh form kukuruzy. [Contents of photosynthetic pigments
in leaves of maize chlorophyll mutants and lines.] - Izv. Akad. Nauk mold.
SSR, Ser. biol. khim. Nauk 1980 (6) : 31 - 37, 1980. [In R.]

50210 - KOMAROVA, Yu.M., DOMAN, N.G., SHAPOSHNIKOV, G.L. : Dve formy karboangidrazy
v khloroplastakh bobov. [Two forms of carbonic anhydrase from broad bean
chloroplasts.] - Biokhimiya 47 : 1027 - 1034, 1982. [In R, ab : E.]

50211 - KOMAROVA, Yu.M., DOMAN, N.G., VAKLINOVA, S.G. : Svetlinna regulyatsiya na
rastitelnata karboankhidraza. [Light regulation of plant carbonic anhydrase.]
- Fiziol. Rast. (Sofiya) 8 (1) : 3 - 9, 1982. [In Bulg., ab : E, R.]

*50212 - KOMISSARENKO, N.F., STUPAKOVA, É.P., POKALN, D.A. : Komponenty Adonis flam-
meus. [Components of Adonis flammeus.] - Khim. prir. Soed. 1981 (2) :
249 - 250, 1981. [Car; in R.]

50213 - KON, T., KOMATSU, Y., KATAOKA, T. : [Influence of abnormal weather on growth
and yield of paddy rice in warmer regions.] - Jap. J. Crop Sci. 51 : 165 -
- 171, 1982. [Dry-matter accumulation; in Jap., ab : E.]

50214 - KONDRAT'EV, K.Ya., FEDCHENKO, P.P., BARMINA, Yu.M. : Opyt opredeleniya soderzhaniya khlorofilla v list'yakh rasteniĭ po tsvetovym koordinatam. [Expe-
rimental determination of chlorophyll content in plant leaves according to
colour coordinates.] - Dokl. Akad. Nauk SSSR 262 : 1022 - 1024, 1982. [In R.]

50215 - KONDRAT'EV, K.Ya., KOZODEROV, V.V., FEDCHENKO, P.P., BARMINA, Yu.M. : K voprosu opredeleniya soderzhaniya khlorofilla v list'yakh rasteniĭ po dannym
spektral'nykh izmereniĭ. [Determination of chlorophyll content in plant
leaves using spectral measurements.] - Dokl. Akad. Nauk SSSR 265 : 1508 -
- 1510, 1982. [In R.]

50216 - KONISHI, T. : High-performance liquid chromatography method for isolation
of membrane proteins from halobacterial membrane. - In : COLOWICK, S.P.,
KAPLAN, N.O. (ed.) : Methods in Enzymology. Vol.88. Pp. 202 - 207. Academic
Press, New York - London - Paris - San Diego - San Francisco - São Paulo -
- Sydney - Tokyo - Toronto 1982. [Bacteriorhodopsin.]

50217 - KONONENKO, A.A., KNOX, P.P., VENEDIKTOV, P.S., GARAB, Gy.I., FALUDI-DANIEL,
A. : Electric polarization and thermoluminescence of chloroplasts. - In :
4th Conference on Luminescence. Conference Digest. Pp. 125 - 132. Szeged
1982.

50218 - KONOPKA, A. : Physiological ecology of a metalimnetic Oscillatoria rubescens
population. - Limnol. Oceanogr. 27 : 1154 - 1161, 1982. [Ps.]

50219 - KONOVALOVA, G.V., TYAPKIN, V.S. : Biomassa fitoplanktona v zalive Pos'eta
Yaponskogo Morya. [Biomass of phytoplankton in the Possyet Bay, Sea of Japan.]
- Biol. Morya 1982 (2) : 12 - 19, 1982. [In Jap., ab : E.]

50220 - KORENBROT, J.I. : The assembly of bacteriorhodopsin-containing planar membranes by the sequential transfer of air-water interface films. - In :
COLOWICK, S.P., KAPLAN, N.O. (ed.) : Methods in Enzymology. Vol.88. Pp. 45 -
- 55. Academic Press, New York - London - Paris - San Diego - San Francisco -
- São Paulo - Sydney - Tokyo - Toronto 1982.

50221 - KORENSTEIN, R., HESS, B. : Analysis of photocycle and orientation in thin
         layers. - In : COLOWICK, S.P., KAPLAN, N.O. (ed.) : Methods in Enzymology.
         Vol.88. Pp. 180 - 193. Academic Press, New York - London - Paris - San Diego -
         - San Francisco - São Paulo - Sydney - Tokyo - Toronto 1982. [Bacteriorho-
         dopsin.]

50222 - KORENSTEIN, R., HESS, B. : Cooperativity of photocycle in purple membranes.
         - In : COLOWICK, S.P., KAPLAN, N.O. (ed.) : Methods in Enzymology. Vol.88.
         Pp. 193 - 201. Academic Press, New York - London - Paris - San Diego - San
         Francisco - São Paulo - Sydney - Tokyo - Toronto 1982. [Bacteriorhodopsin.]

50223 - KORKIN, A.A., IONOV, S.P. : Fiziko-khimicheskie osnovy stroeniya i funktsio-
         nirovaniya reaktsionnykh tsentrov fotosinteziruyushchikh bakteriĭ. II. Orga-
         nizatsiya vzaimodeĭstviya pigmentov v reaktsionnykh tsentrakh. [Physico-che-
         mical principles of the structure and functioning of pigment interaction
         in reaction centres.] - Zh. fiz. Khim. 56 : 513 - 523, 1982. [In R.]

50224 - KÖRNER, C. : CO_2 exchange in the alpine sedge *Carex curvula* as influenced by
         canopy structure, light and temperature. - Oecologia 53 : 98 - 104, 1982.

50225 - KOROLĚVA, O.Ya. : K voprosu o vliyanii perekisi vodoroda na violaksantinovyĭ
         tsikl v list'yakh zelenykh rasteniĭ. [Effect of hydrogen peroxide on the
         violaxanthin cycle in green plant leaves.] - Fiziol. Rast. 29 : 784 - 786,
         1982. [In R.]

50226 - KORTE, C.J., WATKIN, B.R., HARRIS, W. : Use of residual leaf area index and
         light interception as criteria for spring-grazing management of a ryegrass-
         -dominant pasture. - New Zeal. J. agr. Res. 25 : 309 - 319, 1982.

50227 - KORVATOVSKIĬ, B.N., PASHCHENKO, V.Z., RUBIN, A.B., RUBIN, L.B., TUSOV, V.B. :
         Avtomatizirovannyĭ impul'snyĭ fluorometr s vysokoĭ vremennoĭ razreshayush-
         cheĭ sposobnost'yu i chuvstvitel'nost'yu. [Automatic pulsed fluorometer
         with short resolution time and high sensitivity.] - Biol. Nauki 1982 (11) :
         105 - 112, 1982. [In R.]

50228 - KOSAKOVSKAYA, I.V. : Ribulozodifosfatkarboksilaza: Osobennosti biosinteza
         i sub"edinichnoĭ struktury. [Ribulose bisphosphate carboxylase: Peculiarities
         of biosynthesis and subunit structure.] - Fiziol. Biokhim. kul't. Rast. 14 :
         419 - 426, 1982. [In R, ab : E.]

50229 - KOSAKOVSKAYA, I.V., KOMARNITSKIĬ, I.K., CHERNYAD'EV, I.I., DOMAN, N.G. :
         Analiz polipeptidnogo stroeniya i karboksilaznoĭ aktivnosti ribulozodifosfat-
         karboksilazy nekotorykh vidov i gibridov tabaka. [Analysis of polypeptide
         structure and carboxylase activity of ribulose bisphosphate carboxylase in
         certain tobacco species and hybrids.] - Fiziol. Biokhim. kul't. Rast. 14 :
         138 - 142, 200, 1982. [In R, ab : E.]

50230 - KOSOVEL, V., TALARICO, L. : Preliminary ultrastructural characterization of
         R-phycocyanin from the red alga *Gracilaria verrucosa* (HUDS.) PAPENFUSS. -
         Photosynthetica 16 : 373 - 374, 380a, 1982.

50231 - KÖSTER, S., HEBER, U. : Light scattering and quenching of 9-aminoacridine
         fluorescence as indicators of the phosphorylation state of the adenylate
         system in intact spinach chloroplasts. - Biochim. biophys. Acta 680 : 88 -
         - 94, 1982.

50232 - KOUCHKOVSKY, Y. de, HARAUX, F., SIGALAT, C. : Effect of hydrogen-deuterium
         exchange on energy-coupled processes in thylakoids. A new illustration of the
         hypothesis of local proton gradients with the energy-transducing biomembra-
         nes. - FEBS Lett. 139 : 245 - 249, 1982.

50233 - KOVACS, K.L., BAGYINKA, Cs., RAK, E. : Orientation of hydrogenase in the
         photosynthetic membrane of *Thiocapsa roseopersicina*. - Stud. biophys. 90 :
         71 - 72, 1982.

50234 - KOW, Y.W., ERBES, D.L., GIBBS, M. : Chloroplast respiration. A means of sup-
         plying oxidized pyridine nucleotide for dark chloroplastic metabolism. -
         Plant Physiol. 69 : 442 - 447, 1982.

50235 - **KOW, Y.W., GIBBS, M.** : Characterization of a photosynthesizing reconstituted spinach chloroplast preparation. Regulation by primer, adenylates, ferredoxin, and pyridine nucleotides. - Plant Physiol. *69* : 179 - 186, 1982.

50236 - **KOW, Y.W., SMYTH, D.A., GIBBS, M.** : Oxidation of NAD(P)H in a reconstituted spinach chloroplast preparation using ascorbate and hydrogen peroxide. - Plant Physiol. *69* : 740 - 741, 1982.

50237 - **KOYAMA, Y., KITO, M., TAKII, T., SAIKI, K., TSUKIDA, K., YAMASHITA, J.** : Configuration of the carotenoid in the reaction centers of photosynthetic bacteria. Comparison of the resonance Raman spectrum of the reaction center of *Rhodopseudomonas sphaeroides* GIC with those of *cis-trans* isomers of β-carotene. - Biochim. biophys. Acta *680* : 109 - 118, 1982.

*50238 - **KOZHOVA, O.M.** : Nekotorye osobennosti formirovaniya fitoplanktona vodokhranilishch. [Some properties of phytoplankton formation of water reservoirs.] - Vod. Resursy *1978*(3) : 94 - 106, 1978. [Primary production; in R.]

50239 - **KOZHOVA, O.M., PAUTOVA, V.N.** : Pervichnaya produktsiya v Bratskom vodokhranilishche i faktory, ee opredelyayushchie. [Primary production in Bratsk water reservoir and factors determining it.] - Vod. Resursy *1982* (1) : 128 - - 139, 1982. [In R.]

50240 - **KOZHOVA, O.M., ZAGORENKO, F.G.** : O sostoyanii fitoplanktona Baĭkala. [Phytoplankton composition in Baikal.] - Vod. Resursy *1982* (4) : 149 - 157, 1982.

50241 - **KOZLOWSKI, T.T.** : Water supply and tree growth. Part I Water deficits. - Forest Abstr. *43* (2) : 57 - 95, 1982. [Ps.]

50242 - **KOZLOWSKI, T.T.** : Water supply and tree growth. Part II Flooding. - Forest Abstr. *43* (3) : 145 - 161, 1982. [Ps.]

50243 - **KRAMBECK, H.-J.** : Solar energy and quanta in Baltic lakes. - Arch. Hydrobiol. *95* : 197 - 206, 1982.

50244 - **KRAMER, H.J.M., AMESZ, J.** : Anisotropy of the emission and absorption bands of spinach chloroplasts measured by fluorescence polarization and polarized excitation spectra at low temperature. - Biochim. biophys. Acta *682* : 201 - - 207, 1982.

50245 - **KRAMER, H.J.M., KINGMA, H., SWARTHOFF, T., AMESZ, J.** : Prompt and delayed fluorescence in pigment-protein complexes of a green photosynthetic bacterium. - Biochim. biophys. Acta *681* : 359 - 364, 1982.

50246 - **KRASICHKOVA, G.V., GILLER, Yu.E., TURBIN, N.V.** : Sravnitel'naya kharakteristika fotosinteticheskogo apparata razlichnykh form tritikale. [Comparative characteristics of the photosynthetic apparatus in different forms of triticale.] - Fiziol. Rast. *29* : 959 - 963, 1982. [In R, ab : E.]

50247 - **KRASNOVSKIĬ, A.A., BYSTROVA, M.I.** : Samosborka agregirovannykh struktur khlorofilla i ego analogov. [Self-assembly of aggregated structures of chlorophyll and its analogues.] - In : Sborka Predbiologicheskikh i Biologicheskikh Struktur. Pp. 48 - 62, 337. Nauka, Moskva 1982. [In R.]

50248 - **KRASNOVSKIĬ, A.A., MAL'TSEV, S.V.** : Fotovydelenie i fotopogloshchenie vodoroda khloroplastami bez ĕkzogennoĭ gidrogenazy. [Photoproduction and photoconsumption of hydrogen by chloroplasts without exogenous hydrogenase.] - Dokl. Akad. Nauk SSSR *267* : 506 - 509, 1982. [In R.]

50249 - **KRASNOVSKIĬ, A.A., SEMENOVA, A.N., NIKANDROV, V.V.** : Modelirovanie pervoĭ fotosistemy s pomoshch'yu liposom, soderzhashchikh khlorofill. [Modelling of photosystem 1 using liposomes containing chlorophyll.] - Dokl. Akad. Nauk SSSR *262* : 469 - 472, 1982. [In R.]

50250 - **KRASNOVSKIĬ, A.A. (ml.), VENEDIKTOV, E.A., CHERNENKO, O.M.** : Tushenie singletnogo kisloroda khlorofillami i porfirinami. [Quenching of singlet oxygen with chlorophylls and porphyrins.] - Biofizika *27* : 966 - 972, 1982. [In R, ab : E.]

*50251 - KRASNOVSKY, A.A. : Evolution of uphill electron transfer in photosynthesis.
          - In : KAGEYAMA, M., NAKAMURA, K., OSHIMA, T., UCHIDA, T. (ed.) : Science
          and Scientists: Essays by Biochemists, Biologists and Chemists. Pp. 47 - 55.
          Japan Sci. Soc. Press, Tokyo 1981.

 50252 - KRASNOVSKY, A.A., SEMENOVA, A.M., NIKANDROV, V.V. : Chlorophyll-containing
          liposomes: photoreduction of methyl viologen and photoproduction of hydro-
          gen. - Photobiochem. Photobiophys. 4 : 227 - 232, 1982.

 50253 - KRASNOVSKY, A.A.,Jr. : Delayed fluorescence and phosphorescence of plant
          pigments. - Photochem. Photobiol. 36 : 733 - 741, 1982.

 50254 - KRAUSE, G.H., VERNOTTE, C., BRIANTAIS, J.-M. : Photoinduced quenching of
          chlorophyll fluorescence in intact chloroplasts and algae. Resolution into
          two components. - Biochim. biophys. Acta 679 : 116 - 124, 1982.

 50255 - KRAUSE, H., HELML, M., GERHARDT, V., GEBHARDT, W. : In vivo measurements of
          photosynthetically active pigment systems in fresh waters using delayed
          fluorescence. - Arch. Hydrobiol. Beih. (Ergebn. Limnol.) 16 : 47 - 54,
          1982.

 50256 - KRAUSSE, G.W. : Induktion, Selektion und Nutzung von Mutanten bei Soja. 1.
          Mitt. Chlorophyllmutationsraten nach Samenbehandlung mit N-nitroso-N-Methyl-
          harnstoff in Abhängigkeit von der Nachkommenschaftsgröße und der Chimärenna-
          tur der M_1-Pflanzen. - Arch. Züchtungsforsch. 12 : 11 - 22, 1982.

 50257 - KRAWIARZ, K., OLEKSYN, J., KAROLEWSKI, P. : Effect of NO_2 on photosynthetic
          pigments in the leaves of Populus "Hybrida 280". - Arboret. kórnic. 26 :
          163 - 172, 1982.

 50258 - KREBBERS, E.T., LARRINUA, I.M., McINTOSH, L., BOGORAD, L. : The maize chlo-
          roplast genes for the β and ε subunits of the photosynthetic coupling fac-
          tor CF_1 are fused. - Nucl. Acids Res. 10 : 4985 - 5002, 1982.

 50259 - KREMER, B.P., MARKHAM, J.W. : Primary metabolic effects of cadmium in the
          brown alga, Laminaria saccharina. - Z. Pflanzenphysiol. 108 : 125 - 130,
          1982. [Ps.]

 50260 - KRENDELEVA, T.E., KUKARSKIKH, G.P., NIZOVSKAYA, N.V., TULBU, G.V. : O po-
          vrezhdenii reaktsiĭ H^+-ATPaznogo kompleksa khloroplastov nizkimi kontsentra-
          tsiyami glutarovogo al'degida. [Damage of the chloroplast H^+-ATPase complex
          by low concentrations of glutaraldehyde.] - Biokhimiya 47 : 904 - 910, 1982.
          [In R, ab : E.]

 50261 - KRENDELEVA, T.E., TULBU, G.V. : Ob uchastii fotosistemy I i fotosistemy II
          v protsessakh transformatsii énergii na membranakh tilakoidov. [Participation
          of photosystem I and photosystem II in processes of energy transduction in
          thylakoid membranes.] - Biol. Nauki 1982 (6) : 5 - 13, 1982. [In R.]

 50262 - KRETSCHMER, H., EHWALD, R., ERDMANN, B. : Zur Bedeutung des Assimilatstaus
          für die Nachmittagsdepression der Netto-CO_2-Aufnahmerate bei Zuckerrüben. -
          In : HOFFMANN, P., HIEKE, B. (ed.) : Photosynthese: Regulation und Evolution.
          (Colloquia Pflanzenphysiologie Nr.5.) Pp. 240 - 243. Humboldt-Universität,
          Berlin 1982.

 50263 - KRETSCHMER, H., SCHÄFER, W. : Langzeitwirkung von Wasserstress auf Zucker-
          rüben bei einem Feldversuch. - In : UNGER, K., SCHUH, J. (ed.) : Umwelt-
          -Stress. Pp. 245 - 248. Martin-Luther-Universität, Halle 1982.

 50264 - KREWER, J.A., HOLM, H.W. : The phosphorus-chlorophyll a relationship in pe-
          riphytic communities in a controlled ecosystem. - Hydrobiologia 94 : 173 -
          - 176, 1982.

 50265 - KRIEDEMANN, P.E. : Improving photosynthetic efficiency of horticultural
          crops. - In : Proceedings 21st International Horticultural Congress. Volume
          II. Pp. 623 - 632. International Society for Horticultural Science, Wage-
          ningen 1982.

 50266 - KRIEG, D.R. : Sorghum. - In : TEARE, I.D., PEET, M.M. (ed.) : Crop-Water
          Relations. Pp. 351 - 388. John Wiley & Sons, New York - Chichester - Bris-
          bane - Toronto - Singapore 1982. [Ps.]

50267 - KRIMM, S., DWIVEDI, A.M. : Infrared spectrum of the purple membrane: clue to a proton conduction mechanism? - Science *216* : 407 - 408, 1982. [Bacteriorhodopsin.]

50268 - KRINNER, M., HAUSKA, G., HURT, E., LOCKAU, W. : A cytochrome f-b_6 complex with plastoquinol-cytochrome c oxidoreductase activity from *Anabaena variabilis*. - Biochim. biophys. Acta *681* : 110 - 117, 1982.

50269 - KRIZEK, D.T. : Guidelines for measuring and reporting environmental conditions in controlled-environment studies. - Physiol. Plant. *56* : 231 - 235, 1982.

50270 - KRIZEK, D.T. : Plant response to atmospheric stress caused by waterlogging. - In : CHRISTIANSEN, M.N., LEWIS, C.F. (ed.) : Breeding Plants for Less Favorable Environments. Pp. 293 - 334. John Wiley & Sons, New York - Chichester - Brisbane - Toronto - Singapore 1982. [Ps.]

50271 - KROGMANN, D.W. : Evolution of photosynthetic catalysts in cyanobacteria. - In : SCHIFF, J.A., LYMAN, H. (ed.) : On the Origins of Chloroplasts. Pp. 27 - 34. Elsevier/North-Holland, New York - Amsterdam - Oxford 1982.

*50272 - KROT, V.F. : Rost eli v opyte s azotnym udobreniem. [Growth of fir under nitrogen nutrition.] - In : Voprosy Lesovedeniya i Lesovodstva v Karelii. Pp. 93 - 106, 196. Inst. Lesa Karel'skogo Filiala Akademii Nauk SSSR, Petrozavodsk 1975. [In R.]

*50273 - KRSTIĆ, B. : Uticaj abiotskih i biotskih činilaca na fotosintezu pšenice. [Effects of abiotic and biotic factors upon photosynthesis in wheat.] - Posebna Izd. srpska Akad. Nauka Umetn. *536*, Od. prirod.-mat. Nauka *53* : 21 - - 38, 1981. [In Serb., ab : E.]

*50274 - KRSTIĆ, B. : Uticaj nekih činilaca na proces asimilacije CO_2 kod šećerne repe. [Effect of some factors on CO_2 assimilation process in sugar beet.] - Posebna Izd. srpska Akad. Nauka Umetn. *538*, Od. prirod.-mat. Nauka *54* [BELIĆ, J. (ed.) : Fiziologija Šećerne Repe] : 17 - 38, 1981. [In Serb., ab : E.]

50275 - KRSTIĆ, B. : Nove koncepcije D.J.Arnona o fotofosforilaciji. [New concepts of D.I.Arnon on photophosphorylation.] - Savrem. Poljoprivreda *30* : 105 - - 111, 1982. [In Croat., ab : E.]

*50276 - KRUEGER, R.W., MILES, D. : Photosynthesis in fescue. I.High rate of electron transport and phosphorylation in chloroplasts of hexaploid plants. - Plant Physiol. *67* : 763 - 767, 1981.

50277 - KRUPA, Z. : The action of lipases on chloroplast membrane I. The release of plastocyanin from galactolipase-treated thylakoid membranes. - Photosynthesis Res. *3* : 95 - 104, 1982.

50278 - KRUPA, Z., RUSZKOWSKI, M., GILOWSKA-JUNG, E. : The effect of chromate on the synthesis of plastid pigments and lipoquinones in *Zea mays* L. seedlings. - Acta Soc. Bot. Pol. *51* : 275 - 281, 1982.

50279 - KSHETRAPAL, S., TANWAR, T.C., JAIN, U. : On the occurrence of Kranz type of leaf anatomy in certain species of Rajasthan. - Nat. Acad. Sci. Lett. *5* : 39 - - 40, 1982.

50280 - KU, S.B., EDWARDS, G.E. : Quantum requirement of photosynthetic bacteria and plants. - In : ZABORSKY, O.R. (ed.) : CRC Handbook of Biosolar Resources. Vol.I/1. Pp. 37 - 54. CRC Press, Boca Raton 1982.

50281 - KUCEY, R.M.N., PAUL, E.A. : Carbon flow, photosynthesis, and N_2 fixation in mycorrhizal and nodulated faba beans (*Vicia faba* L.). - Soil Biol. Biochem. *14* : 407 - 412, 1982.

50282 - KUDZHMAUSKAS, Sh.P. : Teoreticheskoe issledovanie kinetiki zakhvata lokalizovannykh èksitonov v fotosinteziruyushchikh sistemakh. [Theoretical study of the kinetics of trapping of localized excitons in photosynthesizing systems.] - Liet. Fiz. Rinkinys *22* (5) : 32 - 37, 1982. [In R, ab : E, Lithu.]

50283 - **KUFER, W., KRAUSS, C., SCHEER, H.** : Zwei milde, regioselektive Abbaumethoden von Biliprotein-Chromophoren. - Angew. Chem. *94* : 455 - 456, 1982.

50284 - **KUFNER, R., CZYGAN, F.-C., SCHNEIDER, L.** : Plastiden-Metamorphose und Pigmentgehalt in Kalluskulturen von *Ruta graveolens* L. - Ber. deut. bot. Ges. *95* : 397 - 411, 1982.

*50285 - **KÜHBAUCH, W., IMHOFF, H., VOIGTLÄNDER, G.** : Verteilung der ^{14}C-Aktivität nach $^{14}CO_2$-Applikation auf die Nichtstrukturkohlenhydrate in Stumpfblättrigen Ampfer, Wiesenkerbel und Bärenklau in verschiedenen Wachstumsstadien. - Z. Acker- Pflanzenbau *149* : 488 - 502, 1980.

50286 - **KUJIRA, Y., KANDA, M.** : The relationship between the root growth analysis and the growth analysis in *Lolium multiflorum* LAM. - Jap. J. Crop Sci. *51* : 1 - 7, 1982.

50287 - **KULAKOVA, O.Yu., CHUPROVA, N.A., REPYAKH, S.M., BARABASH, N.D.** : Khimicheskiĭ sostav khvoi kedrovogo stlanika. [Chemical composition of *Pinus pumila* needles.] - Khim. Drev. *1982* (4) : 107 - 108, 1982. [Chl, Car; in R.]

50288 - **KUMAKOV, V.A.** : Fotosinteticheskaya deyatel'nost' rasteniĭ v aspekte selektsii. [Photosynthetic activity of plants in the aspect of selection.] - In : Fiziologiya Fotosinteza. Pp. 283 - 293. Nauka, Moskva 1982. [In R.]

50289 - **KUMAR, A., TABITA, F.R., VAN BAALEN, C.** : Isolation and characterization of heterocysts from *Anabaena* sp. strain CA. - Arch. Microbiol. *133* : 103 - - 109, 1982. [Bil.]

*50290 - **KUMAR, A., WILSON, D., COCKING, E.C.** : Polypeptide composition of fraction 1 protein of the somatic hybrid between *Petunia parodii* and *Patunia parviflora*. - Biochem. Genet. *19* : 255 - 261, 1981.

*50291 - **KUMAR, D., TIESZEN, L.L.** : Photosynthesis in *Coffea arabica*. II. Effects of water stress. - Exp. Agr. *16* : 21 - 27, 1980.

50292 - **KUMAR, K.B., KHAN, P.A.** : Peroxidase & polyphenol oxidase in excised ragi (*Eleusine corocana* cv_PR 202) leaves during senescence. - Indian J. exp. Biol. *20* : 412 - 416, 1982. [Chl.]

50293 - **KUMAR, K.B., KHAN, P.A.** : Effect of insecticides, oxydementon-methyl and dimethoate, an chlorophyll retention & hydrogen peroxide utilization in ragi (*Eleusine coracana* GAERTN. cv_PR 202) leaves during senescence. - Indian J. exp. Biol. *20* : 889 - 893, 1982.

50294 - **KUMAR, N.** : Photosynthetic activity of fruiting organs. 1. Photosystem 2 activity of chloroplasts from flag leaf and developing seeds of barley. - Photosynthetica *16* : 561 - 563, 1982.

50295 - **KUMAR, N., MOHANTY, P.** : Effect of 6-benzyl aminopurine on the stabilization of absorption spectrum and Hill activity of isolated chloroplasts. - Biochem. Physiol. Pflanzen *177* : 137 - 142, 1982.

50296 - **KUMAR, P.A.** : Correlation between photosynthetic rate and other physiological characters in seven tobacco cultivars. - Photosynthetica *16* : 564 - 567, 1982.

*50297 - **KUMAZAWA, K., JANAGISAWA, K.** : [Determination of ^{13}C concentration in samples by means of infrared absorption.] - J. Sci. Soil Manure Jap. *52* : 74 - - 76, 1981. [In Jap.]

50298 - **KUMAZAWA, S., MITSUI, A.** : Hydrogen metabolism of photosynthetic bacteria and algae. - In : ZABORSKY, O.R. (ed.) : CRC Handbook of Biosolar Resources. Vol.I/1. Pp. 299 - 316. CRC Press, Boca Raton 1982.

50299 - **KUME, N., KATOH, T.** : Dissociation kinetics of *Anabaena* phycobilisomes. - Plant Cell Physiol. *23* : 803 - 812, 1982.

50300 - **KUMMEROW, J., AVILA, G., ALJARO, M.-E., ARAYA, S., MONTENEGRO, G.** : Effect of fertilizer on fine root density and shoot growth in Chilean mattoral. - Bot. Gaz. *143* : 498 - 504, 1982. [Dry-matter accumulation.]

50301 - **KUNDU, A.L., CHATTERJEE, B.N.** : Growth analysis of turmeric as a sole crop and in mixture with other crops. - Indian J. agr. Sci. *52* : 584 - 589, 1982.

50302 - **KUNII, H.** : Life cycle and growth of *Potamogeton crispus* L. in a shallow pond, Ojaga-ike. - Bot. Mag. (Tokyo) *95* : 109 - 124, 1982. [Dry-matter production.]

50303 - **KUNKEL, D.D.** : Thylakoid centers: structures associated with the cyanobacterial photosynthetic membrane system. - Arch. Microbiol. *133* : 97 - 99, 1982.

50304 - **KÜPPERS, M., HALL, A.E., SCHULZE, E.-D.** : Effects of day-to-day changes in root temperature on leaf conductance to water vapour and CO_2 assimilation rates of *Vigna unguiculata* L. WALP. - Oecologia *52* : 116 - 120, 1982.

50305 - **KURATA, K., ROTH, R.** : Theoretische Untersuchung der Turbulenz innerhalb eines Pflanzenbestandes (I): Modelierung der Turbulenzintensitätsprofile innerhalb eines Pflanzenbestandes. - Arch. Meteorol. Geophys. Bioklimatol., Ser. B *31* : 99 - 112, 1982.

50306 - **KURETS, V.K.** : Modelirovanie - odin iz êffektivnykh metodov intensifikatsii êkologo-fiziologicheskikh issledovaniĬ. [Modelling - one of effective methods of intensification of ecologo-physiological studies.] - In : Vliyanie Faktorov VneshneĬ Sredy i Fiziologicheski Aktivnykh Veshchestv na Termorezistentnost' i Produktivnost' RasteniĬ. Pp. 6 - 13. Institut Biologii, Karel'skiĬ Filial Akademii Nauk SSSR, Petrozavodsk 1982. [Ps; in R.]

50307 - **KURKIN, K.A., BOGATYREVA, V.V.** : Dinamika prirostov nadzemnoĬ massy oroshaemykh lugovykh fitotsenozov. [Increment dynamics of the aboveground biomass of the irrigated meadow phytocoenoses.] - Bot. Zh. *67* : 1618 - 1627, 1982. [In R, ab : E.]

50308 - **KURODA, K., FUKUCHI, M.** : Vertical distribution of chlorophyll *a* in the Indian sector of the Antarctic Ocean in 1972 - 1973. - Antarctic Record *74* : 127 - 142, 1982.

50309 - **KUROIWA, T., KAWANO, S., NISHIBAYASHI, S., SATO, C.** : Epifluorescent microscopic evidence for maternal inheritance of chloroplast DNA. - Nature *298* : 481 - 483, 1982.

*50310 - **KUROIWA, T., SUZUKI, T.** : Circular nucleoids isolated from chloroplasts in a brown alga *Ectocarpus siliculosus*. - Exp. Cell Res. *134* : 457 - 461, 1981.

50311 - **KUSCHMITZ, D., HESS, B.** : *Trans-cis* isomerisation of the retinal chromophore of bacteriorhodopsin during the photocycle. - FEBS Lett. *138* : 137 - 140, 1982.

50312 - **KUSCHMITZ, D., HESS, B.** : Spectroscopic methods for protonation state determination. - In : COLOWICK, S.P., KAPLAN, N.O. (ed.) : Methods in Enzymology. Vol.88. Pp. 254 - 265. Academic Press, New York - London - Paris - San Diego - - San Francisco - São Paulo - Sydney - Tokyo - Toronto 1982. [Bacteriorhodopsin.]

50313 - **KUWABARA, T., MURATA, N.** : An improved purification method and a further characterization of the 33-kilodalton protein of spinach chloroplasts. - Biochim. biophys. Acta *680* : 210 - 215, 1982.

50314 - **KUWABARA, T., MURATA, N.** : Inactivation of photosynthetic oxygen evolution and concomitant release of three polypeptides in the photosystem II particles of spinach chloroplasts. - Plant Cell Physiol. *23* : 533 - 539, 1982.

50315 - **KUWABARA, T., MURATA, N.** : Evidence against identity of the atrazine receptor protein with the 33-kilodalton protein from spinach chloroplasts. - Plant Cell Physiol. *23* : 663 - 667, 1982.

50316 - **KUZNETSOV, V.N., IL'NITSKIĬ, O.A., RABOTYAGOV, V.D., SEMIN, V.S.** : Nekotorye metodicheskie aspekty analiza shirokopolosnykh spektrov rastitel'nykh materialov. [Some methodical aspects of broad-band spectra analysis of plant materials.] - Byull. gos. nikitsk. bot. Sada *47* : 81 - 87, 1982. [In R, ab : E.]

50317 - KUZNETSOVA, E.A. : Flyuorestsentsiya list'ev vysshikh rastenii pri povyshen-
        nykh temperaturakh. [Fluorescence of higher plant leaves at raised tempera-
        tures.] - Biofizika 27 : 809 - 817, 1982. [In R, ab : E.]

50318 - KUZNETSOVA, E.A., KUKUSHKIN, A.K. : Vliyanie ionnogo sostava sredy na tushe-
        nie flyuorestsentsii 9-aminoakridina i atebrina v khloroplastakh. [Effect of
        ionic composition of medium on quenching of 9-aminoacridine and atebrin fluo-
        rescence in chloroplasts.] - Biofizika 27 : 539 - 541, 1982. [In R, ab : E.]

50319 - KYLE, D.J., HAWORTH, P., ARNTZEN, C.J. : Thylakoid membrane protein phospho-
        rylation leads to a decrease in connectivity between photosystem II reaction
        centers. - Biochim. biophys. Acta 680 : 336 - 342, 1982.

50320 - KYLE, D.J., ZALIK, S. : Development of photochemical activity in relation
        to pigment and membrane protein accumulation in chloroplasts of barley and
        its virescent mutant. - Plant Physiol. 69 : 1392 - 1400, 1982.

50321 - KYLE, D.J., ZALIK, S. : Photosystem II activity, plastoquinone A levels,
        and fluorescence characterization of a virescent mutant of barley. - Plant
        Physiol. 70 : 1026 - 1031, 1982.

*50322 - LAAKKONEN,A., MÄLKKI, P., NIEMI, Å. : Studies on the sinking, degradation
        and sedimentation of organic matter off Hanko peninsula, entrance to the
        Gulf of Finland, in 1979. - Meri 9 : 3 - 42, 1981/2. [Chl.]

*50323 - LAASCH, H., PFISTER, K., URBACH, W. : Comparative binding of photosystem II -
        - herbicides to isolated thylakoid membranes and intact green algae. - Z.
        Naturforsch. 36 C : 1041 - 1049, 1981.

50324 - LADYGIN, V.G., BIL', K.Ya. : Antennaya forma khlorofilla fotosistemy II v
        khloroplastakh. [Antenna form of chlorophyll of the photosystem II in chlo-
        roplasts.] - Biofizika 27 : 37 - 41, 1982. [In R, ab : E.]

50325 - LADYGIN, V.G., BIL', K.Ya., BOZHOK, G.V. : Formy khlorofilla i struktura
        khloroplastov mutantov Pisum sativum s poterei aktivnosti fotosistemy I
        ili fotosistemy II. [Chlorophyll species and chloroplast structure in Pisum
        sativum mutants lacking activities in photosystem I or photosystem II.] -
        Fiziol. Rast. 29 : 479 - 487, 1982. [In R, ab : E.]

50326 - LADYGIN, V.G., SEMENOVA, G.A., CHEMERILOVA, V.I., KVITKO, K.V., STOLBOVA,
        A.V., TAGEEVA, S.V., IROSHNIKOVA, G.A. : Variabel'nost' strukturnoi organi-
        zatsii khloroplastov u allel'nykh svetochuvstvitel'nykh mutantov khlamido-
        monady. [Structural organization of chloroplast variability in light-sensi-
        tive allelic mutants of Chlamydomonas reinhardii.] - Tsitologiya 24 : 391 -
        - 399, 1982. [In R, ab : E.]

50327 - LAHAV, E., TROCHOULIAS, T. : The effect of temperature on growth and dry
        matter production of avocado plants. - Aust. J. agr. Res. 33 : 549 - 558,
        1982. [Leaf resistance.]

50328 - LAINSON, R.A., THORNLEY, J.H.M. : A model for leaf expansion in cucumber. -
        Ann. Bot. 50 : 407 - 425, 1982. [Ps.]

50329 - LAISK, A.Kh. : Sootvetstvie fotosinteziruyushchei sistemy usloviyam sredy.
        [Correspondence of the photosynthetic system to environmental conditions.]
        - In : Fiziologiya Fotosinteza. Pp. 221 - 234. Nauka, Moskva 1982. [In R.]

50330 - LAJOLO, F.M., LANFER MARQUEZ, U.M. : Chlorophyll degradation in a spinach
        system at low and intermediate water activities. - J. Food Sci. 47 : 1995 -
        - 1998, 2003, 1982.

50331 - LAKSO, A.N. : Precautions on the use of excised shoots for photosynthesis
        and water relations measurements of apple and grape leaves. - 'HortScience
        17 : 368 - 370, 1982.

50332 - LAKSO, A.N., PRATT, C., PEARSON, R.C., POOL, R.M., SEEM, R.C., WELSER, M.J.:
        Photosynthesis, transpiration, and water use efficiency of mature grape
        leaves infected with Uncinula necator (powdery mildew). - Phytopathology
        72 : 232 - 236, 1982.

50333 - **LAM, E., FRY, I., PACKER, L., MUKOHATA, Y.** : Comparison of the O_{640} photo-
-intermediate and acid-induced species in membrane patches from *Halobacterium halobium* S_9 and R_1mW strains. - FEBS Lett. *146* : 106 - 110, 1982.

50334 - **LAM, E., MALKIN, R.** : Ferredoxin-mediated reduction of cytochrome b-563 in
a chloroplast cytochrome b-563/f complex. - FEBS Lett. *141* : 98 - 101,
1982.

50335 - **LAM, E., MALKIN, R.** : *In vitro* reconstitution of the electron pathway from
water to cytochrome f of spinach thylakoids. - FEBS Lett. *144* : 190 - 194,
1982.

50336 - **LAM, E., MALKIN, R.** : Photoreactions of cytochrome b_6 in systems using resolved chloroplast electron-transfer complexes. - Biochim. biophys. Acta
682 : 378 - 386, 1982.

50337 - **LAM, E., MALKIN, R.** : Reconstitution of the chloroplast noncyclic electron
transport pathway from water to NADP with three integral protein complexes.
- Proc. nat. Acad. Sci. USA *79* : 5494 - 5498, 1982.

50338 - **LAMBERS, H., SIMPSON, R.J., BEILHARZ, V.C., DALLING, M.J.** : Translocation
and utilization of carbon in wheat (*Triticum aestivum*). - Physiol. Plant.
56 : 18 - 22, 1982.

50339 - **LAMBERT-CASTEL, F., PENOT, M.** : Action des pétroles de l'Amoco Cadiz sur
la croissance et certains aspects du métabolisme d'une algue phytoplancton
nique *Pavlova lutheri* (DROOP) GREEN. - Publ. CNEXO (Actes Colloq.) *14*
(Indices Biochimiques et Milieux Marins) : 411 - 422, 1982. [Ps.]

50340 - **LAMBOU, V.W., HERN, S.C., TAYLOR, W.D., WILLIAMS, L.R.** : Chlorophyll, phosphorus, Secchi disk, and trophic state. - Water Resour. Bull. *18* : 807 -
- 813, 1982.

50341 - **LANARAS, T., CODD, G.A.** : Variations in ribulose 1,5-bisphosphate carboxylase protein levels, activities and subcellular distribution during photoautotrophic batch culture of *Chlorogloeopsis fritschii*. - Planta *154* : 284 -
- 288, 1982.

50342 - **LANDERS, D.H.** : Effects of naturally senescing aquatic macrophytes on nutrient chemistry and chlorophyll a of surrounding waters. - Limnol. Oceanogr.
27 : 428 - 439, 1982.

50343 - **LANDOLT, W.** : Der Einfluß einer praxisnahen SO_2-Begasung auf das $^{14}CO_2$-Fixie-
rungsmuster von Buchen (*Fagus sylvatica* L.). - Europe. J. Forest Pathol.
12 : 331 - 339, 1982.

*50344 - **LANDSBERG, J.J.** : Limits to apple yields imposed by weather. - In : **HURD,
R.G., BISCOE, P.V., DENNIS, C.** (ed.) : Opportunities for Increasing Crop
Yields. Pp. 161 - 180. Pitman Publishing Ltd., London 1979. [Photosyntha-
tes.]

50345 - **LANDSBERG, P.T., TONGE, G.** : Kinetic photochemical and photovoltaic energy
conversion models. - Photochem. Photobiol. *35* : 769 - 781, 1982.

B50346 - **LANGE, O.L., NOBEL, P.S., OSMOND, C.B., ZIEGLER, H.** (ed.) : Physiological
Plant Ecology II. Water Relations and Carbon Assimilation.(Encyclopedia
of Plant Physiology. New Series. Vol. 12 B.) - Springer-Verlag, Berlin -
- Heidelberg - New York 1982.

50347 - **LANGE, O.L., TENHUNEN, J.D.** : Water relations and photosynthesis in desert
lichens. - J. Hattori bot. Lab. *53* : 309 - 313, 1982.

50348 - **LANGE, O.L., TENHUNEN, J.D., BRAUN, M.** : Midday stomatal closure in Mediterranean type sclerophylls under simulated habitat conditions in an environmental chamber. I. Comparison of the behaviour of various European Mediterranean species. - Flora *172* : 563 - 579, 1982. [Ps, Chl.]

50349 - **LANYI, J.K.** : Spectrophotometric determination of halorhodopsin in *Halo-
bacterium halobium* membranes. - In : **COLOWICK, S.P., KAPLAN, N.O.** (ed.) :
Methods in Enzymology. Vol.88. Pp. 439 - 443. Academic Press, New York -
- London - Paris - San Diego - San Francisco - São Paulo - Sydney - Tokyo -
- Toronto 1982.

50350 - **LARKUM, A.W.D., ANDERSON, J.M.** : The reconstitution of a photosystem II protein complex, P-700-chlorophyll a-protein complex and light-harvesting chlorophyll a/b-protein. - Biochim. biophys. Acta *679* : 410 - 421, 1982.

☆50351 - **LARSON, R.A., BOTT, T.L., HUNT, L.L., ROGENMUSER, K.** : Photooxidation products of a fuel oil and their antimicrobial activity. - Environm. Sci. Technol. *13* : 965 - 969, 1979. [Ps.]

☆50352 - **LARSSON, K., PUANG-NGERN, S.** : The aqueous system of monogalactosyl diglycerides and digalactosyl diglycerides - significance to the structure of the thylakoid membrane. - In : APPELQVIST, L.-Å., LILJENBERG, C. (ed.) : Advances in the Biochemistry of Plant Lipids. Pp. 27 - 33. Elsevier/North--Holland Biomedical Press, Amsterdam - New York - Oxford 1979.

50353 - **LARSSON, M., INGEMARSSON, B., LARSSON, C.-M.** : Photosynthetic energy supply for NO_3^- assimilation in *Scenedesmus*. - Physiol. Plant. *55* : 301 - 308, 1982.

50354 - **LARSSON, M., LARSSON, C.M., ULLRICH, W.R.** : Regulation by amino acids of photorespiratory ammonia and glycolate release from *Ankistrodesmus* in the presence of methionine sulfoximine. - Plant Physiol. *70* : 1637 - 1640, 1982.

50355 - **LARSSON, U., HAGSTRÖM, Å.** : Fractionated phytoplankton primary production, exudate release and bacterial production in a Baltic eutrophication gradient. - Mar. Biol. *67* : 57 - 70, 1982.

50356 - **LASTRA, O., GÓMEZ, M., LÓPEZ-GORGÉ, J., DEL RÍO, L.A.** : Catalase activity and isozyme pattern of the metalloenzyme system, superoxide dismutase, as a function of leaf development during growth of *Pisum sativum* L. plants. - Physiol. Plant. *55* : 209 - 213, 1982.

50357 - **LÁSZTITY, B.** : Szezonális változások az öszi árpa tápelemtartalmában. [Seasonal changes in the nutrient content of of winter barley.] - Növénytermelés *31* : 155 - 164, 1982. [Production; in Hung., ab : E.]

50358 - **LAVALLEE, D.K., MCDONOUGH, T.J.,Jr., CIOFFI, L.** : Fluorometric properties of N-methyltetraphenylporphine and several derivatives: evaluation as standards for determination of chlorophyll concentrations. - Appl. Spectrosc. *36* : 430 - 435, 1982.

50359 - **LAVERGNE, J.** : Two types of primary acceptors in chloroplasts photosystem II. I. Different recombination properties. - Photobiochem. Photobiophys. *3* : 257 - 271, 1982.

50360 - **LAVERGNE, J.** : Two types of primary acceptors in chloroplasts photosystem II. II. Reduction in two successive photoacts. - Photobiochem. Photobiophys. *3* : 273 - 285, 1982.

50361 - **LAVERGNE, J.** : Interaction of exogenous benzoquinone with photosystem II in chloroplasts. The semiquinone form acts as a dichlorophenyldimethylurea--insensitive secondary acceptor. - Biochem. biophys. Acta *679* : 12 - 18, 1982.

50362 - **LAVERGNE, J.** : Mode of action of 3-(3,4-dichlorophenyl)-1,1-dimethylurea. Evidence that the inhibitor competes with plastoquinone for binding to a common site on the acceptor side of Photosystem II. - Biochim. biophys. Acta *682* : 345 - 353, 1982.

50363 - **LAVOREL, J., DENNERY, J.-M.** : The slow component of photosystem II luminescence. A process with distributed rate constant? - Biochim. biophys. Acta *680* : 281 - 289, 1982.

50364 - **LAVOREL, J., LAVERGNE, J., ETIENNE, A.-L.** : A reflection on several problems of luminescence in photosynthetic systems. - Photobiochem. Photobiophys. *3* : 287 - 314, 1982.

50365 - **LAVOREL, J., SEIBERT, M.** : Patterns of oxygen emission from active oxygen--evolving photosystem II particles subjected to sequences of flashes. - FEBS Lett. *144* : 101 - 103, 1982.

50366 - **LAW, R.M., MANSFIELD, T.A.** : Oxides of nitrogen and the greenhouse atmosphere. - In : UNSWORTH, M.H., ORMROD, D.P. (ed.) : Effects of Gaseous Air Pollution in Agriculture and Horticulture. Pp. 93 - 112. Butterworth Scientific, London - Boston - Sydney - Wellington - Durban - Toronto 1982. [Stomatal resistance.]

50367 - **LAWN, R.J.** : Response of four grain legumes to water stress in south-eastern Queensland. II. Plant growth and soil water extraction patterns. - Aust. J. agr. Res. *33* : 497 - 509, 1982. [Dry-matter accumulation, leaf area index.]

50368 - **LAWYER, A.L., CORNWELL, K.L., GEE, S.L., BASSHAM, J.A.** : Effects of glycine hydroxamate, carbon dioxide, and oxygen on photorespiratory carbon and nitrogen metabolism in spinach mesophyll cells. - Plant Physiol. *69* : 1136 - - 1139, 1982.

50369 - **LAZAREK, S.** : Structure and productivity of epiphytic algal communities on *Lobelia dortmanna* L. in acidified and limed lakes. - Water,Air,Soil Pollut. *18* : 333 - 342, 1982. [Chl.]

50370 - **LaZERTE, B.D., SZALADOS, J.E.** : Stable carbon isotope ratio of submerged freshwater macrophytes. - Limnol. Oceanogr. *27* : 413 - 418, 1982.

50371 - **LEA, P.J., MILLS, W.R., WALLSGROVE, R.M., MIFLIN, B.J.** : Assimilation of nitrogen and synthesis of amino acids in chloroplasts and cyanobacteria (blue-green algae). - In : SCHIFF, J.A., LYMAN, H. (ed.) : On the Origins of Chloroplasts. Pp. 149 - 178. Elsevier/North-Holland, New York - Amsterdam - Oxford 1982.

50372 - **LEACH, J.E., PARKINSON, K.J., WOODHEAD, T.** : Photosynthesis, respiration and evaporation of a field-grown potato crop. - Ann. appl. Biol. *101* : 377 - - 390, 1982.

50373 - **LEAN, D.R.S., MURPHY, T.P., PICK, F.R.** : Photosynthetic response of lake plankton to combined nitrogen enrichment. - J. Phycol. *18* : 509 - 521, 1982. [Chl.]

50374 - **LEAVITT, S.W., LONG, A.** : Evidence for $^{13}C/^{12}C$ fractionation between tree leaves and wood. - Nature *298* : 742 - 744, 1982.

50375 - **LE BOUTEILLER, A., HERBLAND, A.** : Diel variation of chlorophyll *a* as evidenced from a 13-day station in the equatorial Atlantic Ocean. - Oceanol. Acta *5* : 433 - 441, 1982.

50376 - **LECHOWICZ, M.J.** : The effects of simulated acid precipitation on photosynthesis in the caribou lichen *Cladina stellaris* (OPIZ) BRODO. - Water, Air, Soil Pollut. *18* : 421 - 430, 1982.

50377 - **LECHOWICZ, W.** : Blue and red light as factor in biogenesis of photosynthetic apparatus of maize seedlings. - In : HOFFMANN, P., HIEKE, B. (ed.) : Photosynthese: Regulation und Evolution. (Colloquia Pflanzenphysiologie Nr.5.). Pp. 176. Humboldt-Universität, Berlin 1982.

50378 - **LECLERC, J.C., DÖHLER, G., ROSSLENBROICH, H.-J.** : $^{14}CO_2$-fixation under various limited light conditions in *Porphyridium cruentum*. - Plant Sci.Lett. *24* : 225 - 229, 1982.

50379 - **LECLERCQ, J.-M., DUPUIS, P., SÂNDORFY, C.** : Possibility of a double well potential in the proton bridge of visual pigments and bacteriorhodopsin. - Croat. chem. Acta *55* : 105 - 119, 1982.

50380 - **LEDDET, C., GENEVÈS, L.** : Influence de basses températures (+6 °C et +2 °C) sur l'organisation fine des mitochondries et des chloroplastes dans des tissus d'*Ephedra* en culture *in vitro*. - Ann. Sci. nat., Bot.,13 Sér. *4* : 27 - - 49, 1982.

50381 - **LEDOIGT, G., FREYSSINET, G.** : Plastid ribosome. - Biol. Cell *46* : 215 - 238, 1982.

50382 - **LEDYAÏKINA, N.A., VOZNESENSKIÏ, B.L.** : Vliyanie kisloroda na uglekislotnyĭ gazoobmen pustynnykh rasteniĭ. [Effect of oxygen on CO_2 exchange in desert plants.] - Fiziol. Rast. *29* : 1134 - 1140, 1982. [In R, ab : E.]

50383 - **LEE, C.H., SUGIURA, A., TOMANA, T.** : [Effect of flooding on the growth and some physiological changes of young apple rootstocks.] - J. jap. Soc. hort. Sci. *51* : 270 - 277, 1982. [Chl; in Jap., ab : E.]

50384 - **LEE, C.Y.** : Comparison of two correction methods for the bias due to the logarithmic transformation in the estimation of biomass. - Can. J. Forest Res. *12* : 326 - 331, 1982.

50385 - **LEE, Y.-K., PIRT, S.J.** : Maximum photosynthetic efficiency of biomass growth: A criticism of some measurements. - Biotechnol. Bioeng. *24* : 507 - - 509, 1982.

50386 - **LEECH, R.M., LEESE, B.M.** : Isolation of etioplasts from maize. - In : EDELMAN, M., HALLICK, R.B., CHUA, N.-H. (ed.) : Methods in Chloroplast Molecular Biology. Pp. 221 - 237. Elsevier Biomedical Press, Amsterdam - New York - Oxford 1982.

✺50387 - **LEECH, R.M., THOMSON, W.W., PLATT-ALOIA, K.A.** : Observations on the mechanics of chloroplast division in higher plants. - New Phytol. *87* : 1 - 9, 1981.

50388 - **LEEDALE, G.F.** : Special cytology: morphology and morphogenesis of eukaryotic algal cells. - Progr. Bot. *44* : 32 - 42, 1982. [Chloroplast.]

50389 - **LEEGOOD, R.** : Ribulose-1,5-bisphosphate (RuBP) carboxylase-oxygenase. - In : COOMBS, J., HALL, D.O. (ed.) : Techniques in Bioproductivity and Photosynthesis. Pp. 152 - 158. Pergamon Press, Oxford - New York - Toronto - Sydney - - Paris - Frankfurt 1982.

50390 - **LEEGOOD, R.C., EDWARDS, G.E., WALKER, D.A.** : Chloroplasts and protoplasts. - In : COOMBS, J., HALL, D.O. (ed.) : Techniques in Bioproductivity and Photosynthesis. Pp. 94 - 111. Pergamon Press, Oxford - New York - Toronto - - Sydney - Paris - Frankfurt 1982.

50391 - **LEEGOOD, R.C., KOBAYASHI, Y., NEIMANIS, S., WALKER, D.A., HEBER, U.** : Co- -operative activation of chloroplast fructose-1,6-bisphosphatase by reductant, pH and substrate. - Biochim. biophys. Acta *682* : 168 - 178, 1982.

50392 - **LEEGOOD, R.C., WALKER, D.A.** : Regulation of fructose-1,6-bisphosphatase activity in leaves. - Planta *156* : 449 - 456, 1982.

50393 - **LEE-KADEN, J., SIMONIS, W.** : Amino acid uptake and energy coupling dependent on photosynthesis in *Anacystis nidulans*. - J. Bacteriol. *151* : 229 - 236, 1982.

50394 - **LEFORT-TRAN, M.** : Ontogeny of chloroplastic envelope. - In : AKOYUNOGLOU, G., EVANGELOPOULOS,A.E., GEORGATSOS, J., PALAIOLOGOS, G., TRAKATELLIS, A., TSIGANOS, C.P. (ed.) : Cell Function and Differentiation, Part B. Biogenesis of Energy Transducing Membranes and Membrane and Protein Energetics. Pp. 67 - - 76. Alan R. Liss Inc., New York 1982.

50395 - **LEICKNAM, J.P., ANITOFF, O.E., GALLICE, M.J., HENRY, M., RUTLEDGE, D., TAYEB, A.E.K.** : Interaction de porphyrines de magnesium et de la chlorophylle "a" avec des molecules polaires. II-Complexes formes en solution avec des molécules polaires. Grandeurs thermodynamiques caracteristiques de la formation des complexes entre la pyridine et la tetra-phenyl-porphyrine de magnesium. - J. Chim. phys. *79* : 171 - 180, 1982.

50396 - **LEINA, G.D., YUDINA, O.S.** : Dykhanie i ego rol' v produktsionnom protsesse dvukh sortov yarovoĭ pshenitsy. [Respiration and its role in the production process of two winter wheat varieties.] - Fiziol. Biokhim. kul't. Rast. *14* : 491 - 497, 1982. [In R, ab : E.]

50397 - **LEKHOTSKI, Ė., RAKHIMBERDIEVA, M.G., KARAPETYAN, N.V.** : Spetsifika funktsionirovaniya fotosistemy 2 u khlorelly, vyrashchennoĭ v prisutstvii tserulenina ili piridazinonovykh gerbitsidov. [Specific functioning of photosystem 2 in *Chlorella* grown in presence of cerulenin or pyridazinone herbicides.] - Prikl. Biokhim. Mikrobiol. *18* : 405 - 410, 1982. [In R, ab : E.]

50398 - LEKHOTSKI, É., RAKHIMBERDIEVA, M.G., KARAPETYAN, N.V. : Vliyanie blokirova-
niya piridazinonami sinteza karotinoidov na formirovanie fotosistemy 2 v
list'yakh yachmenya. [Effect of blocking of carotenoid synthesis with pyridazinones on photosystem II formation in barley leaves.] - Fiziol. Rast.
29: 682 - 686, 1982. [In R, ab : E.]

50399 - LEMKE, H.D., BERGMEYER, J., OESTERHELT, D. : Determination of modified po-
sitions in the polypeptide chain of bacteriorhodopsin. - In : COLOWICK, S.
P., KAPLAN, N.O. (ed.) : Methods in Enzymology. Vol.88. Pp. 89 - 98. Academic Press, New York - London - Paris - San Diego - San Francisco - São Paulo -
Sydney - Tokyo - Toronto 1982.

50400 - LEMKE, H.-D., BERGMEYER, J., STRAUB, J., OESTERHELT, D. : Reversible inhibition of the proton pump bacteriorhodopsin by modification of tyrosine 64.
- J. biol. Chem. *257* : 9384 - 9388, 1982.

*50401 - LEMKE, H.-D., OESTERHELT, D. : The role of tyrosine residues in the funct-
ion of bacteriorhodopsin. Specific nitration of tyrosine 26. - Europe. J.
Biochem. *115* : 595 - 604, 1981.

50402 - LEMOINE, Y., DUBACQ, J.-P., ZABULON, G. : Changes in light-harvesting capacities and Δ3-*trans*-hexadecenoic acid content in dark- and light-grown
Picea abies. - Physiol. vég. *20* : 487 - 503, 1982.

50403 - LENCI, F. : Photomovements of microorganisms. - In : HÉLÈNE, C., CHARLIER,M.,
MONTENAY-GARESTIER,T., LAUSTRIAT,G. (ed.): Trends in Photobiology. Pp. 421 -
- 435. Plenum Press, New York - London 1982. [Ps, bacteriorhodopsin.]

50404 - LENDZIAN, K.J. : Gas permeability of plant cuticles. Oxygen permeability.
- Planta *155* : 310 - 315, 1982.

50405 - LENZ, F., ANTOSZEWSKI, R. : Effect of low oxygen on green pepper plants. -
Gartenbauwissenschaft *47* : 1 - 4, 1982. [Ps.]

50406 - LENZ, F., NOGA, G. : Photosynthese und Atmung bei Apfelfrüchten. - Erwerbs-
obstbau *24* : 198 - 200, 1982.

50407 - LEONARDI, S. : Sulla produttivita primaria in colture erbacee della Sicilia.
[Primary production in herbaceous cultures of Sicilia.] - Ecol. méditer.
8 : 143 - 164, 1982. [In Ital., ab : E.]

50408 - LEONARDI, S., RAPP, M. : Phytomasse et minéralomasse d'un taillis de chêne
vert du massif de l'Etna. - Ecol. méditer. *8* : 125 - 138, 1982. [Dry-matter
distribution.]

*50409 - LEPPÄNEN, J.M., TAMELANDER, G. : Composition of particulate matter and its
relation to plankton biomass in the trophogenic layer off Tvärminne, at the
entrance to the Gulf of Finland. - Meri *9* : 56 - 70, 1981/2. [Chl.]

50410 - LERBS, S., WOLLGIEHN, R. : The occurrence of two nicks in the heavy chloroplast ribosomal RNA of *Nicotiana rustica*. - Biochem. Physiol. Pflanzen
177 : 431 - 439, 1982.

50411 - LERMA, C., GÓMEZ-LOJERO, C. : Photosynthetic phosphorylation by a membrane
preparation of the cyanobacterium *Spirulina maxima*. - Biochim. biophys. Acta
680 : 181 - 186, 1982.

50412 - LE TKHI LAN OAN', RUSINOVA, N.G., DOMAN, N.G. : Dissotsiatsiya i samosborka ribulozodifosfat-karboksilazy iz list'ev *Phaseolus aureus*. [Dissociation
and self-assembly of ribulose-1,5-bisphosphate carboxylase from leaves of
Phaseolus aureus.] - In : Sborka Predbiologicheskikh i Biologicheskikh
Struktur. Pp. 310 - 313, 343. Nauka, Moskva 1982. [In R.]

50413 - LETO, K. : Photosynthetic mutants of maize. - In : SHERIDAN, W.F. (ed.) :
Maize for Biological Research. Pp. 317 - 325. Plant mol. Biol. Assoc., Charlottesville 1982.

50414 - LETO, K.J., KERESZTES, A., ARNTZEN, C.J. : Nuclear involvement in the appearance of a chloroplast-encoded 32,000 dalton thylakoid membrane polypeptide
integral to the photosystem II complex. - Plant Physiol. *69* : 1450 - 1458,
1982.

50415 - LEVANON, H., NORRIS, J.R. : Triplet state and chlorophylls. - In : FONG, F.
K. (ed.) : Light Reaction Path of Photosynthesis. Pp. 152 - 195. Springer-
-Verlag, Berlin - Heidelberg - New York 1982.

50416 - LEVAVASSEUR, G., GIRAUD, G. : Modification de la photosynthèse nette d'une
Ulve de Roscoff en fonction de la durée d'éclairement. - Physiol. vég. 20 :
143 - 154, 1982.

50417 - LEVERENZ, J., DEANS, J.D., FORD, E.D., JARVIS, P.G., MILNE, R., WHITEHEAD,
D. : Systematic spatial variation of stomatal conductance in a Sitka spruce
plantation. - J. appl. Ecol. 19 : 835 - 851, 1982.

*50418 - LEVI, Y., BERNER, T., COHEN, Y. : CO_2 exchange and growth rate of the loess
soil crusts algae in the Negev desert of Israel. - In : SHUVAL, H.I. (ed.):
Developments in Arid Zone Ecology and Environmental Quality. Pp. 43 - 48.
Balaban International Science Services, Philadelphia 1981.

50419 - LEW, R., TSUJI, H. : Effect of benzyladenine treatment duration on δ-aminolevulinic acid accumulation in the dark, chlorophyll lag phase abolition,
and long-term chlorophyll production in excised cotyledons of dark-grown
cucumber seedlings. - Plant Physiol. 69 : 663 - 667, 1982.

50420 - LEWIS, A. : Resonance Raman spectroscopy of rhodopsin and bacteriorhodopsin:
An overview. - In : COLOWICK, S.P., KAPLAN, N.O. (ed.) : Methods in Enzymo-
logy. Vol.88. Pp. 561 - 617. Academic Press, New York - London - Paris -
San Diego - San Francisco - São Paulo - Sydney - Tokyo - Toronto 1982.

50421 - LEWIS, A. : Kinetic resonance Raman spectroscopy with microsampling rotating
cells. - In : COLOWICK, S.P., KAPLAN, N.O. (ed.) : Methods in Enzymology.
Vol.88. Pp. 659 - 666. Academic Press, New York - London - Paris - San
Diego - San Francisco - São Paulo - Sydney - Tokyo - Toronto 1982. [Bacterio-
rhodopsin.]

50422 - LEWIS, A., PERREAULT, G.J. : Emission spectroscopy of rhodopsin and bacte-
riorhodopsin. - In : COLOWICK, S.P., KAPLAN, N.O. Methods in Enzymology.
Vol.88. Pp. 217 - 229. Academic Press, New York - London - Paris -San Diego -
San Francisco - São Paulo - Sydney - Tokyo - Toronto 1982.

50423 - LEWIS, M.R., KEMP, W.M., CUNNINGHAM, J.J., STEVENSON, J.C. : A rapid technique for preparation of aquatic macrophyte samples for measuring ^{14}C incorporation. - Aquat. Bot. 13 : 203 - 207, 1982.

50424 - LEY, A.C., MAUZERALL, D.C. : Absolute absorption cross-sections for photosystem II and the minimum quantum requirement for photosynthesis in Chlorella vulgaris. - Biochim. biophys. Acta 680 : 95 - 106, 1982.

50425 - LEY, A.C., MAUZERALL, D.C. : The reversible decline of oxygen flash yields
at high ·flash energies. Evidence for total annihilation of excitations in
photosystem II. - Biochim. biophys. Acta 680 : 174 - 180, 1982.

50426 - LI, J., HOLLINGSHEAD, C. : Formation of crystalline arrays of chlorophyll
a/b-light-harvesting protein by membrane reconstitution. - Biophys. J. 37 :
363 - 370, 1982.

50427 - LI Tong-zhu, SUN Gu-chou, HAO Nai-bin, KUANG Ting-yun : [The chlorophyll
protein complexes resolved from thylakoid membranes of blue-green algae
Anabaena cylindrica.] - Acta Phytophysiol. sin. 8 : 205 - 212, 1982. [In
Chin., ab : E.]

50428 - LI, W.K.W., HARRISON, W.G. : Carbon flow into the end-products of photosynthesis in short and long incubations of a natural phytoplankton population. - Mar. Biol. 72 : 175 - 182, 1982.

50429 - LI, W.K.W., MORRIS, I. : Temperature adaptation in Phaeodactylum tricornutum
BOHLIN : Photosynthetic rate compensation and capacity. - J. exp. mar. Biol.
Ecol. 58 : 135 - 150, 1982.

50430 - LI, W.K.W., PLATT, T. : Distribution of carbon among photosynthetic end-products in phytoplankton of the eastern Canadian Arctic. - J. Phycol. 18 :
466 - 471, 1982.

50431 - LI Xi-jing, YUAN Xiao-hua, WU Xiang-yu : [Comparative studies of some kinetic properties of glyceraldehyde 3-phosphate dehydrogenase and fructose-
-bisphosphatase in chloroplasts from wheat and rice.] - Acta bot. sin. *24* :
39 - 45, 1982. [In Chin., ab : E.]

50432 - LI Xiong-biao, WU Guang-yao :[Changes of ribulose bisphosphate carboxylase-
-oxygenase and malate dehydrogenase activities with leaf age in *Vicia faba*.]
- Acta Phytophysiol. sin. *8* : 197 - 203, 1982. [In Chin., ab : E.]

50433 - LI Yuxiang, LI Jigeng : [Studies on the chlorophyll-protein complexes of
chloroplast mutants in maize.] - Acta genet. sin. *9* : 344 - 349, 1982.
[In Chin., ab : E.]

50434 - LI, Y.-S., UENG, S.-H. : Zn^{2+} or silicotungstate affects steady state chlorophyll *a* fluorescence of chloroplasts suspended in distilled water. - Bot.
Bull. Acad. sin. *23* : 27 - 38, 1982.

50435 - LI, Y.-S., UENG, S.-H. : Silicotungstate induced discrepancy in the rates
of photosynthetic electron transport determined by fluorescence dip and
absorption methods. - Bot. Bull. Acad. sin. *23* : 135 - 144, 1982.

50436 - LI, Y.-S., UENG, S.-H., LIN, B.-Y. : Silicotungstate lowers dramatically the
quantum yield of chlorophyll fluorescence *in situ* without affecting the
rate of electron transport. - Biochim. biophys. Acta *681* : 469 - 473, 1982.

50437 - LIANG, Z., YU, C., HUANG, A.H.C. : Isolation of spinach leaf peroxisomes in
0.25 molar sucrose solution by percoll density gradient centrifugation. -
Plant Physiol. *70* : 1210 - 1212, 1982. [Chl.]

50438 - LIBIK, A., WOJTACZEK, T., MACZEK, W. : Efekty produkcyjne dokarmiania dwu-
tlenkiem węgla pomidorów szklarniowych. [The influence of CO_2 enrichment
on yields of glasshouse tomatoes.] - Zesz. nauk. Akad. rol. Krakow. *171*
(Ogrodnictwo 9) : 75 - 96, 1982. [Dry matter and leaf area formation; in
Pol., ab : E, R.]

50439 - LICHTENTHALER, H.K. : Synthesis of prenyllipids in vascular plants (including chlorophylls, carotenoids, and prenylquinones). - In : ZABORSKY, O.R.
(ed.) : CRC Handbook of Biosolar Resources. Vol.I/1. Pp. 405 - 421. CRC
Press, Boca Raton 1982.

50440 - LICHTENTHALER, H.K., BACH, T.J., WELLBURN, A.R. : Cytoplasmic and plastidic
isoprenoid compounds of oat seedlings and their distinct labelling from
^{14}C-mevalonate. - In : WINTERMANS, J.F.G.M., KUIPER, P.J.C. (ed.) : Biochemistry and Metabolism of Plant Lipids. Pp. 489 - 500. Elsevier Biomedical
Press, Amsterdam 1982. [Chl, Car.]

50441 - LICHTENTHALER, H.K., KUHN, G., PRENZEL, U., BUSCHMANN, C., MEIER, D. :
Adaptation of chloroplast-ultrastructure and of chlorophyll-protein levels
to high-light and low-light growth conditions. - Z. Naturforsch. *37 C* :
464 - 475, 1982.

50442 - LICHTENTHALER, H.K., KUHN, G., PRENZEL, U., MEIER, D. : Chlorophyll-protein
levels and degree of thylakoid stacking in radish chloroplasts from high-
-light, low-light and bentazon-treated plants. - Physiol. Plant. *56* : 183 -
- 188, 1982.

50443 - LICHTENTHALER, H.K., MEIER, D., RETZLAFF, G., HAMM, R. : Distribution and
effects of bentazon in crop plants and weeds. - Z. Naturforsch. *37 C* : 889 -
- 897, 1982.

50444 - LICHTENTHALER, H.K., PRENZEL, U., KUHN, G. : Carotenoid composition of chlo-
rophyll-carotenoid-proteins from radish chloroplasts. - Z. Naturforsch.
37 C : 10 - 12, 1982.

50445 - LIDDELL, P.A., NEMETH, G.A., LEHMAN, W.R., JOY, A.M., MOORE, A.L., BENSASSON,
R.V., MOORE, T.A., GUST, D. : Mimicry of carotenoid function in photosynthe-
sis: synthesis and photophysical properties of a carotenopyropheophorbide.
- Photochem. Photobiol. *36* : 641 - 645, 1982.

50446 - LIEBIG, H.-P. : Einflüsse endogener und exogener Faktoren auf die Ertrags-
        bildung von Salatgurken (*Cucumis sativus* L.). I. Stoffproduktion und Er-
        tragsrhythmik. - Gartenbauwissenschaft *47* : 4 - 13, 1982. [Growth analy-
        sis.]

50447 - LIETH, H., ESSER, G. : Zur Modellierung der Beziehung zwischen globaler
        Netto-Primärproduktivität und Umweltfaktoren. - In : UNGER, K., SCHUH, J.
        (ed.) : Umwelt-Stress. Pp. 303 - 321. Martin-Luther-Universität, Halle
        1982.

50448 - LIEW, K.Y., TAN, S.H., MORSINGH, F., KHOO, L.E. : Adsorption of β-carotene:
        II. On cation exchanged bleaching clays. - J. amer. Oil Chem. Soc. *59* :
        480 - 484, 1982.

50449 - LILLEY, R.McC., STITT, M., MADER, G., HELDT, H.W. : Rapid fractionation
        of wheat leaf protoplasts using membrane filtration. The determination of
        metabolite levels in the chloroplasts, cytosol, and mitochondria. - Plant
        Physiol. *70* : 965 - 970, 1982.

50450 - LIMAR', R.S., MATVIENKO, I.I. : Transport assimilyatov v kolos'ya yarovoĭ
        pshenitsy. [Assimilate transport into ears of spring wheat.] - Tr. prikl.
        Bot., Genet. Selektsii *72* (2) : 12 - 17, 1982. [In R, ab : E.]

*50451 - LIN, C.K., SIMMONS, M.S. : Effects of pentachlorobiphenyl on growth of
        nutrient enriched phytoplankton from Lake Michigan. - J. Great Lakes Res.
        7 : 481 - 485, 1981.

50452 - LIN, W.C., MOLNAR, J.M. : Supplementary lighting and CO_2 enrichment for
        accelerated growth of selected wood ornamental seedlings and rooted cut-
        tings. - Can. J. Plant Sci. *62* : 703 - 707, 1982.

50453 - LIN Zhen-wu, WANG Yu-qin, WU Shao-bo, TANG Yu-wei : [Enhancement of chloro-
        plast development by 6-benzylaminopurine in etiolated wheat leaves I. Ef-
        fect of red light on chlorophyll formation.] - Acta Phytophysiol. sin. *8* :
        399 - 406, 1982. [In Chin., ab : E.]

50454 - LIN, Z.F., EHLERINGER, J. : Effects of leaf age on photosynthesis and water
        use efficiency of papaya. - Photosynthetica *16* : 514 - 519, 1982.

50455 - LIN, Z.F., EHLERINGER, J. : Changes in spectral properties of leaves as re-
        lated to chlorophyll content and age of papaya. - Photosynthetica *16* : 520 -
        - 525, 1982.

50456 - LIN Zhi-fang, EHLERINGER, J.R. : [The effects of light, temperature, water
        vapour pressure deficit and carbon dioxide on photosynthesis in papaya.] -
        Acta Phytophysiol. sin. *8* : 363 - 372, 1982. [In Chin., ab : E.]

50457 - LINDEBOOM, H.J., DE BREE, B.H.H. : Daily production and consumption in an
        eelgrass (*Zostera marina*) community in saline Lake Greveningen: discrepan-
        cies between the O_2 and ^{14}C method. - Neth. J. Sea Res. *16* : 362 - 379,
        1982.

50458 - LINDEBOOM, H.J., DE KLERK, H.A.J., DRIESSCHE, V.D., SANDEE, A.J.J. : Pro-
        duction and decomposition of eelgrass (*Zostera marina* L.) in saline Lake
        Greveningen. - Hydrobiol. Bull. *16* : 93 - 102, 1982.

50459 - LINDENMAYER, A. : Rules of growth: Some comments on Erickson's models of
        plant growth. - Acta biotheor. *31A* (Suppl.) : 152 - 161, 1982. [Dry-matter
        accumulation.]

50460 - LINDLEY, E.V., HOWLES, P.N., MacDONALD, R.E. : Demonstration of primary
        sodium transport activity in *Halobacterium halobium* envelope vesicles. -
        In : COLOWICK, S.P., KAPLAN, N.O. (ed.) : Methods in Enzymology. Vol.88.
        Pp. 426 - 433. Academic Press, New York - London - Paris - San Diego -
        - San Francisco - São Paulo - Sydney - Tokyo - Toronto 1982. [Bacterio-
        rhodopsin.]

50461 - LINDSKOG, S. : Carbonic anhydrase. - In : EICHHORN, G.L., MARZILLI, L.G.
        (ed.) : Advances in Inorganic Biochemistry. Vol.4. Pp. 115 - 170. Elsevier
        Biomedical, New York - Amsterdam - Oxford 1982.

50462 - LING, A.H., ROBERTSON, G.W. : Reflection coefficient of some tropical vegetation covers. - Agr. Meteorol. *27* : 141 - 144, 1982.

50463 - LIPINSKY, E.S., KRESOVICH, S. : Sugar crops as a solar energy converter. -
Experientia *38* : 13 - 18, 1982. [Photosynthates.]

*50464 - LIPSCHULTZ, C.A., GANTT, E. : Association of phycoerythrin and phycocyanin:
in vitro formation of a functional energy transferring phycobilisome complex
of *Porphyridium sordidum*. - Biochemistry *20* : 3371 - 3376, 1981.

50465 - LITSCHER, T., WHITEMAN, P.C. : Light transmission and pasture composition
under smallholder coconut plantations in Malaita, Solomon Islands. - Exp.
Agr. *18* : 383 - 391, 1982.

50466 - LITTLEJOHN, R.O.,Jr., GREEN, J.M., WILLIAMS, G.J.,III : A morning peak in
acidity during phase II Crassulacean acid metabolism in *Opuntia erinacea,
Echinocereus viridiflorus,* and *Mammillaria vivipara*. - Plant Sci. Lett. *27*:
43 - 49, 1982.

50467 - LITTLER, M.M., ARNOLD, K.E. : Primary productivity of marine macroalgal
functional-form groups from southwestern North America. - J. Phycol. *18* :
307 - 311, 1982.

50468 - LITVIN, F.F., SINESHCHEKOV, V.A., BOĬCHENKO, V.A. : Sootnoshenie biofizi-
cheskikh i fiziologicheskikh zakonomernosteĭ nachal'nykh stadiĭ fotosinteza. [Relationship between biophysical and physiological regularities of
primary stages of photosynthesis.] - In : Fiziologiya Fotosinteza. Pp.
34 - 54. Nauka, Moskva 1982. [In R.]

*50469 - LIU Zhen-chee : [A study on the photosynthetic characters of different plant
types of rice.] - Zhongguo Nongye Kexue (Beijing) *3* : 6 - 10, 1980. [In
Chin., ab : E.]

50470 - LIU Zhenchee, LIU Zenye, ZENG Shufen, MA Dapeng : [A study on some photosynthetic characters of rice.] - Zhongguo Nongye Kexue (Beijing) *5* : 33 - 39,
1982. [In Chin., ab : E.]

50471 - LIVORNESS, J., SMITH, T.D. : The role of manganese in photosynthesis. -
In : Structure and Bonding. Vol.48. Pp. 2 - 44. Springer-Verlag, Berlin -
- Heidelberg 1982.

50472 - LJUBEŠIČ, N. : Phytoferritin accumulation in chromoplasts of *Sorbus aucupa-
ria* L. fruits. - Acta bot. croat. *41* : 29 - 32, 1982. [Chloroplasts.]

*50473 - LLAMA, M.J., SERRA, J.L., RAO, K.K., HALL, D.O. : Characterization of hydro-
genases from blue-green algae. - Biochem. Soc. Trans. *6* : 1337 - 1339,
1978.

50474 - LLOYD, E.J. : Mobilization of carbon during senescence in detached leaves
of *Lolium temulentum*. - Plant Sci. Lett. *25* : 313 - 319, 1982. [Chl.]

*50475 - LLOYD, N.D.H., McLACHLAN, J.L., BIDWELL, R.G.S. : A rapid infra-red carbon
dioxide analysis screening technique for predicting growth and productivity
of marine algae. - In : LEVRING, T. (ed.) : Xth International Seaweed Symposium. Proceedings. Pp. 461 - 466. Walter de Gruyter, Berlin - New York
1981.

B50476 - LOCKERETZ, W. (ed.) : Agriculture as a Producer and Consumer of Energy.
(AAAS Selected Symposium 78.) - Westview Press, Boulder 1982. [Biomass,
energy content.]

50477 - LOEBLICH, A.R.,III : General characteristics of phytoplankton: *Dinoflagel-
lates*. - In : ZABORSKY, O.R. (ed.) : CRC Handbook of Biosolar Resources.
Vol.I/2. Pp. 37 - 47. CRC Press, Boca Raton 1982. [Chl, Car.]

50478 - LOEBLICH, L.A. : Photosynthesis and pigments influenced by light intensity
and salinity in the halophile *Dunaliella salina (Chlorophyta)*. - J. mar.
Biol. Assoc. UK *62* : 493 - 508, 1982.

50479 - LOEHLE, C. : Growth and maintenance respiration: a reconciliation of Thornley's model and the traditional view. - Ann. Bot. *51* : 741 - 747, 1982.
[Ps.]

50480 - LOESCHER, V.H., MARLOW, G.C., KENNEDY, R.A. : Sorbitol metabolism and sink-
-source interconversions in developing apple leaves. - Plant Physiol. *70* :
335 - 339, 1982. [Ps, Chl.]

50481 - LØKEN, Ø., SIREVÅG, R. : Evidence for the presence of the glyoxylate cycle
in *Chloroflexus*. - Arch. Microbiol. *132* : 276 - 279, 1982.

*50482 - LOKTEVA, T.N., LUK'YANOVA, L.M. : Issledovaniya CO_2-gazoobmena rasteniĭ
Khibin. [CO_2 gas exchange of plants in the region of Khibiny.] - In : Raz-
vitie Botanicheskikh Issledovaniĭ na Kol'skom Severe. Pp. 111 - 120, 152.
Akademiya Nauk SSSR, Apatity 1981. [In R.]

50483 - LOLAS, P., GALOPOULOS, A. : [Effect of herbicides on growth, chemical composition, nitrate reductase activity and chlorophyll of tobacco.] - Zizanio-
logy *1* : 11 - 15, 1982. [In Greek, ab : E.]

50484 - LONDON, E., KHORANA, H.G. : Denaturation and renaturation of bacteriorhodopsin in detergents and lipid-detergent mixtures. - J. biol. Chem. *257* :
7003 - 7011, 1982.

50485 - LONG, S.P. : Measurement of photosynthetic gas exchange. - In : COOMBS, J.,
HALL, D.O. (ed.) : Techniques in Bioproductivity and Photosynthesis. Pp.
25 - 36. Pergamon Press, Oxford - New York - Toronto - Sydney - Paris -
- Frankfurt 1982.

50486 - LONGUENESSE, J.-J. : Température nocturne et photosynthése. III. Influence
de la température appliquée pendant une nuit sur les échanges gazeux de la
tomate (*Lycopersicon esculentum* MILL.) - Agronomie *2* : 805 - 811, 1982.

50487 - LONGUENESSE, J.-J., CONUS, G., SARROUY, C. : Température nocturne et photosynthèse. II. Une chambre d'assimilation climatisée pour la mesure des
échanges gazeux de plantes entières. - Agronomie *2* : 777 - 781, 1982.

50488 - LONGWORTH, J.W. : The trends and future of photobiology: physical and bio-
physical aspects. - In : HÉLÈNE, C., CHARLIER, M., MONTENAY-GARESTIER, T.,
LAUSTRIAT, G. (ed.) : Trends in Photobiology. Pp. 147 - 152. Plenum Press,
New York - London 1982. [Chl.]

50489 - LORENC-PLUCIŃSKA, G. : Effect of sulphur dioxide on CO_2 exchange in SO_2-to-
lerant and SO_2-susceptible Scots pine seedlings. - Photosynthetica *16* : 140 -
- 144, 1982.

50490 - LORENC-PLUCIŃSKA, G. : Influence of SO_2 on CO_2 assimilation and carbon metabolism in photosynthetic processes in Scots pine. - Arboretum kórnickie
27 : 285 - 310, 1982.

50491 - LORENC-PLUCIŃSKA, G., OLEKSYN, J. : Effect of HF on photosynthesis, photorespiration and dark respiration in Scotch pine. - Fluoride *15* : 149 - 156,
1982.

50492 - LORIMER, G.H. : Activities of RuBP carboxylase-oxygenase. - In : EDELMAN,
M., HALLICK, R.B., CHUA, N.-H. (ed.) : Methods in Chloroplast Molecular Biology. Pp. 803 - 808. Elsevier Biomedical Press, Amsterdam - New York - Oxford 1982.

50493 - LORIMER, G.H. : Respiration and photorespiration in plants. - In : ZABORSKY,
O.R. (ed.) : CRC Handbook of Biosolar Resources. Vol.I/1. Pp. 323 - 327.
CRC Press, Boca Raton 1982.

50494 - LÖSCH, R., TENHUNEN, J.D., PEREIRA, J.S., LANGE, O.L. : Diurnal courses of
stomatal resistance and transpiration of wild and cultivated mediterranean
perrenials at the end of the summer dry season in Portugal. - Flora *172* :
138 - 160, 1982.

*50495 - LOSEV, A.P., LYAL'KOVA, N.D., SAGUN, E.I. : Ėnergiya nizhnikh tripletnykh
sostoyaniĭ ryada karotinoidov i bakteriokhlorofil'nykh pigmentov. [Energy
of the lower triplet states of a series of carotenoids and bacteriochlorophyll pigments.] - Zh. prikl. Spektrosk. *35* : 671 - 677, 1981. [In R, ab :
E.]

*50496 - LOSEV, A.P., ZEN'KEVICH, É.I., SAGUN, E.I. : Izuchenie kontsentratsionnykh zavisimostei vykhodov fluorestsentsii i tripletnykh sostoyanii pigmentov tipa khlorofilla i fluorestseina. [Study of concentration dependences of fluorescence yields and triplet states of pigments of chlorophyll and fluorescein types.] - Izv. Akad. Nauk SSSR, Ser. fiz. *44* : 783 - 788, 1980. [In R.]

50497 - LOTH, P. : Verwendung von Glasfaserpapier in der Wasseranalytik. Teil 2. Seston- und Chlorophyllbestimmung sowie die Bestimmung des partikulären Kohlenstoffes. - Acta hydrochim. hydrobiol. *10* : 323 - 337, 1982.

50498 - LOTSHAW, W.T., ALBERTE, R.S., FLEMING, G.R. : Low-intensity subnanosecond fluorescence study of the light-harvesting chlorophyll *a/b* protein. - Biochim. biophys. Acta *682* : 75 - 85, 1982.

50499 - LOU Shi-qing, ZHANG Qi-de, TANG Chong-qin, LIN Shi-ching, HAO Nai-bin, LI Tung-shu, KUANG Ting-yun : [Structure and function of chloroplast membranes. IX. Isolation of the light-harvesting chlorophyll *a/b*-protein complex from spinach chloroplast membrane.] - Acta biochim. biophys. sin. *14* : 431 - 439, 1982. [In Chin., ab : E.]

50500 - LOVGREN, T., PEACOCK, R., LAVI, J., KARP, M., RAUNIO, R. : The bioluminescent assay of NADH and NADPH. - Int. Lab. *12* (6) : 58, 60, 62, 64, 66-67, 1982.

50501 - LOWE, R.H., SHEEN, S.J. : Accumulation of soluble proteins and nitrogenous compounds in the leaf of bright and burley tobaccos during the growing season. - Beitr. Tabakforsch. Int. *11* : 161 - 169, 1982. [Chl.]

50502 - LOWMAN, M.D. : Effects of different rates and methods of leaf area removal on rain forest seedlings of coachwood (*Ceratopetalum apetalum*). - Aust. J. Bot. *30* : 477 - 483, 1982.

50503 - LOZIER, R.H. : Rapid kinetic optical absorption spectroscopy of bacteriorhodopsin phytocycles. - In : COLOWICK, S.P., KAPLAN, N.O. (ed.) : Methods in Enzymology. Vol.88. Pp. 133 - 162. Academic Press, New York - London - Paris - San Diego - San Francisco - São Paulo - Sydney - Tokyo - Toronto 1982.

50504 - LUBIÁN, L.M. : *Nannochloropsis gaditana* sp. nov., una nueva *Eustigmatophyceae* marina. [*Nannochloropsis gaditana* sp. nov., a new marine *Eustigmatophyceae*.] - Lazaroa *4* : 287 - 293, 1982. [Chloroplast, Car; in Span., ab : E.]

50505 - LUBIÁN, L.M., ESTABLIER, R. : Estudio comparativo de la composición de pigmentos en varias cepas de *Nannochloropsis (Eustigmatophyceae)*. [Comparative study of pigment composition of some *Nannochloropsis* strains *(Eustigmatophyceae)*.] - Invest. Pesquera *46* : 379 - 389, 1982. [In Span., ab : E.]

50506 - LUCAS, W.J. : Mechanism of acquisition of exogenous bicarbonate by internodal cells of *Chara corallina*. - Planta *156* : 181 - 192, 1982. [Ps.]

50507 - LUCIER, A.A., HINCKLEY, T.M. : Phenology, growth and water relations of irrigated and non-irrigated black walnut. - Forest Ecol. Manage. *4* : 127 - 142, 1982.

50508 - LUDDEN, P.W., PRESTON, G.G., DOWLING, T.E. : Comparison of active and inactive forms of iron protein from *Rhodospirillum rubrum*. - Biochem. J. *203* : 663 - 668, 1982.

50509 - LUDLOW, M.M. : Measurement of solar radiation, temperature and humidity. - In : COOMBS, J., HALL, D.O. (ed.) : Techniques in Bioproductivity and Photosynthesis. Pp. 5 - 16. Pergamon Press, Oxford - New York - Toronto - Sydney - Paris - Frankfurt 1982.

50510 - LUDLOW, M.M. : Measurement of stomatal conductance and plant water status. - In : COOMBS, J., HALL, D.O. (ed.) : Techniques in Bioproductivity and Photosynthesis. Pp. 44 - 57. Pergamon Press, Oxford - New York - Toronto - - Sydney - Paris - Frankfurt 1982.

50511 - LUDLOW, M.M., STOBBS, T.H., DAVIS, R., CHARLES-EDWARDS, D.A. : Effect of
        sward structure of two tropical grasses with contrasting canopies on light
        distribution, net photosynthesis and size of bite harvested by grazing
        cattle. - Aust. J. agr. Res. 33 : 187 - 201, 1982.

50512 - LUGANSKAYA, A.N., LEBEDEV, N.N., BRIN, G.P., KRASNOVSKIĬ, A.A. : Spektral'-
        nye i fotokhimicheskie svoĭstva khlorofilla, adsorbirovannogo na neorgani-
        cheskikh nositelyakh. Fotovosstanovlenie kisloroda. [Spectral and photo-
        chemical properties of chlorophyll adsorbed on inorganic carriers. Photo-
        reduction of oxygen.] - Zh. fiz. Khim. 56 : 1753 - 1757, 1982. [In R.]

50513 - LUKASHEV, E.P., VOZARI, É., KONONENKO, A.A., RUBIN, A.B., ABDULAEV, N.G. :
        Regulyatsiya tsikla fotokhromnykh prevrashcheniĭ bakteriorodopsina élektri-
        cheskim polem. [Control of bacteriorhodopsin photochromic cycle by electric
        field.] - Bioorg. Khim. 8 : 1173 - 1179, 1982. [In R, ab : E.]

50514 - LUK'YANOVA, L.M. : Ékologo-fiziologicheskie aspekty izucheniya pigmentnoĭ
        sistemy rasteniĭ. I. Vliyanie vneshnikh faktorov, sezonnaya i sutochnaya
        dinamika. [Eco-physiological aspects of study of plant pigment system. I.
        The effects of environmental factors, seasonal and diurnal dynamics.] -
        Bot. Zh. 67 : 265 - 277, 1982. [In R, ab : E.]

50515 - LUK'YANOVA, L.M. : Ékologo-fiziologicheskie aspekty izucheniya pigmentnoĭ
        sistemy rasteniĭ. II. Vliyanie ékologo-geograficheskikh usloviĭ i sistema-
        ticheskoĭ prinadlezhnosti rasteniĭ. [Eco-physiological aspects of study of
        plant pigment system. II. The influence of eco-geographical factors and
        systematical confinement of plants.] - Bot. Zh. 67 : 409 - 418, 1982. [In
        R, ab : E.]

*50516 - LUK'YANOVA, L.M., LOKTEVA, T.N. : Fiziologicheskie issledovaniya v Polyarno-
        -al'piĭskom botanicheskom sadu. [Physiological studies in the Arctic Alpine
        Botanical Garden.] - In : Razvitie Botanicheskikh Issledovaniĭ na Kol'skom
        Severe. Pp. 93 - 111. Akad. Nauk SSSR, Kol'skiĭ Filial, Apatity 1981. [Ps,
        Chl; in R.]

*50517 - LUK'YANOVA, N.M., OSIPOV, A.V. : Izuchenie svetolyubiya vechnozelenykh ras-
        teniĭ s ispol'zovaniem lazernogo svetovozbuzhdeniya. [Study of evergreen
        plants photophily by means of laser light induction.] - Byull. gos. nikitsk.
        bot. Sada 3 (40) : 77 - 81, 1979. [In R, ab : E.]

50518 - LUPATOV, V.M. : Analiz modeleĭ mekhanizma vydeleniya kisloroda pri fotosin-
        teze. [Analysis of oxygen evolution models in photosynthesis.] - Stud. bio-
        phys. 91 : 133 - 140, 1982. [In R, ab : E.]

50519 - LUPATOV, V.M. : Dva reaktsionnykh tsentra v sisteme razlozheniya vody pri
        fotosinteze. [Two reaction centres in water splitting system in photosyn-
        thesis.] - Stud. biophys. 91 : 141 - 148, 1982. [In R, ab : E.]

50520 - LÜTKE-BRINKHAUS, F., LIEDVOGEL, B., KREUZ, K., KLEINIG, H. : Phytoene syn-
        thase and phytoene dehydrogenase associated with envelope membranes from
        spinach chloroplasts. - Planta 156 : 176 - 180, 1982. [Chl, Car.]

50521 - LÜTTKE, A., BONOTTO, S. : Chloroplasts and chloroplast DNA of Acetabularia
        mediterranea: Facts and hypotheses. - Int. Rev. Cytol. 77 : 205 - 242,
        1982.

50522 - LÜTTKE, A., BONOTTO, S. : Starch accumulation in Acetabularia mediterranea
        after X-irradiation. - Environm. exp. Bot. 22 : 293 - 299, 1982. [Chl.]

*50523 - LÜTZ, C., BENZ, J., RÜDIGER, W. : Esterification of chlorophyllide in pro-
        lamellar body (PLB) and prothylakoid (PT) fractions from Avena sativa etio-
        plasts. - Z. Naturforsch. 36 C : 58 - 61, 1981.

50524 - LUTZ, M., CHINSKY, L., TURPIN, P.-Y. : Triplet states of carotenoids bound
        to reaction centres of photosynthetic bacteria: Time-resolved resonance
        Raman spectroscopy. - In : LASCOMBE, J., HUONG, P.V. (ed.) : Raman Spectro-
        scopy Linear and Nonlinear. Pp. 789 - 790. J.Wiley & Sons, Chichester - New
        York - Brisbane - Toronto - Singapore 1982.

50525 - LUTZ, M., CHINSKY, L., TURPIN, P.-Y. : Triplet states of carotenoids bound to reaction centers of photosynthetic bacteria: time-resolved resonance Raman spectroscopy. - Photochem. Photcbiol. *36* : 503 - 515, 1982.

50526 - LUTZ, M., HOFF, A.J., BREHAMET, L. : Bacteriochlorophyll *a*-protein inter- actions in a complex from *Prosthecochloris aestuarii*. A resonance Raman study. - Biochim. biophys. Acta *679* : 331 - 341, 1982.

50527 - LYMAN, H. : Rapporteur's summary: Molecular biology and control of plastid development. - In : SCHIFF, J.A., LYMAN, H. (ed.) : On the Origins of Chlo- roplasts. Pp. 319 - 322. Elsevier/North-Holland, New York - Amsterdam - Oxford 1982.

50528 - LYNCH, D.V., THOMPSON, G.A.,Jr. : Low temperature-induced alterations in the chloroplast and microsomal membranes of *Dunaliella salina*. - Plant Physiol. *69* : 1369 - 1375, 1982. [Chl.]

50529 - LYON, C.K., KOHLER, G.O. : Stabilization of carotene and xanthophyll in alfalfa leaf protein concentrates. - J. agr. Food Chem. *30* : 934 - 937, 1982.

50530 - LYUTOVA, M.I. : Ustoĭchivost' reaktsii Khilla k razlichnym povrezhdayush- chim vozdeĭstviyam u dvukh sortov pshenitsy *Triticum aestivum (Poaceae)*. [The resistance of Hill reaction to different injurious agents in two cul- tivars of wheat *Triticum aestivum (Poaceae)*.] - Bot. Zh. *67* : 433 - 439, 1982. [In R, ab : E.]

50531 - MacCOLL, R. : Phycobilisomes and biliproteins. - Photochem. Photobiol. *35* : 899 - 904, 1982.

50532 - MACHADO, E.C., PEREIRA, A.R., FAHL, J.I., ARRUDA, H.V., CIONE, J. : Índices biométricos de duas variedades de cana-de-açúcar. [Biometric indices of two sugar cane cultivars.] - Pesq. agropec. bras. *17* : 1323 - 1329, 1982. [Growth analysis; in Port., ab : E.]

50533 - MACHADO, E.C., PEREIRA, A.R., FAHL, J.I., ARRUDA, H.V., DA SILVA, W.J., TEIXEIRA, J.P.F. : Análise quantitativa de crescimento de quatro variedades de milho em três densidades de plantio, através de funções matemáticas aju- stadas. [Quantitative growth analysis of four maize cultivars grown in three plant densities, with the use of fitted mathematical functions.] - Pesq. agropec. brasil. *17* : 825 - 833, 1982. [In Port., ab : E.]

50534 - MACHEREL, D., RAVANEL, P., TISSUT, M. : Effects of herbicidal carbamates on mitochondria and chloroplasts. - Pestic. Biochem. Physiol. *18* : 280 - - 288, 1982.

*50535 - MACHOWICZ, E. : Aparat fotosyntetyczny w starzejącym się liściu. [Photosyn- thetic apparatus in senescing leaf.] - Zesz. nauk. Uniw. jagielloń. *596* (Prace Biol. mol. 7) : 247 - 253, 1980. [In Pol., ab : E.]

50536 - MADER, P., NAUŠ, J., MAKOVEC, P., SOFROVÁ, D., VACEK, J., GRÉC, L., MASOJÍ- DEK, J. : Effects of ontogeny and additional fertilization with nitrogen on composition and activity of the spring barley photosynthetic apparatus. - Photosynthetica *16* : 161 - 175, 1982.

50537 - MADGWICK, H.A.I., OLIVER, G., HOLTEN-ANDERSON, P. : Above-ground biomass, nutrients, and energy content of trees in a second-growth stand of *Agathis australis*. - New Zeal. Forest. Sci. *12* : 3 - 6, 1982.

50538 - MADHAVAN, S., SMITH, B.N. : Localization of ribulose bisphosphate carboxy- lase in the guard cells by an indirect, immunofluorescence technique. - Plant Physiol. *69* : 273 - 277, 1982.

50539 - MADIGAN, M., COX, J.C., GEST, H. : Photopigments in *Rhodopseudomonas capsu- lata* cells grown anaerobically in darkness. - J. Bacteriol. *150* : 1422 - - 1429, 1982.

50540 - MADIGAN, M., GEST, H. : Nitrogen fixation in photosynthetic bacteria. - In :
ZABORSKY, O.R. (ed.) : CRC Handbook of Biosolar Resources. Vol.I/1. Pp. 245 -
- 250. CRC Press, Boca Raton 1982. [Ps.]

50541 - MADORE, M., WEBB, J.A. : Stachyose synthesis in isolated mesophyll cells
of *Cucurbita pepo*. - Can. J. Bot. *60* : 126 - 130, 1982. [Photosynthates.]

50542 - MAE, T., OHIRA, K. : Relation between leaf age and nitrogen incorporation
in the leaf of the rice plant (*Oryza sativa* L.). - Plant Cell Physiol. *23* :
1019 - 1024, 1982. [Chl.]

50543 - MAEDA, A., TAKEUCHI, Y., YOSHIZAWA, T. : Absorption spectral properties
of acetylated bacteriorhodopsin in purple membrane depending on pH. - Biochemistry *21* : 4479 - 4483, 1982.

50544 - MAEHL, P. : Phytoplankton production in relation to physico-chemical conditions in a small, oligotrophic subarctic lake in South Greenland. - Holarctic
Ecol. *5* : 420 - 427, 1982.

50545 - MAGDE, D., BERENS, S.J., BUTLER, W.L. : Picosecond fluorescence in spinach
chloroplasts. - Proc. SPIE (Int. Soc. Opt. Eng.) *322* : 80 - 86, 1982.

50546 - MAGNUSON, C.E., FARES, Y., GOESCHL, J.D., NELSON, C.E., STRAIN, B.R.,
JAEGER, C.H., BILPUCH, E.G. : An integrated tracer kinetics system for stu-
dying carbon uptake and allocation in plants using continuously produced
$^{11}CO_2$. - Radiat. environm. Biophys. *21* : 51 - 65, 1982.

50547 - MAGOMEDOV, I.M., AGAEV, M.G. : Nekotorye ėvolyutsionnye aspekty C_4-foto-
sinteza. [Some evolutionary aspects of C_4-photosynthesis.] - Bot. Zh. *67* :
1094 - 1100, 1982. [In R.]

50548 - MAHALL, B.E., SCHLESINGER, W.H. : Effects of irradiance on growth, photosynthesis, and water use efficiency of seedlings of the chaparral shrub,
Geanothus megacarpus. - Oecologia *54* : 291 - 299, 1982.

50549 - MAHNKEN, R., WILHM, J. : Diel variation in species composition and diversity, density, and chlorophyll content of phytoplankton in an intermittent
stream in Oklahoma. - Southwest. Naturalist *27* : 79 - 86, 1982.

50550 - MAHON, J.D. : Field evaluation of growth and nitrogen fixation in peas selected for high and low photosynthetic CO_2 exchange. - Can. J. Plant Sci.
62 : 5 - 17, 1982.

50551 - MAHRO, B., GRIMME, L.H. : H_2-photoproduction by green-algae: The significance of anaerobic pre-incubation periods and of high light intensities
for H_2-photoproductivity of *Chlorella fusca*. - Arch. Microbiol. *132* : 82 -
- 86, 1982.

50552 - MAKASHARIPOVA, K.A., ZELENSKIĬ, M.I. : Tenevynoslivost' pshenits razlichnoĭ vidovoĭ i ėkologicheskoĭ prinadlezhnosti. [Shade-tolerance in wheats
of different species and ecological origin.] - Tr. prikl. Bot., Genet. Selektsii *72* (2) : 77 - 83, 1982. [In R, ab : E.]

*50553 - MÄKELÄ, A., HARI, P., KELLOMÄKI, S. : A model for the effect of air pollu-
tion on forest growth.   - Silva fenn. *15* : 481 - 482, 1981. [Ps.]

*50554 - MAKOVSKIĬ, V.I., KRAVCHENKO, L.Yu. : Sezonnaya dinamika nadzemnoĭ fitomassy
travyannogo i mokhovogo yarusov bolotnykh fitotsenozov (Il'menskiĭ zapoved-
nik). [Seasonal dynamics of aboveground biomass of herb and moss layers on
bog phytocenoses (Il'mensk reservation).] - Tr. Inst. Ėkol. Rast. Zhivot.
130 (Strukturno-Funktsional'nye Vzaimosvyazi v Biogeotsenozakh Yuzhnogo
Urala) : 91 - 97, 1979. [In R.]

50555 - MAKSYMOWYCH, R., MAKSYMOWYCH, A.B. : Petiole development and xylem differentiation in *Xanthium* represented by the plastochron index. - Amer. J. Bot.
69 : 23 - 30, 1982.

50556 - MALHOTRA, K., OELZE-KAROW, H., MOHR, H. : Action of light on accumulation
of carotenoids and chlorophylls in the milo shoot (*Sorghum vulgare* PERS.). -
Planta *154* : 361 - 370, 1982.

50557 - MALHOTRA, O.P., SRIVASTAVA, D.K. : Steady-state kinetic studies of green
gram glyceraldehyde 3-phosphate dehydrogenase. - Indian J. Biochem. Biophys.
19 : 95 - 101, 1982.

50558 - MALIK, N.S.A. : Senescence in detached oat leaves I. Changes in free amino
acid levels. - Plant Cell Physiol. *23* : 49 - 57, 1982. [Chl.]

*50559 - MALKIN, R. : Redox properties of the DBMIB-Rieske iron-sulfur complex in
spinach chloroplast membranes. - FEBS Lett. *131* : 169 - 172, 1981.

50560 - MALKIN, R. : Interaction of photosynthetic electron transport inhibitors and
the Rieske iron-sulfur center in chloroplasts and the cytochrome b_6-f complex. - Biochemistry *21* : 2945 - 2950, 1982.

50561 - MALKIN, R. : Photosystem I. - Annu. Rev. Plant Physiol. *33* : 455 - 479,
1982.

50562 - MALKIN, R. : Redox properties and functional aspects of electron carriers
in chloroplast photosynthesis. - In : BARBER, J. (ed.) : Electron Transport
and Photophosphorylation. Pp. 1 - 47. Elsevier Biomedical Press, Amsterdam -
- New York - Oxford 1982.

50563 - MALKIN, R., CROWLEY, R. : Quinone analogs and their interaction with the
chloroplast electron-transport chain. - In : TRUMPOVER, B.L. (ed.) : Function
of Quinones in Energy Conserving Systems. Pp. 453 - 462. Academic Press,
New York 1982.

50564 - MALKINA, I.S. : Vliyanie osveshchennosti i vozrasta dereva na assimilyatsion-
nuyu sposobnost' khvoi sosny obyknovennoĭ. [Effect of illuminance and tree
age on CO_2 assimilation in pine needles.] - Fiziol. Rast. *29* : 465 - 470,
1982. [In R, ab : E.]

50565 - MALKINA, I.S., ERPERT, S.D. : Assimilyatsionnaya sposobnost' krony derev'ev
s razlichnym tipom vetvleniya. [Assimilation capacity of tree crowns with
different type of branching.] - Zh. obshch. Biol. *43* : 109 - 113, 1982.
[In R, ab : E.]

50566 - MALONE, T.C. : Phytoplankton photosynthesis and carbon-specific growth:
light-saturated rates in a nutrient-rich environment. - Limnol. Oceanogr.
27 : 226 - 235, 1982.

50567 - MAL'TSEV, S.V., KRASNOVSKIĬ, A.A. : Vydelenie i pogloshchenie molekulyar-
nogo vodoroda izolirovannymi khloroplastami vysshikh rastenii. [Evolution
and absorption of molecular hydrogen by chloroplasts isolated from higher
plants.] - Fiziol. Rast. *29* : 951 - 958, 1982.

50568 - MAL'YAN, A.N. : Kineticheskoe issledovanie Mg^{2+}-zavisimoĭ CF_1-ATPazy v prisutstvii stimulyatorov. [Kinetic study of Mg^{2+}-dependent CF_1-ATPase in the
presence of stimulators.] - Biokhimiya *47* : 540 - 545, 1982. [In R, ab :
E.]

50569 - MALYAN, A.N., KUZMIN, A.N., VITSEVA, O.I. : On the participation of the
"regulatory" site of CF_1-ATPase in photophosphorylation. - Biochem. Int.
5 : 325 - 328, 1982.

50570 - MANDOLI, D.F., BRIGGS, W.R. : Optical properties of etiolated plant tissues.
- Proc. nat. Acad. Sci. USA *79* : 2902 - 2906, 1982.

50571 - MANETAS, Y. : Changes in properties of phosphoenolpyruvate carboxylase from
the CAM plant *Sedum praealtum* D.C. upon dark/light transition and their stabilization by glycerol. - Photosynthesis Res. *3* : 321 - 333, 1982.

50572 - MANETAS, Y., GAVALAS, N.A. : Evidence for essential sulfhydryl group(s) in
photosynthetic phosphoenolpyruvate carboxylase: protection by substrate,
metal-substrate and glucose-6-phosphate against p-chloromercuribenzoate
inhibition. - Photosynthetica *16* : 59 - 66, 1982.

*50573 - MANGAT, B.S. : The effect of diuron on the soluble nucleotide pool and
growth of bean and corn seedlings. - Pest. Biochem. Physiol. *10* : 251 -
- 258, 1979. [Chl.]

50574 - **MANNER, R.** : Biomass production: A plant breeder's view. - In : Improvement of Oil-Seed and Industrial Crops by Induced Mutations. Pp. 265 - 281. Int. Atomic Energy Agency, Vienna 1982.

50575 - **MANNERS, J.M., GAY, J.L.** : Transport, translocation and metabolism of ^{14}C-photosynthates at the host-parasite interface of *Pisum sativum* and *Erysiphe pisi*. - New Phytol. *91* : 221 - 244, 1982.

50576 - **MANOLOV, P.** : Sintez, transport i metaboliz"m na sorbitola i svobodnite zakhari pri narastvaneto i uzryavaneto na plodovete ot cheresha i sliva. [Sorbitol and free sugars synthesis, transport and metabolism in the course of cherry and plum fruit growth and ripening.] - Gradin. lozar. Nauka *19* (2) : 23 - 29, 1982. [Photosynthates; in Bulg., ab : E, R.]

50577 - **MANSFIELD, R., BARBER, J.** : Manganese levels associated with inside-out thylakoid membranes in relation to oxygen evolution. - FEBS Lett. *140* : 165 - 168, 1982.

50578 - **MANSFIELD, R.W., NAKATANI, H.Y., BARBER, J., MAURO, S., LANNOYE, R.** : Charge density on the inner surface of pea thylakoid membranes. - FEBS Lett. *137* : 133 - 136, 1982.

50579 - **MÄNTELE, W., SIEBERT, F., KREUTZ, W.** : Kinetic properties of rhodopsin and bacteriorhodopsin measured by kinetic infrared spectroscopy (KIS). - In : COLOWICK, S.P., KAPLAN, N.O. (ed.) : Methods in Enzymology. Vol.88. Pp. 729 - 740. Academic Press, New York - London - Paris - San Diego - San Francisco - São Paulo - Sydney - Tokyo - Toronto 1982.

50580 - **MAR, T., PICOREL, R., GINGRAS, G.** : Photosynthetic unit size and electron-transport chain in a photoreaction center-depleted mutant of *Rhodospirillum rubrum*. - Biochim. biophys. Acta *682* : 354 - 363, 1982.

*50581 - **MARANO, B.** : Relationships between nutrient supply and meteorological factors in optimizing maize hybrid productivity. - In : Proceedings of the International Seminar on Soil Environment and Fertility Management in Intensive Agriculture. Pp. 525 - 534. Tokyo 1977. [Growth analysis.]

*50582 - **MARANO, B.** : Analisi della crescita e della produttività di due ibridi di mais di diversa classe di maturità. Azione della competizione e della concimazione. [Growth analysis and productivity of two hybrids of maize of different maturity. Action of competition and fertilization.] - Ann. Fac. Sci. agr. Univ. Napoli Portici, Ser. IV *13* : 294 - 312, 1979. [In Ital., ab : E.]

*50583 - **MARANO, B., MATTEI, F.** : Behaviour of sorghum hybrid (*Sorghum bicolor* (L.) MOENCH × *Sorghum sudanese* (PIPER) STAPF) in presence of high plant density. II - Chemical composition and nutritive value. - Agrochimica *21* : 370 - 378, 1977. [Growth analysis.]

*50584 - **MARANO, B., MATTEI, F.** : Behaviour of a sorghum hybrid (*Sorghum bicolor* L., MOENCH × *Sorghum sudanese* PIPER, STAPF.) in presence of high plant density. I - Competition and productivity. - Ann. Fac. Sci. agr. Univ. Napoli Portici, Ser. IV *12* : 217 - 230, 1978.

*50585 - **MARANO, B., PALMIERI, G.** : Analisi della crescita e produttività di un ibrido di girasole per effetto della competizione e della concimazione. [Competition and fertilization action on the growth and productivity of a sunflower hybrid.] - In : Convegno sugli Aspetti Genetici Agronomici e Patologici del Girasole e sulle Caratteristiche Industriali, Alimentari e Commerciali del Prodotto. Pp. 1 - 17. Consiglio Nazionale della Ricerche, Pisa 1978. [In Ital., ab : E.]

*50586 - **MARANO, B., PALMIERI, G.** : Azione della competizione e della disponibilità idrica sulla crescita, produttività e valore nutritivo della granella di due mais ibridi a diverso ciclo di maturità. Prova condotta nell'ambiente pedoclimatico del basso Volturno. [Action of competition and water availability on growth, productivity and nutritive value of grains of two maize hybrids of different maturity cycle. Test in pedoclimatic conditions of the Volturno basin.] - Ann. Fac. Sci. agr. Univ. Napoli Portici, Ser. IV *12* : 145 - 158, 1978. [In Ital., ab : E.]

50587 - MARANO, B., PALMIERI, G. : Effects of bentazon and phosphate fertilization on some growth and productivity parameters related to chlorophyll and phosphorus content of two soybean cultivars. - Agrochimica *26* : 431 - 441, 1982.

50588 - MARCHYULENENE, D.P., MOTEYUNENE, É.B., SHULIENE, R.Yu., VROBLYAVICHYUTE, R.R., LEGIN, V.K. : Vliyanie kompleksona ÉDTA na vodorosli. [Effect of EDTA on algae.] - Gidrobiol. Zh. *18* (5) : 88 - 95, 1982. [Ps; in R, ab : E.]

50589 - MARCUS, Y., ZENVIRTH, D., HAREL, E., KAPLAN, A. : Induction of HCO_3^- transporting capability and high photosynthetic affinity to inorganic carbon by low concentration of CO_2 in *Anabaena variabilis*. - Plant Physiol. *69* : 1008 - 1012, 1982.

50590 - MARDER, J.B., MATTOO, A.K., EDELMAN, M. : Photosynthesis and the rapidly-metabolised 32 kilodalton chloroplast membrane protein. - In : AKOYUNOGLOU, G., EVANGELOPOULOS, A.E., GEORGATSOS, J., PALAIOLOGOS, G., TRAKATELLIS, A., TSIGANOS, C.P. (ed.) : Cell Function and Differentiation, Part B. Biogenesis of Energy Transducing Membranes and Membrane and Protein Energetics. Pp. 91 - 100. Alan R. Liss Inc., New York 1982.

50591 - MARGARIS, N.S. : Quality and quantity of latex which can be produced from natural vegetation in Greece. - In : GRASSI, G., PALZ, W. (ed.) : Energy from Biomass. Vol.3. Pp. 49 - 53. D.Reidel Publ. Co., Dordrecht - Boston - London 1982. [Plant age and biomass.]

50592 - MARIANI, P., RASCIO, N. : Plastid ultrastructure in etiolated seedlings of *"Ginkgo biloba"* L. - Caryologia *35* : 390 - 391, 1982.

*50593 - MARIANI COLOMBO, P., RASCIO, N. : Ruthenium red staining for electron microscopy of plant material. - J. Ultrastruct. Res. *60* : 135 - 139, 1977. [Chloroplast.]

50594 - MARINI, R.P., BARDEN, J.A. : Net photosynthesis, dark respiration, transpiration, and stomatal resistance of young and mature apple trees as influenced by summer or dormant pruning. - J. amer. Soc. hort. Sci. *107* : 170 - - 174, 1982.

50595 - MARKAROVA, E.N., SULEĬMANOVA, Sh.S., VESELOVSKIĬ, V.A., MINEEVA, L.A. : Vliyanie sveta vysokoĭ intensivnosti na fotosinteticheskie reaktsii sinezelenykh vodorosleĭ. [Effect of high light intensity on photosynthetic reactions in blue-green algae.] - Fiziol. Rast. *29* : 578 - 585, 1982. [In R, ab : E.]

50596 - MARKER, A.F.H., JINKS, S. : The spectrophotometric analysis of chlorophyll a and phaeopigments in acetone, ethanol and methanol. - Arch. Hydrobiol. Beih. Ergeb. Limnol. *16* : 3 - 17, 1982.

50597 - MARKOVSKAYA, E.F., OBSHATKO, L.A. : O roli temperaturnogo faktora na rannikh ètapakh ontogeneza ogurechnogo rasteniya. [Role of temperature factor in early phases of ontogeny of a cucumber plant.] - In : Vliyanie Faktorov Vneshneĭ Sredy i Fiziologicheski Aktivnykh Veshchestv na Termorezistentnost' i Produktivnost' Rasteniĭ. Pp. 104 - 108, 157 - 158. Inst. Biol. karel'. Filiala Akad. Nauk SSSR, Petrozavodsk 1982. [Dry-matter production; in R.]

50598 - MARKWELL, J.P., BAKER, N.R., THORNBER, J.P. : Metabolic regulation of the thylakoid protein kinase. - FEBS Lett. *142* : 171 - 174, 1982.

50599 - MARKWELL, J.P., THORNBER, J.P. : Treatment of the thylakoid membrane with surfactants. Assessment of effectiveness using the chlorophyll a absorption spectrum. - Plant Physiol. *70* : 633 - 636, 1982.

50600 - MARRA, J., HEINEMANN, K. : Photosynthesis response by phytoplankton to sunlight variability. - Limnol. Oceanogr. *27* : 1141 - 1153, 1982.

50601 - MARRA, J., HOUGHTON, R.W., BOARDMAN, D.C., NEALE, P.J. : Variability in surface chlorophyll a at a shelf-break front. - J. mar. Res. *40* : 575 - 591, 1982.

50602 - **MARRS, B.L.** : Genetic analysis of carotenogenesis in *Rhodopseudomonas capsulata*. - In : BRITTON, G., GOODWIN, T.W. (ed.) : Carotenoid Chemistry and Biochemistry. Pp. 273 - 277. Pergamon Press, Oxford - New York - Toronto - - Sydney - Paris - Frankfurt 1982.

50603 - **MARSH, D., WATTS, A.** : Diffusible spin labels used to study lipid-protein interactions with rhodopsin and bacteriorhodopsin. - In : COLOWICK, S.P., KAPLAN, N.O. (ed.) : Methods in Enzymology. Vol.88. Pp. 762 - 772. Academic Press, New York - London - Paris - San Diego - San Francisco - São Paulo - Sydney - Tokyo - Toronto 1982.

50604 - **MARSHALL, B., BISCOE, P.V.** : Environmental and physiological factors affecting assimilate supply during grain growth. - In : HAWKINS, A.F., JEFFCOAT, B. (ed.) : Opportunities for Manipulation of Cereal Productivity. Monograph 7. Pp. 179 - 192. British Plant Growth Regulator Group, Wantage 1982. [Ps.]

*50605 - **MARSHALL, H.G., NESIUS, K.K., CIBIK, S.J.** : Phytoplankton studies within the Virginia Barrier Islands II.Seasonal study of phytoplankton within the Barrier Island channels. - Castanea *46* : 89 - 99, 1981.

50606 - **MARTIN, B., ORT, D.R.** : Insensitivity of water-oxidation and photosystem II activity in tomato to chilling temperatures. - Plant Physiol. *70* : 689 - - 694, 1982.

50607 - **MARTIN, C.E.** : Translocation of nocturnally fixed ^{14}C in the Crassulacean Acid Metabolism epiphyte *Tillandsia usneoides* L. - Bot. Gaz. *143* : 1 - 4, 1982.

50608 - **MARTIN, C.E., CHURCHILL, S.P.** : Chlorophyll concentrations and a/b ratios in mosses collected from exposed and shaded habitats in Kansas. - J. Bryol. *12* : 297 - 304, 1982.

50609 - **MARTIN, C.E., LUBBERS, A.E., TEERI, J.A.** : Variability in Crassulacean Acid Metabolism: a survey of North Carolina succulent species. - Bot. Gaz. *143* : 491 - 497, 1982.

50610 - **MARTIN, S.S.** : Sugarbeet. - In : TEARE, I.D., PEET, M.M. (ed.) : Crop Water Relations. Pp. 423 - 444. John Wiley & Sons, New York - Chichester - Brisbane - Toronto - Singapore 1982. [Ps.]

50611 - **MARTIN, W.L., SHARIK, T.L., ODERWALD, R.G., SMITH, D.W.** : Phytomass: Structural relationships for woody plant species in the understory of an Appalachian oak forest. - Can. J. Bot. *60* : 1923 - 1927, 1982.

50612 - **MARTINOIA, E., DALLING, M.J., MATILE, P.** : Catabolism of chlorophyll: Demonstration of chloroplast localized peroxidative and oxidative activities. - Z. Pflanzenphysiol. *107* : 269 - 279, 1982.

50613 - **MARTÍ VIUDES, M.** : Estudio de los factores que afectan la asimilación y la excreción de *Phaeodactylum tricornutum* BOHLIN en cultivos de volumen limitado. [Factors affecting assimilation and excretion rates in *Phaeodactylum tricornutum* BOHLIN in batch cultures.] - Invest. Pesq. *46* : 91 - 119, 1982. [In Span., ab : E.]

50614 - **MARUYAMA, K., OTSUKI, T.** : [Advances of photosynthesis mechanism in organic chemistry.] - Kagaku (Kyoto) *37* : 329 - 333, 1982. [In Jap.]

50615 - **MASKIEWICZ, R., BIELSKI, B.H.J.** : Kinetics of some electron-transfer reactions in biological photosystems II. Study of intramolecular electron- -transfer rates between ferredoxin-NADP reductase, NADP⁺ radical and oxidized ferredoxin. - Biochim. biophys. Acta *680* : 297 - 303, 1982.

50616 - **MASLOV, V.G.** : Primenenie spektroskopii vyzhiganiya provalov dlya issledovaniya pervichnykh protsessov fotosinteza. [Application of hole-burning spectroscopy to the study of primary processes of photosynthesis.] - Eesti NSV Tead. Akad. Toim., Füüs., Mat. *31* : 166 - 169, 1982. [In R, ab : E, Est.]

50617 - **MASLOV, V.G., CHUNAEV, A.S.** : Issledovanie pervichnykh fotoprotsessov v fotosisteme 2 mutantov khlamidomonady metodom spektroskopii vyzhiganiya provalov. [Study of primary photoprocesses in photosystem 2 of *Chlamydomonas* mutant strains by hole-burning spectroscopy.] - Mol. Biol. *16* : 604 - - 611, 1982. [In R, ab : E.]

50618 - **MASLOVA, T.G., ZELENSKIĬ, M.I., SAPOZHNIKOV, D.I.** : Izuchenie reaktsii violaksantinovogo tsikla v svyazi s vydeleniem kisloroda pri fotosinteze khlorelly. [Study of violaxanthin cycle reactions with respect to photosynthetic oxygen evolution in *Chlorella*.] - Fiziol. Rast. *29* : 697 - 704, 1982. [In R, ab : E.]

50619 - **MASSIG, G., STOCKBURGER, M., GÄRTNER, W., OESTERHELT, D., TOWNER, P.** : Structural conclusion on the Schiff base group of retinylidene chromophores in bacteriorhodopsin from characteristic vibrational bands in the resonance Raman spectra of BR_{570} (all-*trans*), BR_{603} (3-*dehydroretinal*) and BR_{548} (13-*cis*). - J. Raman Spectrosc. *12* : 287 - 294, 1982.

50620 - **MATHESON, I.B.C., RODGERS, M.A.J.** : Crocetin, a water-soluble carotenoid monitor for singlet molecular oxygen. - Photochem. Photobiol. *36* : 1 - 4, 1982.

50621 - **MATHIES, R.** : Resonance Raman spectroscopy of rhodopsin and bacteriorhodopsin isotopic analogs. - In : COLOWICK, S.P., KAPLAN, N.O. (ed.) : Methods in Enzymology. Vol.88. Pp. 633 - 643. Academic Press, New York - London - Paris - San Diego - San Francisco - São Paulo - Sydney - Tokyo - Toronto 1982.

50622 - **MATHIEU, Y.** : pH-dependence of phosphoenolpyruvate carboxylase from *Acer pseudoplatanus* cell suspensions. - Plant Sci. Lett. *28* : 111 - 119, 1982.

50623 - **MATHIS, P., SCHENCK, C.C.** : The functions of carotenoids in photosynthesis. - In : BRITTON, G., GOODWIN, T.W. (ed.) : Carotenoid Chemistry and Biochemistry. Pp. 339 - 351. Pergamon Press, Oxford - New York - Toronto - Sydney - Paris - Frankfurt 1982.

50624 - **MATHIS, P., SCHENCK, C.** : The mechanism of photosynthetic oxygen evolution: Primary photoreactions and electron carriers. - In : HALL, D.O., PALZ, W. (ed.) : Photochemical, Photoelectrochemical and Photobiological Processes. Vol.1. Pp. 129 - 133. D. Reidel Publ. Co., Dordrecht - Boston - London 1982.

50625 - **MATOCHA, J.E., PENNINGTON, D.** : Effects of plant iron recycling on iron chlorosis of grain sorghum grown on calcareous soils. - J. Plant Nutr. *5* : 869 - 882, 1982.

50626 - **MATOH, T., TAKAHASHI, E.** : Changes in the activities of ferredoxin- and NADH-glutamate synthase during seedling development of peas. - Planta *154* : 289 - 294, 1982.

50627 - **MATORIN, D.N., GROSSER, M., BORNER, T., KHERRMANN, F.N., GOSTIMSKIĬ, S.A.** : Funktsional'naya aktivnost' i polipeptidnyĭ sostav tilakoidnykh membran mutantov gorokha s neaktivnoĭ fotosistemoĭ II. - [Functional activity and polypeptide composition of thylakoid membranes of pea mutants with inactive photosystem II.] - Biol. Nauki *1982* (1) : 35 - 38, 1982. [In R.]

50628 - **MATORIN, D.N., ORTOIDZE, T.V., NIKOLAEV, G.M., VENEDIKTOV, P.S., RUBIN, A.B.** : Effects of dehydration on electron transport activity in chloroplasts. - Photosynthetica *16* : 226 - 233, 1982.

50629 - **MATSUDA, Y.** : [Sex differentiation and chloroplast genetics in *Chlamydomonas*.] - Iden *36* (10) : 11 - 17, 1982. [In Jap.]

*50630 - **MATSUMOTO, Y., MORIYA, T., HANDA, S., HANAMI, Y., KAKUBARI, Y., NEGISI, K.** : [A handy assimilation chamber following up fluctuations in ambient air temperature.] - Seibutsu Kankyo Tosetsu [Environm. Contr. Biol.] *19* : 17 - - 23, 1981. [In Jap., ab : E.]

50631 - **MATSUMOTO, Y., NEGISI, K.** : [Photosynthesis and respiration in advanced growth of *Abies veitchii* growing under a deep canopy and in the open.] - J. jap. Forest. Soc. *64* : 165 - 176, 1982. [In Jap., ab : E.]

50632 - **MATSUMOTO, Y., SUKIGARA, N., KAKUBARI, Y., NEGISI, K.** : [A system for measuring photosynthesis and respiration rates of standing trees and its application to a *Fagus crenata* stand.] - J. jap. Forest. Soc. *64* : 149 - 154, 1982. [In Jap.]

50633 - MATSUMURA-KADOTA, H., MUTO, S., MIYACHI, S. : Light-induced conversion of NAD$^+$ to NADP$^+$ in *Chlorella* cells. - Biochim. biophys. Acta *679*: 300 - 307, 1982.

50634 - MATSUNAGA, T., MITSUI, A. : Seawater-based hydrogen production by immobilized marine photosynthetic bacteria. - Biotechnol. Bioeng. Symp. *12* : 441 - - 450, 1982.

50635 - MATTOO, A.K., MARDER, J.B., GRESSEL, J., EDELMAN, M. : Presence of the rapidly-labelled 32 000-dalton chloroplast membrane protein in triazine-resistant biotypes. - FEBS Lett. *140* : 36 - 40, 1982.

50636 - MATUSZKIEWICZ, J.M., ROO-ZIELIŃSKA, E.M., SOLON, J. : Relation between the size of the above-ground and the underground phytomass of plant communities in the region of Gurwan-Turuu in Central Mongolia. - Acta Soc. Bot. Pol. *51* : 323 - 330, 1982.

50637 - MAUK, A.G., BORDIGNON, E., GRAY, H.B. : Analysis of the kinetics of electron transfer between blue copper proteins and inorganic redox agents. Reactions involving bis(dipicolinate) complexes of cobalt(III) and iron(II) and stellacyanin, plastocyanin, and azurin. - J. amer. chem. Soc. *104* : 7654 - 7657, 1982.

50638 - MAUNEY, J.R., SZAREK, S.R. : Drought-tolerant plants: dry matter production of crop and native species. - In : ZABORSKY, O.R. (ed.) : CRC Handbook of Biosolar Resources. Vol.I/2. Pp. 379 - 386. CRC Press, Boca Raton 1982.

50639 - MAURING, K.Kh., AVARMAA, R.A. : Mekhanizmy vyzhiganiya provalov v spectrakh khlorofilla i feofitina. [Mechanisms of hole-burning in the spectra of chlorophyll and pheophytin.] - Eesti NSV Tead. Akad. Toim., Füüs., Mat. *31* : 155 - 160, 1982. [In R, ab : E, Est.]

50640 - MAUSETH, J.D. : A morphometric study of the ultrastructure of *Echinocereus engelmannii (Cactaceae)*. IV. Leaf and spine primordia. - Amer. J. Bot. *69* : 546 - 550, 1982. [Chloroplast.]

50641 - MAUZERALL, D. : Statistical theory of the effect of mutiple excitation in photosynthetic systems. - In : ALFANO, R.R. (ed.) : Biological Events Probed by Ultrafast Laser Spectroscopy. Pp. 215 - 235. Academy Press, New York - London - Paris - San Diego - San Francisco - São Paulo - Sydney - Tokyo - Toronto 1982.

50642 - MAVOUNGOU,Z.C., BEUNARD, P., TRUONG, B. : Test de résistance à la transpiration de quatre variétés de soja en culture aéroponique. - Agron. trop. *37* : 288 - 294, 1982. [Stomatal resistance.]

50643 - MAYER, H. : Synthesis of optically active carotenoids with ε-end groups. - In : BRITTON, G., GOODWIN, T.W. (ed.) : Carotenoid Chemistry and Biochemistry. Pp. 55 - 70. Pergamon Press, Oxford - New York - Toronto - Sydney - Paris - Frankfurt 1982.

50644 - MAYNE, B.C., POOLE, R.E. : Nitrogen fixation by the unicellular cyanobacterium *Gloeothece*. Differential effect of uncouplers of phosphorylation on photosynthesis and acetylene reduction. - Annu. Rep. C.F.Kettering Res. Lab. *1982*: 15, 1982.

50645 - MAZUR, B.J., CHUI, C.-F. : Sequence of the gene coding for the β-subunit of dinitrogenase from the blue-green alga *Anabaena*. - Proc. nat. Acad. Sci. USA *79* : 6782 - 6786, 1982. [Chloroplast.]

50646 - MAZZOLINI, A.P., ANDERSON, C.A., LADIGES, P.Y., LEGGE, G.J.F. : Distribution of mineral elements in green and chlorotic leaf tissue of *Eucalyptus obliqua* L'HÉRIT., determined with a scanning proton microprobe. - Aust. J. Plant Physiol. *9* : 261 - 269, 1982.

50647 - McCARTHY, S.A., MATTHEIS, J.R., REBEIZ, C.A. : Chloroplast biogenesis: biosynthesis of protochlorophyll(ide) via acidic and fully esterified biosynthetic branches in higher plants. - Biochemistry *21* : 242 - 247, 1982.

50648 - McCARTY, R.E., CARMELI, C. : Proton translocating ATPases of photosynthetic membranes. - In : GOVINDJEE (ed.) : Photosynthesis. Vol.1. Pp. 647 - 695. Academic Press, New York - London - Paris - San Diego - San Francisco - São Paulo - Sydney - Tokyo - Toronto 1982.

50649 - MCCRACKEN, J.L., FRANK, H.A., SAUER, K. : Radical pair interactions in spinach chloroplasts. - Biochim. biophys. Acta *679* : 156 - 168, 1982.

*50650 - McCREE, K.J. : Photosynthetically active radiation. - In : LANGE, O.L., NOBEL, P.S., OSMOND, C.B., ZIEGLER, H. (ed.) : Physiological Plant Ecology I. Pp. 41 - 55. Springer-Verlag, Berlin - Heidelberg - New York 1981.

50651 - McCREE, K.J. : The role of respiration in crop production. - Iowa State J. Res. *56* : 291 - 306, 1982.

50652 - McCREE, K.J. : Maintenance requirements of white clover at high and low growth rates. - Crop Sci. *22* : 345 - 351, 1982.

50653 - McCREE, K.J., AMTHOR, M.E. : Effects of diurnal variation in temperature on the carbon balances of white clover plants. - Crop Sci. *22* : 822 - 827, 1982.

50654 - McDANIEL, M.E., BROWN, J.C. : Differential iron chlorosis of oat cultivars - - a review. - J. Plant Nutr. *5* : 545 - 552, 1982.

50655 - McDANIEL, R.G. : The physiology of temperature effects on plants. - In : CHRISTIANSEN, M.N., LEWIS, C.F. (ed.) : Breeding Plants for Less Favorable Environments. Pp. 13 - 45. A Wiley-Interscience Publication, John Wiley & Sons, New York - Chichester - Brisbane - Toronto - Singapore 1982. [Chloroplast.]

50656 - MCDONNEL, A., STAEHELIN, L.A. : Reconstitution of a chlorophyll-protein complex into liposomes. - In : EDELMAN, M., HALLICK, R.B., CHUA, N.-H. (ed.): Methods in Chloroplast Molecular Biology. Pp. 857 - 861. Elsevier Biomedical Press, Amsterdam - New York - Oxford 1982.

50657 - McEWAN, A.G., GEORGE, C.L., FERGUSON, S.J., JACKSON, J.B. : A nitrate reductase activity in *Rhodopseudomonas capsulata* linked to electron transfer and generation of a membrane potential. - FEBS Lett. *150* : 277 - 280, 1982.

50658 - McINTYRE, G.I., HSIAO, A.I. : Influence of nitrogen and humidity on rhizome bud growth and glyphosate translocation in quackgrass (*Agropyron repens*). - Weed Sci. *30* : 655 - 660, 1982. [Photosynthates.]

B50659 - McLAREN, J.S. (ed.) : Chemical Manipulation of Crop Growth and Development. - Butterworth Scientific, London - Boston - Durban - Singapore - Sydney - Toronto - Wellington 1982. [Ps, Chl, photorespiration.]

50660 - McLAUGHLIN, S.B., McCONATHY, R.K., DUVICK, D., MANN, L.K. : Effects of chronic air pollution stress on photosynthesis, carbon allocation, and growth of white pine trees. - Forest Sci. *28* : 60 - 70, 1982.

50661 - McLEAN, M.B., SAUER, K. : The dependence of reaction center and antenna triplets on the redox state of Photosystem I. - Biochim. biophys. Acta *679* : 384 - 392, 1982.

50662 - McMICHAEL, B.L., HESKETH, J.D. : Field investigations of the response of cotton to water deficits. - Field Crops Res. *5* : 319 - 333, 1982. [Ps.]

50663 - McMILLAN, C., SMITH, B.N. : Comparison of $\delta^{13}C$ values for seagrasses in experimental cultures and in natural habitats. - Aquat. Bot. *14* : 381 - 387, 1982.

50664 - McWILLIAM,J.R., KRAMER, P.J., MUSSER, R.L. : Temperature-induced water stress in chilling-sensitive plants. - Aust. J. Plant Physiol. *9* : 343 - - 352, 1982. [Ps.]

50665 - MÉALLIER, P., TISSUT, M., BASTIDE, J. : Relation spectre ultraviolet - activité inhibitrice des herbicides sur la photosynthèse. 1. Cas des benzamides. - Chemosphère *11* : 459 - 463, 1982.

50666 - **MEDINA, E.** : Temperature and humidity effects on dark CO_2 fixation by *Kalanchoe pinnata*. - Z. Pflanzenphysiol. *107*: 251 - 258, 1982.

50667 - **MEESON, B.W., CHANG, S.S., SWEENEY, B.M.** : Characterization of peridinin--chlorophyll *a*-proteins from the marine dinoflagellate *Ceratium furca*. - Bot. mar. *25* : 347 - 350, 1982.

50668 - **MEHLHORN, R.J., PROBST, I.** : Light-induced pH gradients measured with spin--labeled amine and carboxylic acid probes: Application to *Halobacterium halobium* cell envelope vesicles. - In : COLOWICK, S.P., KAPLAN, N.O. (ed.) : Methods in Enzymology. Vol.88. Pp. 334 - 344. Academic Press, New York - London - Paris - San Diego - San Francisco - São Paulo - Sydney - Tokyo - Toronto 1982.

50669 - **MEI Zhenan** : [Initial process of bacterial photosynthesis.] - Shengwu Huaxue Yu Shengwu Wuli Jinzhan *47* : 13 - 21, 1982. [In Chin.]

50670 - **MEIER, D., LICHTENTHALER, H.K.** : Special senescence stages in chloroplast ultrastructure of radish seedlings induced by the photosystem II-herbicide bentazon. - Protoplasma *110*: 138 - 142, 1982.

50671 - **MEINDL, U., KIERMAYER, O.** : Über die Kern- und Chloroplastenmigration von *Micrasterias denticulata* BRÉB. II. Die Chloroplastenmigration und ihre Veränderung durch verschiedene Stoffe. - Phyton *22* : 213 - 231, 1982.

50672 - **MEINHARDT, S.W., CROFTS, A.R.** : The site and mechanism of action of myxothiazol as an inhibitor of electron transfer in *Rhodopseudomonas sphaeroides*. - FEBS Lett. *149* : 217 - 222, 1982.

50673 - **MEINHARDT, S.W., CROFTS, A.R.** : Kinetics and thermodynamic resolution of cytochrome c_1 and cytochrome c_2 from *Rhodopseudomonas sphaeroides*. - FEBS Lett. *149* : 223 - 227, 1982.

50674 - **MEINZER, F.C.** : The effect of vapor pressure on stomatal control of gas exchange in Douglas fir (*Pseudotsuga menziesii*) saplings. - Oecologia *54* : 236 - 242, 1982.

50675 - **MEINZER, F.C.** : The effect of light on stomatal control of gas exchange in Douglas fir (*Pseudotsuga menziesii*) saplings. - Oecologia *54* : 270 - 274, 1982.

50676 - **MEINZER, F.C.** : Models of steady-state and dynamic gas exchange responses to vapor pressure and light in Douglas fir (*Pseudotsuga menziesii*) saplings. - Oecologia *55* : 403 - 408, 1982.

50677 - **MEISTER, A.** : Limits for decomposition of spectra into single bands. - J. theor. Biol. *94* : 541 - 553, 1982.

50678 - **MEISTER, A., BRECHT, E., JANK, H.-W.** : Zerlegung von Spektren in ihre Komponenten II. Spektrenzerlegung mit dem FORTRAN-Programm RESO. - Kulturpflanze *30* : 141 - 154, 1982.

50679 - **MELACK, J.M.** : Photosynthetic activity and respiration in an equatorial African soda lake. - Freshwater Biol. *12* : 381 - 399, 1982.

50680 - **MELANDRI, B.A., ZANNONI, D., DE SANTIS, A., CASADIO, R.** : Hydrogen photometabolism in *Rhodopseudomonas capsulata*. - In : HALL, D.O., PALZ, W. (ed.): Photochemical, Photoelectrochemical and Photobiological Processes. Vol.1. Pp. 160 - 164. D.Reidel Publ.Co., Dordrecht - Boston - London 1982.

*50681 - **MELCHIONNA, M., DE MASI, F.** : The fine structure of the vegetative cells of *Erythrocystis montagnei*, a symbiotic red alga. - Cytobios *20* : 113 - - 119, 1978. [Chloroplast.]

50682 - **MELIS, A., OW, R.A.** : Photoconversion kinetics of chloroplast photosystems I and II. Effect of Mg^{2+}. - Biochim. biophys. Acta *682* : 1 - 10, 1982.

50683 - **MELIS, A., ZEIGER, E.** : Chlorophyll *a* fluorescence transients in mesophyll and guard cells. Modulation of guard cell photophosphorylation by CO_2. - Plant Physiol. *69* : 642 - 647, 1982.

50684 - MELLER, E., GASSMAN, M.L. : Biosynthesis of 5-aminolevulinic acid: two pathways in higher plants. - Plant Sci. Lett. *26* : 23 - 29, 1982.

50685 - MENCZEL, L., GALIBA, G., NAGY, F., MALIGA, P. : Effect of radiation dosage on efficiency of chloroplast transfer by protoplast fusion in *Nicotiana*. - Genetics *100* : 487 - 495, 1982.

50686 - MENGEL, K., ARNEKE, W.-W. : Effect of potassium on the water potential, the pressure potential, the osmotic potential and cell elongation in leaves of *Phaseolus vulgaris*. - Physiol. Plant. *54* : 402 - 408, 1982.

B50687 - MENGEL, K., KIRKBY, E.A. : Principles of Plant Nutrition. 3rd Edition. - International Potash Institute, Worblaufen - Bern 1982. [Ps, Chl.]

50688 - MENOUX, Y., FERRON, F. : Effets d'une déshydratation sur l'ouverture stomatique et le métabolisme carbone à la lumière des feuilles de Lin (*Linum usitatissimum* L. cv.Hera). - Photosynthetica *16* : 526 - 532, 1982.

50689 - MERINO, J., FIELD, C., MOONEY, H.A. : Construction and maintenance costs of Mediterranean-climate evergreen and deciduous leaves. I.Growth and CO_2 exchange analysis. - Oecologia *53* : 208 - 213, 1982. [Growth analysis.]

50690 - MERRETT, M.J., ARMITAGE, T.L. : The effect of oxygen concentration on photosynthetic biomass production by algae. - Planta *155* : 95 - 96, 1982.

50691 - MERRITT, R.H., KOHL, H.C.,Jr.: Effect of root temperature and photoperiod on growth and crop productivity efficiency of *Petunia*. - J. amer. Soc. hort. Sci. *107* : 997 - 1000, 1982. [Growth analysis.]

50692 - MESSIER, J., GUCKERT, A. : Relations entre fertilisation azotée et activité photosynthétique chez la Betterave sucrière. - Bull. E.N.S.A.I.A. *24* : 81 - - 96, 1982.

*50693 - METEÏKO, T.Ya., MONASTYRETSKAYA, E.V. : Vliyanie sestona pri ispol'zovanii v kachestve udobreniĭ na dinamiku pigmentov u rasteniĭ. [Effect of seston used as fertilizer on pigment dynamics in plants.] - In : Formirovanie i Kontrol' Kachestva Poverkhnostnykh Vod. Pp. 84 -   . Naukova Dumka, Kiev 1976. [Chl, in R.]

50694 - METS, L. : Purification and characterization of chloroplast ribosomes and ribosomal subunits from *Chlamydomonas reinhardii*. - In : EDELMAN, M., HALLICK, R.B., CHUA, N.-H. (ed.) : Methods in Chloroplast Molecular Biology. Pp. 629 - 635. Elsevier Biomedical Press, Amsterdam - New York - Oxford 1982.

50695 - METZ, J.G., MILES, D. : Use of nuclear mutant of maize to identify components of Photosystem II. - Biochim. biophys. Acta *681* : 95 - 102, 1982.

50696 - METZGER, U. : Funktion und Eigenschaften des B-Proteins im photosynthetischen Elektronentransport. - In : HOFFMANN, P., HIEKE, B. (ed.) : Photosynthese: Regulation und Evolution. (Colloquia Pflanzenphysiologie Nr.5.).Pp. 93 - 97. Humboldt-Universität, Berlin 1982.

50697 - METZGER, U., KLOTZ, S. : Resistenz von Unkräutern gegenüber Harnstoffherbiziden. - In : HOFFMANN, P., HIEKE, B. (ed.) : Photosynthese: Regulation und Evolution. (Colloquia Pflanzenphysiologie Nr.5.). Pp. 172 - 175. Humboldt--Universität, Berlin 1982.

50698 - METZIG, G., RASCHKE, E. : The determination of chlorophyll from satellite measurements of the ocean color. - Ann. Meteorol. *18* : 45 - 47, 1982.

50699 - METZNER, H., FISCHER, K., BAZLEN, O. : Tracer studies on the oxygen evolution in photosynthesis. - In : SCHMIDT, H.-L., FÖRSTEL, H., HEINZINGER, K. (ed.) : Stable Isotopes. Pp. 517 - 527. Elsevier, Amsterdam 1982.

50700 - MEYER, H.-U., BIEHL, B. : Relation between photosynthetic and phenolase activities in spinach chloroplasts. - Phytochemistry *21* : 9 - 12, 1982.

50701 - MEYERS, S.P., MOLIN, W.T., SELMAN, B.R., SCHRADER, L.E. : Ploidy effects in isogenic populations of alfalfa III.Chloroplast thylakoid-bound coupling factor 1 in protoplasts and leaves. - Plant Physiol. *70* : 1715 - 1717, 1982.

50702 - MEYERS, S.P., NICHOLS, S.L., BAER, G.R., MOLIN, W.T., SCHRADER, L.E. :
Ploidy effects in isogenic populations of alfalfa I.Ribulose-1,5-bisphos-
phate carboxylase, soluble protein, chlorophyll, and DNA in leaves. - Plant
Physiol. 70 : 1704 - 1709, 1982.

*50703 - MEZHUNTS, B.Kh. : Osobennosti nekotorykh fiziologicheskikh funktsiĭ u raste-
niĭ rozovoĭ gerani i tabaka pri gidroponicheskom vyrashchivanii. [Characte-
ristics of some physiological functions of rose geranium and tobacco plants
grown in open-air hydroponics.] - Soobsh. Inst. agrokhim. Probl. Gidroponiki
Akad. Nauk arm. SSR 20 : 17 - 25, 1980. [Ps; in R, ab : E.]

50704 - MGALOBLISHVILI, M.P., LITVINOV, A.I., KALANDADZE, A.N., SANADZE, G.A. :
Deĭstvie tsiklogeksimida, khloramfenikola i diurona na fotosintez i izopre-
novyĭ èffekt protoplastov mezofilla list'ev topolya. [Effect of cycloheximide, chloramphenicol and DCMU on photosynthesis and isoprene biosynthesis
in mesophyll protoplasts of poplar leaves.] - Fiziol. Rast. 29 : 372 - 377,
1982. [In R, ab : E.]

50705 - MICHAEL, G. : Photosynthese unter SO$_2$- und Frost-stress bei Koniferen. - In :
UNGER, K., SCHUH, J. (ed.) : Umwelt-Stress. Pp. 281 - 284. Martin-Luther-
-Universität, Halle 1982.

50706 - MICHAEL, G., FEILER, S., RANFT, H., TESCHE, M. : Der Einfluß von Schwefeldi-
oxid und Frost auf Fichten (Picea abies (L.) KARST.). - Flora 172 : 317 -
- 326, 1982. [Ps.]

50707 - MICHALSKI, W.P., KANIUGA, Z. : Photosynthetic apparatus of chilling-sensitive
plants. XI. Reversibility by light of cold- and dark-induced inactivation of
cyanide-sensitive superoxide dismutase activity in tomato leaf chloroplasts.
- Biochim. biophys. Acta 680: 250 - 257, 1982.

50708 - MICHEL, H. : Three-dimensional crystals of a membrane protein complex. The
photosynthetic reaction centre from Rhodopseudomonas viridis. - J. mol.
Biol. 158 : 567 - 572, 1982.

50709 - MICHEL, H. : Characterization and crystal packing of three-dimensional bacteriorhodopsin crystals. - EMBO J. 1 : 1267 - 1271, 1982.

50710 - MICHEL, H., OESTERHELT, D. : Preparation of new two- and three-dimensional
crystal forms of bacteriorhodopsin. - In : COLOWICK, S.P., KAPLAN, N.O.
(ed.) : Methods in Enzymology. Vol.88. Pp. 111 - 117. Academic Press, New
York - London - Paris - San Diego - San Francisco - São Paulo - Sydney -
Tokyo - Toronto 1982.

50711 - MICHELSON, A.M. : Bioluminescence and its application. - In : HÉLÈNE, C.,
CHARLIER, M., MONTENAY-GARESTIER, T., LAUSTRIAT, G. (ed.) : Trends in Photobiology. Pp. 133 - 146. Plenum Press, New York - London 1982. [ATP determination.]

50712 - MIKHAĬLOVA, T.P. : Peredvizhenie ^{14}C-assimilyatov iz otdel'nykh list'ev
tabaka. [Translocation of ^{14}C-labelled photosynthates from single leaves
in tobacco.] - Fiziol. Rast. 29 : 32 - 37, 1982. [In R, ab : E.]

*50713 - MIKHEEVA, S.A. : O deĭstvii nekotorykh khimicheskikh agentov na svetozavisi-
muyu bioèlektricheskuyu aktivnost' list'ev rasteniĭ. [Effect of some chemical
factors on light-dependent bioelectric activity of plant leaves.] - In :
Svetozavisimaya Bioèlektricheskaya Aktivnost' List'ev Rasteniĭ. Pp. 63 - 76,
118. Ural'. Gos. Universitet, Sverdlovsk 1977. [Ps; in R, ab E.]

*50714 - MIKHEEVA, S.A., RYBIN, I.A., SHAVNIN, S.A. : O svetoindutsirovannykh izmene-
niyakh potentsiala lista kukuruzy. [Light-induced changes in potential of
maize leaves.] - In : Svetozavisimaya Bioèlektricheskaya Aktivnost' List'-
ev Rasteniĭ. Pp. 23 - 32, III. Ural'. Gos. Universitet, Sverdlovsk 1980.
[Ps; in R, ab : E.]

50715 - MIKULOVICH, T.P., KUKINA, I.M. : RNA synthesis in intact chloroplasts from
excised pumpkin cotyledons: some characteristics of transcription and the
effect of phytohormones. - Biochem. Physiol. Pflanzen 177 : 419 - 429,
1982.

50716 - MIKULSKA, E., DAMSZ, B., ŻOŁNIEROWICZ, H. : Structural and functional polymorphism of plastids in leaves of *Clivia miniata* RGL. II. Ontogenesis of
plastids in mesophyll cells and in cells surrounding vascular bundles. -
Acta Soc. Bot. Pol. *51* : 157 - 166, 1982.

50717 - MILBORROW, B.V. : Stereochemical aspects of carotenoid biosynthesis. - In :
BRITTON, G., GOODWIN, T.W. (ed.) : Carotenoid Chemistry and Biochemistry.
Pp. 279 - 295. Pergamon Press, Oxford - New York - Toronto - Sydney - Paris -
- Frankfurt 1982.

50718 - MILCHUNAS, D.G., LAUENROTH, W.K., DODD, J.L. : The effect of SO_2 on ^{14}C
translocation in *Agropyron smithii* RYDB. - Environm. exp. Bot. *22* : 81 - 91,
1982.

50719 - MILES, D. : The use of mutations to probe photosynthesis in higher plants. -
In : EDELMAN, M., HALLICK, R.B., CHUA, N.-H. (ed.) : Methods in Chloroplast
Molecular Biology. Pp. 75 - 107. Elsevier Biomedical Press, Amsterdam - New
York - Oxford 1982.

50720 - MILIVOJEVIĆ, D., KRSTIĆ, B., SARIĆ, M. : The effect of the deficiency of some
mineral nutrient ions on pigment content and chloroplast ultrastructure in
sunflower. - Period. Biol. *84* : 160 - 162, 1982.

*50721 - MILIVOJEVIĆ, D., STANKOVIĆ, Ž. : Grada i funkcija hloroplasta šećerne repe.
[Structure and function of sugar beet chloroplast.] - Posebna Izd. srpska
Akad. Umetn. *538*, Od. prirod.-mat. Nauka *54* [BELIĆ, J. (ed.) : Fiziologija
Šećerne Repe] : 1 - 15, 1981. [In Serb., ab : E.]

50722 - MILLAN-NUÑEZ, R., ALVAREZ-BORREGO, S., NELSON, D.M. : Effects of physical
phenomena on the distribution of nutrients and phytoplankton productivity
in a coastal lagoon. - Estuar. coast. Shelf Sci. *15* : 317 - 335, 1982.

*50723 - MILLER, B.L., HUFFAKER, R.C. : Hydrolysis of ribulose-1,5-bisphosphate carbo-
xylase by partially purified endoproteinases of senescing primary barley
leaves. - In : LYONS, J.M., VALENTINE, R.C., PHILLIPS, D.A., RAINS, D.W.,
HUFFAKER, R.C. (ed.) : Genetic Engineering of Symbiotic Nitrogen Fixation
and Conservation of Fixed Nitrogen. Pp. 293 - 303. Plenum Press, New York -
London 1981.

50724 - MILLER, B.L., HUFFAKER, R.C. : Hydrolysis of ribulose-1,5-bisphosphate carboxylase by endoproteinases from senescing barley leaves. - Plant Physiol.
69 : 58 - 62, 1982.

50725 - MILLER, M., COX, R.P. : Rapid equilibration of added Mn^{2+} across the chloroplast thylakoid membrane. - Photobiochem. Photobiophys. *4* : 243 - 248,
1982.

50726 - MILLER, M.E., PRICE, C.A. : Protein synthesis by developing plastids isolated
from *Euglena gracilis*. - FEBS Lett. *147* : 156 - 160, 1982.

50727 - MILLERO, F.J., MORSE, J.W. : The carbonate system in seawater. - In : ZABOR-
SKY, O.R. (ed.) : CRC Handbook of Biosolar Resources. Vol.I/2. Pp. 517 -
- 529. CRC Press, Boca Raton 1982.

50728 - MILLS, J.D., MITCHELL, P. : Modulation of coupling factor ATPase activity
in intact chloroplasts. Reversal of thiol modulation in the dark. - Biochim.
biophys. Acta *679* : 75 - 83, 1982.

50729 - MILLS, J.D., MITCHELL, P. : Thiol modulation of CF_0-CF_1 stimulates acid/
/base dependent phosphorylation of ADP by broken pea chloroplast. - FEBS
Lett. *144* : 63 - 67, 1982.

50730 - MINAS, H.J., CODISPOTI, L.A., DUGDALE, R.C. : Nutrients and primary production in the upwelling region off Northwest Africa. - Rapp. Proc.-verb.
Réun. Cons. int. Explor. Mer *180*: 148 - 183, 1982.

50731 - MINAS, Kh.Zh. : Diagramnyĭ analiz gidrologicheskikh i khimicheskikh faktorov
i ego primenenie v issledovanii sistemy produktsiya-regeneratsiya. [Diagramme
analysis of hydrological and chemical factors, and its use in the study of
the system production-regeneration.] - In : Pervichnaya i Vtorichnaya Produktsiya Morskikh Organizmov. Pp. 17 - 34. Kiev 1982. [Ps; in R.]

50732 - MINAS, M., ABBOUD, M., SLAWYK, G. : Production primaire et relations entre certains paramètres de la biomasse phytoplanctonique dans le Dôme de Guinée. - Rapp. Proc.-verb. Réun. Cons. int. Explor. Mer *180* : 214 - 218, 1982.

50733 - MINCHIN, P.E.H., THORPE, M.R. : Evidence for a flow of water into sieve tubes associated with phloem loading. - J. exp. Bot. *33* : 233 - 240, 1982. [Photosynthates.]

50734 - MINOT, R., MEUNIER, J.-C., BUC, J., RICARD, J. : The role of pH and magnesium concentration in the light activation of chloroplastic fructose bisphosphatase. - FEBS Lett. *142* : 118 - 120, 1982.

*50735 - MISHRA, S.D., GAUR, B.K., BEDEKAR, V.W., SINGH, B.B. : Paramagnetics changes in metallic ions during leaf senescence. - Acta bot. indica *4* : 1 - 5, 1976. [Chl.]

*50736 - MISHRA, S.D., GAUR, B.K., BEDEKAR, V.W., SINGH, B.B. : Isolation, identification and significance of free radical in senescing leaves. - Acta bot. indica *4* : 131 - 138, 1976. [Chl.]

50737 - MISRA, A.N., BISWAL, U.C. : Changes in the content of plastid macromolecules during aging of attached and detached leaves, and of isolated chloroplasts of wheat seedlings. - Photosynthetica *16* : 22 - 26, 1982.

50738 - MISRA, A.N., BISWAL, U.C. : Differential changes in the electron transport properties of thylakoid membranes during aging of attached and detached leaves, and of isolated chloroplasts. - Plant Cell Environm. *5* : 27 - 30, 1982.

*50739 - MISZALSKI, Z. : Photochemical activity of chloroplasts isolated from SO_2--fumigated spinach leaves in relation to changes in RuDPC activity. - Bull. Acad. pol. Sci., Sér. Sci. biol. *29* : 483 - 488, 1981.

50740 - MITSUI, A. : Nitrogen and hydrogen metabolism in marine tropical photosynthetic prokaryotes. - In : STANIER, R.Y., COHEN-BAZIRE, G. (ed.) : 5th International Symposium on Photosynthetic Prokaryotes. Pp. B66-1 - B66-2. Bombannes - Bordeaux 1982.

50741 - MITSUI, A., MURRAY, R., ENTENMANN, B., MIYAZAWA, K., POLK, E. : Utilization of marine blue-green algae and macroalgae in warm water mariculture. - In : SAN PIETRO, A. (ed.) : Biosaline Research. A Look to the Future. Pp. 216 - 225. Plenum Press, New York 1982.

50742 - MIURA, Y., YAGI, K., SHOGA, M., MIYAMOTO, K. : Hydrogen production by a green alga, *Chlamydomonas reinhardtii*, in an alternating light/dark cycle. - Biotechnol. Bioeng. *24* : 1555 - 1563, 1982.

50743 - MIZIORKO, H.M., BEHNKE, C.E., HOUKOM, E.C. : Protein liganding to the activator cation of ribulosebisphosphate carboxylase. - Biochemistry *21* : 6669 - 6674, 1982.

50744 - MIZUTANI, H., WADA, E. : Effect of high atmospheric CO_2 concentration on $\delta^{13}C$ of algae. - Origins Life *12* : 377 - 390, 1982.

50745 - MOELLER, C.H., MUDD, J.B. : Localization of filipin-sterol complexes in the membranes of *Beta vulgaris* roots and *Spinacia oleracea* chloroplasts. - Plant Physiol. *70* : 1554 - 1561, 1982.

50746 - MOHR, H. : Phytochrome and gene expression. - In : HÉLÈNE, C., CHARLIER, M., MONTENAY-GARESTIER, T., LAUSTRIAT, G. (ed.) : Trends in Photobiology. Pp. 515 - 530. Plenum Press, New York - London 1982. [Chloroplast.]

50747 - MOKRONOSOV, A.T. : Donorno-aktseptornye otnosheniya v ontogeneze rasteniĭ. [Source-sink relations in plant ontogeny.] - In : Fiziologiya Fotosinteza. Pp. 235 - 250. Nauka, Moskva 1982. [Photosynthates; in R.]

*50748 - MOLCHANOV, M.I., KOTOVSKAYA, A.P., TRUSOVA, V.M., SHAPOSHNIKOV, G.L. : Aminokislotnyĭ sostav belkovykh komponentov membran pri biogeneze khloroplastov. [Amino-acid composition of protein components of membranes in the chloroplast biogenesis.] - Dokl. Akad. Nauk SSSR *231* : 499 - 502, 1976. [In R.]

*50749 - MOLCHANOV, M.I., TRUSOVA, V.M., SERGIENKO, I.Z., KOTOVSKAYA, A.P. : Gliko-lipoproteidy membran pri biogeneze khloroplastov. [Glycolipoproteins of membranes in the chloroplast biogenesis.] - Dokl. Akad. Nauk SSSR *231*: 752 - 755, 1976. [In R.]

50750 - MOLIN, W.T., MEYERS, S.P., BAER, G.R., SCHRADER, L.E. : Ploidy effects in isogenic populations of alfalfa II. Photosynthesis, chloroplast number, ribulose-1,5-bisphosphate carboxylase, chlorophyll, and DNA in protoplasts. - Plant Physiol. *70* : 1710 - 1714, 1982.

50751 - MOLL, R.A., STOERMER, E.F. : A hypothesis relating trophic status and subsurface chlorophyll maxima of lakes. - Arch. Hydrobiol. *94* : 425 - 440, 1982.

50752 - MONDAL, R., CHOUDHURI, M.A. : Regulation of senescence of excised leaves of some C_3 and C_4 species by endogenous H_2O_2. - Biochem. Physiol. Pflanzen *177* : 403 - 417, 1982.

*50753 - MONOTTI, M., BORGHI, B., GHIDUCCI, M., BOGGINI, G., GAMBELLI, M.L. : Effects of irrigation and other agronomic practices on wheat grain yield. - In : Limites de Potentialité de Production du Blé dans Différents Systèmes de Culture et dans Différentes Zônes Mediterranéennes. Pp. 139 - 150. Ist. Agron. Gen. Coltiv. Erb., Univ. Bologna, Bologna 1981. [Growth analysis.]

50754 - MONSON, R.K., LITTLEJOHN, R.O.,Jr., WILLIAMS, G.J.,III : The quantum yield for CO_2 uptake in C_3 and C_4 grasses. - Photosynthesis Res. *3* : 153 - 159, 1982.

50755 - MONSON, R.K., STIDHAM, M.A., WILLIAMS, G.J.,III, EDWARDS, G.E., URIBE, E.G.: Temperature dependence of photosynthesis in *Agropyron smithii* RYDB. I. Factors affecting net CO_2 uptake in intact leaves and contribution from ribulose-1,5-bisphosphate carboxylase measured *in vivo* and *in vitro*. - Plant Physiol. *69* : 921 - 928, 1982.

50756 - MONSON, R.K., WILLIAMS, C.J.,III : A correlation between photosynthetic temperature adaptation and seasonal phenology patterns in the shortgrass prairie. - Oecologia *54* : 58 - 62, 1982.

50757 - MOOK, J.H., VAN DER TOORN, J. : The influence of environmental factors and management on stands of *Phragmites australis* II. Effects on yield and its relationships with shoot density. - J. appl. Ecol. *19* : 501 - 517, 1982. [Growth analysis.]

50758 - MOONEY, H.A., GULMON, S.L. : Constraints on leaf structure and function in reference to herbivory. - BioScience *32* : 198 - 206, 1982. [Ps.]

50759 - MOORE, A.L., JOY, A., TOM, R., GUST, D., MOORE, T.A., BENSASSON, R.V., LAND, E.J. : Photoprotection by carotenoids during photosynthesis: motional dependence of intramolecular energy transfer. - Science *216* : 982 - 984, 1982.

50760 - MOORE, C.J., LUFF, S.E., HALLAM, N.D. : Fine structure and physiology of the desiccation-tolerant mosses, *Barbula torquata* TAYL. and *Triquetrella papillata* (HOOK.F. and WILS.) BROTH., during desiccation and rehydration. - Bot. Gaz. *143* : 358 - 367, 1982. [Ps.]

50761 - MOORE, P.D. : Evolution of photosynthetic pathways in flowering plants. - Nature *295* : 647 - 648, 1982.

50762 - MORALES, D., JIMENEZ, M.S., IRIARTE, J., GIL, F. : Altitudinal effects on chlorophyll and carotenoid concentrations in gymnosperm leaves. - Photosynthetica *16* : 362 - 372, 1982.

50763 - MORAN, R. : Formulae for determination of chlorophyllous pigments extracted with N,N-dimethylformamide. - Plant Physiol. *69* : 1376 - 1381, 1982.

50764 - MOREHART, A.L., MELCHIOR, G.L. : Influence of water stress on verticillium wilt of yellow-poplar. - Can. J. Bot. *60* : 201 - 209, 1982. [Stomatal resistance.]

50765 - MORELAND, D.E., HUBER, S.C., NOVITZKI, W.P. : Interaction of herbicides with cellular and liposome membranes. - In : MORELAND, D.E., St.JOHN, J.B., HESS, F.D. (ed.) : Biochemical Responses Induced by Herbicides. Pp. 79 - 96. American Chemical Society, Washington, D.C. 1982. [Ps.]

*50766 - MORGAN, D.C., SMITH, H. : Non-photosynthetic responses to light quality. - In : LANGE, O.L., NOBEL, P.S., OSMOND, C.B., ZIEGLER, H. (ed.) : Physiological Plant Ecology I. Pp. 109 - 134. Springer-Verlag, Berlin - Heidelberg - - New York 1981. [Ps, Chl.]

50767 - MORITA, S. : Response of photosynthetic bacteria to light quality, light intensity, temperature, CO_2, HCO_3^-, O_2, and pH. - In : ZABORSKY, O.R. (ed.): CRC Handbook of Biosolar Resources. Vol.I/2. Pp. 147 - 152. CRC Press, Boca Raton 1982. [Ps.]

50768 - MORITA, S., ITOH, S., NISHIMURA, M. : Correlation between the activity of membrane-bound ATPase and the decay rate of flash-induced 515-nm absorbance change in chloroplasts in intact leaves, assayed by means of rapid isolation of chloroplasts. - Biochim. biophys. Acta 679 : 125 - 130, 1982.

50769 - MORONEY, J.V., McCARTY, R.E. : Effect of proteolytic digestion on the Ca^{2+}--ATPase activity and subunits of latent and thiol-activated chloroplast coupling factor 1. - J. biol. Chem. 257: 5910 - 5914, 1982.

50770 - MORONEY, J.V., McCARTY, R.E. : Light-dependent cleavage of the γ subunit of coupling factor 1 by trypsin causes activation of Mg^{2+}-ATPase activity and uncoupling of photophosphorylation in spinach chloroplasts. - J. biol. Chem. 257 : 5915 - 5920, 1982.

50771 - MORONEY, J.V., WARNCKE, K., McCARTY, R.E. : The distance between thiol groups in the γ subunit of coupling factor 1 influences the proton permeability of thylakoid membranes. - J. Bioenerg. Biomembranes 14 : 347 - 359, 1982.

50772 - MORRIS, I. : Primary production of the oceans. - In : BURNS, R.G., SLATER, J.H. (ed.) : Experimental Microbial Ecology. Pp. 239 - 252. Blackwell Scientific Publications, Oxford 1982.

50773 - MORRIS, P., HALL, D.O. : The inherent stability of Chenopodium quinoa chloroplasts. - Plant Sci. Lett. 25 : 353 - 357, 1982. [Ps.]

50774 - MORRIS, P., NASH, G.V., HALL, D.O. : The stability of electron transport in in vitro chloroplast membranes. - Photosynthesis Res. 3 : 227 - 240, 1982.

50775 - MORRISON, L.E., SCHELHORN, J.E., COTTON, T.M., BERING, C.L., LOACH, P.A. : Electrochemical and spectral properties of ubiquinone and synthetic analogs: relevance to bacterial photosynthesis. - In : TRUMPOWER, B.L. (ed.) : Function of Quinones in Energy Conserving Systems. Pp. 35 - 58. Academic Press, New York 1982.

*50776 - MORRISON, S.L., HUFFAKER, R.C., GUIDARA, C.R. : Light interaction with nitrate reduction. - In : LYONS, J.M., VALENTINE, R.C., PHILLIPS, D.A., RAINS, D.W., HUFFAKER, R.C. (ed.) : Genetic Engineering of Symbiotic nitrogen Fixation and Conservation of Fixed Nitrogen. Pp. 547 - 560. Plenum Press, New York - London 1981. [Ps.]

50777 - MÖRSCHEL, E. : Accessory polypeptides in phycobilisomes of red algae and cyanobacteria. - Planta 154 : 251 - 258, 1982.

50778 - MORTIMER, D.C., CZUBA, M. : Structural damage to leaf chloroplasts of Elodea densa caused by methylmercury accumulated from water. - Ecotoxicol. environm. Safety 6 : 193 - 195, 1982.

50779 - MOSER, L.E., VOLENEC, J.J., NELSON, C.J. : Respiration, carbohydrate content, and leaf growth of tall fescue. - Crop Sci. 22 : 781 - 786, 1982.

50780 - MOSKALENKO, A.A., SULEĬMANOV, S.Yu., KUZNETSOVA, N.Yu., EROKHIN, Yu.E., ALIEV, D.A. : Spektral'nye kharakteristiki i osobennosti sostoyaniya pigmentov v svetosobirayushchem komplekse iz khloroplastov pshenitsy. [Spectral characteristics and states of pigments in the light-harvesting complex from wheat chloroplasts.] - Fiziol. Rast. 29 : 687 - 696, 1982. [In R, ab : E.]

50781 - MOSLEMI, A.A. : Tree biomass. - In : SMITH, W.R. (ed.) : Energy from Forest
Biomass. Pp. 55 - 65. Academic Press, New York - London 1982.

50782 - MOTOMURA, Y. : Incorporation of ^{14}C-assimilates into GA-treated and -untreat-
ed inflorescences following assimilation of ^{14}CO$_2$ by individual leaves in
grape shoot. - Tohoku J. agr. Res. *33* : 1 - 13, 1982.

50783 - MOTT, K.A., GIBSON, A.C., O'LEARY, J.W. : The adaptive significance of am-
phistomatic leaves. - Plant Cell Environm. *5* : 455 - 460, 1982. [Ps.]

50784 - MOTTO, M.G., NAKANISHI, K. : Heavy atom labeling of retinal in bacteriorhodopsin. - In : COLOWICK, S.P., KAPLAN, N.O. (ed.) : Methods in Enzymology.
Vol.88. Pp. 178 - 180. Academic Press, New York - London - Paris - San Diego - San Francisco - São Paulo - Sydney - Tokyo - Toronto 1982.

50785 - MOUTONNET, P., BOIS, J.-F. : Efficience comparée de l'eau transpirée par le
maïs, le tournesol et le riz. - Agr. Meteorol. *27* : 209 - 215, 1982. [Ps.]

50786 - MOYSE, A. : La respiration des Végétaux verts à l'obscurité: les effets de
la lumière sur cette respiration. - Bull. Soc. bot. Fr., Actual. bot. *129*:
53 - 72, 1982. [Ps.]

*50787 - MRÁČEK, Z. : Zápoj v kulturách borovice (*Pinus silvestris* L.) založených
s různou počáteční hustotou sazenic. [Canopy in pine plantations (*Pinus
silvestris* L.) established with various plant density.] - Práce VÚLHM [Rep.
Forest. Game Managem. Res. Inst.] *50*: 115 - 127, 1977. [Canopy development;
in Czech, ab : E, R.]

50788 - MUALLEM, A., HALL, D.O. : Ascorbate as a substrate for photoproduction of
hydrogen by photosystem I of chloroplasts. - Plant Physiol. *69* : 1116 -
- 1120, 1982.

50789 - MUCHOW, R.C., CHARLES-EDWARDS, D.A. : An analysis of the growth of mung
beans at a range of plant densities in tropical Australia I.Dry matter
production. - Aust. J. agr. Res. *33* : 41 - 51, 1982.

50790 - MUCHOW, R.C., CHARLES-EDWARDS, D.A. : An analysis of the growth of mung
beans at a range of plant densities in tropical Australia II. Seed production. - Aust. J. agr. Res. *33* : 53 - 61, 1982.

50791 - MUCHOW, R.C., COATES, D.B., WILSON, G.L., FOALE, M.A. : Growth and productivity of irrigated *Sorghum bicolor* (L. MOENCH) in northern Australia. I
Plant density and arrangement effects on light interception and distribution,
and grain yield, in the hybrid Texas 610SR in low and medium latitudes. -
Aust. J. agr. Res. *33* : 773 - 784, 1982.

50792 - MUDD, J.B. : Effects of oxidants on metabolic function. - In : UNSWORTH,
M.H., ORMROD, D.P. (ed.) : Effects of Gaseous Air Pollution in Agriculture
and Horticulture. Pp. 189 - 203, Butterworth Scientific, London - Boston -
- Sydney - Wellington - Durban - Toronto 1982. [Chloroplast.]

50793 - MUDD, J.B. : Lipid metabolism. - In : SCHIFF, J.A., LYMAN, H. (ed.) : On
the Origins of Chloroplasts. Pp. 131 - 148. Elsevier/North-Holland, New
York - Amsterdam - Oxford 1982. [Chloroplast.]

50794 - MUKOHATA, Y. : Energy conversion in photosynthetic membranes and organelles.
a. H$^+$ pumping and ATP synthesis. - Recent Progr. nat. Sci. Jap. *7* : 41 - 43,
1982.

50795 - MUKOHATA, Y., SUGIYAMA, Y. : Isolation of the white membrane of crystalline
bacterio-opsin from *Halobacterium halobium* R$_1$mW lacking carotenoid. - In :
COLOWICK, S.P., KAPLAN, N.O. (ed.) : Methods in Enzymology. Vol.88. Pp. 407 -
- 411. Academic Press, New York - London - Paris - San Diego - San Francisco -
- São Paulo - Sydney - Tokyo - Toronto 1982.

50796 - MULDOON, D.K., PEARSON, C.J., WHEELER, J.L. : The effect of temperature
on growth and development of *Echinochloe* millets. - Ann. Bot. *50* : 665 -
- 672, 1982. [Growth analysis.]

50797 - MÜLLER, D., KLUGE, M., GRÖSCHEL-STEWART, U. : Comparative studies on immunological and molecular properties of phosphoenolpyruvate carboxylase in species of *Sedum* and *Kalanchoë* performing crassulacean acid metabolism (CAM). - Plant Cell Environm. *5* : 223 - 230, 1982.

*50798 - MÜLLER, E., HATTENBACH, A., GÜNDEL, J., HERMANN, G. : Biomembranes and light. — In : Biophysics of Membrane Transport. Part II. Pp. 311 - 336. Wroclaw 1981. [Chl.]

50799 - MÜLLER, H., NEUFANG, H., KNOBLOCH, K. : Purification and properties of the coupling-factor ATPases F_1 from *Rhodopseudomonas palustris* and *Rhodopseudomonas sphaeroides*. - Europe. J. Biochem. *127* : 559 - 566, 1982.

50800 - MÜLLER, J. : Auswirkung von Stress auf die CO_2-Aufnahmerate - Abschätzung mit einem Modell. - In : UNGER, K., SCHUH, J. (ed.) : Umwelt-Stress. Pp. 212 - 216. Martin-Luther-Universität, Halle 1982.

50801 - MÜLLER, J. : Analyse der ontogenetischen Veränderungen der CO_2-Aufnahme von Blättern bei Winterweizen in der Kornfüllungsperiode mit einem Modell für den CO_2-Gaswechsel. - In : HOFFMANN, P., HIEKE, B. (ed.): Photosynthese: Regulation und Evolution. (Colloquia Pflanzenphysiologie Nr.5.) Pp. 247 - - 249. Humboldt-Universität, Berlin 1982.

50802 - MÜLLER, M., NISIUS, A. : Protochlorophyllide photoconversion initiates the transformation of reaggregated prolamellar body tubules *in vitro*. - Z. Naturforsch. *37 C* : 476 - 480, 1982.

50803 - MÜLLER, R.K., BERNHARD, K., VECCHI, M. : Recent advances in the synthesis and analysis of 3,4-oxygenated xanthophylls. - In : BRITTON, G., GOODWIN, T.W. (ed.) : Carotenoid Chemistry and Biochemistry. Pp. 27 - 54. Pergamon Press, Oxford - New York - Toronto - Sydney - Paris - Frankfurt 1982.

50804 - MULLET, J.E., GROSSMAN, A.R., CHUA, N.-H. : Synthesis and assembly of the polypeptide subunits of photosystem I. - Cold Spring Harbor Symp. quant. Biol. *46* : 979 - 984, 1982.

50805 - MUNAWAR, M., MUNAWAR, I.F., ROSS, P.E., DAGENAIS, A. : Microscopic evidence of phytoplankton passing through glass-fibre and its implications for chlorophyll analysis. - Arch. Hydrobiol. *94* : 520 - 528, 1982.

50806 - MUNDA, I.M., MARKHAM, J.W. : Seasonal variations of vegetation patterns and biomass constituents in the rocky eulittoral of Helgoland. - Helgol. Meeresuntersuch. *35* : 131 - 151, 1982.

50807 - MURAKAMI, S. : [Surface structure of photosynthetic membranes.] - In : YOSHIDA, M. (ed.) : Fuoto Baiorojii: Koseiri Gensho No Shoki Katei. Pp. 194 - 208. Kodansha Saientifiku, Tokyo 1982. [In Jap.]

50808 - MURAKAMI, S., IKEUCHI, M. : Biochemical characterization and localization of the 36,000-dalton NADPH: protochlorophyllide oxidoreductase in squash etioplasts. - In : AKOYUNOGLOU, G., EVANGELOPOULOS, A.E., GEORGATSOS, J., PALAIOLOGOS, G., TRAKATELLIS, A., TSIGANOS, C.P. (ed.) : Cell Function and Differentiation, Part B. Biogenesis of Energy Transducing Membranes and Membrane and Protein Energetics. Pp. 13 - 23. Alan R. Liss Inc., New York 1982.

*50809 - MURAKAMI, T. : [Studies on the apparent photosynthetic rate of mulberry plant II. Profile of the photosynthetic rate according to the leaf order at different growth stages.] - Bull. sericult. Exp. Sta. *28* : 575 - 598, 1981. [In Jap., ab : E.]

*50810 - MURAKAMI, T. : [Studies on the apparent photosynthetic rate of mulberry plant III. Change in relationship between the photosynthetic rate and the aging of the leaf of mulberry plants.] - Bull. sericult. Exp. Sta. *28* : 599 - 612, 1981. [In Jap., ab : E.]

50811 - MURAKAMI, T. : Characteristics of photosynthesis in mulberry leaves. - Jap. agr. Res. quart. *16* : 46 - 50, 1982.

50812 - MURAKAMI, T., INAYAMA, M. : Effects of growing temperature on translocation of ^{14}C-photosynthates in cucumber seedlings. - Jap. agr. Res. quart. *16* : 105 - 113, 1982.

50813 - **MURATA, N.** : [Lipid structure of photosynthesis membranes.] - In : YOSHIDA, M. (ed.) : Fuoto Baiorojii: Koseri Gensho No Shoki Katei. Pp. 178 - 194. Kodansha Saientifiku, Tokyo 1982. [In Jap.]

50814 - **MURATA, N., MURAKAMI, S.** : Photosynthetic membranes. - Recent Progr. nat. Sci. Jap. *7* : 9 - 14, 1982.

*50815 - **MURATA, Y., YAMAGUCHI, J.** : [Photosynthesis and dry matter production.] - Tanpakushitsu Kakusan Koso Bessatsu *21* : 203·- 213, 1979. [In Jap.]

50816 - **MURAV'EVA, A.S., GAZIZOVA, I.I.** : Izmenenie biosinteza pigmentov v list'-yakh bobov pod vliyaniem rogora. [Changes in biosynthesis of pigments in bean leaves under the effect of rogor.] - Fiziol. Biokhim. kul't. Rast. *14* : 590 - 595, 1982. [In R, ab : E.]

50817 - **MUREĬ, I.A., VELICHKOV, D.K., SHUL'GIN, I.A.** : Gazoobmen podsolnechnika i kukuruzy posle vklyucheniya sveta. [Gas exchange in sunflower and maize following switching on the light.] - Fiziol. Rast. *29* : 70 - 79, 1982. [In R, ab : E.]

50818 - **MURPHY, D.J.** : The importance of non-planar bilayer regions in photosynthetic membranes and their stabilisation by galactolipids. - FEBS Lett. *150* : 19 - 26, 1982.

50819 - **MURPHY, D.J., WALKER, D.A.** : The properties of transketolase from photosynthetic tissue. - Planta *155* : 316 - 320, 1982.

50820 - **MURPHY, D.J., WALKER, D.A.** : Acetyl coenzyme A biosynthesis in the chloroplast. What is the physiological precursor? - Planta *156* : 84 - 88, 1982.

50821 - **MURPHY, E.M., FERRELL, W.K.** : Diurnal and seasonal changes in leaf conductance, xylem water potential, and abscisic acid of Douglas-fir [*Pseudotsuga menziesii* (MIRB.) FRANCO] in five habitat types. - Forest Sci. *28* : 627 - 638, 1982.

*50822 - **MURTY, K.S., NAYAK, S.K., SAHU, G.** : Effect of low light stress on rice crop. - Crop Plant Response of Environmental Stresses. (Proc. Symp.) Pp. 74 - 84. Central Rice Research Institute, Cuttack 1976. [Photosynthates.]

50823 - **MURTY, P.S.S., MURTY, K.S.** : Efficiency of ^{14}C sucrose translocation in high and low sterile rice varieties under normal and low light conditions. - J. nuclear agr. Biol. *11* : 100 - 101, 1982.

50824 - **MURZAEVA, S.V., KRAVCHENKO, V.V.** : Fotofosforiliruyushchaya aktivnost' khloroplastov, vydelennykh iz sifonovoĭ zelenoĭ vodorosli *Cadium fragile*. [Photophosphorylation activity in chloroplasts isolated from the siphonous green alga *Codium fragile*.] - Fiziol. Rast. *29* : 591 - 596, 1982.[In R, ab : E.]

50825 - **MUSCATINE, L.** : Establishment of photosynthetic eukaryotes as endosymbionts in animal cells. - In : SCHIFF, J.A., LYMAN, H. (ed.) : On the Origins of Chloroplasts. Pp. 77 - 92. Elsevier/North-Holland, New York - Amsterdam - Oxford 1982.

50826 - **MUSGRAVE, S.C., KERBY, N.W., CODD, G.A., STEWART, W.D.P.** : Sustained ammonia production in immobilized filaments of the nitrogen-fixing cyanobacterium *Anabaena* 27893. - Biotechnol. Lett. *4* : 647 - 652, 1982.

50827 - **MUSTÁRDY, L.A., VU, T.T., FALUDI-DÁNIEL, Á.** : Stomatal response and photosynthetic capacity of maize leaves at low temperature. A study on varietal differences in chilling sensitivity. - Physiol. Plant. *55* : 31 - 34, 1982.

50828 - **MUTO, S.** : Distribution of calmodulin within wheat leaf cells. - FEBS Lett. *147* : 161 - 164, 1982. [Chloroplast.]

50829 - **MUTO, S.** : [Calmodulin and physiological functions of plants.] - Tanpakushitsu Kakusan Koso *27* : 2189 - 2200, 1982. [Chloroplast; in Jap.]

50830 - **MUTO, S., IZAWA, S., MIYACHI, S.** : Light-induced Ca^{2+} uptake by intact chloroplasts. - FEBS Lett. *139* : 250 - 254, 1982.

50831 - MYERS, J., GRAHAM, J.-R., WANG, R.T. : Protochlorophyll(ide) in a blue-green
        alga. - Plant Physiol. *69* : 549 - 550, 1982.

50832 - NABEDRYK, E., TIEDE, D.M., DUTTON, P.L., BRETON, J. : Conformation and ori-
        entation of the protein in the bacterial photosynthetic reaction center. -
        Biochim. biophys. Acta *682* : 273 - 280, 1982.

*50833 - NACHMIAS, A., BARASH, I., SOLEL, Z., STROBEL, G.A. : Effect of mal seco to-
        xin on lemon leaf cells. - Phytoparasitica *8* : 51 - 60, 1980. [Ps.]

50834 - NAGASAKI, T., HAGIWARA, H., INADA, Y. : Fibrin membrane endowed with biolo-
        logical function. Immobilization of spinach chloroplasts and rat liver micro-
        somes. - FEBS Lett. *142* : 159 - 161, 1982.

50835 - NAIRN, J.A., HAEHNEL, W., REISBERG, P., SAUER, K. : Picosecond fluorescence
        kinetics in spinach chloroplasts at room temperature. Effects of Mg^{2+}. -
        Biochim. biophys. Acta *682* : 420 - 429, 1982.

50836 - NAKAMOTO, H., KU, M.S.B., EDWARDS, G.E. : Inhibition of C_4 photosynthesis by
        (benzamidooxy)acetic acid. - Photosynthesis Res. *3* : 293 - 305, 1982.

50837 - NAKAMURA, Y. : [Control of metabolism of photosynthesized starch. Effect
        of temperature.] - Kagaku to Seibutsu *20* : 5 - 6, 1982. [In Jap.]

50838 - NAKAMURA, Y., MIYACHI, S. : Change in starch photosynthesized at different
        temperatures in *Chlorella*. - Plant Sci. Lett. *27* : 1 - 6, 1982.

50839 - NAKAMURA, Y., MIYACHI, S. : Effect of temperature on starch degradation in
        Chlorella vulgaris 11h cells. - Plant Cell Physiol. *23* : 333 - 341, 1982.
        [Ps.]

50840 - NAKAMURA, Y., TSUZUKI, M., MIYACHI, S. : Regulation of carbon metabolism
        by environmental factors. - Rec. Progr. nat. Sci. Jap. *7* : 61 - 66, 1982.

50841 - NAKASEKO, K., GOTOH, K. : [Comparative studies on dry matter production
        plant type and productivity in soybean, azuki bean and kidney bean VI. Dif-
        ference in the early vegetative growth in relation to seed weight and cha-
        racter of cotyledon.] -Jap. J. Crop Sci. *51* : 110 - 116, 1982. [In Jap.,
        ab : E.]

*50842 - NAKATANI, H.Y., BARBER, J. : Cholate extraction of a heme-protein from spi-
        nach thylakoids and its possible involvement in PS-II oxygen evolution. -
        Photobiochem. Photobiophys. *2* : 69 - 78, 1981.

50843 - NAKHUTSRISHVILI, G.Sh., CHERNUSKA, A. : Ėkologicheskiĭ analiz vliyaniya vypa-
        sa na vysokogornye luga Tsentral'nogo Kavkaza. [Ecological analysis of the
        influence of grazing on alpine meadows in Central Kaukasus.] - Dokl. Akad.
        Nauk SSSR *267* : 503 - 505, 1982. [Resistances, canopy structure; in R.]

50844 - NASH, T.H.III, LANGE, O.L., KAPPEN, L. : Photosynthetic patterns of Sonoran
        Desert lichens. II. A multivariate laboratory analysis. - Flora *172* : 419 -
        - 426, 1982.

50845 - NASH, T.H.III, MOSER, T.J., BERTKE, C.C., LINK, S.O., SIGAL, L.L., WHITE,
        S.L., FOX, C.A. : Photosynthetic patterns of Sonoran Desert lichens I. Envi-
        ronmental considerations and preliminary field measurements. - Flora *172*:
        335 - 345, 1982.

50846 - NASSI, L., CILENTO, G. : Excitation of chloroplasts induced by phenylacetal-
        dehyde. - Photochem. Photobiol. *36* : 121 - 123, 1982.

*50847 - NASYROV, Yu.S. : Genetic modification of the CO_2 carboxylation reactions
        as a factor improving efficiency of photosynthesis. - Indian J. Plant Phy-
        siol. *24* : 26 - 36, 1981.

50848 - NASYROV, Yu.S. : Geneticheskaya regulyatsiya formirovaniya i aktivnosti
        fotosinteticheskogo apparata. [Genetical regulation of formation and acti-
        vity of photosynthetic apparatus.] - In : Fiziologiya Fotosinteza. Pp. 146 -
        - 164. Nauka, Moskva 1982. [In R.]

50849 - **NASYROV, Yu.S.** : Genetika fotosinteza i selektsiya. [Genetics of photosynthesis and breeding.] - Novoe Zhizni Nauke Tekh., Ser. Biol. *1982* (12) :
1 - 60, 1982. [In R.]

50850 - **NASYROV, Yu.S.** : Genetika fotosinteza v svyazi s problemami selektsii. [Ge-
netics of photosynthesis as connected with breeding problems.] - Sel'skokhoz. Biol. *17* : 834 - 840, 1982. [In R, ab : E.]

50851 - **NATARAJAN, L.V., BLANKENSHIP, R.E.** : Linear dichroism of the 740nm absorbing
form of chlorophyll a. - Spectrosc. Lett. *15* : 527 - 532, 1982.

50852 - **NATHAN, K.K.** : Note on relationship between photosynthetically active radiation and total radiation. - Ann. Arid Zone *21* : 259 - 261, 1982.

50853 - **NATO, A., HOARAU, J., BOURDU, R.** : The contribution of photosynthetic acti-
vity in the growth of *Nicotiana tabacum* green cell suspension culture. -
In : FUJIWARA, A. (ed.) : Plant Tissue Culture 1982. Pp. 253 - 254. Maruzen,
Tokyo 1982.

50854 - **NAUMOVA, T.V.** : Izmenchivost' fiziologicheskikh pokazateleĭ i èlementov produktivnosti yarovoĭ pshenitsy v svyazi s mestom reproduktsii semyan. [Vari-
ation in physiological indices and productivity elements in spring wheat
as related to the place of seed reproduction.] - Tr. prikl. Bot.Genet. Selek.
72 (2) : 90 - 93, 1982. [In R, ab : E.]

50855 - **NAVARRO, S., ALMELA, L., GARCÍA, A.L.** : Application of derivative spectroscopy to the quantitative determination of chlorophylls and related pigments.
1. Simultaneous determination of chlorophylls a and b. - Photosynthetica
16 : 134 - 139, 1982.

50856 - **NAYLOR, A.W., GILES, L.J.** : Growth, pigment synthesis, and ultrastructural
responses of *Phaseolus vulgaris* L. cv. Blue Lake to intermittent and flash-
ing light. - Plant Physiol. *70* : 257 - 263, 1982.

50857 - **NEILD, R.E.** : Temperature and rainfall influences on the phenology and yield
of grain sorghum and maize: a comparison. - Agr. Meteorol. *27* : 79 - 88,
1982. [Production model.]

*50858 - **NELSON, N.** : Proton ATPase of chloroplasts. - Curr. Topics Bioenerg. *11* :
1 - 33, 1981.

50859 - **NELSON, N.** : Isolation of chloroplast Photosystem I reaction center. - In :
EDELMAN, M., HALLICK, R.B., CHUA, N.-H. (ed.) : Methods in Chloroplast Mo-
lecular Biology. Pp. 907 - 915. Elsevier Biomedical Press, Amsterdam - New
York - Oxford 1982.

50860 - **NELSON, N.** : Structure and function of the higher plant coupling factor. -
In : BARBER, J. (ed.) : Electron Transport and Photophosphorylation. Pp.
81 - 104. Elsevier Biomedical Press, Amsterdam - New York - Oxford 1982.

50861 - **NELSON, N., DICKMANN, D.I., GOTTSCHALK, K.W.** : Autumnal photosynthesis in
short-rotation intensively cultured *Populus* clones. - Photosynthetica *16* :
321 - 333, 1982.

50862 - **NELSON, N.D., MICHAEL, D.** : Photosynthesis, leaf conductance, and specific
leaf weight in long and short shoots of *Populus* 'Tristis #1' grown under
intensive culture. - Forest Sci. *28* : 737 - 744, 1982.

50863 - **NEMCHENKO, O.A., MUSATENKO, L.I.** : Modelirovanie rosta i metabolizma rasteniĭ na rannykh ètapakh ontogeneza. [Modelling of growth and metabolism of
plants et early stages of ontogenesis.] - Fiziol. Biokhim. kul't. Rast. *14*:
439 - 445, 1982. [Photosynthates; In R, ab : E.]

50864 - **NEORI, A., HOLM-HANSEN, O.** : Effect of temperature on rate of photosynthesis
in antarctic phytoplankton. - Polar Biol. *1* : 33 - 38, 1982.

50865 - **NESBITT, D.M., BERG, S.P.** : The influence of spinach thylakoid lumen volume
and membrane proximity on the rotational motion of the spin label tempamine.
- Biochim. biophys. Acta *679* : 169 - 174, 1982.

50866 - NETZEL, T.L. : Electron transfer reactions in reaction centers of photosyn-
thetic bacteria and in reaction centers models. - In : ALFANO, R.R. (ed.) :
Biological Events Probed by Ultrafast Laser Spectroscopy. Pp. 79 - 117.
Academic Press, New York - London - Paris - San Diego - San Francisco - São
Paulo - Sydney - Tokyo - Toronto 1982.

50867 - NETZEL, T.L., BERGKAMP, M.A., CHANG, C.K. : Benzoquinone quenching of diporphyrin excited states: kinetic evidence for distinguishing electron-
-transfer photoproducts from (π-π*) states. - J. amer. chem. Soc. *104* : 1952 -
- 1957, 1982.

50868 - NEUFANG, H., MÜLLER, H., KNOBLOCH, K. : On the quantitation of bacteriochlorophyll in chromatophore suspensions from purple bacteria. - Biochim. biophys.
Acta *681* : 327 - 329, 1982.

50869 - NEUGEBAUER, D.-C. : Metal decoration of the purple membrane. - In : COLOWICK,
S.P., KAPLAN, N.O. (ed.) : Methods in Enzymology. Vol.88. Pp. 235 - 241.
Academic Press, New York - London - Paris - San Diego - San Francisco - São
Paulo - Sydney - Tokyo - Toronto 1982.

50870 - NEUMANN, K.-H., BENDER, L., KUMAR, A., SZEGOE, M. : Photosynthesis and pathway of carbon in tissue cultures of *Daucus* and *Arachis*. - In : FUJIWARA, A.
(ed.) : Plant Tissue Culture 1982. Pp. 251 - 252. Maruzen, Tokyo 1982.

50871 - NEWELL, R.I.E. : An evaluation of the wet oxidation technique for use in determining the energy content of seston samples. - Can. Bull. Fish. aquat.
Sci. *39* : 1383 - 1388, 1982.

50872 - NEWELL, S.J. : Translocation of ^{14}C-photoassimilate in two stoloniferous
Viola species. - Bull. Torrey bot. Club *109* : 306 - 317, 1982.

50873 - NEWSOME, R.D., KOZLOWSKI, T.T., TANG, Z.C. : Responses of *Ulmus americana*
seedlings to flooding of soil. - Can. J. Bot. *60* : 1688 - 1695, 1982. [Growth
analysis, stomatal resistance.]

50874 - NEWTON, J.W. : The relationship between photosynthesis and nitrogen fixation
revisited. - In : KAPLAN, N.O., ROBINSON, A. (ed.) : From Cyclotrons to Cytochromes. Pp. 391 - 396. Academic Press, New York 1982.

*50875 - NICHIPOROVICH, A.A. : Fotosintez i biosfera. [Photosynthesis and the biosphe-
re.] - Tr. biogeokhim. Lab. Akad. Nauk SSSR *17* : 84 - 90, 1979. [In R.]

*50876 - NICHIPOROVICH, A.A. : Fotosintez i problemy ènergetiki. [Photosynthesis and
problems of energetics.] - In : Preobrazovanie Solnechnoĭ Énergii. Pp. 142 -
- 157. Akad. Nauk SSSR, Chernogolovka 1981. [In R.]

50877 - NICHIPOROVICH, A.A. : Fiziologiya fotosinteza i produktivnost' rasteniĭ.
[Physiology of photosynthesis and plant productivity.] - In : Fiziologiya
Fotosinteza. Pp. 7 - 33. Nauka, Moskva 1982. [In R.]

B50878 - NICHOLLS, D.G. : Bioenergetics. - Academic Press, London - New York - Paris -
San Diego - San Francisco - São Paulo - Sydney - Tokyo - Toronto 1982. [Ps.]

50879 - NICHOLLS, J.M., FOULDS, I.F., LAMBERT, G., CARR, N.G. : The amplification
of hydrogenase activity in photosynthetic prokaryotes by nutrition and by
application of recombinants DNA technology and cloning procedures. - In :
HALL, D.O., PALZ, W. (ed.) : Photochemical, Photoelectrochemical and Photobiological Processes. Vol.1. Pp. 170 - 173. D.Reidel Publ. Co., Dordrecht -
Boston - London 1982.

50880 - NICKLISCH, A. : Regulation der Primärproduktion auf der Ebene von Biozönosen.
- In : HOFFMANN, P., HIEKE, B. (ed.) : Photosynthese: Regulation und Evolution. (Colloquia Pflanzenphysiologie Nr.5.) Pp. 218 - 228. Humboldt-Universität, Berlin 1982.

50881 - NICOLAS, P., HEIZMANN,P., NIGON, V. : Les différents phénotypes de mutants
non photosynthétiques (phot⁻) chez *Euglena gracilis*: fréquence de formation
par irradiation ultraviolette. - Compt. rend. Acad. Sci. Paris, Sér. III
294 : 145 - 148, 1982.

50882 - NICOLAS, P., NIGON, V. : Production of plastidial antibiotic-resistant mutants
by ultraviolet irradiation of *Euglena gracilis*. - Mol.gen. Genet. *185* : 184 -
- 185, 1982.

50883 - NICOLAY, K., HELLINGWERF, K.J., VAN GEMERDEN, H., KAPTEIN, R., KONINGS,W.N.: ^{31}P NMR studies of photophosphorylation in intact cells of *Chromatium vinosum*. - FEBS Lett. *138* : 249 - 254, 1982.

*50884 - NIEBUR, W.S., FEHR, W.R. : Agronomic evaluation of soybean genotypes resistant to iron deficiency chlorosis. - Crop Sci. *21* : 551 - 554, 1981.

50885 - NIELL, F.X., FERNANDEZ, C. : Medida de la concentración de pigmentos en algas rojas. Interferencia de la extracción de pigmentos hidrosolubles en la estimación de clorofila *a*. [Estimation of photosynthetic pigments in red algae. Interference of hydrosoluble pigments extraction on chlorophyll *a* estimation.] - Invest. Pesquera *46* : 185 - 189, 1982. [In Span., ab : E.]

50886 - NIERZWICKI, S.A., MARATEA, D., BALKWILL, D.L., HARDIE, L.P., MEHTA, V.B., STEVENS, S.E.,Jr. : Ultrastructure of the cyanobacterium, *Mastigocladus laminosus*. - Arch. Microbiol. *133* : 11 - 19, 1982. [Chromatophores.]

50887 - NILSEN, S., JOHNSEN, Ø.: Effect of CO_2, O_2 and diamox on photosynthesis and photorespiration in *Chlamydomonas reinhardtii* (green alga) and *Anacystis nidulans* (cyanobacterium, blue-green alga). - Physiol. Plant. *56* : 273 - - 280, 1982.

50888 - NILSSON, L.-O. : Determination of current energy forest growth and biomass production. - Projekt Energiskogsodling tekn. Rapp. *27* : 1 - 36, 1982.

50889 - NILWIK, H.J.M., GOSIEWSKI, W., BIERHUIZEN, J.F. : The influence of irradiance and external CO_2-concentration on photosynthesis of different tomato genotypes. - Sci. Hort. *16* : 117 - 123, 1982.

*50890 - NIMMO, I.A., BAUERMEISTER, A., DALE, J.E. : Evaluation of the jacknife technique for fitting multiexponential functions to biochemical data. - Anal. Biochem. *110* : 407 - 411, 1981. [^{14}C analysis.]

50891 - NISHIDA, K. : Photorespiration and various types of CO_2 fixation. c. CAM (Crassulacean acid metabolism). - Recent Progr. nat. Sci. Jap. *7* : 59 - 60, 1982.

50892 - NISHIDA, K. : Relationship between CO_2 concentration within the leaf and photosynthesis during deacidification in a CAM plant, *Kalanchoë pinnatum*. - Physiol. Plant. *54* : 451 - 454, 1982.

50893 - NISHIMURA, M. : Energy conversion in photosynthetic membranes and organelles. b. Energy conversion and changes in physical parameters of microenvironment. - Recent Progr. nat. Sci. Jap. *7* : 44 - 46, 1982.

50894 - NISHIMURA, M., AKAZAWA, T. : Purification of ribulose-1,5-bisphosphate carboxylase and its subunits from higher plants. - In : EDELMAN, M., HALLICK, R.B., CHUA, N.-H. (ed.) : Methods in Chloroplasts Molecular Biology. Pp. 751 - 765. Elsevier Biomedical Press, Amsterdam - New York - Oxford 1982.

50895 - NISHIMURA, M., BHUSAWANG, P., STRZALKA, K., AKAZAWA, T. : Developmental formation of glutamine synthetase in greening pumpkin cotyledons and its subcellular localization. - Plant Physiol. *70* : 353 - 356, 1982.

50896 - NISHIMURA, M., DOUCE, R., AKAZAWA, T. : Isolation and characterization of metabolically competent mitochondria from spinach leaf protoplasts. - Plant Physiol. *69* : 916 - 920, 1982. [Photorespiration.]

50897 - NISHIZAWA, A.N., YEE, B.C., BUCHANAN, B.B. : Chloroplast fructose-1,6-bisphosphatase from spinach leaves. - In : EDELMAN, M., HALLICK, R.B., CHUA, N.-H. (ed.) : Methods in Chloroplast Molecular Biology. Pp. 707 - 713. Elsevier Biomedical Press, Amsterdam - New York - Oxford 1982.

*50898 - NITISEWOJO, P., ISHAK, S. : The effects of pH and magnesium on the colour of dried spinach leaves. - Sains malaysiana *6* : 37 - 42, 1977. [Chl.]

50899 - NIXDORF, B. : Untersuchungen zur Tagesdynamik der Primärproduktion des Phytoplanktons im Müggelsee. - In : HOFFMANN, P., HIEKE, B. (ed.) : Photosynthese: Regulation und Evolution. (Colloquia Pflanzenphysiologie Nr.5.) Pp. 237 - 239. Humboldt-Universität, Berlin 1982.

50900 - NIZIOŁEK, S., BIELAWSKI, W., NALBORCZYK, E. : Photosynthesis in detached
rye leaves at normal and low oxygen concentration. II. Incorporation of
$^{14}CO_2$ into amino acids. - Acta biochim. pol. 29 : 339 - 347, 1982.

50901 - NOACK, K. : Circular dichroism of carotenoids and its use in investigations
of their structures, configurations and conformations. - In : BRITTON, G.,
GOODWIN, T.W. (ed.) : Carotenoid Chemistry and Biochemistry. Pp. 135 - 153.
Pergamon Press, Oxford - New York - Toronto - Sydney - Paris - Frankfurt
1982.

50902 - NOBEL, P.S. : Orientation, PAR interception, and nocturnal acidity increases
for terminal cladodes of a widely cultivated cactus, Opuntia ficus-indica. -
Amer. J. Bot. 69 : 1462 - 1469, 1982.

50903 - NOGA, G., LENZ, F. : Einfluss von verschiedenen Klimafaktoren auf den CO_2-
-Gaswechsel von Äpfeln während der Licht- und Dunkelperiode. - Gartenbauwissenschaft 47 : 193 - 197, 1982.

50904 - NOORUDEEN, A.M., KULANDAIVELU, G. : On the possible site of inhibition
of photosynthetic electron transport by ultraviolet-B (UV-B) radiation. -
Physiol. Plant. 55 : 161 - 166, 1982.

50905 - NORBY, R.J., KOZLOWSKI, T.T. : The role of stomata in sensitivity of Betula
papyrifera seedlings to SO_2 at different humidities. - Oecologia 53 : 34 -
- 39, 1982. [Growth analysis.]

50906 - NORRIS, I.B. : Soil moisture and growth of contrasting varieties of Lolium,
Dactylis and Festuca species. - Grass Forage Sci. 37 : 273 - 283, 1982.
[Growth analysis.]

50907 - NORRIS, I.B., THOMAS, H. : Recovery of ryegrass species from drought. - J.
agr. Sci. 98 : 623 - 628, 1982. [Energy utilization.]

50908 - NORRIS, J.R., BOWMAN, M.K., BUDIL, D.E., TANG, J., WRAIGHT, C.A., CLOSS, G.L.:
Magnetic characterization of the primary state of bacterial photosynthesis.
- Proc. nat. Acad. Sci. USA 79 : 5532 - 5536, 1982.

*50909 - NORRIS, J.R., THURNAUER, M.C., BOWMAN, M.K., TRIFUNAC, A.D. : Electron spin
echo spectroscopy and photosynthesis. - Front. biol. Energ. 1 : 581 - 591,
1978.

50910 - NORSTOG, K. : Experimental embryology of gymnosperms. - In : JOHRI, B.M.
(ed.) : Experimental Embryology of Vascular Plants. Pp. 25 - 51. Springer-
-Verlag, Berlin - Heidelberg - New York 1982. [Chl.]

50911 - NOTT, D.L., OSMOND, C.B. : Purification and properties of phosphoenolpyruvate
carboxylase from plants with crassulacean acid metabolism. - Aust. J. Plant
Physiol. 9 : 409 - 422, 1982.

*50912 - NOVAKOVA, A.A., USPENSKAYA, N.Ya., ALEKSANDROV, A.Yu., KONONENKO, A.A.,
KUZMIN, R.N., RUBIN, A.B. : Effects of heat denaturation of intact cells
and chromatophores of Chromatium minutissimum on their Mössbauer spectra. -
Stud. biophys. 80 : 55 - 58, 1980.

50913 - NUGENT, J.H.A., EVANS, M.C.W. : Identification of components in the photosystem II reaction centre. - Biochem. Soc. Trans. 10 : 406 - 407, 1982.

50914 - NUGENT, J.H.A., EVANS, M.C.W., DINER, B.A. : Characteristics of the photosystem II reaction centre. II. Electron donors. - Biochim. biophys. Acta
682 : 106 - 114, 1982.

*50915 - NUMATA, M. : Primary producers in meadows. - In : COUPLAND, R.T. (ed.) :
Grassland Ecosystems of the World. Vol.18. Analysis of Grasslands and their
Uses. Pp. 127 - 138. Cambridge University Press, Cambridge 1979. [Growth
analysis.]

50916 - NURMI, A., VAPAAVUORI, E. : Chlorophyll-protein complexes in Salix sp. "aquatica gigantea" under strong and weak light I. Spectral characterization of
the chlorophyll-protein complexes. - Plant Cell Physiol. 233 : 785 - 790,
1982.

*50917 - OBOLONSKIĬ, V.V., KOPTSEVA, L.G. : Vliyanie kombinirovannogo krasnogo sveta
na svetoindutsiruemuyu bioèlektricheskuyu reaktsiyu list'ev rasteniĭ. [Act-
ion of the combined red light on a light-induced bioelectric response of
plant leaves.] - In : Svetozavisimaya Bioèlektricheskaya Aktivnost' List'-
ev Rasteniĭ. Pp. 69 - 83, III. Ural'. Gos. Univ., Sverdlovsk 1980. [Ps;
in R, ab : E.]

*50918 - OBOLONSKIĬ, V.V., RYBIN, I.A. : Bioèlektricheskie èffekty perekhoda iz od-
noĭ spektral'noĭ oblasti v druguyu na list'yakh kukuruzy i bobov. [Bioelec-
tric effects of a transition from one spectral range to another in leaves
of maize and broad bean.] - In : Svetozavisimaya Bioèlektricheskaya Aktivnost' List'ev Rasteniĭ. Pp. 89 - 98, 118. Ural'. Gos. Univ., Sverdlovsk
1977. [In R, ab : E.]

50919 - OCHIAI, H., SHIBATA, H., SAWA, Y., KATOH, T. : Properties of the chloroplast
film electrode immobilized on an SnO_2-coated glass plate. - Photochem. Photobiol. 35 : 149 - 155, 1982.

50920 - OELZE-KAROW, H., MOHR, H. : Phytochrome action on chlorophyll synthesis. A
study of the escape from photoreversibility. - Plant Physiol. 70 : 863 -
- 866, 1982.

50921 - OELZE-KAROW, H., MOHR, H. : Phytochrome and the development of photophosphorylation capacity - an appraisal of the experimental approach. - Photochem.
Photobiol. 35 : 223 - 227, 1982.

*50922 - OESTERHELT, D. : Isoprenoids and bacteriorhodopsin in halobacteria.-Progr.mol.
subcell. Biol. 4 : 133 - 166, 1976.

50923 - OESTERHELT, D. : Lichtenergieumwandlung in Halobacterien - ein zweiter Weg
der Natur zur Photosynthese. - Nova Acta leopoldina NF 55 (246) : 21 - 38,
1982.

50924 - OESTERHELT, D. : Reconstitution of the retinal proteins bacteriorhodopsin
and halorhodopsin. - In : COLOWICK, S.P., KAPLAN, N.O. (ed.) : Methods in
Enzymology. Vol.88. Pp. 10 - 17. Academic Press, New York - London - Paris -
San Diego - San Francisco - São Paulo - Sydney - Tokyo - Toronto 1982.

50925 - OESTERHELT, D. : Photophosphorylation and reconstitution of photophospho-
rylation in halobacterial cells. - In : COLOWICK, S.P., KAPLAN, N.O. (ed.):
Methods in Enzymology. Vol.88. Pp. 349 - 355. Academic Press, New York -
London - Paris - San Diego - San Francisco - São Paulo - Sydney - Tokyo -
Toronto 1982.

50926 - OETTMEIER, W., JOHANNINGMEIER, U., TREBST, A. : Inhibitors of plastoquinone
function as tools for identification of its binding proteins in chloroplasts.
- In : TRUMPOWER, B.L. (ed.) : Function of Quinones in Energy Conserving Systems. Pp. 425 - 441. Academic Press, New York - London 1982.

50927 - OETTMEIER, W., MASSON, K. : Picrate as an inhibitor of photosystem II in
photosynthetic electron transport. - Europe. J. Biochem. 122 : 163 - 167,
1982.

50928 - OETTMEIER,W., MASSON, K., FEDTKE, C., KONZE, J., SCHMIDT, R.R. : Effect of
different photosystem II inhibitors on chloroplasts isolated from species
either susceptible or resistant toward s-triazine herbicides. - Pestic. Biochem. Physiol. 18 : 357 - 367, 1982.

*50929 - OETTMEIER, W., MASSON, K., GODDE, D. : Inhibition of photosynthetic electron
transport by the quinone antagonist UHDBT. - Z. Naturforsch. 36 C : 272 -
- 275, 1981.

50930 - OETTMEIER, W., MASSON, K., JOHANNINGMEIER, U. : Evidence for two different
herbicide-binding proteins at the reducing side of photosystem II. - Biochim.
biophys. Acta 679 : 376 - 383, 1982.

50931 - OETTMEIER, W., MASSON, K., SOLL, H.-J., HURT, E., HAUSKA, G. : Photoaffinity
labelling of plastoquinone binding sites in chloroplast cytochrome b_6/f-com-
plex. - FEBS Lett. 144 : 313 - 317, 1982.

50932 - OGAWA, S. : Disintegration of chloroplasts during zygote formation in *Spirogyra verruculosa*. - Bot. Mag. (Tokyo) *95* : 249 - 260, 1982.

50933 - OGAWA, T. : Simple oscillations in photosynthesis of higher plants. - Biochim. biophys. Acta *681* : 103 - 109, 1982.

*50934 - OGAWA, T., AIBA, S. : Bioenergetic analysis of mixotrophic growth in *Chlorella vulgaris* and *Scenedesmus acutus*. - Biotechnol. Bioeng. *23* : 1121 - - 1132, 1981. [Ps.]

50935 - OGAWA, T., GRANTZ, D., BOYER, J., GOVINDJEE : Effects of cations and abscisic acid on chlorophyl *a* fluorescence in guard cells of *Vicia faba*. - Plant Physiol. *69* : 1140 - 1144, 1982.

50936 - OGBUEHI, S.N., BRANDLE, J.R. : Influence of windbreak-shelter on soybean growth, canopy structure, and light relations. - Crop Sci. *22* : 269 - 273, 1982. [Growth analysis.]

*50937 - OGINO, C. : [Photosynthesizing bacteria as feed for fish.] - Hakko to Kogyu *36* : 836 - 841, 1978. [In Jap.]

50938 - OGREN, W.L., CHOLLET, R. : Photorespiration. - In : GOVINDJEE (ed.) : Photosynthesis. Vol.II. Pp. 191 - 230. Academic Press, New York - London - Paris - San Diego - San Francisco - São Paulo - Sydney - Tokyo - Toronto 1982.

50939 - OHAD, I., DREWS, G. : Biogenesis of the photosynthetic apparatus in prokaryotes and eukaryotes. - In : GOVINDJEE (ed.) : Photosynthesis. Vol.II. Pp. 89 - - 140. Academic Press, New York - London - Paris - San Diego - San Francisco - São Paulo - Tokyo - Toronto 1982.

50940 - OH-HAMA, T. : [Photosynthetic carbon assimilation in C_3- and C_4-plants - tracer experiments using ^3H, ^{14}C, ^{13}C and ^{18}O.] - Radioisotopes *31* : 480 - 489, 1982. [In Jap.]

50941 - OH-HAMA, T., SETO, H., OTAKE, N., MIYACHI, S. : ^{13}C-NMR evidence for the pathway of chlorophyll biosynthesis in green algae. - Biochem. biophys. Res. Commun. *105* : 647 - 652, 1982.

50942 - OHKI, K. : Soybean nitrate reductase activity and photosynthesis related to manganese status determined by plant analysis. - In : SCAIFE, A. (ed.) : Plant Nutrition. Vol.2. Pp. 448 - 453. Commonwealth Agricultural Bureaux, Farnham Royal 1982.

50943 - OHKI, K., WATANABE, M., FUJITA, Y. : Action of near UV and blue light on the photocontrol of phycobiliprotein formation; a complimentary chromatic adaptation. - Plant Cell Physiol. *23* : 651 - 656, 1982.

50944 - OHMANN, E. : Autotrophie-Typen und ihre evolutionäre Entwicklung. - In : HOFFMANN, P., HIEKE, B. (ed.) : Photosynthese: Regulation und Evolution. (Colloquia Pflanzenphysiologie Nr.5.) Pp. 25 - 32. Humboldt-Universität, Berlin 1982.

50945 - OHMORI, M. : Assimilation of nitrate, nitrite, and ammonia in algae. - In : ZABORSKY, O.R. (ed.) : CRC Handbook of Biosolar Resources. Vol.I/1. Pp. 231 - - 235. CRC Press, Boca Raton 1982. [Ps.]

*50946 - OHNISHI, J., YAMADA, M. : Development of chloroplasts in etiolated *Avena* leaves during greening. - Sci. Pap. Coll. gen. Educ. Univ. Tokyo *30* : 167 - - 174, 1980.

50947 - OHNISHI, J.-I., YAMADA, M. : Glycerolipid synthesis in *Avena* leaves during greening of etiolated seedlings III. Synthesis of α-linolenoyl-manogalactosyl diacylglycerol from liposomal linolenoyl-phosphatidylcholine by *Avena* plastids in the presence of phosphatidylcholine-exchange protein. - Plant Cell Physiol. *23* : 767 - 773, 1982.

50948 - OHNISHI, O. : Population genetics of cultivated common buckwheat, *Fagopyrum esculentum* MOENCH. I.Frequency of chlorophyll-deficient mutants in Japanese populations. - Jap. J. Genet. *57* : 623 - 639, 1982.

50949 - **OHSUGI, R., MURATA, T., CHONAN, N.** : C₄ syndrome of the species in the *Dichotomiflora* group of the genus *Panicum (Gramineae)*. - Bot. Mag. (Tokyo) *95* : 339 - 347, 1982.

50950 - **OHYA, T., NAITO, K., SUZUKI, H.** : Combined effect of benzyladenine and potassium on the level of light-harvesting chlorophyll a/b protein in detached cucumber cotyledons. - Z. Pflanzenphysiol. *108* : 39 - 47, 1982.

50951 - **OHYAMA, K., WETTER, L.R., YAMANO, Y., FUKUZAWA, H., KOMANO, T.** : A simple method for isolation of chloroplast DNA from *Marchantia polymorpha* L. suspension cultures. - Agr. biol. Chem. *46* : 237 - 242, 1982. [RuBPC.]

50952 - **OI, V.T., GLAZER, A.N., STRYER, L.** : Fluorescent phycobiliprotein conjugates for analyses of cells and molecules. - J. Cell Biol. *93* : 981 - 986, 1982.

50953 - **OIE, T., MAGGIORA, G.M., CHRISTOFFERSEN, R.E.** : Structural characterization of a special-pair chlorophyll dimer model of P700. - Int. J. Quantum Chem., quantum Biol. Symp. *9* : 157 - 171, 1982.

50954 - **OKAMURA, M.Y., FEHER, G., NELSON, N.** : Reaction centers. - In : GOVINDJEE (ed.) : Photosynthesis. Vol.1. Pp. 195 - 272. Academic Press, New York - London - Paris - San Diego - San Francisco - São Paulo - Sydney - Tokyo - Toronto 1982.

*50955 - **OKAMURA, M.Y., FREDKIN, D.R., ISAACSON, R.A., FEHER, G.** : Magnetic interactions and electron transfer kinetics of the reduced intermediate acceptor in reaction centers (RCs) of *Rhodopseudomonas sphaeroides* R-26. Evidence for thermally induced tunneling. - In : CHANCE, B., DEVAULT, D.C., FRAUENFELDER, H., MARCUS, R.A., SCHRIEFFER, J.R., SUTIN, N. (ed.) : Tunneling in Biological Systems. Pp. 729 - 743. Academic Press, New York - San Francisco - London 1979.

*50956 - **OKAYAMA, S.** : [Role of quinones in photosynthetic systems.] - Tanpakushitsu Kakusan Koso, Bessatsu *21* : 96 - 103, 1979. [In Jap.]

50957 - **O'KELLY, C.J.** : Chloroplast pigments in selected marine *Chaetophoraceae* and *Chaetosiphonaceae (Chlorophyta)*: The occurrence and significance of siphonaxanthin. - Bot. mar. *25* : 133 - 137, 1982.

50958 - **OKU, T.** : Photoactivation of O_2-evolving system accompanying thiol generation in spruce chloroplasts. - Photobiochem. Photobiophys. *4* : 275 - 281, 1982.

50959 - **OLDFIELD, E., KINSEY, R.A., KINTANAR, A.** : Recent advances in the study of bacteriorhodopsin dynamic structure using high-field solid-state nuclear magnetic resonance spectroscopy. - In : COLOWICK, S.P., KAPLAN, N.O. (ed.): Methods in Enzymology. Vol.88. Pp. 310 - 325. Academic Press, New York - London - Paris - San Diego - San Francisco - São Paulo - Sydney - Tokyo - Toronto 1982.

50960 - **O'LEARY, M.H.** : Phosphoenolpyruvate carboxylase: an enzymologist's view. - Annu. Rev. Plant Physiol. *33* : 297 - 315, 1982.

50961 - **O'LEARY, M.H., DÍAZ, E.** : Phosphoenol-3-bromopyruvate. A mechanism-based inhibitor of phosphoenolpyruvate carboxylase from maize. - J. biol. Chem. *257* : 14603 - 14605, 1982.

*50962 - **OLECH, K., STANEK, R., BLAMOWSKI, Z.K.** : Przebieg fotosyntezy i oddychanie u wygłodzonych siewek fasoli. [The course of photosynthesis and respiration in starving bean seedlings.] - Ann. Univ. M. Curie-Skłodowska (Lublin) Sect. E *31* : 147 - 156, 1976. [In Pol., ab : E, R.]

50963 - **OLIVEIRA, L.** : The development of chloroplasts in root meristematic tissue of *Secale cereale* L. seedlings. - New Phytol. *91* : 263 - 275, 1982.

50964 - **OLIVER, R.P., GRIFFITHS, W.T.** : Pigment-protein complexes of illuminated etiolated leaves. - Plant Physiol. *70* : 1019 - 1025, 1982.

50965 - **OLSEN, L.F.** : Transient kinetics of the electron transfer between P-700, plastocyanin and cytochrome f in chloroplasts suspended in fluid media at sub-zero temperatures. - Biochim. biophys. Acta *682* : 482 - 490, 1982.

50966 - OLSEN, L.F., COX, R.P. : Transient kinetics of the reaction between cytochrome c-552 or plastocyanin and P-700 in subchloroplast particles. - Biochim.
biophys. Acta 679 : 436 - 443, 1982.

50967 - OLSEN, R.A., BROWN, J.C., BENNETT, J.H., BLUME, D. : Reduction of Fe^{3+} as it
relates to Fe chlorosis. - J. Plant Nutr. 5 : 433 - 445, 1982.

50968 - OLUFAJO, O.O., DANIELS, R.W., SCARISBRICK, D.H. : The effect of pod removal
on the translocation of ^{14}C photosynthate from leaves in *Phaseolus vulgaris*
L. cv. Lochness. - J. hort. Sci. 57 : 333 - 338, 1982.

*50969 - OMASA, K., ABO, F., NATORI, T., TOTSUKA, T. : Analysis of air pollutant
sorption by plants (3) Sorption under fumigation with NO_2, O_3 or NO_2 + O_3. -
Res. Rep. nat. Inst. environm. Stud. 11 (Studies on the Effects of Air Pollutants on Plants and Mechanism of Phytotoxicity) : 213 - 224, 1980. [Sto-
matal resistance.]

*50970 - ONDOK, J.P., GLOSER, J. : Fotosyntetický model porostu rákosu (*Phragmites
communis* TRIN.). [Model of photosynthesis in a reed stand (*Phragmites communis* TRIN.).] - Acta ecol. 8 (18) : 43 - 69, 1978. [In Czech, ab : E, R.]

50971 - ONDOK, J.P., POKORNÝ, J. : Model of diurnal regime of O_2 and CO_2 in stands of
submerged aquatic vegetation. - Ekológia (ČSSR) 1 : 381 - 394, 1982. [Ps.]

50972 - O'NEAL, S.W., PRINCE, J.S. : Relationship between seasonal growth, photosynthetic production and apex mortality of *Caulerpa paspaloides (Chlorophyceae)*. - Mar. Biol. 72 : 61 - 67, 1982.

50973 - O'NEIL, K.J., CARROW, R.N. : Kentucky bluegrass growth and water use under
different soil compaction and irrigation regimes. - Agron. J. 74 : 933 - 936,
1982. [Primary production.]

50974 - ONISHI, J.C., NIEDERMAN, R.A. : *Rhodopseudomonas sphaeroides* membranes: alterations in phospholipid composition in aerobically and phototrophically grown
cells. - J. Bacteriol. 149 : 831 - 839, 1982. [Chl.]

50975 - ONO, T., INOUE, Y. : Photoactivation of the water-oxidation system in isolated intact chloroplasts prepared from wheat leaves grown under intermittent
flash illumination. - Plant Physiol. 69 : 1418 - 1422, 1982.

50976 - ONO, Y. : [Effects of leaf area index, specific leaf area, nitrogen content
in leaves and distribution ratio of dry matter to pod on net assimilation
rate of peanut plants in the first half of fruiting stage.] - Jap. J. Crop
Sci. 51 : 287 - 292, 1982. [In Jap., ab : E.]

50977 - ONOUE, Y., HIRAKI, K., NISHIKAWA, Y. : [Studies on the delayed fluorometric
analysis of metalloporphyrins.] - Bunseki Kagaku 31 : 169 - 174, 1982. [In
Jap., ab : E.]

50978 - ONOUE, Y., KOTANI, M., HIRAKI, K., SHIGEMATSU, T., NISHIKAWA, Y. : Delayed
fluorometric determination of chlorophyll c in natural waters. - Bunsei Kagaku 31 : E45 - E47, 1982.

50979 - OOTA, T., TAKENAGA, H., KANEKI, Y. : [Effect of X-ray irradiation on the
photosynthesis of maize.] - Tokyo Nogyo Daigaku [J.agr.Sci.] 26 : 305 - 310,
1982. [In Jap., ab : E.]

50980 - OOTAKE, Y. : [Effect of temperature on the internal morphology and development of leaves in Chinese cabbage (*Brassica campestris* L.).] - J. jap. Soc.
hort. Sci. 51 : 329 - 337, 1982. [In Jap., ab : E.]

50981 - OPARINA, L.A. : Mnozhestvennye molekulyarnye formy "malik"-ènzima v list'-
yakh *Zea mays*. [Multiple molecular forms of malic enzyme from the leaves of
Zea mays.] - Biokhimiya 47 : 1035 - 1038, 1982. [In R, ab : E.]

50982 - OPRITOV, V.A., PYATYGIN, S.S., RETIVIN, V.G. : Vozniknovenie potentsialov
deĭstviya u vysshikh rasteniĭ v otvet na neznachitel'noe lokal'noe okhlazh-
denie. [Generation of action potential in higher plants in response to slight
local cooling.] - Fiziol. Rast. 29 : 338 - 344, 1982. [Chloroplast; in R,
ab : E.]

50983 - **OQUIST, G.** : Seasonally induced changes in acyl lipids and fatty acids of chloroplast thylakoids of *Pinus silvestris*. A correlation between the level of unsaturation of monogalactosyldiglyceride and the rate of electron transport. - Plant Physiol. *69* : 869 - 875, 1982.

50984 - **ÖQUIST, G., BRUNES, L., HÄLLGREN, J.-E.** : Photosynthetic efficiency of *Betula pendula* acclimated to different quantum flux densities. - Plant Cell Environm. *5* : 9 - 15, 1982.

50985 - **ÖQUIST, G., BRUNES, L., HÄLLGREN, J.-E.** : Photosynthetic efficiency during ontogenesis of leaves of *Betula pendula*. - Plant Cell Environm. *5* : 17 - 21, 1982.

50986 - **ÖQUIST, G., FORK, D.C.** : Effects of desiccation on the excitation energy distribution from phycoerythrin to the two photosystems in the red alga *Porphyra perforata*. - Physiol. Plant. *56* : 56 - 62, 1982.

50987 - **ÖQUIST, G., FORK, D.C.** : Effects of desiccation on the 77 °K fluorescence properties of the liverwort *Porella navicularis* and the isolated lichen green alga *Trebouxia pyriformis*. - Physiol. Plant. *56* : 63 - 68, 1982.

50988 - **ÖQUIST, G., HAGSTRÖM, A., ALM, P., SAMUELSSON, G., RICHARDSON, K.** : Chlorophyll *a* fluorescence, an alternative method for estimating primary production. - Mar. Biol. *68* : 71 - 75, 1982.

50989 - **ORLYUK, A.P., LAVRINENKO, Yu.A.** : Nasledovanie pokazateleĭ fotosinteticheskoĭ deyatel'nosti gibridami yarovoĭ pshenitsy v usloviyakh orosheniya. [Inheritance of photosynthetic characteristics by spring wheat hybrids under irrigation.] - Sel'skokhoz. Biol. *17* : 309 - 314, 1982. [In R, ab : E.]

50990 - **ORMROD, D.P.** : Air pollutant interactions in mixtures. - In : **UNSWORTH,M.H., ORMROD, D.P.** (ed.) : Effects of Gaseous Air Pollution in Agriculture and Horticulture. Pp. 307 - 331. Butterworth Scientific, London - Boston - Sydney - Wellington - Durban - Toronto 1982.[Ps.]

50991 - **ORMROD, D.P.** : Effects of pollutants on horticultural plants. - In : Proceedings 21st International Horticultural Congrass. Volume II. Pp. 646 - 656. International Society for Horticultural Science, Wageningen 1982. [Ps.]

50992 - **ORON, G., SHELEF, G.** : Maximizing algal yield in high-rate oxidation ponds. - J. Environm. Eng. Div.-ASCE *108* : 730 - 738, 1982.

50993 - **ORR, G.L., HESS, F.D.** : Proposed site(s) of action of new diphenyl ether herbicides. - In : **MORELAND, D.E., St.JOHN, J.B., HESS, F.D.** (ed.) : Biochemical Responses Induced by Herbicides. Pp. 131 - 152. American Chemical Society, Washington, D.C. 1982. [Car.]

50994 - **ORT, D.R., MELANDRI, B.A.** : Mechanism of ATP synthesis. - In : **GOVINDJEE** (ed.) : Photosynthesis. Vol.1. Pp. 537 - 587. Academic Press, New York - London - Paris - San Diego - San Francisco - São Paulo - Sydney - Tokyo - Toronto 1982.

*50995 - **OSAKI, M., TANAKA, A.** : ["Current photosynthates" and "Storage substances" as the respiratory substrates in the rice plant.] - J. Sci. Soil Manure *50* : 540 - 546, 1979. [In Jap.]

50996 - **OSBORNE, L.L.** : Acute metabolic responses of lotic epilithic communities to total residual chlorine. - Bull. environm. Contam. Toxicol. *28* : 524 - 529, 1982. [Primary production.]

*50997 - **OSIPOV, L.F., SAKEVICH, A.I.** : Poluchenie khlorofill-karotinovoĭ pasty iz sinezelenykh vodorosleĭ-vozbuditeleĭ "tsveteniya" vody. [Production of chlorophyll-carotene paste from blue-green algae - the cause of blooming in water.] - In : Formirovanie i Kontrol' Kachestva Poverkhnostnykh Vod. Pp. 80 - - 82. Naukova Dumka, Kiev 1976. [In R.]

50998 - **OSMAN, M.E.H., METZNER, H., FISCHER, K.** : Effects of nitrate on thylakoid reactions. 1. Influence on photosynthetic electron transport. - Photosynthetica *16* : 7 - 12, 1982.

50999 - OSMOND, C.B., WINTER, K., ZIEGLER, H. : Functional significance of different
         pathways of CO_2 fixation in photosynthesis. - In : LANGE, O.L., NOBEL, P.S.,
         OSMOND, C.B., ZIEGLER, H. (ed.) : Physiological Plant Ecology II. Water
         Relations and Carbon Assimilation. Pp. 479 - 547. Springer-Verlag, Berlin -
         Heidelberg - New York 1982.

51000 - OSTROVSKAYA, L.K. : Ul'trastrukturnaya i funktsional'naya spetsifichnost'
         organizatsii khloroplastov i elektrontransportnoĭ tsepi fotosinteza. [Ultra-
         structure and functional specificity of chloroplast organization and electron
         transport chain of photosynthesis.] - In : Fiziologiya Fotosinteza. Pp. 76 -
         - 88. Nauka, Moskva 1982. [In R.]

51001 - OSZLÁNYI, J. : Wood, bark, needles, leaves and roots energy values of *Pinus
         silvestris* L., *Picea excelsa* LINK. and *Fagus silvatica* L. - Ekológia (ČSSR)
         1 : 289 - 296, 1982.

51002 - O'TOOLE, J.C., TOMAR, V.S. : Transpiration, leaf temperature and water poten-
         tial of rice and barnyardgrass in flooded fields. - Agr. Meteorol. *26* : 285 -
         - 296, 1982. [Stomatal resistance.]

51003 - OTTOLENGHI, M. : Molecular aspects of the photocycles of rhodopsin and bac-
         teriorhodopsin: A comparative overview. - In : COLOWICK, S.P., KAPLAN, N.O.
         (ed.) : Methods in Enzymology. Vol.88. Pp. 470 - 491. Academic Press, New
         York - London - Paris - San Diego - San Francisco - São Paulo - Sydney -
         Tokyo - Toronto 1982.

51004 - OUTLAW, W.H.,Jr. : Carbon metabolism in guard cells. - In : CREASY, L.L.,
         HRAZDINA, G. (ed.) : Cellular and Subcellular Localization in Plant Metabo-
         lism. Pp. 185 - 222. Plenum Press, New York - London 1982.

51005 - OUTLAW, W.H.,Jr., TARCZYNSKI, M.C., ANDERSON, L.C. : Taxonomic survey for
         the presence of ribulose-1,5-bisphosphate carboxylase activity in guard
         cells. - Plant Physiol. *70* : 1218 - 1220, 1982.

51006 - OVCHINNIKOV, Yu.A. : Rhodopsin and bacteriorhodopsin: structure-function
         relationships. - FEBS Lett. *148* : 179 - 191, 1982.

51007 - OVCHINNIKOV, Yu.A., ABDULAEV, N.G., DERGACHEV, A.E., DRACHEV, A.L., DRACHEV,
         L.A., KAULEN, A.D., KHITRINA, L.V., LAZAROVA, Z.P., SKULACHEV, V.P. : Photo-
         electric and spectral responses of bacteriorhodopsin modified by carbodiimide
         and amine derivatives. - Europe. J. Biochem. *127* : 325 - 332, 1982.

51008 - OVCHINNIKOV, Yu.A., ABDULAEV, N.G., MODYANOV, N.N. : Structural basis of
         proton-translocating protein function. - Annu. Rev. Biophys. Bioenerg. *11* :
         445 - 463, 1982. [Ps, Chl.]

51009 - OVERDIECK, D. : Tagesgang des CO_2-Gaswechsels und der Saccharid-Produktion
         von Sonnenblumenblättern (*Helianthus annuus* L.). - Flora *172* : 511 - 531,
         1982.

*51010 - OVIATT, C.A., NIXON, S.W., PEREZ, K.T., BUCKLEY, B. : On the season and na-
         ture of perturbations in microcosm experiments. - In : DAME, R.F. (ed.) :
         Marsh-Estuarine Systems Simulation. Pp. 143 - 164. University of South Caro-
         lina Press, Columbia 1979. [Chl.]

51011 - OWENS, G.C., OHAD, I. : Phosphorylation of *Chlamydomonas reinhardi* chloro-
         plast membrane proteins *in vivo* and *in vitro*. - J. Cell Biol. *93* : 712 - 718,
         1982.

51012 - OWENS, G.C., WETTERN, M., LAVINTMAN, N., ISH-SHALOM, D., SCHUSTER, G., KIRI-
         LOVSKY, D., OHAD, I. : Aspects of the organization and development of the
         photosystem II unit in *Chlamydomonas reinhardi*. - In : AKOYUNOGLOU, G.,
         EVANGELOPOULOS, A.E., GEORGATSOS, J., PALAIOLOGOS, G.,TRAKATELLIS, A.,
         TSIGANOS, C.P. (ed.) : Cell Function and Differentiation, Part B. Biogenesis
         of Energy Transducing Membranes and Membrane and Protein Energetics. Pp.
         223 - 231. Alan R. Liss Inc., New York 1982.

51013 - OWENS, T.G., FALKOWSKI, P.G. : Enzymatic degradation of chlorophyll *a* by ma-
         rine phytoplankton *in vitro*. - Phytochemistry *21* : 979 - 984, 1982.

51014 - OWONUBI, J.J., KANEMASU, E.T. : Water use efficiency of three height iso-
lines of sorghum. - Can. J. Plant Sci. *62* : 35 - 46, 1982. [Stomatal resistance.]

51015 - PACKHAM, N.K., DUTTON, P.L., MUELLER, P. : Direct measurement of light induced currents and potentials generated by bacterial reaction centers. - In :
HÉLÈNE, C., CHARLIER, M., MONTENAY-GARESTIER, T., LAUSTRIAT, G. (ed.) :
Trends in Photobiology. Pp. 571 - 577. Plenum Press, New York - London
1982.

51016 - PACKHAM, N.K., MANSFIELD, R.W., BARBER, J. : Action of cyanide on the photosynthetic water-splitting process. - Biochim. biophys. Acta *681* : 538 -
- 541, 1982.

51017 - PAERL, H.W. : Feasibility of ^{55}Fe autoradiography as performed on N_2-fixing
Anabaena spp. populations and associated bacteria. - Appl. environm. Microbiol. *43* : 210 - 217, 1982. [Ps.]

*51018 - PAGNI, P.G.S., WALNE, P.L., PAGNI, R.M. : On the occurrence of α-carotene
in isolated stigmata of *Euglena gracilis* var. *bacillaris*. - Phycologia *20* :
431 - 434, 1981.

51019 - PAILLOTIN, G., VERMEGLIO, A., BRETON, J. : Organization of the photosynthetic
pigments and transfer of the excitation energy. - In : HÉLÈNE, C., CHARLIER,
M., MONTENAY-GARESTIER, T., LAUSTRIAT, G. (ed.) : Trends in Photobiology.
Pp. 539 - 547. Plenum Press, New York - London 1982.

*51020 - PAIS, I., FEHÉR, M., FARKAS, E., CORNIDES, I. : A titán hatása különböző
növények leveleinek klorofilltartalmára. [Effect of titanium on the chlorophyll content of different plants.] - Kertészetiegy. Közl. *41* : 135 - 140,
1978. [In Hung., ab : E.]

51021 - PAKSHINA, E.V., LEBEDEV, N.N., SHAPOSHNIKOVA, M.G., KRASNOVSKIĬ, A.A. :
Obratimoe deĭstvie intensivnogo sveta na fotobiokhimicheskie svoĭstva
khromatophorov *Rhodospirillum rubrum*. [Reversible effect of intensive light
on photobiochemical properties of *Rhodospirillum rubrum* chromatophores.] -
Biokhimiya *47* : 534 - 539, 1982. [In R, ab : E.]

51022 - PALESS, Gy., E.TÓTH, É., SZIGETI, Z. : 3,5-Diszubsztituált-4-hidroxi-benzo-
nitrilek hatása a fotoszintetikus apparátusra. [Effect of 3,5-disubstituted-
-4-hydroxybenzonitriles on the photosynthetic system.] - Magy. Kem. Lapja
37 : 426 - 429, 1982. [In Hung., ab : E, R.]

51023 - PALIT, P., BHATTACHARYYA, A.C. : Source-sink control of jute productivity. -
Indian J. Plant Physiol. *25* : 187 - 200, 1982. [Growth analysis.]

51024 - PALLAS, J.E., Jr., KAYS, S.J. : Inhibition of photosynthesis by ethylene -
A stomatal effect. - Plant Physiol. *70* : 598 - 601, 1982.

51025 - PALMER, J.D. : Physical and gene mapping of chloroplast DNA from *Atriplex
triangularis* and *Cucumis sativa*. - Nucl. Acids Res. *10* : 1593 - 1605, 1982.

51026 - PALMER, J.D., EDWARDS, H., JORGENSEN, R.A., THOMPSON, W.F. : Evolutionary
variation in transcription an location of chloroplast genes. - Carnegie
Inst. Washington Year Book *81* : 94 - 95, 1982.

51027 - PALMER, J.D., STEIN, D.B. : Chloroplast DNA from the fern *Osmunda cinnamo-
mea:* physical organization, gene localization and comparison to angiosperm
chloroplast DNA. - Curr. Genet. *5* : 165 - 170, 1982.

51028 - PALMER, J.D., THOMPSON, W.F. . Chloroplast DNA rearrangements are more fre-
quent when a large inverted repeat sequence is lost. - Cell *29* : 537 - 550,
1982.

51029 - PALMER, J.D., ZAMIR, D. : Chloroplast DNA evolution and phylogenetic relationships in *Lycopersicon*. - Proc. nat. Acad. Sci. USA *79* : 5006 - 5010,
1982.

51030 - **PALTA, J.A.** : Gas exchange of four cassava cultivars in relation to light intensity. - Exp. Agr. *18* : 375 - 382, 1982.

51031 - **PAN, D., TAN, K.H.** : Regulation of thermostable phosphoenolpyruvate acid carboxylase from maize leaves by malate, citrate (activators) and fructose 1,6-bisphosphate, acetyl-CoA (inhibitors). - Plant Sci. Lett. *27* : 69 - 75, 1982.

51032 - **PAN, R.L., FAN, I.-J., BHARDWAJ, R., GROSS, E.L.** : A photosynthetic photo-electrochemical cell using flavin mononucleotide as the electron acceptor. - Photochem. Photobiol. *35* : 655 - 664, 1982.

51033 - **PANDEV, S., STANEV, V., K"DREV, T.** : Vliyanie na nedostiga na makroelementi-te v khranitelniya raztvor v"rkhu izmenenieto na nyakoi fotosintetichni po-kazateli na sl"nchogleda. [Effect of macroelement deficit in the nutrient solution on the changes of some photosynthetic indices in sunflower.] - Fi-ziol. Rast. (Sofia) *8*(2) : 45 - 51, 1982. [In Bulg., ab : E, R.]

51034 - **PAPAGEORGIOU, G.C., DEMOSTHENOPOULOU-KARAOULANI, E.** : Stabilization of the morphology and the photosynthetic function of isolated intact chloroplasts with glutaraldehyde. - Z. Pflanzenphysiol. *105* : 210 - 210, 1982.

51035 - **PAPAGEORGIOU, G.C., ISAAKIDOU, J.** : The pH dependence of the photosynthetic electron transport in glutaraldehyde-treated thylakoids. - FEBS Lett. *138* : 19 - 24, 1982.

51036 - **PARAMONOVA, L.I., NAUSH, Ya., KRESLAVSKIĬ, V.D., STOLOVITSKIĬ, Yu.M.** : Tushenie fluorestsentsii khlorofilla i feofitina fukoksantinom. [Quenching of chlorophyll and pheophytin fluorescence by fucoxanthin.] - Biofizika *27* : 197 - 201, 1982. [In R, ab : E.]

51037 - **PARKHURST, D.F.** : Stereological methods for measuring internal leaf structu-re variables. - Amer. J. Bot. *69* : 31 - 39, 1982. [Ps.]

51038 - **PARKKHERST, D.F.** : Adaptatsionnoe znachenie raspredeleniya ust'its na pover-khnostyakh list'ev. [Adaptation significance of stomata distribution on the leaf surface.] - Fiziol. Biokhim. kul't. Rast. *14* : 315 - 326, 1982. [Ps; in R, ab : E.]

51039 - **PARSON, W.W.:** Light-induced volume changes. - In : **COLOWICK, S.P., KAPLAN, N.O.** (ed.) : Methods in Enzymology. Vol.88. Pp. 272 - 276. Academic Press, New York - London - Paris - San Diego - San Francisco - São Paulo - Sydney - Tokyo - Toronto 1982. [Purple membrane.]

51040 - **PARSON, W.W.** : Photosynthetic bacterial reaction centers: interactions among the bacteriochlorophylls and bacteriopheophytins. - Annu. Rev. Biophys. Bio-eng. *11* : 57 - 80, 1982.

51041 - **PARSON, W.W., KE, B.** : Primary photochemical reactions. - In : **GOVINDJEE** (ed.) : Photosynthesis. Vol.1. Pp. 331 - 385. Academic Press, New York - London - Paris - San Diego - San Francisco - São Paulo - Sydney - Tokyo - Toronto 1982.

51042 - **PARSONS, A.J., ROBSON, M.J.** : Seasonal changes in the physiology of S24 pe-rennial ryegrass (*Lolium perenne* L.). 4.Comparison of the carbon balance of the reproductive crop in spring and the vegetative crop in autumn. - Ann. Bot. *50* : 167 - 177, 1982.

51043 - **PARSONS, L.R.** : Plant responses to water stress. - In : **CHRISTIANSEN, M.N., LEWIS, C.F.** (ed.) : Breeding Plants for Less Favorable Environments. Pp. 175 - 192. John Wiley & Sons, New York - Chichester - Brisbane - Toronto - Singapore 1982. [Ps.]

51044 - **PARSONS, T.R.** : The future of controlled ecosystem enclosure experiments. - In : **GRICE, G.D., REEVE, M.R.** (ed.) : Marine Mesocosms. Biological and Chemical Research in Experimental Ecosystems. Pp. 411 - 418. Springer-Verlag, New York - Heidelberg - Berlin 1982.

*51045 - **PARTHIER, B.** : Compartment cooperation in the synthesis of chloroplast pro-teins. - In : **SCHÜTTE, H.R., GROSS, D.** (ed.) : Proceedings of the Internatio-nal Conference on Regulation of Developmental Processes in Plants. Pp. 136 - -158. VEB G. Fischer, Jena 1978.

51046 - **PARTHIER, B.** : RNA-Sequenzen und Organellen-Evolution. - In : HOFFMANN, P.,
HIEKE, B. (ed.) : Photosynthese: Regulation und Evolution. (Colloquia Pflanzenphysiologie Nr.5.) Pp. 33 - 48, Humboldt-Universität, Berlin 1982. [Chlo-
roplast.]

51047 - **PARTHIER, B.** : The cooperation of nuclear and plastid genomes in plastid
biogenesis and differentiation. - Biochem. Physiol. Pflanzen *177* : 283 -
- 317, 1982.

51048 - **PARTHIER, B.** : Light-induced chloroplast differentiation. - In : NOVER, L.,
LUCKNER, M., PARTHIER, B. (ed.) : Cell Differentiation. Pp. 280 - 304.
Springer-Verlag, Berlin - Heidelberg - New York 1982.

51049 - **PASSIOURA, J.B.** : Water in the soil-plant-atmosphere continuum. - In :
LANGE, O.L., NOBEL, P.S., OSMOND, C.B., ZIEGLER, H. (ed.) : Physiological
Plant Ecology II. Water Relations and Carbon Assimilation. Pp. 5 - 33.
Springer-Verlag, Berlin - Heidelberg - New York 1982. [Ps.]

51050 - **PASSOVA, R.** : Primenenie metoda disk-DSN-èlektroforeza dlya razdeleniya
membrannykh belkov khloroplastov. [Fractionation of chloroplast membrane
proteins by disc electrophoresis in sodium dodecyl sulfate.] - Eesti NSV
Tead. Akad. Toimet., Biol. *31* : 292 - 297, 1982. [In R, ab : E, Est.]

51051 - **PATEL, K.R., SINGH, Y.D.** : Retarding effect of distilled water infiltration
on chlorophyll breakdown. - Biochem. Physiol. Pflanzen *177* : 275 - 277,
1982.

51052 - **PATERSON, D., KENWORTHY, J.B.** : An investigation of the effects of fluoride
on selected moss species. - In : UNSWORTH, M.H., ORMROD, D.P. (ed.) : Effects
of Gaseous Air Pollution in Agriculture and Horticulture. Pp. 486 - 488.
Butterworths, London 1982. [Ps.]

51053 - **PATERSON, D.R., ARNTZEN, C.J.** : Detection of altered inhibition of photosystem II reactions in herbicide-resistant plants. - In : EDELMAN, M.,
HALLICK, R.B., CHUA, N.-H. (ed.) : Methods in Chloroplast Molecular Biology.
Pp. 109 - 118. Elsevier Biomedical Press, Amsterdam - New York - Oxford
1982.

51054 - **PATHAN, S.N., NIMBALKAR, J.D.** : Photosynthesis in *Alternanthera (Amaranthaceae)* species differing in carbon dioxide fixation pathways. - Photosynthetica *16* : 119 - 122, 1982.

51055 - **PATNAIK, G., COCKING, E.C.** : A new enzyme mixture for the isolation of leaf
protoplasts. - Z. Pflanzenphysiol. *107* : 41 - 45, 1982.

51056 - **PATTENDEN, G.** : Stereocontrolled synthesis of polyene isoprenoids. - In :
BRITTON, G., GOODWIN, T.W. (ed.) : Carotenoid Chemistry and Biochemistry.
Pp. 87 - 96. Pergamon Press, Oxford - New York - Toronto - Sydney - Paris -
Frankfurt 1982. [Car.]

51057 - **PATTERSON, D.T., FLINT, E.P.** : Interacting effects of CO_2 and nutrient concentration. - Weed Sci. *30* : 389 - 394, 1982. [Chl.]

*51058 - **PÄTZOLD, H., BAUER, U.** : Účinnost závlahy postřikem čistou vodou na výnos,
kontinuitu produkce a jakost píce různých travních porostů v severních
oblastech NDR. [Efficiency of sprinkler irrigation in yield, continuity of
production and quality of fodder of various grass covers in northern G.D.R.]
- Acta Univ. Agr., Fac. agron. (Brno) *A 28*(3-4): 275 - 283, 1980. [In Czech,
ab : G.]

51059 - **PAULECHOVÁ, K., PAULECH, C.** : Patofyziologické zmeny v listoch slivky napad-
nutej vírusom zakrpatenosti slívky. [Pathophysiological changes in plum
leaves infected with prune dwarf virus.] - Acta Inst. bot. Acad. Sci. slov.
Ser. B *1* : 321 - 336, 1975. [Ps, Chl; in Slovak, ab : E, R.]

51060 - **PAUTOVA, L.A., KONOVALOVA, G.V.** : Letne-osenniǐ fitoplankton proliva Starka
Yaponskogo morya. [Summer-autumn phytoplankton of the Stark Strait, Sea of
Japan.] - Biol. Morya *1982*(5) : 20 - 28, 1982. [In R, ab : E.]

51061 - PEARCY, R.W., OSTERYOUNG, K., RANDALL, D. : Carbon dioxide exchange characteristics of C_4 Hawaiian *Euphorbia* species native to diverse habitats. - Oecologia *22* : 333 - 341, 1982.

51062 - PEAREN, J.R., HUME, D.J. : Non-destructive estimation of ^{14}C in soybeans
immediately after labelling. - Crop Sci. *22* : 669 - 671, 1982.

51063 - PEARLSTEIN, R.M. : Chlorophyll singlet excitons. - In : GOVINDJEE (ed.) :
Photosynthesis. Vol.1. Pp. 293 - 330. Academic Press, New York - London -
Paris - San Diego - San Francisco - São Paulo - Sydney - Tokyo - Toronto
1982.

51064 - PEARLSTEIN, R.M. : Exciton migration and trapping in photosynthesis. - Photochem. Photobiol. *35* : 835 - 844, 1982.

51065 - PEARLSTEIN, R.M., DAVIS, R.C., DITSON, S.L. : Giant circular dichroism of
high molecular weight chlorophyllide-apomyoglobin complexes. - Proc. nat.
Acad. Sci. USA *79* : 400 - 402, 1982.

51066 - PEAT, J.R., JEFFCOAT, B. : The potential for increasing soyabean yield with
plant growth regulators. - In : McLAREN, J.S. (ed.) : Chemical Manipulation
of Crop Growth and Development. Pp. 237 - 249. Butterworth Scientific, London - Boston - Durban - Singapore - Sydney - Toronto - Wellington 1982.
[Ps.]

51067 - PEEL, E. : Photoautotrophic growth of suspension cultures of *Asparagus offi-
cinalis* L. cells in turbidostat. - Plant Sci. Lett. *24* : 147 - 155, 1982.

51068 - PEINE, G., HOFFMANN, P., WIEDER, G., SCHILLING, G. : Die Pyridinnucleotid-
gehalte in Weizenkeimpflanzen unter besonderer Berücksichtigung der Photo-
synthesebedingungen. - In : HOFFMANN, P., HIEKE, B. (ed.) : Photosynthese:
Regulation und Evolution. (Colloquia Pflanzenphysiologie Nr.5.) Pp. 79 - 82.
Humboldt-Universität, Berlin 1982.

51069 - PEINERT, R., SAURE, A., STEGMANN, P., STIENEN, C., HAARDT, H., SMETACEK, V.:
Dynamics of primary production and sedimentation in a coastal ecosystem. -
Neth. J. Sea Res. *16* : 276 - 289, 1982.

51070 - PEISER, G.D., LIZADA, M.C.C., YANG, S.F. : Sulfite-induced lipid peroxidation
in chloroplasts as determined by ethane production. - Plant Physiol. *70* :
994 - 998, 1982. [Ps.]

51071 - PEISKER, M. : The effect of CO_2 leakage from bundle sheath cells on carbon
isotope discrimination in C_4 plants. - Photosynthetica *16* : 533 - 541,
1982.

51072 - PEISKER, M. : Theoretische Analyse der Änderungen des $\delta^{13}C$-Wertes bei der
Evolution des C_4-pathway der Photosynthese. - In : HOFFMANN, P., HIEKE, B.
(ed.) : Photosynthese: Regulation und Evolution. (Colloquia Pflanzenphysiologie Nr.5.) Pp. 163 - 167. Humboldt-Universität, Berlin 1982.

51073 - PEISKER, M., VÁCLAVÍK, J. : Die Beziehung zwischen Transpiration und CO_2-Auf-
nahme bei Maisblättern nach Unterbrechung der Wasserversorgung. - In :
UNGER, K., SCHUH, J. (ed.) : Umwelt-Stress. Pp. 221 - 224. Martin-Luther-Universität, Halle 1982.

51074 - PELLEGRINO, F., ALFANO, R.R. : Time-resolved fluorescence spectroscopy. -
In : ALFANO, R.R. (ed.) : Biological Events Probed by Ultrafast Laser Spec-
troscopy. Pp. 27-53. Academic Press, New York - London - Paris - San Diego -
San Francisco - São Paulo - Sydney - Tokyo - Toronto 1982. [Ps, Chl.]

51075 - PENNY, M.G., BAYFIELD, N.G. : Photosynthesis in desiccated shoots of *Poly-
trichum*. - New Phytol. *91* : 637 - 645, 1982.

51076 - PERCHOROVICZ, J.T., RAYNES, D.A., JENSEN, R.G. : Measurement and preservation
of the *in vivo* activation of ribulose 1,5-bisphosphate carboxylase in leaf
extracts. - Plant Physiol. *69* : 1165 - 1168, 1982.

51077 - PEREIRA, A.R., MACHADO, E.C., DE CAMARGO, M.B.P. : Solar radiation regime in
three cassava (*Manihot esculenta* CRANTZ) canopies. - Agr. Meteorol. *26* : 1 -
- 10, 1982.

*51078 - PERES EÏRIS, M., ROMANENKO, V.I., PUBIENES, M.A. : Potreblenie êÏkhornieï *Eichhornia crassipes* (MART.) SOLMS. anionov karbonata cherez kornevuyu siste-mu. [Use of carbonate anions through root system in *Eichhornia crassipes* (MART.) SOLMS.] - Biol. vnutr. Vod (Leningrad) *48* : 16 - 18, 1980. [Ps; in R.]

51079 - PERESYPKIN, V.F., VORONKOV, L.A., GORDIENKO, T.K. : Fotokhimicheskaya aktiv-nost'khloroplastov ozimoï pshenitsy pri zarazhenii buroï rzhavchinoï.[Photo-chemical activity of chloroplasts in winter wheat infected with brown rust.] - Dokl. vsesoyuz. Akad. sel'skokhoz. Nauk Im.V.I.Lenina *1982* (12) : 17 - 19, 1982. [In R.]

51080 - PERRIN, P.W. : Poststorage effect of light, temperature and nutrient spray treatments on chlorophyll development in cabbage. - Can. J. Plant Sci. *62* : 1023 - 1026, 1982.

51081 - PERROT-RECHENMANN, C., VIDAL, J., BRULFERT, J., BURLET, A., GADAL, P. : A comparative immunocytochemical localization study of phosphoenolpyruvate carboxylase in leaves of higher plants. - Planta *155* : 24 - 30, 1982.

*51082 - PERSTNEVA, T.A. : Struktura listovogo pologa i intensivnost' solnechnoï radiatsii v nasazhdeniyakh yabloni na podvoe M IX. [Canopy structure and so-lar irradiance in stands of apple tree on M IX rootstock.] - In : Voprosy Intensifikatsii Plodovodstva. Pp. 34 - 38. Kishinev. sel'skokhoz. Inst., Kishinev 1978. [In R.]

*51083 - PERSTNEVA, T.A. : Kontsentratsiya pigmentov i opticheskie svoïstva list'ev yabloni na slaboroslom podvoe v usloviyakh intensivnoï kul'tury. [Concentra-tion of pigments and optical properties of apple tree leaves on poorly-grow-ing rootstocks during intensive cultivation.] - In : Voprosy Intensifikatsii Plodovodstva. Pp. 39 - 43. Kishinev. sel'skokhoz. Inst., Kishinev 1978. [In R.]

51084 - PESCHEK, G.A., MUCHL, R., KIENZL, P.F., SCHMETTERER, G. : Characteristic temperature dependences of respiratory and photosynthetic electron-transport activities in membrane preparations from *Anacystis nidulans* grown at differ-ent temperatures. - Biochim. biophys. Acta *679* : 35 - 43, 1982.

51085 - PESCHEK, G.A., SCHMETTERER, G. : Evidence for plastoquinol-cytochrome f/b-563 reductase as a common electron donor to $P700$ and cytochrome oxidase in cyano-bacteria. - Biochem. biophys. Res. Commun. *108* : 1188 - 1195, 1982.

51086 - PESCHL, A. : Reaktion des CO_2- und Wasserdampfgaswechsels abgehärteter Pflan-zen von *Picea abies*, *Pseudotsuga menziesii* und *Abies grandis* bei Frosttrock-nis. - Flora *172* : 427 - 447, 1982.

51087 - PESCHL, A. : Der Einfluß der Kalium- und Stickstoffversorgung auf den Gas-wechsel junger Pflanzen von *Picea abies*, *Pseudotsuga menziesii* und *Abies grandis* bei Frosttrocknis. - Flora *172* : 449 - 461, 1982.

*51088 - PEŠKA, J. : Simple equipment for the exposure of plant assimilating organs in the $^{14}CO_2$-labelled atmosphere under field conditions. - Acta Univ. Agr., Fac. agron. (Brno) *24* : 587 - 592, 1976.

*51089 - PETERS, G.A., ITO, O., TYAGI, V.V.S., KAPLAN, D. : Physiological studies on N_2-fixing *Azolla*. - In : LYONS, J.M., VALENTINE, R.C., PHILLIPS, D.A., RAINS, D.W., HUFFAKER, R.C. (ed.) : Genetic Engineering of Symbiotic Nitrogen Fixation and Conservation of Fixed Nitrogen. Pp. 343 - 362. Plenum Press, New York - London 1981. [Ps, Chl.]

51090 - PETERS, R., CHERRY, R.J. : Lateral and rotational diffusion of bacteriorho-dopsin in lipid bilayers: Experimental test of the Saffman-Delbrück equations. - Proc. nat. Acad. Sci. USA *79* : 4317 - 4321, 1982.

51091 - PETERSON, K.M., BILLINGS, W.D. : Growth of alpine plants under controlled drought. - Arctic alpine Res. *14* : 189 - 194, 1982. [Growth analysis.]

51092 - PETERSON, N.C., BLESSINGTON, T.M. : Antitranspirant influences on the post-harvest quality of *Ficus benjamina* L. - Res. Highlights *45* (7) : 7 - 8, 1982. [Chl.]

51093 - **PETERSON,R.B.** : Enhanced incorporation of tritium into glycolate during
photosynthesis by tobacco leaf tissue in the presence of tritiated water. -
Plant Physiol. *69* : 192 - 197, 1982.

51094 - **PETERSON, R.B., ZELITCH, I.** : Relationship between net CO_2 assimilation and
dry weight accumulation in field-grown tobacco. - Plant Physiol. *70* : 677 -
- 685, 1982.

51095 - **PETKE, J.D., MAGGIORA, G.M., SHIPMAN, L.L., CHRISTOFFERSEN, R.E.** : Stereo-
electronic properties of photosynthetic and related systems. XI. *Ab initio*
quantum mechanical characterization of the electronic structure and spectra
of the chlorophyllide *a* and pheophorbide *a* anion radicals. - Photochem.
Photobiol. *36* : 383 - 393, 1982.

51096 - **PETKOV, P.S.** : Dinamika na natrupvane na plastidnite pigmenty v listata
na pshenitsa sort Sadovo 1 v zavisimost ot reaktsiyata mu k"m predshestveni-
ka, posevnata norma i fona na torene. [Dynamics of plastid pigment accumulation in leaves of wheat cv. Sadovo 1, depending on its reaction to pre-
decessor seeding rate and fertilizer background.] - Rasteniev. Nauki *19* (2):
3 - 11, 1982. [In Bulg., ab : E, R.]

51097 - **PETROV, A., MANOLOV, P., MANSUR, A.** : Zakonomernosti v razpredelyaneto na
produktite na fotosintezata v yab"lkovoto d"rvo. [Regularities of photosynthate distribution in apple trees.] - Gradin. lozar. Nauka *19* (1) : 16 -
- 24, 1982. [In Bulg., ab : E, R.]

51098 - **PETRUKHIN, Yu.A., KONSTANTINOVA, L.M.** : Fotosinteticheskiĭ metabolism ugleroda v list'yakh khlorofituma. [Photosynthetic metabolism of carbon in the
leaves of *Chlorophytum*.] - Biol. Nauki *1982* (7) : 95 - 99, 1982. [In R.]

51099 - **PFANDER, H., SCHURTENBERGER, H.** : Biosynthesis of C_{20}-carotenoids in *Crocus
sativus*. - Phytochemistry *21* : 1039 - 1042, 1982.

51100 - **PFENNIG, N.** : Chlorophylls of photosynthetic bacteria. - In : **ZABORSKY, O.R.**
(ed.) : CRC Handbook of Biosolar Resources. Vol.I/1. Pp. 59 - 62. CRC Press,
Boca Raton 1982.

51101 - **PFENNIG, N.** : Sources and collections of photosynthetic bacteria (anoxyphotobacteria; phototrophic bacteria). - In : **ZABORSKY, O.R.** (ed.) : CRC Handbook of Biosolar Resources. Vol.I/1. Pp. 591 - 592. CRC Press, Boca Raton
1982.

*51102 - **PFISTER, K., LICHTENTHALER, H.K., BURGER, G., MUSSO, H., ZAHN, M.** : The
inhibition of photosynthetic light reactions by halogenated naphthoquinones.
- Z. Naturforsch. *36 C* : 645 - 655, 1981.

51103 - **PFISTERER, J., LACHMANN, P., KLOPPSTECH, K.** : Transport of proteins into
chloroplasts. Binding of nuclear-coded chloroplast proteins to the chloroplast envelope. - Europe. J. Biochem. *126* : 143 - 148, 1982.

51104 - **PHAM THI, A.T., PIMENTEL, C., VIEIRA DA SILVA, J.** : Effects of water stress
on photosynthesis and photorespiration of *Atriplex nummularia*, a C_4 plant.
- Photosynthetica *16* : 334 - 342, 1982.

51105 - **PHELAN, D.M., TAYLOR, R.M., FRICKE, S.** : A maintenance-free dissolved oxygen monitor. - Int. Lab. *12* (7) : 60, 62-64, 66, 68, 70, 72, 74-75, 1982.

51106 - **PHILIPP, J.E.** : Maximum likelyhood estimation of primary productivity co-
efficients. - Radiat. environm. Biophys. *20* : 301 - 310, 1982.

*51107 - **PHILLIPS, R.** : Deacidification in a plant with crassulacean acid metabolism
associated with anion-cation balance. - Nature *287* : 727 - 728, 1980.

51108 - **PHIPPS, D.W.,Jr., PARDY, R.L.** : Host enhancement of symbiont photosynthesis
in the hydra-algae symbiosis. - Biol. Bull. *162* : 83 - 94, 1982.

51109 - **PHLIPS, E.J.,MITSUI,A.** : Light intensity preference and tolerance of aquatic photosynthetic macroalgae. - In : **ZABORSKY, O.R.** (ed.) : CRC Handbook
of Biosolar Resources. Vol.I/2. Pp. 309 - 333. CRC Press, Boca Raton
1982.

51110 - PHLIPS, E.J., MITSUI, A. : Light intensity preference and tolerance of aquatic photosynthetic microorganisms. - In : ZABORSKY, O.R. (ed.) : CRC Handbook of Biosolar Resources. Vol.I/2. Pp. 257 - 308. CRC Press, Boca Raton
1982.

51111 - PHLIPS, E.J., MITSUI, A. : Temperature preference and tolerance of aquatic
photosynthetic microorganisms. - In : ZABORSKY, O.R. (ed.) : CRC Handbook
of Biosolar Resources. Vol.I/2. Pp. 335 - 361. CRC Press, Boca Raton 1982.

51112 - PHLIPS, E.J., MITSUI, A. : Temperature preference and tolerance of aquatic
photosynthetic macroalgae. - In : ZABORSKY, O.R. (ed.) : CRC Handbook of
Biosolar Resources. Vol.I/2. Pp. 363 - 377. CRC Press, Boca Raton 1982.

51113 - PIAZZA, G.J., SMITH, M.G., GIBBS, M. : Characterization of the formation
and distribution of photosynthetic products by *Sedum praealtum* chloroplasts.
- Plant Physiol. *70* : 1748 - 1758, 1982.

51114 - PICAUD, A., ACKER, S., DURANTON, J. : A simple step separation of PS 1,
PS 2 and chlorophyll-antenna particles from spinach chloroplasts. - Photosynthesis Res. *3* : 203 - 213, 1982.

51115 - PICCIONI, R., BELLEMARE, G., CHUA, N.-H. : Methods of polyacrylamide gel
electrophoresis in the analysis and preparation of plant polypeptides. -
In : EDELMAN, M., HALLICK, R.B., CHUA, N.-H. (ed.) : Methods in Chloroplast
Molecular Biology. Pp. 985 - 1014. Elsevier Biomedical Press, Amsterdam -
New York - Oxford 1982.

51116 - PICK, U. : Isolation of the ATPase complex (CF_0-CF_1). - In : EDELMAN, M.,
HALLICK, R.B., CHUA, N.-H. (ed.) : Methods in Chloroplast Molecular Biology. Pp. 873 - 880. Elsevier Biomedical Press, Amsterdam - New York - Oxford
1982.

51117 - PICK, U., BASSILIAN, S. : Activation of magnesium ion specific adenosine-
triphosphatase in chloroplast coupling factor 1 by octyl glucoside. - Biochemistry *21* : 6144 - 6152, 1982.

51118 - PICK, U., CONRAD, P.L., CONRAD, J.M., DURBIN, R.D., SELMAN, B.R. : Syner-
gistic activation of an Mg-specific ATPase activity in chloroplast coupling
factor by octylglucoside and tentoxin. - Biochim. biophys. Acta *682* : 55 -
- 58, 1982.

51119 - PICKARD, W.F. : Why is the substomatal chamber as large as it is? - Plant
Physiol. *69* : 971 - 974, 1982. [Stomatal resistance.]

51120 - PIERRE, M., QUIEIROZ, O. : Modulation by leaf age and SO_2 concentration
of the enzymic response to subnecrotic SO_2 pollution. - Environm. Pollut.
A *28* : 209 - 217, 1982. [Ps.]

51121 - PIETARINEN, I., KANNINEN, M., HARI, P., KELLOMÄKI, S. : A simulation model
for daily growth of shoots, needles, and stem diameter in Scots pine trees.
- Forest Sci. *28* : 573 - 581, 1982. [Ps, photosynthates.]

51122 - PINEAU, B. : Biosynthesis of ribulose-1,5-bisphosphate carbyxylase in greening cells of *Euglena gracilis*. The accumulation of ribulose-1,5-bisphosphate
carboxylase and of its subunits. - Planta *156* : 117 - 128, 1982.

51123 - PINGREE, R.D., HOLLIGAN, P.M., MARDELL, G.T., HARRIS, R.P. : Vertical distribution of plankton in the Skagerrak in relation to doming of the seasonal thermocline. - Continental Shelf Res. *1* : 209 - 219, 1982. [Chl.]

51124 - PINGREE, R.D., MARDELL, G.T., HOLLIGAN, P.M., GRIFFITHS, D.K., SMITHERS,J.:
Celtic Sea and Armorian current structure and the vertical distributions
of temperature and chlorophyll. - Continental Shelf Res. *1* : 99 - 116,
1982.

51125 - PINTÉR, L. : Trends of above-ear and below-ear leaf areas and of grain
yield per unit area in maize (*Zea mays* L.) hybrids with different genotypes. - Acta agron. Acad. Sci. hung. *31* : 201 - 206, 1982.

51126 - PINTÉR, L., KOROM, S. : Az ötven centiméteres sortávolság hatása a kukorica
(*Zea Mays* L.) víz-, tápanyag- és fényellátottságára. [Water, nutrient and
light supply as affected by 50 cm row spacing in maize (*Zea mays* L.).] -
Növénytermelés *31* : 35 - 40, 1982. [In Hung., ab : E.]

*51127 - PINTILIE, C., BERCA, M., CARAMETE, A. : Studii asupra formării producţiei
la floarea-soarelui în condiţiile administrării unor erbicide cu mod de
actiune diferit. [Formation of sunflower production under administering
some herbicides with different effects.] - Lucr. ştiint. Inst. agron. "Nicolae Bălcescu"(Bucureşti), Ser.A *18-19* : 81 - 86, 1975/6 (1978). [Chl;
in Roum., ab : E.]

*51128 - PINTILIE, C., CARAMETE, A., BERCA, M. : Efectul erbicidelor inhibitoare
de fotosinteza asupra porumbului. [Effect of photosynthesis inhibiting herbicides on maize.] - Lucr. ştiint. Inst. agron. "Nicolae Bălcescu" (Bucureşti), Ser.A *18-19* : 87 - 91, 1975/6 (1978). [In Roum., ab : E.]

51129 - PIP, E., ROBINSON, G.G.C. : A study of the seasonal dynamics of three phyco-
periphytic communities using nuclear track autoradiography. I. Inorganic
carbon uptake. - Arch. Hydrobiol. *94* : 341 - 371, 1982.

51130 - PISKORNIK, Z., MYCZKOWSKI, J., ZALASIŃSKI, J., KORFEL, J. : Akumulacja
skrobi w liściach roślin pomidora uprawianych przy wyższych stężeniach CO_2.
[Starch accumulation in leaves of tomato plants grown at increased CO_2 concentrations.] - Zesz. nauk. Akad. rol. Krakow. *171* (Ogrodnictwo 9) : 25 -
- 42, 1982. [Chloroplast; in Pol., ab : E, R.]

51131 - PISSAIA, A., COLASANTE, L.O., COSTA, J.A. : Efeitos de desfolhamentos arti-
ficiais sobre a produção e acúmulo de matéria seca em duas cultivares de
soja. [Effect of artificial defoliation on dry matter production and accumulation in two soybean cultivars.] - Pesq. agropec. brasil. *17* : 873 - 881,
1982. [In Port., ab : E.]

51132 - PISTORIUS, E.K., VOSS, H. : Presence of an amino acid oxidase in photosystem II of *Anacystis nidulans*. - Europe. J. Biochem. *126* : 203 - 209, 1982.

51133 - PITCAIRN, C.E.R., GRACE, J. : The effect of wind and a reduced supply of
phosphorus and nitrogen on the growth and water relations of *Festuca arundinacea* SCHREB. - Ann. Bot. *49* : 649 - 660, 1982. [Growth analysis.]

51134 - PITTOCK, A.B., SALINGER, M.J. : Towards regional scenarios for a CO_2-warmed
Earth. - Climatic Change *4* : 23 - 40, 1982.

51135 - PJON, C.-J. : Effects of metal chelators on leaf senescence in maize and
hydrangea. - Plant Cell Physiol. *23* : 1427 - 1433, 1982. [Chl.]

51136 - PLANCHON, C., FESQUET, J. : Effect of the D genome and of selection on photosynthesis in wheat. - Theor. appl. Genet. *61* : 359 - 365, 1982.

51137 - PLATT, S.G., RAND, L. : Methionine sulfoximine effects on C_4 plant leaf
discs: comparison with C_3 species. - Plant Cell Physiol. *23* : 917 - 921,
1982. [Ps, photorespiration.]

*51138 - PLATT, T. : Spectral analysis of spatial structure in phytoplankton populat-
ions. - In: STEELE,J.H.(ed.): Spatial Pattern in Plankton Communities. Vol.
3. Pp. 73 - 84. Plenum Press, New York 1978. [Chl.]

51139 - PLATT, T., HARRISON, W.G., IRWIN, B., HORNE, E.P., GALLEGOS, C.L. : Photosynthesis and photoadaptation of marine phytoplankton in the Arctic. - Deep-
Sea Res. *29 A* : 1159 - 1170, 1982.

51140 - PLEKHANOV, S.E., MAKSIMOVA, I.V., BOLDYREVA, L.S., TAMBIEV, A.Kh., SHELYAS-
TINA, N.N. : Fiziologicheskie osobennosti razvitiya nakopitel'noɣ kul'tury
Scenedesmus quadricauda. [Physiological peculiarities of *Scenedesmus quadricauda* accumulative culture development.] - Biol. Nauki *1982* (2) : 77 -
- 81, 1982. [Ps, Chl; in R.]

51141 - POCINKI, A.G., BLANKENSHIP, R.E. : Kinetics of electron transfer in duro-
quinone-reconstituted reaction centers from photosynthetic bacteria. - FEBS
Lett. *147* : 115 - 119, 1982.

51142 - POLESSKAYA, O.G., MAL'TSEV, S.V., KRASNOVSKIĬ, A.A. : Obrazovanie i pogloshchenie vodoroda kul'turami i izolirovannymi geterotsistami tsianobakteriĭ.
[Hydrogen production and consumption by cultures and isolated heterocysts
of cyanobacteria.] - Prikl. Biokhim. Mikrobiol. *18* : 316 - 323, 1982. [In
R, ab : E.]

*51143 - POLLINGHER, U., BERMAN, T. : Biomass and chlorophyll. - In : SERRUYA, C.
(ed.) : Lake Kinneret. Pp. 247 - 252. Dr. W. Junk, Publ., The Hague 1978.

51144 - POOLE, D.K., MILLER, P.C. : Carbon dioxide flux from three arctic tundra
types in north-central Alaska, U.S.A. - Arctic alp. Res. *14* : 27 - 32,
1982. [Production.]

51145 - POPOV, Ė.G., BEZDENEZHNYKH, V.A., TALANOV, A.V. : Optimizatsiya protsessa
formirovaniya produktivnosti teplichnoĭ kul'tury ogurtsa. [Optimization
of the process of formation of productivity of a glasshouse cucumber crop.]
- In : Vliyanie Faktorov Vneshneĭ Sredy i Fiziologicheski Aktivnykh Veshchestv na Termorezistentnost' i Produktivnost' Rasteniĭ. Pp. 116 - 126, 158.
Inst. Biol. karel'. Filiala Akad. Nauk SSSR, Petrozavodsk 1982. [Ps; in R.]

51146 - POPOV, P., STANKOVA, P., STANKOV, I. : Fotosintezirashcha ploshch i chista
produktivnost na fotosintezata pri kheksaploidni formi tritikale. [Photo-
synthesizing area and net photosynthetic productivity of some hexaploid
triticale forms.] - Rasteniev. Nauki *19* : 83 - 87, 1982. [In Bulg., ab :
E, R.]

*51147 - POPOV, V.F. : Soderzhanie khlorofilla i produktivnost' fotosinteza yabloni
v zavisimosti ot sposobov soderzhaniya pochvy v pal'mettnom sadu. [Content
of chlorophyll and photosynthetic rate of apple trees in relation to soil
maintenance methods in orchards.] - In : Voprosy Intensifikatsii Plodovodstva. Pp. 44 - 46. Kishinev. sel'skokhoz. Inst., Kishinev 1978. [In R.]

51148 - POPOVA, I.A., MASLOVA, T.G., POPOVA, O.F., MIROSLAVOV, E.A., TSAR'KOVA, V.A.:
Osobennosti fotosinteticheskogo apparata krapivy dvudomnoĭ, proizrastayush-
cheĭ v razlichnykh svetovykh usloviyakh. [Some properties of photosynthetic
apparatus in great nettle growing under various light conditions.] - Fiziol.
Rast. *29* : 1102 - 1108, 1982. [In R, ab : E.]

51149 - POPOVA, L.P. : State and activity of ribulosediphosphate carboxylase *in vitro*. - Dokl. bolg. Akad. Nauk *35* : 509 - 512, 1982.

51150 - POPOVA, L.P. : Ferredoxin-dependent activation of the ribulosediphosphate
carboxylase from spinach. - Dokl. bolg. Akad. Nauk *35* : 1287 - 1290, 1982.

51151 - POPOVA, L.P., DIMITROVA, O.D., VAKLINOVA, S.G. : Effect of GA_3 on the chlorophyll content, on the intensity of photosynthetic CO_2 fixation and on
the activity of carboxylating enzymes in C_3 and C_4 plants. - Dokl. bolg.
Akad. Nauk *35* : 797 - 800, 1982.

51152 - POPOVA, L.P., DIMITROVA, O.D., VAKLINOVA, S.G. : Effect of ABA on the chlorophyll content, the intensity of the photosynthetic CO_2 fixation and the
activity of the carboxylating enzymes in C_3 and C_4 plants. - Dokl. bolg.
Akad. Nauk *35* : 1291 - 1294, 1982.

51153 - POPOVIĆ, Ž. : Prinosnost gajenih biljaka i njena veza sa pojedinim fizio-
loškim procesima. [Productivity of cultivated plants and its connection
with some physiological processes.] - Agrohemija *1982*: 389 - 438, 1982.
[Ps, Chl; in Croat., ab : E.]

51154 - POPP, M., OSMOND, C.B., SUMMONS, R.E. : Pathway of malic acid synthesis
in response to ion uptake in wheat and lupin roots: evidence from fixation
of ^{13}C and ^{14}C. - Plant Physiol. *69* : 1289 - 1292, 1982.

51155 - PORRA, R.J., KLEIN, O., WRIGHT, P.E. : ^{13}C-NMR studies of chlorophyll biosynthesis in higher plants: an unequivocal proof of the participation of
the C_5 pathway and evidence of a new route for the incorporation of glycine. - Biochem. Int. *5* : 345 - 350, 1982.

51156 - PORTER, J.R., ALLEN, M.F., LANE, L.C., BOOSALIS, M.G. : Platte Valley Yel-
lows, a chlorotic condition of soybeans: Symptoms and preliminary analyses.
- Plant Soil *68* : 283 - 287, 1982.

51157 - PORTER, J.R., BRAGG, P.L., RAYNER, J.H., WEIR, A.H., LANDSBERG, J.J. : The ARC winter wheat simulation model - principles and progress. - In : HAWKINS, A.F., JEFFCOAT, B. (ed.) : Opportunities for Manipulation of Cereal Productivity. Monograph 7. Pp. 97 - 108. British Plant Growth Regulator Group, Wantage 1982. [Ps.]

51158 - PORTIS, A.R.,Jr. : Effects of the relative extrachloroplastic concentrations of inorganic phosphate, 3-phosphoglycerate, and dihydroxyacetone phosphate on the rate of starch synthesis in isolated spinach chloroplasts. - Plant Physiol. *70* : 393 - 396, 1982.

51159 - PORTIS, A.R.,Jr. : Introduction to photosynthesis: Carbon assimilation and plant productivity. - In : GOVINDJEE (ed.) : Photosynthesis. Vol.II. Pp. 1 - - 12. Academic Press, New York - London - Paris - San Diego - San Francisco - São Paulo - Sydney - Tokyo - Toronto 1982.

51160 - POSPÍŠILOVÁ, J., SOLÁROVÁ, J., JANDA, J. : Responses of epidermal diffusive conductance to simultaneous changes in two factors: determination of interactions. - Biol. Plant. *24* : 155 - 157, 1982.

51161 - POST, E., GOEYENS, L., VANDENHOUDT, A. : Determination of marine phytoplanktonic biomass. - Biol. Oceanogr. Committee C.M. *1982/L:28* : 1 - 14, 1982.

51162 - POTTER, J.W., BLACK, C.C.,Jr. : Differential protein composition and gene expression in leaf mesophyll cells and bundle sheath cells of the C_4 plant *Digitaria sanguinalis* (L.) SCOP. - Plant Physiol. *70* : 590 - 597, 1982.

51163 - POTTOSIN, I.I., SHINKAREV, V.P., RUBIN, A.B. : Kinetika fotoindutsirovannykh redoks-prevrashcheniĭ vysokopotentsial'nogo tsitokhroma v khromatoforakh sernykh purpurnykh bakteriĭ *Ectothiorhodospira shaposhnikovii.* [Kinetics of light-induced redox changes of high-potential cytochrome in the chromatophores of sulphur purple bacteria *Ectothiorhodospira shaposhnikovii.*] - Biofizika *27* : 977 - 982, 1982. [In R, ab : E.]

51164 - POULSEN, C. : Isolation and purification of ribulosebisphosphate carboxylase and its subunits by gel chromatography. - In : EDELMAN, M., HALLICK, R.B., CHUA, N.-H. (ed.) : Methods in Chloroplast Molecular Biology. Pp. 767 - 781. Elsevier Biomedical Press, Amsterdam - New York - Oxford 1982.

51165 - POWLES, S.B., BJÖRKMAN, O. : High light and water stress effects on photosynthesis in *Nerium oleander.* I. Photoinhibition of photosynthesis: effect on chlorophyll fluorescence at 77 K in intact leaves and in chloroplast membranes. - Carnegie Inst. Washington Year Book *81* : 74 - 76, 1982.

51166 - POWLES, S.B., BJÖRKMAN, O. : Photoinhibition of photosynthesis: effect on chlorophyll fluorescence at 77 K in intact leaves and in chloroplast membranes of *Nerium oleander.* - Planta *156* : 97 - 107, 1982.

51167 - POWLES, S.B., CHAPMAN, K.S.R., WHATLEY, F.R. : Effect of photoinhibitory treatments on the activity of light-activated enzymes of C_3 and C_4 photosynthetic carbon metabolism. - Plant Physiol. *69* : 371 - 374, 1982.

*B51168 - PREBBLE, J.N. : Mitochondria, Chloroplasts and Bacterial Membranes. - Longman, London - New York 1981.

*51169 - PRECHEUR, R., GREIG, J.K., ARMBRUST, D.V. : The effects of wind and wind--plus-sand on tomato plants. - J. amer. Soc. hort. Sci. *103* : 351 - 355, 1978. [Ps.]

51170 - PREISS, J. : Biosynthesis of fructose, sucrose, and starch. - In : ZABORSKY, O.R. (ed.) : CRC Handbook of Biosolar Resources. Vol.I/1. Pp. 331 - 350. CRC Press, Boca Raton 1982.

*51171 - PREISS, J., KAPPEL, W.K., GREENBERG, E. : Comparative regulation of α1,4--glucan synthesis in photosynthetic and non-photosynthetic systems. - In : ATKINSON, D.E., FOX, C.F. (ed.) : Modulation of Protein Function. Pp. 161 - - 184. Academic Press, New York - London 1979.

51172 - PRENZEL, U., LICHTENTHALER, H.K. : Localization of β-carotene in chlorophyll a-proteins and changes in its levels during short-term high-light
exposure of plants. - In : WINTERMANS, J.F.G.M., KUIPER, P.J.C. (ed.) :
Biochemistry and Metabolism of Plant Lipids. Pp. 565 - 572. Elsevier Biomedical Press, Amsterdam - New York - Oxford 1982.

51173 - PRICE, C.A., REARDON, E.M. : Isolation of chloroplasts for protein synthesis
from spinach and Euglena gracilis by centrifucation in silica sols. - In :
EDELMAN, M., HALLICK, R.B., CHUA, N.-H. (ed.) : Methods in Chloroplast
Molecular Biology. Pp. 189 - 209. Elsevier Biomedical Press, Amsterdam -
New York - Oxford 1982.

51174 - PRIESTLE, J.P.,Jr., RHYNE, R.H.,Jr., SALMON, J.B., HACKERT, M.L. : Phycobiliproteins: comparison of solution and single crystal fluorescence for
C-phycocyanin and B-phycoerythrin. - Photochem. Photobiol. 35 : 827 - 834,
1982.

51175 - PRIEUR, P., LOUGUET, P. : Comparaison d'activités enzymatiques en relation
avec le métabolisme de l'acide malique, et des teneurs en acides organiques
et en glucides solubles de lambeaux d'épidermes à stomates ouverts ou fer-
més, chez Pelargonium x hortorum et Vicia faba. - Compt. rend. Acad. Sci.
Paris, Sér. III 294 : 845 - 848, 1982. [Ps.]

51176 - PRINCE, R.C., MATSUURA, K., HURT, E., HAUSKA, G., DUTTON, P.L. : Reduction
of cytochromes b_6 and f in isolated plastoquinol-plastocyanin oxidoreductase driven by photochemical reaction centers from Rhodopseudomonas sphaeroides. - J.biol.Chem. 257:3379 - 3381, 1982.

51177 - PRINCE, R.C., O'KEEFE, D.P., DUTTON, P.L. : The organization of the cyclic
electron transfer system in photosynthetic bacterial membranes. Is this
the hardware of a chemiosmotic system? - In : BARBER, J. (ed.) : Electron
Transport and Photophosphorylation. Pp. 197 - 248. Elsevier Biomedical
Press, Amsterdam - New York - Oxford 1982.

51178 - PRINS, H.B.A., O'BRIEN, J., ZANSTRA, P.E. : Bicarbonate utilization in
aquatic angiosperms. pH and CO_2 concentrations at the leaf surface. - In :
SYMOENS, J.J., HOOPER, S.S.,COMPÈRE, P. (ed.) : Studies on Aquatic Vascular
Plants. Pp. 112 - 119. Royal Botanical Society of Belgium, Brussels 1982.

51179 - PRINS, H.B.A., SNEL, J.F.H., ZANSTRA, P.E. : The mechanism of photosynthetic bicarbonate utilization. - In : SYMOENS, J.J., HOOPER, S.S., COMPÈRE,P.
(ed.) : Studies on Aquatic Vascular Plants. Pp. 120 - 126. Royal Botanical
Society of Belgium, Brussels 1982.

51180 - PRINS, H.B.A., SNEL, J.F.H., ZANSTRA, P.E., HELDER, R.J. : The mechanism
of bicarbonate assimilation by the polar leaves of Potamogeton and Elodea.
CO_2 concentrations at the leaf surface. - Plant Cell Environm. 5 : 207 -
- 214, 1982.

51181 - PRIOUL, J.-L. : Limiting factors in photosynthesis - from the chloroplast
to the plant canopy. - In : HÉLÈNE, C., CHARLIER, M., MONTENAY-GARESTIER,
T., LAUSTRIAT, G. (ed.) : Trends in Photobiology. Pp. 633 - 643. Plenum
Publ. Corp., New York - London 1982.

51182 - PRIOUL, J.-L., SILSBURY, J.H. : A physiological analysis of the effect of
sowing density on the rate of subterranean clover. - Aust. J. agr. Res.
33 : 213 - 222, 1982. [Growth analysis.]

51183 - PRISCU, J.C., AXLER, R.P., CARLTON, R.G., REUTER, J.E., ARNESON, P.A.,
GOLDMAN, C.R. : Vertical profiles of primary productivity, biomass and phy-
sico-chemical properties in meromictic Big Soda Lake, Nevada, U.S.A. -
Hydrobiologia 96 : 113 - 120, 1982.

51184 - PRISCU, J.C., VERDUIN, J., DEACON, J.E. : Primary productivity and nutrient
balance in a lower Colorado River reservoir. - Arch. Hydrobiol. 94 : 1 - 23,
1982.

51185 - PROCOPIOU, J., WALLACE, A. : Mineral composition of two populations of leaves - green and iron chlorotic - of the same age all from the same tree.
- J. Plant Nutr. 5 : 811 - 820, 1982.

51186 - PROCTOR, J.T.A., BODNAR, J.M., BLACKBURN, W.J., WATSON, R.L. : Analysis of
the effects of the spotted tentiform leafminer (*Phyllonorycter blancardella*)
on the photosynthetic characteristics of apple leaves. - Can. J. Bot. *60* :
2734 - 2740, 1982.

51187 - PROHASKA, K.R., FEHR, W.R. : Recurrent selection for resistance to iron deficiency chlorosis in soybeans. - Crop Sci. *21* : 524 - 526, 1981.

51188 - PROKHARCHYK, R.A., LYAGENCHANKA, B.I., VALYNETS, A.P., SHALAMITSKAYA, L.M. :
Paraŭnal'naya atsènka fotasintetychnaĭ aktyŭnastsi roznykh sartoŭ lyutsèr-
ny. [Comparison of photosynthetic activity of different cultivars of alfalfa.] - Vestsi Akad. Navuk belarus.SSR, Ser. biyal. Navuk *1982* (1) : 14 - 18,
121, 1982. [In Belorus., ab : E, R.]

51189 - PROKHOROVA, S.A. : Deĭstvie nefteproduktov na fotosintez buroĭ vodorosli
Fucus vesiculosus L. [Effect of oil products on photosynthesis of *Fucus vesiculosus* L.] - Biol. Nauki *1982* (6) : 69 - 72, 1982. [In R.]

*51190 - PRONINA, N.B. : Adenilatkinaznaya aktivnost' belkov raznoĭ stepeni vossta-
novlennosti, vydelennykh iz khloroplastov gorokha. [Adenylate-kinase activity of proteins with different degree of reduction, separated from pea
chloroplasts.] - Izv. Akad. Nauk SSSR, Ser. biol. *1981* : 873 - 880, 1981.
[In R, ab : E.]

51191 - PROTASOVA, N.N., KEFELI, V.I. : Fotosintez i rost vysshikh rasteniĭ, ikh
vzaimosvyazi i korrelyatsii. [Photosynthesis and growth of higher plants,
its relations and correlations.] - In : Fiziologiya Fotosinteza. Pp. 251 -
- 270. Nauka, Moskva 1982. [In R.]

51192 - PROTSENKO, M.A., RADZIEVSKAYA, M.G. : Ul'trastruktura protoplastov, vydelennykh iz mezofilla lista probirochnykh rasteniĭ kartofelya (*Solanum tuberosum* L.). [Fine structure of mesophyll protoplasts isolated from axenic
dwarf plantlets of potato (*Solanum tuberosum* L.)] - Fiziol. Rast. *29* : 134 -
- 141, 1982. [Chloroplast; in R, ab : E.]

51193 - PRUDHOMME, T.I. : The effect of defoliation history on photosynthetic rates
in mountain birch. - Rep. Kevo subarctic Res. Sta. *18* : 5 - 9, 1982.

*51194 - PUCHER-PETKOVIĆ, T., MARASOVIĆ, I. : Importance relative du phytoplancton
de tailles différentes dans les eaux du large de l'Adriatique Centrale
(Stončica). - Rapp. Commun. int. Mer médit. *27* (7) : 87 - 89, 1981. [Prima-
ry production.]

51195 - PUKACKI, P., GIERTYCH, M. : Seasonal changes in light transmission by bud
scales of spruce and pine. - Planta *154* : 381 - 383, 1982.

51196 - PULICH, W.M.,Jr. : Edaphic factors related to shoalgrass (*Halodule wrightii*
ASCHERS.) production. - Bot. mar. *25* : 467 - 475, 1982. [Dry-matter production.]

51197 - PUROHIT, K., McFADDEN, B.A., SALUJA, A. : Stoichiometry in the assay of ribulose bisphosphate oxygenase and carboxylase. - Anal. Biochem. *124* : 158 -
- 166, 1982.

51198 - PUROHIT, S.S. : Monocarpic senescence in *Helianthus annuus* L. 1.Relation
of fruit induced senescence, chlorophyll and chlorophyllase activity. -
Photosynthetica *16* : 542 - 545, 1982.

51199 - PUROHIT, S.S. : Prevention by kinetin of ethylene-induced chlorophyllase
activity in senescing detached leaves of *Helianthus annuus*. - Biochem. Physiol. Pflanzen *177* : 625 - 627, 1982.

51200 - PUSHNIK, J.C., MILLER, G.W. : The effects of iron and light treatments on
chloroplast composition and ultrastructure in iron-deficient barley leaves.
- J. Plant Nutr. *5* : 311 - 321, 1982.

*51201 - QUEIROZ, O. : Programmation de l'adaptation saisonnière: un modèle basé
sur les rythmes métaboliques circadiens. - Bull. Soc. bot. Fr. *128* : 35 -
- 42, 1981. [Ps.]

51202 - QUEIROZ, O., BRULFERT, J. : Photoperiod-controlled induction and enhancement of seasonal adaptation to drought. - In : TING, I.P., GIBBS, M. (ed.):
Crassulacean Acid Metabolism. Pp. 208 - 229. Waverly Press, Baltimore
1982.

51203 - QUINONES, F.A., SHAW, S.T., CARDENAS, M. : Attempts to predict forage
yield and protein content of individual plants of tall wheatgrass [*Agropyron elongatum* (HOST) BEAUV.]. - Field Crops Res. *5* : 365 - 371, 1982.

51204 - QUIZENBERRY, J.E. : Breeding for drought resistance and plant water use
efficiency. - In : CHRISTIANSEN, M.N., LEWIS, C.F. (ed.) : Breeding Plants
for Less Favorable Environments. Pp. 193 - 212. Wiley-Interscience Pub.,
John Wiley & Sons, New York - Chichester - Brisbane - Toronto - Singapore
1982. [Resistances.]

51205 - RABE, R., NOBEL, W., KOHLER, A. : Effects of sodium chloride on photosynthesis and some enzyme activities of *Potamogeton alpinus*. - Aquat. Bot. *14* :
159 - 165, 1982.

51206 - RABE, R., SCHUSTER, H., KOHLER, A. : Effects of copper chelate on photosynthesis and some enzyme activities of *Elodea canadensis*. - Aquat. Bot.
14 : 167 - 175, 1982.

51207 - RABINOWITCH, H.D., SKLAN, D., BUDOWSKI, P. : Photo-oxidative damage in the
ripening tomato fruit: Protective role of superoxide dismutase. - Physiol.
Plant. *54* : 369 - 374, 1982. [Chl.]

51208 - RACKOVSKY, S., SCHER, H. : Effect of neighboring charges and external
fields on photosynthetic electron transfer. - Biochim. biophys. Acta *681* :
152 - 160, 1982.

*51209 - RADEMAKER, H., HOFF, A.J. : The balance between primary forward and back
reactions in bacterial photosynthesis. - Biophys. J. *34* : 325 - 344,
1981.

51210 - RADHAKRISHNA, K., BHATTATHIRI, P.M.A., DEVASSY, V.P. : Chlorophyll *a*,
phaeopigments & particulate organic carbon in the northern & western Bay
of Bengal. - Indian J. mar. Sci. *11* : 287 - 291, 1982.

51211 - RADICS, L., MENYHÉRT, Z., ÁNGYÁN, J. : A kukorica vetísidejének és tőtávol-
ságának hatása az *Amaranthus retroflexus* L. növekedésére. [Growth of *Amaranthus retroflexus* L. as affected by date of sowing and spacing of maize.]
- Növénytermelés *31* : 237 - 242, 1982. [Growth analysis.]

51212 - RADIN, J.W., PARKER, L.L., GUINN, G. : Water relations of cotton plants
under nitrogen deficiency V. Environmontal control of abscisic acid accumulation and stomatal sensitivity to abscisic acid. - Plant Physiol. *70* :
1066 - 1070, 1982. [Stomatal resistance.]

51213 - RADMER, R., OLLINGER, O. : Nitrogen and oxygen evolution by hydroxylamine-
-treated chloroplasts. - FEBS Lett. *144* : 162 - 166, 1982.

51214 - RAGHAVENDRA, A.S., VALLEJOS, R.H. : Regulation of phosphoenolpyruvate carboxylase from C_4 plants: involvement of thiol groups in catalytic reaction
of enzyme from *Amaranthus viridis* L. - Indian J. exp. Biol. 20 : 619 - 622,
1982.

51215 - RAI, H. : Primary production of various size fractions of natural phytoplankton communities in a North German lake. - Arch. Hydrobiol. *95* : 395 -
- 412, 1982.

51216 - RAISON, J.K., ROBERTS, J.K.M., BERRY, J.A. : Correlations between the thermal stability of chloroplast (thylakoid) membranes and the composition and
fluidity of their polar lipids upon acclimation of the higher plant, *Nerium oleander*, to growth temperature. - Biochim. biophys. Acta *688* : 218 -
- 228, 1982.

51217 - RAJENDRUDU, G., DAS, V.S.R. : Biomass production of two species of *Cleome* exhibiting C_3 and C_4 photosynthesis. - Biomass *2* : 223 - 227, 1982.

51218 - RAJENDRUDU, G., DAS, V.S.R. : The carboxylating enzymes in leaves of *Cleome gynandra*, a C_4 dicot plant. - Plant Sci. Lett. *26* : 285 - 291, 1982.

51219 - RAJORA, O.P., RAWAT, M.S. : Natural chlorophyll mutants in *Albizzia lebbek* BENTH. - Indian Forester *108* : 289 - 292, 1982.

51220 - RAKHIMBERDIEVA, M.G., LEKHOTSKI, E., KARAPETYAN, N.V., KRASNOVSKIĬ, A.A. : Vliyanie piridazinonov i tserulenina na biosintez i funktsional'noe sosto-yanie fotosistemy 2 v list'yakh yachmenya. [Effects of pyridazinones and cerulenin on biosynthesis and functional state of photosystem 2 in barley leaves.] - Biokhimiya *47* : 637 - 646, 1982. [In R, ab : E.]

51221 - RAKHTSEENKA, I.N., KAPITULA, A.M. : Intensiŭnasts' fotasintezu elki eŭra-peĭskaĭ i listoŭnitsy sibirskaĭ u chystykh i zmeshanykh kul'turfitatsenozakh. [Net photosynthetic rate of *Picea abies* and *Larix sibirica* in pure and mixed plant associations.] - Vestsi Akad. Nauk belarus. SSR, Ser. biyal. Navuk *1982* (5) : 14 - 17, 1982. [In Belorus., ab : E.]

51222 - RAM, H., SINGH, R. : Chlorophyll content, photosynthetic rates and related enzyme activities in ear parts of two wheat cultivars differing in grain yield. - Plant Physiol. Biochem. *9* : 94 - 102, 1982.

51223 - RAMACHANDRA REDDY, A., DAS, V.S.R. : Pyruvate Pi dikinase activity in different groups of crassulacean acid metabolism plants. - Plant Sci. Lett. *25* : 155 - 159, 1982.

51224 - RAMAMURTHY, S., LÜDDERS, P. : Effect of ammonium and nitrate nutrition on net-photosynthetic rate and carbohydrate content in calamondine (*Citrus madurensis* LOUR.). - Gartenbauwissenschaft *47* : 168 - 173, 1982.

51225 - RAMANI, S., KANNAN, S. : Inadaptive changes in pH with Zn-stress tolerance in some cultivars of cotton and peanut. - J. Plant Nutr. *5* : 207 - 217, 1982. [Chl.]

*51226 - RAMIREZ, F., OKAZAKI, H., TU, S. : Effect of phospholipid composition on activities of bacteriorhodopsin in reconstituted purple membrane. - FEBS Lett. *135* : 123 - 126, 1981.

51227 - RAMOS, C., HALL, A.E. : Relationships between leaf conductance, intercellular CO_2 partial pressure and CO_2 uptake rate in two C_3 and two C_4 plant species. - Photosynthetica *16* : 343 - 355, 1982.

51228 - RAMOS, C., HOFFMAN, G.J., HALL, A.E. : Evaluation of a dual-radioisotope instrument for measuring leaf conductance and photosynthesis. - Agron. J. *74* : 709 - 715, 1982.

51229 - RAMOS, J.L., GUERRERO, M.G., LOSADA, M. : Optimization of conditions for photoproduction of ammonia from nitrate by *Anacystis nidulans*. - Appl. environm. Microbiol. *44* : 1013 - 1019, 1982.

51230 - RAMOS, J.L., GUERRERO, M.G., LOSADA, M. : Photoproduction of ammonia from nitrate by *Anacystis nidulans* cells. - Biochim. biophys. Acta *679* : 323 - - 330, 1982. [Chl.]

51231 - RAMOS, J.L., GUERRERO, M.G., LOSADA, M. : Sustained photoproduction of ammonia from nitrate by *Anacystis nidulans*. - Appl. environm. Microbiol. *44* : 1020 - 1025, 1982.

51232 - RAMSHAW, J.A.M., FELTON, A.A. : The amino acid sequence of plastocyanin from *Cucumis sativus*. - Phytochemistry *21* : 1317 - 1320, 1982.

51233 - RAND, R.H., STORTI, D.W., UPADHYAYA, S.K., COOKE, J.R. : Dynamics of coupled stomatal oscillators. - J. math. Biol. *15* : 131 - 149, 1982.

51234 - RANTY, B., CAVALIÉ, G. : Photosynthetic characteristics of mesophyll cells isolated from sunflower (*Helianthus annuus* L.) leaves. - Photosynthesis Res. *3* : 59 - 67, 1982.

51235 - **RANTY, B., CAVALIÉ, G.** : Purification and properties of ribulose 1,5-bis-phosphate carboxylase from sunflower leaves. - Planta *155* : 388 - 391, 1982.

51236 - **RAO, A.N., DAS, V.S.R.** : Chlorophyll content and photochemical activities of three *Sorghum* cultivars grown under three irradiances. - Photosynthetica *16* : 145 - 147, 1982.

51237 - **RAO, K.K., BRUCE, D.L., GISBY, P.E., MUALLEM, A., HALL, D.O.** : Biophotolysis of water for hydrogen production via natural and artificial catalytic systems. - In : HALL, D.O., PALZ, W. (ed.) : Photochemical, Photoelectrochemical and Photobiological Processes. Vol.1. Pp. 195 - 202. D.Reidel Publ. Co., Dordrecht - Boston - London 1982.

51238 - **RAO, K.K., HALL, D.O.** : Photorespiration. - J. biol. Educ. *16* : 167 - 172, 1982.

51239 - **RAO, K.K., MUALLEM, A., BRUCE, D.L., SMITH, G.D., HALL, D.O.** : Immobilization of chloroplasts, algae and hydrogenases in various solid supports for the photoproduction of hydrogen. - Biochem. Soc. Trans. *10* : 527 - 528, 1982.

51240 - **RAO, L.V.M., DATTA, N., GUHA-MUKHERJEE, S., SOPORY, S.K.** : The effect of blue light on the induction of nitrate reductase of etiolated excised maize leaves. - Plant Sci. Lett. *28* : 39 - 47, 1982. [Ps, Chl.]

51241 - **RAO, P.S.** : Protoplast culture. - In : JOHRI, B.M. (ed.) : Experimental Embryology of Vascular Plants. Pp. 231 - 262. Springer-Verlag, Berlin - Heidelberg - New York 1982. [Chloroplast.]

*51242 - **RASCIO, N., CASADORO, G., ORSENIGO, M., GAVAZZI, G., RACCHI, M.L.** : Ultrastructural features of a proline requiring mutant in *Zea mays* L. - Maydica *25* : 95 - 104, 1980. [Chloroplast.]

*51243 - **RASCIO, N., CASADORO, G., ORSENIGO, M., TONELLI, C.** : Ultrastructural study of an olive necrotic mutant of maize. - Protoplasma *105* : 241 - 249, 1981.

*51244 - **RASCIO, N., ORSENIGO, M., CASADORO, G., RACCHI, M.L., GAVAZZI, G.** : Ultrastructural responses of the *pro 1-1* maize mutant to increase proline concentrations. - Maydica *26* : 75 - 84, 1981. [Chloroplast.]

51245 - **RASKIN, V.I., KASTSYUKEVICH, G.S.** : Paslyadoŭnasts' peratvarènnyaŭ spektral'nykh form khlarafilidu ŭ izalyavanykh membranakh ètyyaplastaŭ. [Sequence of changes of spectral forms of chlorophyllide in isolated membranes of etioplasts.] - Vestsi Akad. Navuk belarus. SSR, Ser. biyal. Navuk *1982* (4) : 38 - 41, 124 - 125, 1982. [In Belorus., ab : E, R.]

51246 - **RASULOV, B.Kh.** : Reassimilyatsiya uglekisloty, obrazovannoĭ v rezul'tate dykhatel'nykh protsessov na svetu, v ontogeneze lista u dvukh vidov khlopchatnika. [Reassimilation of carbon dioxide, the product of respiration processes in the light, during leaf ontogeny in two cotton species.] - Dokl. Akad. Nauk tadzh. SSR *25* : 250 - 254, 1982. [In R.]

51247 - **RASULOV, B.Kh., ASROROV, K.A.** : Ontogeneticheskie izmeneniya fotosinteticheskikh kharakteristik i aktivnost' fotosinteziruyushchego apparata u razlichnykh vidov khlopchatnika. [Ontogenetic changes in photosynthetic characteristics and the activity on the photosynthesizing apparatus in various cotton species.] - Fiziol. Rast. *29* : 378 - 386, 1982. [In R, ab : E.]

51248 - **RASULOV, B.Kh., ASROROV, K.A.** : Zavisimost' intensivnosti fotosinteza razlichnykh vidov khlopchatnika ot udel'noĭ poverkhnostnoĭ plotnosti lista. [Dependence of photosynthetic rate of various cotton species on the specific space density of the leaf.] - In : Fiziologiya Fotosinteza. Pp. 270 - 283. Nauka, Moskva 1982. [In R.]

51249 - **RASULOV, B.Kh., OYA, V.M.** : Opredelenie komponentov dykhaniya na svetu s uchetom ostatochnoĭ kontsentratsii kisloroda. [Estimation of components of respiration in the light considering the presence of residual oxygen.] - Fiziol. Rast. *29* : 616 - 622, 1982. [In R, ab : E.]

51250 - RAULIN, F. : Round table summary: Prebiotic photochemistry and photochemical reactions in space. - In : HÉLÈNE, C., CHARLIER, M., MONTENAY-GARESTIER,T., LAUSTRIAT, G. (ed.) : Trends in Photobiology. Pp. 123 - 132. Plenum Press, New York - London 1982. [Model systems of Ps.]

51251 - RAVEN, J.A., BEARDALL, J., JOHNSTON,A.M. : Inorganic carbon transport in relation to H^+ transport at the plasmalemma of photosynthetic cells. - In : MARMÉ, D., MARRÉ, E., HERTEL, R. (ed.) : Plasmalemma and Tonoplast: their Functions in the Plant Cell. Pp. 41 - 47. Elsevier Biomedical Press, Amsterdam - New York - Oxford 1982.

51252 - RAVINDRAN, P.N., MENON, M.A. : Laboratory screening of cocoa genotypes for drought tolerance. - Planter (Kuala Lumpur) 58 : 334 - 339, 1982. [Chl.]

51253 - RAWSON, H.M., LOVE, D.C. : A chamber for rapid measurements of cereal leaf gas exchange. - Photosynthetica 16 : 67 - 70, 1982.

51254 - RAWSON, H.M., TURNER, N.C. : Recovery from water stress in five sunflower (Helianthus annuus L.) cultivars. I. Effects of the timing of water application on leaf area and seed production. - Aust. J. Plant Physiol. 9 : 437 - 448, 1982. [Growth analysis.]

51255 - RAWSON, H.M., TURNER, N.C. : Recovery from water stress in five sunflower (Helianthus annuus L.) cultivars. II. The development of leaf area. - Aust. J. Plant Physiol. 9 : 449 - 460, 1982.

51256 - RAZJIVIN, A.P., DANIELIUS, R.V., GADONAS, R.A., BORISOV, A.Yu., PISKARSKAS, A.S. : The study of excitation transfer between light-harvesting antenna and reaction center in chromatophores from purple bacterium Rhodospirillum rubrum by selective picosecond spectroscopy. - FEBS Lett. 143 : 40 - 44, 1982.

51257 - RAZUMOVA, N.A., MAKSIMOV, G.B., LEPNĚV, G.P. : Potentsiometricheskie CO_2-datchiki dlya izucheniya dykhatel'nogo i fotosinteticheskogo gazoobmena v vozdushnykh i vodnykh sredakh. [Potentiometric CO_2 sensors for measuring respiratory and photosynthetic gas exchange in atmospheric and water environment.] - Fiziol. Rast. 29 : 189 - 197, 1982. [In R, ab : E.]

51258 - REBEIZ, C.A. : Chlorophyll: anatomy of a discovery. - Chemtech 12 (1) : 52 - - 63, 1982.

51259 - REBEIZ, C.A., DANIELL, H., MATTHEIS, J.R. : Chloroplast bioengineering: The greening of chloroplasts in vitro. - Biotechnol. Bioeng. Symp. 12 : 413 - - 439, 1982.

51260 - REBEIZ, C.A., LASCELLES, J. : Biosynthesis of pigments in plants and bacteria. - In : GOVINDJEE (ed.) : Photosynthesis. Vol.1. Pp. 699 - 780. Academic Press, New York - London - Paris - San Diego - San Francisco - São Paulo - Sydney - Tokyo - Toronto 1982.

51261 - RECALCATI, L.M., BASSO, B., ALBERGONI, F.G., RADICE, M. : On the determination of ^{14}C-labelled photosynthesis products by liquid scintillation counting. - Plant Sci. Lett. 27 : 21 - 27, 1982.

51262 - RÉDEI, E. : Hazai dohányok klorofill és aminosav tartalmának változása a szárítás során. [Changes in the content of chlorophyll and amino acids of Hungarian tobaccoes during drying.] - Élelmiszervizsgálati Közl. 28 : 201 - - 206, 1982. [In Hung., ab : E, G, R.]

51263 - RED'KO, T.P., SHMELEVA, V.L., IVANOV, B.N., MUKHIN, E.N. : Sootnoshenie mezhdu netsiklicheskim i psevdotsiklicheskim transportom élektronov v khloroplastakh gorokha v zavisimosti ot kontsentratsii ferredoksina. [Ferredoxin-dependent correlation between non-cyclic and pseudocyclic electron transport in pea chloroplasts.] - Biokhimiya 47 : 1695 - 1699, 1982. [In R, ab : E.]

51264 - REDLINGER, T., GANTT, E. : A M_r-95,000 polypeptide in Porphyridium cruentum phycobilisomes and thylakoids and in energy transfer. - Proc. nat. Acad. Sci. USA 79 : 5542 - 5546, 1982.

51265 - REHNBERG, B.G., SCHULTZ, D.A., RASCHKE, R.L. : Limitations of electronic particle counting in reference to algal assays. - J. Water Pollut. Control Fed. *54* : 181 - 186, 1982. [Chl.]

51266 - REIBACH, P.H., BENEDICT, C.R. : Biosynthesis of starch in protoplastids of germinating *Ricinus communis* endosperm tissue. - Plant Physiol. *70* : 252 - - 256, 1982. [RuBPC.]

51267 - REISBERG, P., NAIRN, J.A., SAUER, K. : Picosecond fluorescence kinetics in spinach chloroplasts at low temperature. - Photochem. Photobiol. *36* : 657 - - 661, 1982.

51268 - REISS-HUSSON, F., AGALIDIS, I. : Studies on carotenes in bacterial reactions centers. - In : KAPLAN, N.O., ROBINSON, A. (ed.) : From Cyclotrons to Cytochromes. Pp. 381 - 389. Academic Press, New York 1982.

*51269 - REĬTER, B.G., YUDKIN, L.Yu. : Ul'trastruktura khloroplastov pshenitsy pri porazhenii buroĭ rzhavchinoĭ. [Ultrastructure of wheat chloroplasts under leaf rust infection.] - Sel'skokhoz. Biol. *16* : 87 - 91, 1981. [In R, ab : E.]

51270 - REMIŠ, D., BULYCHEV, A.A., KURELLA, G.A. : Kinetics of the increase of electric potential difference across the thylakoid membrane upon chloroplast illumination. - Biológia (Bratislava) *37* : 1057 - 1061, 1982.

51271 - REMISON, S.U., LUCAS, E.O. : Effects of planting density on leaf area and productivity of two maize cultivars in Nigeria. - Exp. Agr. *18* : 93 - 100, 1982.

51272 - REMY, R., AMBARD-BRETTEVILLE, F., VEDEL, F. : EcoRI analysis of chloroplastic DNAs and polypeptidic composition of thylakoids from wheat and related species. - Plant Sci. Lett. *25* : 261 - 270, 1982.

51273 - REMY, R., TRÉMOLIÈRES, A., DUVAL, J.-C., AMBARD-BRETEVILLE, F., DUBACQ,J.-P.: Study of the supramolecular organization of light-harvesting complex (LHCP). Conversion of the oligometric form into the monomeric one by phospholipase A_2 and reconstitution with liposomes. - FEBS Lett. *137* : 271 - 275, 1982.

51274 - RENAUDIN, S., VIDAL, J., LARHER, F. : Characterization of phosphoenolpyruvate carboxylase in a range of parasitic phanerogames. - Z. Pflanzenphysiol. *106* : 229 - 237, 1982.

51275 - RENGER, G. : Biologische Energiekonservierung. - In : HOPPE, W., LOHMANN, W., MARKL, H., ZIEGLER, H. (ed.) : Biophysik. 2nd Ed. Pp. 360 - 385. Springer--Verlag, Berlin - Heidelberg - New York 1982. [Ps.]

51276 - RENGER, G. : Photosynthese. - In : HOPPE, W., LOHMANN, W., MARKL, H., ZIEGLER, H. (ed.) : Biophysik. 2nd Ed. Pp. 532 - 561. Springer-Verlag, Berlin - Heidelberg - New York 1982.

51277 - RENGER, G., REUTER, R. : The destabilization of oxidizing redox equivalents of system II by ADRY reagents in normal and Tris-washed chloroplasts. - Photobiochem. Photobiophys. *3* : 317 - 325, 1982.

51278 - RENGER, G., VOELKER, M. : Studies on the proton release pattern of the donor side of system II. Correlation between oxidation and deprotonization of donor D_1 in Tris-washed inside-out thylakoids. - FEBS Lett. *149* : 203 - 207, 1982.

51279 - RENGER, G., WEISS, W. : The detection of intrinsic 320 nm absorption changes reflecting the turnover of the water-splitting enzyme system Y which leads to oxygen formation in trypsinized chloroplasts. - FEBS Lett. *137* : 217 - 221, 1982.

51280 - RENQUIST, A.R., BREEN, P.J., MARTIN, L.W. : Stomatal behavior and leaf water status of strawberry in different growth environments. - Sci. Hort. *18* : 101 - 110, 1982/83. [Stomatal resistance.]

*51281 - REYMOND, O. : Contribution à l'étude de *Desmatractum* WEST & WEST (*Chlorophyceae, Chlorococcales*) au microscope électronique à transmission. - Arch. Sci. Genève *34* : 259 - 263, 1981. [Chloroplast.]

✻51282 - **REYMOND, O., KOUWETS, F.** : Note sur l'écologie, l'ultrastructure et la ta-
xonomie de l'algue unicellulaire *Desmatractum bipyramidatum* (CHODAT)
PASCHER (= *Bernardinella bipyramidata*, CHODAT), *Chlorophyceae, Chlorococcales*. - Arch. Sci. Genève *34* : 409 - 416, 1981.

51283 - **REYNOLDS, J.F., THORNLEY, J.H.M.** : A shoot:root partitioning model. - Ann.
Bot. *49* : 585 - 597, 1982. [Photosynthates.]

51284 - **RHEE, G.-Y.** : Effects of environmental factors and their interactions on
phytoplankton growth. - In : MARSHALL, K.C. (ed.) : Advances in Microbial
Ecology. Vol.6. Pp. 33 - 74. Plenum Publishing Corp., New York - London
1982. [Ps.]

✻51285 - **RIAUX, C.** : Tidal variations in phytoplankton biomass and seston in a Northern Brittany estuary: "ebb-flood" asymmetry. - Kieler Meeresforsch. *1981*
(Sonderh.5) : 274 - 277, 1981. [Primary production.]

51286 - **RIAUX, C.** : La chlorophylle *a* dans un sédiment estuarien de Bretagne Norde.
- Ann. Inst. océanogr. (Paris) *58* : 185 - 203, 1982.

51287 - **RIAUX, C., GRALL, J.-R.** : Hydrologie et biomasse phytoplanctonique dans un
estuaire de Bretagne Nord, La Penzé. - Int. Rev. ges. Hydrobiol. *67* : 387 -
- 404, 1982.

51288 - **RICH, M., BRODY, S.S.** : Role of various carotenoids in mediating electron
transfer sensitized by chlorophyll and pheophytin. - FEBS Lett. *143* : 45 -
- 48, 1982.

51289 - **RICHARDS, J.H., TEERI, J.A.** : Re-evaluation of proposed C_4 photosynthetic
characteristics in the genus *Larix*. - Physiol. Plant. *55* : 117 - 120,
1982.

51290 - **RICHTER, G., BECKMANN, J., GROß, M., HUNDRIESER, J., SCHNEIDER, C.** :
Blue light-induced synthesis of chloroplast proteins in cultured plant
cells. - In : AKOYUNOGLOU, G., EVANGELOPOULOS, A.E., GEORGATSOS, J., PALA-
IOLOGOS, G., TRAKATELLIS, A., TSIGANOS, C.P. (ed.) : Cell Function and
Differentiation, Part B. Biogenesis of Energy Transducing Membranes and
Membrane and Protein Energetics. Pp. 267 - 276. Alan R. Liss Inc., New York
1982.

51291 - **RICHTER, R., HIEKE, B.** : Zur Bestimmung der Größe der photosynthetischen
Einheit mit intermittierender Belichtung. - In : HOFFMANN, P., HIEKE, B.
(ed.) : Photosynthese: Regulation und Evolution. (Colloquia Pflanzenphysiol.
Nr.5.) Pp. 157 - 159. Humboldt-Universität, Berlin 1982.

51292 - **RIDLEY, S.M.** : Carotenoids and herbicide action. - In : BRITTON, G., GOOD-
WIN, T.W. (ed.) : Carotenoid Chemistry and Biochemistry. Pp. 353 - 369.
Pergamon Press, Oxford - New York - Toronto - Sydney - Paris - Frankfurt
1982.

51293 - **RIEMANN, B., ERNST, D.** : Extraction of chlorophylls *a* and *b* from phytoplankton using standard extraction techniques. - Freshwater Biol. *12* : 217 -
- 223, 1982.

51294 - **RINDT, K.-P.** : Einheit und Vielfalt in der Ausbildung des Photosyntheseapparates bei Prokaryonten. - In : HOFFMANN, P., HIEKE, B.(ed.) : Photosynthese: Regulation und Evolution. (Colloquia Pflanzenphysiologie Nr.5.)
Pp. 135 - 143. Humboldt-Universität, Berlin 1982.

51295 - **RISCH, S.J., HANSEN, M.K.** : Plant growth, flowering phenologies, and yields
of corn, beans and squash grown in pure stands and mixtures in Costa Rica.
- J. appl. Ecol. *19* : 901 - 916, 1982. [Dry-matter accumulation.]

51296 - **ROBERTS, J., PITMAN, R.M., WALLACE, J.S.** : A comparison of evaporation
from stands of Scots pine and Corsican pine in Thetford Chase, East Anglia.
- J. appl. Ecol. *19* : 859 - 872, 1982. [Growth analysis.]

51297 - **ROBERTS, P.L., WOOD, K.R.** : Effects of a severe (P6) and a mild (W) strain
of cucumber mosaic virus on tobacco leaf chlorophyll, starch and cell
ultrastructure. - Physiol. Plant Pathol. *21* : 31 - 37, 1982.

51298 - **ROBERTSON, D.S.** : Chlorophyll and carotenoid mutants. - In : **SHERIDAN, W.F.**
(ed.) : Maize for Biological Research. Pp. 313 - 315. Plant mol. Biol.
Assoc., Charlottesville 1982.

51299 - **ROBERTSON, D.S., FALUDI-DANIEL, A.** : Tests establishing the relationship
between carotenoid biosynthetic mutants of maize studied in Hungary and the
United-States. - J. Hered. *73* : 473, 1982.

51300 - **ROBERTSON, J.D., SCHREIL, W., REEDY, M.** : *Halobacterium halobium* I : A thin-
-sectioning electron microscopic study. - J. Ultrastructure *80* : 148 - 162,
1982.

51301 - **ROBINSON, J.M., GIBBS, M.** : Hydrogen peroxide synthesis in isolated spinach
chloroplast lamellae. An analysis of the Mehler reaction in the presence
of NADP reduction and ATP formation. - Plant Physiol. *70* : 1249 - 1254,
1982.

51302 - **ROBINSON, S.J., DeROO, C.S., YOCUM, C.F.** : Photosynthetic electron transfer
in preparations of the cyanobacterium *Spirulina platensis*. - Plant Physiol.
70 : 154 - 161, 1982.

51303 - **ROBINSON, S.P.** : Light stimulates glycerate uptake by spinach chloroplasts.
- Biochem. biophys. Res. Commun. *106* : 1027 - 1034, 1982.

51304 - **ROBINSON, S.P.** : 3-phosphoglycerate phosphatase activity in chloroplast
preparations as a result of contamination by acid phosphatase. - Plant
Physiol. *70* : 645 - 648, 1982.

51305 - **ROBINSON, S.P.** : Transport of glycerate across the envelope membrane of
isolated spinach chloroplasts. - Plant Physiol. *70* : 1032 - 1038, 1982.

51306 - **ROBSON, M.J.** : The growth and carbon economy of selection lines of *Lolium
perenne* cv. S23 with differing rates of dark respiration. 1. Grown as simulated swards during a regrowth period. - Ann. Bot. *49* : 321 - 329, 1982.
[Ps.]

51307 - **ROBSON, M.J.** : The growth and carbon economy of selection lines of *Lolium
perenne* cv. S23 with differing rates of dark respiration. 2. Grown as young
plants from seed. - Ann. Bot. *49* : 331 - 339, 1982. [Ps.]

*51308 - **ROCHAIX, J.D.** : Organization, function and expression of the chloroplast
DNA of *Chlamydomonas reinhardii*. - Experientia *37* : 323 - 332, 1981. [Ps.]

51309 - **ROCHAIX, J.-D.** : Electron microscope localization of chloroplast genes by
R-loop analysis. - In : **EDELMAN, M., HALLICK, R.B., CHUA, N.-H.**   (ed.) :
Methods in Chloroplast Molecular Biology. Pp. 469 - 476. Elsevier Biomedi-
cal Press, Amsterdam - New York - Oxford 1982.

51310 - **ROCHAIX, J.-D., MALNOË, P.** : Use of DNA-RNA hybridization for locating
chloroplast genes and for estimating the size and abundance of chloroplast
DNA transcripts. In : **EDELMAN, M., HALLICK, R.B., CHUA, N.-H.** (ed.) :
Methods in Chloroplast Molecular Biology. Pp. 477 - 490. Elsevier Biomedical Press, Amsterdam - New York - Oxford 1982.

51311 - **ROMANOVSKIĬ, Yu.V.** : Proyavlenie mezhmolekulyarnykh vzaimodeĭstviĭ v tonko-
strukturnykh spektrakh khlorofilla i ego analogov. [Manifestation of inter-
molecular interactions in the fine-structure spectra of chlorophyll and its
analogues.] - Eesti NSV Tead. Akad. Toim., Füüs., Mat. *31* : 139 - 144, 1982.
[In R, ab : E, Est.]

51312 - **ROMANOWSKA, E., PARYS, E.** : CO_2 exchange rates of seedlings as influenced
by light quality and gibberellic acid. - In : **HOFFMANN, P., HIEKE, B.** (ed.):
Photosynthese: Regulation und Evolution. (Colloquia Pflanzenphysiologie
Nr.5.) Pp. 210 - 213. Humboldt-Universität, Berlin 1982.

*51313 - **RONNEBERGER, D.** : Plankton, Seston und Sauerstoffeintrag des Phytoplanktons
der mittleren Saale. - Limnologica *10* : 73 - 95, 1976.

*51314 - **ROSA, L.** : The rapid activation *in vitro* of the chloroplast fructose 1,6-
-bisphosphatase followed using a new assay procedure. - FEBS Lett. *134* :
151 - 154, 1981.

51315 - ROSCOE, T.J., ELLIS, R.J. : Two-dimensional gel electrophoresis of chloroplast proteins. - In : EDELMAN, M., HALLICK, R.B., CHUA, N.-H. (ed.) :
Methods In Chloroplast Molecular Biology. Pp. 1015 - 1028. Elsevier Biomedical Press, Amsterdam - New York - Oxford 1982.

51316 - ROSE, M.E. : Metastable ion techniques for the analysis of carotenoids iso-
mers. - In : BRITTON, G., GOODWIN, T.W. (ed.) : Carotenoid Chemistry and
Biochemistry. Pp. 167 - 174. Pergamon Press, Oxford - New York - Toronto -
- Sydney - Paris - Frankfurt 1982.

51317 - ROSE, R.J., LINDBECK, A.G.C. : Morphological studies on the transcription
of spinach chloroplast DNA. - Z. Pflanzenphysiol. 106 : 129 - 137, 1982.
[Chloroplast vesicles.]

51318 - ROSENBACH, V., GOLDBERG, R., GILON, C., OTTOLENGHI, M. : On the role of tyrosine in the photocycle of bacteriorhodopsin. - Photochem. Photobiol. 36 :
197 - 201, 1982.

51319 - ROSENBERG, G., RAMUS, J. : Ecological growth strategies in the seaweeds
Gracilaria foliifera (Rhodophyceae) and Ulva sp. (Chlorophyceae): photosynthesis and antenna composition. - Mar. Ecol.-Progr. Ser. 8 : 233 - 241,
1982.

51320 - ROSHCHINA, V.V., BOZHOK, G.V., GOSTIMSKIĬ, S.A. : Issledovanie fotokhimi-
cheskoĭ aktivnosti mutantov gorokha s narusheniem fotosistem. [Photochemi-
cal activity of pea mutants with damaged photosystems.] - Biokhimiya 47 :
1512 - 1521, 1982. [In R, ab : E.]

51321 - ROSHCHINA, V.V., SOLOMATKIN, V.P., MUTUSKIN, A.A. : Membranoaktivnye ingi-
bitory êlektronnogo transporta v khloroplastakh. [Membrane-active inhibitors of electron transport in chloroplasts.] - Biokhimiya 47 : 937 - 944,
1982. [In R, ab : E.]

51322 - ROSINGER, C.H., WILSON, J.M., KERR, M.W. : Changes in the temperature response of Hill-reaction activity of chilling-sensitive and chilling-resis-
tant plants after hardening. - J. exp. Bot. 33 : 321 - 331, 1982.

51323 - ROSSET, R., CAUDE, M., SASSIAT, P., DUTANG, M. : Continuous measurement of
total organic carbon in water by a potentiometric method: An industrial
analyser. - Int. J. environm. anal. Chem. 13 : 19 - 28, 1982.

51324 - ROSSLENBROICH, H.-J., DÖHLER, G. : Einfluß von Ammonium und Nitrat auf die
photosynthetische CO_2-Fixierung von Bellerochea yucatanensis v.STOSCH. -
Biochem. Physiol. Pflanzen 177 : 363 - 373, 1982.

51325 - ROSSMANN, M.G., HENDERSON, R. : Phasing electron diffraction amplitudes
with the molecular replacement method. - Acta crystallogr. A 38: 13 - 20,
1982. [Bacteriorhodopsin.]

51326 - ROTHSCHILD, K.J., ARGADE, P.V., EARNEST, T.N., HUANG, K.-S., LONDON, E.,
LIAO, M.-J., BAYLEY, H., KHORANA, H.G., HERZFELD, J. : The site of attach-
ment of retinal In bacteriorhodopsin. A resonance Raman study. - J. biol.
Chem. 257 : 8592 - 8595, 1982.

51327 - ROTHSCHILD, K.J., MARRERO, H. : Infrared evidence that the Schiff base of
bacteriorhodopsin is protonated: bR570 and K intermediates. - Proc. nat.
Acad. Sci. USA 79 : 4045 - 4049, 1982.

51328 - ROTHSCHILD, K.J., SANCHES, R., CLARK, N.A. : Infrared absorption of photo-
receptor and purple membrane. - In : COLOWICK, S.P., KAPLAN, N.O. (ed.) :
Methods in Enzymology. Vol.88. Pp. 696 - 714. Academic Press, New York -
- London - Paris - San Diego - San Francisco - São Paulo - Sydney - Tokyo -
Toronto 1982.

*51329 - ROUSSEL, C., DUBERTRET, G., LEFORT-TRAN, M. : Effet de la température sur
la distribution des particles intramembranaires dans les thylacoïdes d'Épi-
nards cryofracturés. - Biol. cell. 37 : 67 - 72, 1980.

*51330 - ROUX, E. : La photosynthèse. Historique - Généralités sur les relations:
Structure-Fonction dans les chloroplastes. - Sci. Rech. 1981(Num. spéc.) :
2 - 5, 1981.

51331 - ROUX, E. : Le devenir de l'énergie lumineuse au cours de la photosynthèse. - Compt. rend. Acad. Agr. Fr. *68* : 834 - 845, 1982.

51332 - ROY, H., BLOOM, M., MILOS, P., MONROE, M. : Studies on the assembly of large subunits of ribulose bisphosphate carboxylase in isolated pea chloroplasts. - J. Cell Biol. *94* : 20 - 27, 1982.

51333 - ROY, J., MOONEY, H.A. : Physiological adaptation and plasticity to water stress of coastal and desert populations of *Heliotropium curassavicum* L. - Oecologia *52* : 370 - 375, 1982. [Ps.]

51334 - RÓZSA, Z., DEMETER, S. : Effect of inactivation of the oxygen-evolving system on the thermoluminescence of isolated chloroplasts. - Photochem. Photobiol. *36* : 705 - 708, 1982.

51335 - RUBIN, L.B., PASHCHENKO, V.Z. : Pikosekundnaya fluorometriya protsessa diffuzii éksitona v pigmentnom apparate khloroplastov vysshikh rasteniĭ. [Picosecond fluorometry of the exciton diffusion process in chloroplast pigment apparatus of higher plants.] - Eesti NSV Tead. Akad. Toim., Füüs., Mat. *31* : 192 - 199, 1982. [In R, ab : E, Est.]

*51336 - RUBINOV, A.N., ZEN'KEVICH, É.I., NEMKOVICH, N.A., TOMIN, V.I. : Napravlennyĭ perenos énergii v rastvorakh fotosinteticheskikh pigmentov, vyzvannyĭ orientatsionnym ushireniem urovneĭ énergii. [Directed energy transfer in solutions of photosynthetic pigments, induced by oriented distribution of energy levels.] - Optika Spektrosk. *51* : 848 - 854, 1981. [In R.]

51337 - RUDENKO, T.I., GERTS, S.M., KLIMOV, A.A., SHMELEVA, V.L., MAKAROV, A.D. : Issledovanie konformatsionnykh kolebaniĭ fotofosforiliruyushchikh khloroplastov. [Conformational oscillations in chloroplasts performing photophosphorylation.] - Fiziol. Rast. *29* : 1095 - 1101, 1982. [In R, ab : E.]

*51338 - RÜDIGER, W., BRANDLMEIER, T., BLOS, I., GOSSAUER, A., WELLER, J.-P. : Isolation of the phytochrome chromophore. The cleavage reaction with hydrogen bromide. - Z. Naturforsch. *35 C* : 763 - 769, 1980. [Bil.]

51339 - RUDOĬ, A.B., VEZITSKIĬ, A.Yu., SHLYK, A.A. : Izuchenie fermentnoĭ sistemy prevrashcheniya khlorofillida v khlorofill v étiolirovannykh list'yakh s pomoshch'yu ékzogennykh substratov. [Study of the enzymatic system transforming chlorophyllide into chlorophyll in etiolated leaves, using exogenous substrates.] - Biokhimiya *47* : 733 - 739, 1982. [In R, ab : E.]

*51340 - RUETER, J.G., McCARTHY, J.J., CARPENTER, E.J. : The toxic effect of copper on *Oscillatoria (Trichodesmium) theibautii*. - Limnol. Oceanogr. *24* : 558 - - 562, 1979. [Ps.]

51341 - RUFFNER, H.P. : Metabolism of tartaric and malic acids in *Vitis*: A review - Part B. - Vitis *21* : 346 - 358, 1982. [Photosynthates.]

51342 - RUNDEL, P.W. : Water uptake by organs other than roots. - In : LANGE, O.L., NOBEL, P.S., OSMOND, C.B., ZIEGLER, H. (ed.) : Physiological Plant Ecology II. Water Relations and Carbon Assimilation. Pp. 111 - 134. Springer-Verlag, Berlin - Heidelberg - New York 1982. [Ps.]

51343 - RURAINSKI, H.J., GERHARDT, R., MADER, G. : Spectrophotometric isolation of kinetically different pools of P-700 and their correlation to the reduction of NADP by isolated chloroplasts. I. The effect of light quality and intensity. - Z. Naturforsch. *37 C* : 31 - 39, 1982.

51344 - RUSCKOWSKI, M., ZILINSKAS, B.A. : Allophycocyanin I and the 95 kilodalton polypeptide. Bridge between phycobilisomes and membranes. - Plant Physiol. *70* : 1055 - 1059, 1982.

51345 - RUSSELL, G.K., LYMAN, H. : Production of chloroplast mutants in *Euglena*. - In : EDELMAN, M., HALLICK, R.B., CHUA, N.H. (ed.) : Methods in Chloroplast Molecular Biology. Pp. 39 - 50. Elsevier Biomedical Press, Amsterdam 1982.

51346 - RUTHERFORD, A.W., CROFTS, A.R., INOUE, Y. : Thermoluminescence as a probe of Photosystem II photochemistry. The origin of the flash-induced glow peaks. - Biochim. biophys. Acta *682* : 457 - 465, 1982.

51347 - **RUTHERFORD, A.W., THURNAUER, M.C.** : Radical pair state in photosystem II. - Proc. nat. Acad. Sci. USA *79* : 7283 - 7287, 1982.

51348 - **RÜTTIMANN, A.** : Synthesis and stereochemistry of red pepper carotenoids. - In : BRITTON, G., GOODWIN, T.W. (ed.) : Carotenoid Chemistry and Biochemistry. Pp. 71 - 86. Pergamon Press, Oxford - New York - Toronto - Sydney - Paris - Frankfurt 1982.

51349 - **RUYTERS, G.** : Effects of blue-light on pyruvate kinase activity during chloroplast development of unicellular green algae. - Photochem. Photobiol. *35* : 229 - 231, 1982.

B51350 - **RYABOV, A.K., SIRENKO, L.A.** : Iskusstvennaya Aëratsiya Prirodnykh Vod. [Artificial Aeration of Natural Waters.] - Naukova Dumka, Kiev 1982. [Ps; in R.]

51351 - **RYBERG, M., SUNDQVIST, C.** : Characterization of prolamellar bodies and pro-thylakoids fractionated from wheat etioplasts. - Physiol. Plant. *56* : 125 - - 132, 1982.

51352 - **RYBERG, M., SUNDQVIST, C.** : Spectral forms of protochlorophyllide in pro-lamellar bodies and prothylakoids fractionated from wheat etioplasts. - Physiol. Plant. *56* : 133 - 138, 1982.

*51353 - **RYBIN, I.A.** : Fenomenologiya i proiskhozhdenie svetozavisimoĭ bioëlektricheskoĭ aktivnosti. [Display and origin of light-dependent bioelectric activity.] - In : Svetozavisimaya Bioëlektricheskaya Aktivnost' List'ev Rasteniĭ. Pp. 5 - 22, III. Ural'. Gos. Universitet, Sverdlovsk 1980. [Ps; in R, ab : E.]

*51354 - **RYBIN, I.A., OBOLONSKIĬ, V.V.** : O svyazi statsionarnogo potentsiala lista rasteniĭ s fotosintezom v usloviyakh estestvennogo osveshcheniya. [Relationship between stationary leaf potential and photosynthesis under natural irradiance.] - In : Svetozavisimaya Bioëlektricheskaya Aktivnost' List'ev Rasteniĭ. Pp. 102 - 108, IV. Ural'. Gos. Universitet, Sverdlovsk 1980. [In R, ab : E.]

51355 - **RYBIŃSKI, W., PATYNA, H.** : Ocena poziomu uszkodzeń somatycznych i często-tliwości występowania mutacji chlorofilowych u roślin jęczmienia jarego po traktowaniu MNUA i prędkimi neutronami. [Estimation of somatic injury and chlorophyll mutation frequency in barley plants treated with MNUA and fast neutrons.] - Hodowla Rośl. Aklimat. Nasienn. *26* : 135 - 144, 1982. [In Pol., ab : E, R.]

51356 - **RYBKINA, G.V., KOMPANIETS, I.I., BIGLOVA, S.G.** : O kharaktere svetozavisi-mykh ob"emnykh izmeneniĭ khloroplastov pri potere kletkoĭ razlichnykh kolichestv vody. [Light-independent changes in the chloroplast volume during dehydration of plant cells.] - Fiziol. Rast. *29* : 471 - 478, 1982. [In R, ab : E.]

51357 - **RYČ, M., LEWAK, S.** : Hormone interactions in the formation of the photo-synthetic apparatus in dormant and stratified apple embryos. - Z. Pflanzen-physiol. *107* : 15 - 24, 1982.

51358 - **RYDÉN, L.** : Model of the active site in the blue oxidases based on the ceru-loplasmin-plastocyanin homology. - Proc. nat. Acad. Sci. USA *79* : 6767 - - 6771, 1982.

51359 - **SABALE, A.B.** : Note on organic and inorganic constituents in *Cymbopogon martnii* (ROXB.) WATS. during different stages of growth. - Indian J. agr. Sci. *52* : 199 - 201, 1982. [Chl.]

51360 - **SABALE, A.B.** : Photosynthetic enzymes and photorespiration characteristics during leaf ontogeny in a C_4 fodder grass *Cymbopogon martinii* ROXB. - Photosynthetica *16* : 481 - 485, 1982.

51361 - SABINSKI, F., BARCKHAUS, R.H., FROMME, H.G., SPENER, F. : Dynamics of galactolipids and plastids in nonphotosynthetic cells of *Glycine max* suspension cultures. - Plant Physiol. *70* : 610 - 615, 1982. [Chloroplast.]

51362 - SACHER, R.M. : Strategies to discover plant growth regulators for agronomic
crops. - In : McLAREN, J.S. (ed.) : Chemical Manipulation of Crop Growth
and Development. Pp. 13 - 15. Butterworth Scientific, London - Boston -
Durban - Singapore - Sydney - Toronto - Wellington 1982. [Ps.]

51363 - SADCHIKOV, A.P. : Produktsiya fitoplanktona Glubokogo ozera v 1976 i 1978 gg.
[Phytoplankton production of the Lake Glubokoe in 1976 and 1978.] - Biol.
Nauki *1982* (2) : 59 - 65, 1982. [In R.]

51364 - SADLER, D.M., WORCESTER, D.L. : Neutron diffraction studies of oriented
photosynthetic membranes. - J. mol. Biol. *159* : 467 - 484, 1982.

51365 - SADLER, D.M., WORCESTER, D.L. : Neutron scattering studies of photosynthetic membranes in aqueous dispersion. - J. mol. Biol. *159* : 485 - 499, 1982.

51366 - SAGER, J.C., EDWARDS, J.L., KLEIN, W.H. : Light energy utilization efficiency for photosynthesis. - Trans. ASAE *25* : 1737 - 1746, 1982.

51367 - SAGUN, E.I., GURINOVICH, G.P., LOSEV, A.P., ZEN'KEVICH, É.I. : Fotonika
khlorofilla i ego analogov v kontsentrirovannykh rastvorakh i assotsiatakh.
[Photophysics of chlorophyll and its analogues in concentrated solutions
and associates.] - Eesti NSV Tead. Akad. Toim., Füüs., Mat. *31* : 170 - 173,
1982. [In R, ab : E, Est.]

51368 - SAGUN, E.I., KOCHUBEEVA, N.D., LOSEV, A.P. : Vliyanie vyazkosti rastvorite-
lya na éffektivnost' *T-T* perenosa énergii s khlorofilla na β-karotin. [Ef-
fect of solvent viscosity on the efficiency of *T-T* energy transfer from
chlorophyll to β-carotene.] - Zh. prikl. Spektroskop. *36* : 434 - 441, 1982.
[In R, ab : E.]

*51369 - SAITO, H. : [Seasonal fluctuations of litterfall in evergreen coniferous
(*Chamaecyparis obtusa* SIEB. et ZUCC.) plantation in Mt. Watamuki-yama, Shiga.] - Jap. J. Ecol. *30* : 377 - 384, 1980. [In Jap., ab : E.]

51370 - SAITO, H. : [Primary production over 10 years in evergreen coniferous (*Chamaecyparis obtusa* SIEB. et ZUCC.) plantation in Mt. Watamuki-yama, Shiga.]
- Jap. J. Ecol. *32* : 87 - 98, 1982. [In Jap., ab : E.]

51371 - SAITO, H., FURUNO, T. : [Dry matter production in *Chamaecyparis obtusa*
plantations in Owase, Mie prefecture and in Kamikitayama, Nara prefecture.]
- J. jap. Forest. Soc. *64* : 209 - 219, 1982. [Growth analysis; in Jap.,
ab : E.]

51372 - SAKA, H., CHISAKA, H. : [Determination and comparison of photosynthesis
inhibition by herbicides with oxygen electrode.] - Weed Res. *27* : 217 -
- 224, 1982. [In Jap., ab : E.]

51373 - SAKHAROVA, O.V., MOLODYKH, L.V. : Fotofosforilirovanie izolirovannykh
khloroplastov korotkostebel'nykh sortov pshenitsy razlichnoĭ produktivnosti.
[Photophosphorylation in chloroplasts isolated from short-strawed wheats
differing in productivity.] - Tr. prikl. Bot. Genet. Selek. *72* : 102 - 108,
1982. [In R, ab : E.]

51374 - SAKS, N.M. : Primary production and release of assimilated carbon by *Chlamydomonas provasolii* in culture. - Mar. Biol. *70* : 205 - 208, 1982.

51375 - SAKURAI, H., LIEN, S., SAN PIETRO, A. : Determination of acid-labile sulfide
and zero-valence sulfur in subchloroplast particles in the presence of sodium dodecyl sulfate. - Anal. Biochem. *119* : 372 - 377, 1982.

51376 - SALA, O.E., LAUENROTH, W.K. : Small rainfall events: An ecological role
in semiarid regions. - Oecologia *53* : 301 - 304, 1982. [Production, resistances.]

51377 - SALA, O.E., LAUENROTH, W.K., PARTON, W.J. : Plant recovery following prolonged drought in a shortgrass steppe. - Agr. Meteorol. *27* : 49 - 58, 1982.
[Resistances.]

51378 - **SALA, O.E., LAUENROTH, W.K., REID, C.P.P.** : Water relations: A new dimension for niche separation between *Bouteloua gracilis* and *Agropyron smithii* in North American semi-arid grasslands. - J. appl. Ecol. *19* : 647 - 657, 1982. [Stomatal resistance.]

51379 - **SALAJ, J.** : Vplyv nízkoteplotného stresu na ultraštruktúru chloroplastov *Picea abies* [L.] KARST. [The effect of low-temperature stress upon the chloroplast ultrastructure in *Picea abies* [L.] KARST.] - Biológia (Bratislava) *37* : 425 - 432, 1982. [In Slov., ab : E, R.]

51380 - **SALCHEVA, G.S., POPOVA, L.P.** : Intensity of photosynthesis and activity of carboxylating enzymes in wheat and rye grown on water-logged soil. - Dokl. bolg. Akad. Nauk *35* : 973 - 976, 1982.

*51381 - **SALEMA, R., SANTOS, I.** : Effect of colchicine and vinblastine on chloroplast microtubules of cultured green-cell lines from *Sedum telephium*. - In : RREDEROO, P., DE PRIESTER, W. (ed.) : Electron Microscopy 1980. Proc. 7th Europe.Congress. Vol.2. Pp. 242 - 243. Electron Microsc. Found., Leiden 1980.

51382 - **SALUJA, A.K., McFADDEN, B.A.** : Modification of active site histidine in ribulosebisphosphate carboxylase/oxygenase. - Biochemistry *21* : 89 - 95, 1982.

51383 - **SALZER, J.** : Untersuchungen über die Nettophotosynthese bei Apfelsorten 3. Mitt. Einfluß der Frucht auf die Mobilisierung der Photosynthesereserven bei der Sorte 'Golden Delicious'. - Arch. Züchtungsforsch. *12* : 385 - - 389, 1982.

51384 - **SAMIEV, Kh.S., MARFINA, K.G.** : Élektroforeticheskiĭ analiz belkov khloroplastov kak kriteriĭ ustoĭchivosti khlopchatnika k zasukhe. [Electrophoretic analysis of chloroplast proteins as an index for drought resistance of cotton plants.] - Dokl. vses. Akad. sel'skokhoz. Nauk im. V.I.Lenina *1982* (6) : 19 - 21, 1982. [In R.]

51385 - **SAMIEV, Kh.S., MARFINA, K.G.** : Ul'trastruktura khloroplastov khlopchatnika pri vodnom defitsite. [Ultrastructure of cotton chloroplasts under water deficit.] - Fiziol. Biokhim. kul't. Rast. *14* : 346 - 349, 1982. [In R.]

51386 - **SAMS, C.E., FLORE, J.A.** : The influence of age, position, and environmental variables on net photosynthetic rate of sour cherry leaves. - J. amer. Soc. hort. Sci. *107* : 339 - 344, 1982.

51387 - **SAMUELSSON, G., RICHARDSON, K.** : Photoinhibition at low quantum flux densities in a marine dinoflagellate (*Amphidium carterae*). - Mar. Biol. *70* : 21 - - 26, 1982. [Ps, Chl.]

51388 - **SAMUILOV, V.D.** : Konversiya énergii v membranakh fotosinteziruyushchikh bakteriĭ. [Energy conversion in membranes of photosynthetic bacteria.] - Usp. sovrem. Biol. *93* (1) : 46 - 63, 1982. [In R.]

51389 - **SANADA, Y., NISHIDA, K.** : The presence of pyruvate, orthophosphate dikinase in CAM plants. - Z. Pflanzenphysiol. *105*: 189 - 192, 1982.

51390 - **SANADZE, G.A., BAAZOV, D.I.** : Éffekt usileniya fotosinteza (Émersona) u list'ev topolya. [The Emerson enhancement effect of photosynthesis in poplar leaves.] - Fiziol. Rast. *29* : 901 - 907, 1982. [In R, ab : E.]

*51391 - **SANCES, F.V., TOSCANO, N.C., JOHNSON, M.W., LaPRÉ, L.F.** : Pesticides may reduce lettuce yield. - Calif. Agr. *35* (11/12) : 4 - 5, 1981. [Ps.]

51392 - **SÁNCHEZ-DÍAZ, M., APARICIO-TEJO, P., GONZÁLEZ-MURÚA, C., PEÑA, J.I.** : The effect of NaCl salinity and water stress with polyethylene glycol on nitrogen fixation, stomatal response and transpiration of *Medicago sativa, Trifolium repens* and *Trifolium brachycalycinum* (subclover). - Physiol. Plant. *54* : 361 - 366, 1982. [Stomatal resistance.]

51393 - **SANDBERG, G., JENSEN, E., CROZIER, A.** : Biosynthesis of indole-3-acetic acid in protoplasts, chloroplasts and a cytoplasmic fraction from barley (*Hordeum vulgare* L.). - Planta *156* : 541 - 545, 1982.

51394 - **SANDELIUS, A.S., LILJENBERG, C.** : Light-induced changes in the lipid composition and ultrastructure of plastids from potato tubers. - Physiol. Plant.
56 : 266 - 272, 1982.

51395 - **SAND-JENSEN, K., PRAHL, C.** : Oxygen exchange with the lacunae and across
leaves and roots of the submerged vascular macrophyte, *Lobelia dortmanna* L.
- New Phytol. *91* : 103 - 120, 1982.

51396 - **SANDMANN, G., BÖGER, P.** : Volatile hydrocarbons from photosynthetic membranes containing different fatty acids. - Lipids *17* : 35 - 41, 1982. [Chlo-
roplast.]

51397 - **SANDMANN, G., BÖGER, P.** : Bleaching activity of new 2-phenylpyridazinones :
structure-activity relationship. - Z. Naturforsch. *37 C* : 1092 - 1094, 1982.

51398 - **SANDMANN, G., BÖGER, P.** : Formation and degradation of photosynthetic membranes determined by ^{35}S-labeled sulfolipid. - Plant Sci. Lett. *24* : 347 -
- 352, 1982. [Chl.]

51399 - **SANDMANN, G., BÖGER, P.** : Mode of action of herbicidal bleaching. - In :
MORELAND, D.E., St.JOHN, J.B., HESS, F.D. (ed.) : Biochemical Responses
induced by Herbicides. Pp. 111 - 130. Amer. Chem. Soc., Washington 1982.
[Ps, Chl.]

51400 - **SAN JOSE, J.J., MAYOBRE, F.** : Quantitative growth relationships of cassava
(*Manihot esculenta* CRANTZ): Crop development in a savana wet season. - Ann.
Bot. *50* : 309 - 316, 1982. [Growth analysis.]

51401 - **SANKOFF, D., CEDERGREN, R.J., McKAY, W.** : A strategy for sequence phylogeny
research. - Nucl. Acids Res. *10* : 421 - 431, 1982. [Chloroplast.]

51402 - **SANTARIUS, K.A.** : The mechanism of cryoprotection of biomembrane systems
by carbohydrates. - In : **LI, P.H., SAKAI, A.** (ed.) : Plant Cold Hardiness
and Freezing Stress. Mechanisms and Crop Implications. Vol.2. Pp. 475 -
- 486. Academic Press, New York - London - Paris - San Diego - San Francisco
- São Paulo - Sydney - Tokyo - Toronto 1982. [Ps.]

51403 - **SANTARIUS, K.A.** : Die Wirkung extremer Temperaturen (Frost, Hitze) auf die
Membranen von Pflanzenzellen. - In : **UNGER, K., SCHUH, J.** (ed.) : Umwelt-
-Stress. Pp. 161 - 170. Martin-Luther-Universität, Halle 1982. [Chloro-
plast.]

51404 - **SANTARIUS, K.A.** : Cryoprotection of spinach chloroplast membranes by dex-
trans. - Cryobiology *19* : 200 - 210, 1982.

51405 - **SANTHA, I.M., MEHTA, S.L., KOUNDAL, K.R., SINHA, S.K.** : Photosynthesis and
translocation rate in high lysin mutant barley. - Phytochemistry *26* : 1183 -
- 1187, 1982.

51406 - **SANTILLAN, C.** : Mass production of *Spirulina*. - Experientia *38* : 40 - 43,
1982. [Chl, Car.]

*51407 - **SANTOS, I., SALEMA, R.** : Cytochemical localization of malic dehydrogenase
in chloroplasts of *Sedum telephium*. - In : **BREDEROO, P., DE PRIESTER, W.**
(ed.) : Electron Microscopy 1980. Proc. 7th Europe. Congress. Vol.2. Pp.
246 - 247. Electron Microsc. Found., Leiden 1980.

*51408 - **SANTOS, I., SALEMA, R.** : Chloroplast microtubules in some CAM-plants. -
Bol. Soc. Brot.,Sér.2 *53* : 1115 - 1122, 1981.

51409 - **SANTOS, R.B., SUTTON, B.G.** : Effect of defoliation on Virginia Bunch peanuts
at Camden, N.S.W. - Aust. J. agr. Res. *33* : 1037 - 1048, 1982. [Dry-matter
accumulation.]

51410 - **SARAI, I.** : Efficiency of electron transport processes. - J. theor. Biol.
99 : 341 - 355, 1982. [Ps.]

51411 - **SARANIN, K.I., STAROVOÏTOV, A.A.** : Formirovanie i produktivnost' listovogo
apparata zernovykh kul'tur v zavisimosti ot priemov obrabotki pochvy. [For-
mation and productivity of leaf apparatus in cereal crops depending on
soil cultivation methods.] - Sel'skokhoz. Biol. *1982* (4) : 495 - 499, 1982.
[In R, ab : E.]

*51412 - **SARIĆ, M.R.** : Fiziološke i morfološke osobine idiotipa šećerne repe. [Physiological and morphological features of sugar beet ideotype.] - Posebna Izd. srpska Akad. Nauka Umetn. *538*, Od. prirod.-mat. Nauka *54* [BELIĆ, J. (ed.) : Fiziologija Šećerne Repe] : 223 - 238, 1981. [Photosynthates; In Serb., ab : E.]

*51413 - **SARIĆ, M.R., MILIVOJEVIĆ, D., KRSTIĆ, B.** : Effects of green and yellow light on ultrastructure of chloroplasts and content of pigments in maize and beans. - Bull. Acad. serbe Sci. Arts, Classe Sci. nat. math. *72* (20) : 29 - 34, 1980.

51414 - **SARKAR, H.K., SONG, P.-S., PARK, S.C., LEE, E.** : Model for photosynthetic light harvesting system: energy transfer in anthracene- and biphenyl-chlorophyll derivatives. - J. Luminesc. *26* : 347 - 358, 1982.

51415 - **SARMA, N.P., PATNAIK, A.** : Streptomycin induced nuclear & cytoplasmic chloroplast mutations in rice. - Indian J. exp. Biol. *20* : 177 - 178, 1982.

51416 - **SAROJA, G., BOSE, S.** : Effects of methyl parathion on the growth, cell size, pigment and protein content of *Chlorella protothecoides*. - Environm. Pollut. A *27* : 297 - 308, 1982.

51417 - **SASAHARA, T.** : Changes in size and number of mesophyll cells, nitrogen content and photosynthesis with leaf order in *Brassica* spp. - Ann. Bot. *50* : 379 - 383, 1982.

51418 - **SASAHARA,T.,SENGOKU,T.,SANO,Y.:** CO_2 and water vapour exchange as related to shade tolerance of *Oryza punctata* KOTSCHY. - Photosynthetica *16* : 356 - - 361, 1982.

51419 - **SASAHARA, T., TAKAHASHI, M., KAMBAYASHI, M.** : [Studies on structure and function of the rice ear III. Final ear weight, and increasing rate of ear weight and decreasing rate of straw weight at the maximum increasing period of ear weight.] - Jap. J. Crop Sci. *51* : 18 - 25, 1982. [In Jap., ab : E.]

51420 - **SATHYANARAYANA, B., SUBHASH, K.** : Studies of EMS induced mutations in egg plant. - Geobios *9* : 137 - 141, 1982. [Chl.]

51421 - **SATO, K., IKEDA, T.** : [The growth responses of soybean plant to photoperiod and temperature. VI. Starch accumulation and chloroplast fine structure in soybean leaf as affected by temperature.] - Jap. J. Crop Sci. *51* : 546 - - 552, 1982. [In Jap., ab : E.]

51422 - **SATO, K., PARK, K.B.** : [On the low temperature damage in rice seedling III. Ultrastructural changes of leaf blade chloroplasts caused by low temperature, and their varietal differences.] - Jap. J. Crop Sci. *51* : 205 - - 214, 1982. [In Jap., ab : E.]

51423 - **SATO, K., PARK, K.B.** : [On the low temperature damage in rice seedling IV. Effect of low temperature on electrical conductivity and mineral element contents in water effusate of leaf blades, and their varietal differences.] - Jap. J. Crop Sci. *51* : 215 - 220, 1982. [In Jap., ab : E.]

51424 - **SATOH, H., KOMAI, T., KITAHARA, K., AOYAGI, R., HONDA, T., NOGUCHI, F.** : [Systematic separation of carotenoid pigments from citrus flesh by high--performance liquid chromatography.] - Bull. Tokyo med. Coll.*1982* (2) : 13 - 21, 1982. [In Jap., ab : E.]

51425 - **SATOH, K.** : Mechanism of photoactivation of electron transport in intact *Bryopsis* chloroplasts. - Plant Physiol. *70* : 1413 - 1416, 1982.

51426 - **SATOH, K.** : Fractionation of thylakoid-bound chlorophyll-protein complexes by isoelectric focussing. - In : EDELMAN, M., HALLICK, R.B., CHUA, N.-H. (ed.) : Methods in Chloroplast Molecular Biology. Pp. 845 - 856. Elsevier Biomedical Press, Amsterdam - New York - Oxford 1982.

51427 - **SATOH, K., FORK, D.C.** : Photoinhibition of reaction centers of photosystems I and II in intact *Bryopsis* chloroplasts under anaerobic conditions. - Plant Physiol. *70* : 1004 - 1008, 1982.

51428 - **SATOH, K., FORK, D.C.** : The light-induced decline of chlorophyll fluorescence as an indication of photoinhibition in intact *Bryopsis* chloroplasts illuminated under anaerobic conditions. - Photobiochem. Photobiophys. *4* : 153 - - 162, 1982.

51429 - **SATOH, K., FORK, D.C.** : The participation of cyclic electron flow around photosystem I in state II to state I transitions in the blue-green alga *Synechococcus lividus*. - Carnegie Inst. Washington Year Book *81* : 50 - 54, 1982.

51430 - **SATOH, K., FORK, D.C.** : The identification of photodestruction of reaction centers of photosystems I and II as the early event of photoinhibition. - Carnegie Inst. Washington Year Book *81* : 58 - 61, 1982.

51431 - **SATOH, K., FORK, D.C.** : Fluorescence transients in *Bryopsis* chloroplasts as indicators of the redox state of plastoquinone. - Carnegie Inst. Washington Year Book *81* : 61 - 64, 1982.

51432 - **SATOH, K., MATHIS, P.** : Photosystem II, chlorophyll *a*-protein complex: a study by flash absorption spectroscopy. - Photobiochem. Photobiophys. *2* : 189 - - 198, 1981.

51433 - **SATOH, M., OHYAMA, K.** : [Effect of girdling on translocation of ^{14}C-photosynthetic product and photosynthetic rate in mulberry.] - J. sericult. Sci. Jap. *51* : 469 - 473, 1982. [In Jap., ab : E.]

51434 - **SAUER, A., HEISE, K.-P.** : On the occurrence of monoacylglycerol derivatives in lipid metabolism of chloroplasts. - Z. Naturforsch. *37 C* : 218 - 225, 1982.

51435 - **SAUNDERS, V.A.** : Genetics and molecular biology of photosynthetic bacteria and cyanobacteria. - In : GOVINDJEE (ed.) : Photosynthesis. Vol.II. Pp. 17 - - 42. Academic Press, New York - London - Paris - San Diego - San Francisco - São Paulo - Sydney - Tokyo - Toronto 1982.

51436 - **SAUNDERS, V.A., BUETOW, D.E.** : Introduction to genetics and molecular biology of photosynthetic bacteria, cyanobacteria, and chloroplasts. - In : GOVINDJEE (ed.) : Photosynthesis. Vol.II. Pp. 13 - 15. Academic Press, New York - London - Paris - San Diego - San Francisco - São Paulo - Sydney - Tokyo - Toronto 1982.

51437 - **SAVER, B.G., KNOWLES, J.R.** : Ribulose-1,5-bisphosphate carboxylase: Enzyme--catalyzed appearance of solvent tritium at carbon 3 of ribulose 1,5-bisphosphate reisolated after partial reaction. - Biochemistry *21* : 5398 - - 5403, 1982.

*51438 - **SAVIDGE, G.** : Photosynthesis of marine phytoplankton in fluctuating light regimes. - Mar. Biol. Lett. *1* : 295 - 300, 1980.

51439 - **SAVOYE, D., LECLERC, J.C.** : Apports de la spectroscopie d'absorption à haute résolution à l'étude taxonomique et écologique de quelques *Trebouxia* (Chlorophycées, Chlorococcales). - Cryptogamie: Algologie *3* (2) : 113 - 120, 1982.

51440 - **SAWA, Y., KANAYAMA, K., OCHIAI, H.** : Photosynthetic regeneration of ATP using a strain of thermophilic blue-green algae. - Biotechnol. Bioeng. *24* : 305 - 315, 1982.

51441 - **SAWADA, S., IGARASHI, T., MIYACHI, S.** : Effects of nutritional levels of phosphate on photosynthesis and growth studied with single, rooted leaf of dwarf bean. - Plant Cell Physiol. *23* : 27 - 33, 1982.

51442 - **SAYER, P., GOUTERMAN, M., CONNELL, C.R.** : Metalloid porphyrins and phthalocyanines. - Accounts chem. Res. *15* : 73 - 79, 1982. [Chl.]

51443 - **SAYRE, R.T., CHENIAE, G.M.** : Studies on the reconstitution of O_2-evolution of chloroplasts. - Plant Physiol. *69* : 1084 - 1095, 1982.

*51444 - **SCALA, A., QUAGLIETTA CHIARANDA, F.** : Effect of water potential on *Rhizoctonia* damping-off of cauliflower. - In : Proceedings of the Fifth Congress of the Mediterranean Phytopathological Union. Pubbl.N.48. Pp. 132 - 135. Consiglio Nazionale delle Ricerche, Napoli-Ponticelli 1980. [Stomatal resistance.]

51445 - SCARPONI, L., PERUCCI, P., GIUSQUIANI, P.L., TAFURI, F. : Il ruolo del
ferro nella nutrizione del fagiolino in idroponica. [Role of iron in bean
nutrition in hydroponics.] - Agrochimica *26* : 270 - 279, 1982. [Chl; in
Ital., ab : E, F, G.]

51446 - SCHAAFSMA, T.J., MICHEL-BEYERLE, M. : *In vitro* simulation of sequential
electron transfer processes in photosynthetic reaction centers. - In :
HALL, D.O., PALZ, W. (ed.) : Photochemical, Photoelectrochemical and Photobiological Processes. Vol.1. Pp. 150 - 153. D.Reidel Publ.Co., Dordrecht -
Boston - London 1982.

51447 - SCHACHLER, G., MATSCHKE, J., MAASS, I., SCHÖNBORN, H.-J. : Beziehung zwischen den Wegen der C-Aufnahme und SO_2-Resistenz bei Koniferen. - Biochem.
Physiol. Pflanzen *177* : 651 - 658, 1982.

51448 - SCHACHSCHNEIDER, R. : Ertrag und Ertragsstruktur verschiedener Winterwei-
zengenotypen unter Stressbedingungen. - In : UNGER, K.,SCHUH, J. (ed.) :
Umwelt-Stress. Pp. 207 - 211. Martin-Luther-Universität, Halle 1982.

51449 - SCHAFFERNICHT, H., JUNGE, W. : Deconvolution of the red P700 difference
spectrum based on a set of three Gaussian components: further evidence from
literature spectra. - Photochem. Photobiol. *36* : 111 - 115, 1982.

51450 - SCHANZ, F., WÄLTI, K. : Primary productivity in freshwater environments. -
In : ZABORSKY, O.R. (ed.) : CRC Handbook of Biosolar Resources. Vol.I/2.
Pp. 389 - 394. CRC Press, Boca Raton 1982.

51451 - SCHEER, A., PARTHIER, B. : Dark-induced chloroplast dedifferentiation in
Euglena gracilis. - Planta *156* : 274 - 281, 1982.

51452 - SCHEER, H. : Phycobiliproteins: Molecular aspects of photosynthetic antenna
system. - In : FONG, F.K. (ed.) : Light Reaction Path of Photosynthesis.
Pp. 7 - 45. Springer-Verlag, Berlin - Heidelberg - New York 1982.

51453 - SCHEER, H., FORMANEK, H., SCHNEIDER, S. : Theoretical studies of biliprotein
chromophores and related bile pigments by molecular orbital and Ramachandran
type calculations. - Photochem. Photobiol. *36* : 259 - 272, 1982.

51454 - SCHENCK, C.C., BLANKENSHIP, R.E., PARSON, W.W. : Radical pair decay kinetics,
triplet yields and delayed fluorescence from bacterial reaction centers. -
Biochim. biophys. Acta *680* : 44 - 59, 1982.

51455 - SCHENCK, C.C., DINER, B., MATHIS, P., SATOH, K. : Flash-induced carotenoid
radical cation formation in photosystem II. - Biochim. biophys. Acta *680* :
216 - 227, 1982.

51456 - SCHERER, H.W., SCHUBERT, S., MENGEL, K. : Einfluss der Kaliumernährung
auf die Wachstumsrate, den Gehalt an Kohlenhydraten und die Wasserretention
von jungen Weizenpflanzen. - Z. Pflanzenern. Bodenk. *145* : 237 - 245, 1982.
[Growth analysis.]

51457 - SCHERER, S., BÖGER, P. : Respiration of blue-green algae in the light. -
Arch. Microbiol. *132* : 329 - 332, 1982.

51458 - SCHERER, S., STÜRZL, E., BÖGER, P. : Interaction of respiratory and photosynthetic electron transport in *Anabaena variabilis* KÜTZ. - Arch. Microbiol.
132 : 333 - 337, 1982.

51459 - SCHIEWER, U., AL-SAADI, H.A., HAMEED, H.A. : On the diel rhythm of phytoplankton productivity in Shatt al-Arab at Basrah, Iraq. - Arch. Hydrobiol.
93 : 158 - 172, 1982.

*51460 - SCHIFF, J.A. : Evolution of the control of pigment and plastid development
in photosynthetic organisms. - BioSystems *14* : 123 - 147, 1981.

51461 - SCHIFF, J.A. : Additional comments: Evolution of the regulation of plastid
development. - In : SCHIFF, J.A., LYMAN, H. (ed.) : On the Origins of Chloroplasts. Pp. 315 - 317. Elsevier/North-Holland, New York - Amsterdam -
Oxford 1982.

51462 - SCHIFF, J.A., CUNNINGHAM, F.X.,Jr., GREEN, M.S. : Carotenoids in relation
to chloroplasts and other organelles. - In : BRITTON, G., GOODWIN, T.W.
(ed.) : Carotenoid Chemistry and Biochemistry. Pp. 329 - 338. Pergamon
Press, Oxford - New York - Toronto - Sydney - Paris - Frankfurt 1982.

B51463 - SCHIFF, J.A., LYMAN, H. (ed.) : On the Origins of Chloroplasts. - Elsevier/ /North-Holland, New York - Amsterdam - Oxford 1982.

51464 - SCHIMPF, C., GOVINDARAJAN, A.G., PARTHIER, B. : Influence of preillumination (potentiation) and carbon substrates on plastid aminoacyl-tRNA synthetases in greening *Euglena* cells. - Biochem. Physiol. Pflanzen *177* : 777 - 788, 1982. [Ps, Chl.]

51465 - SCHIMZ, A., SPERLING, W., HILDERBRAND, E., KÖHLER-HAHN, D. : Bacteriorhodopsin and the sensory pigment of the photosystem 565 in *Halobacterium halobium*. - Photochem. Photobiol. *36* : 193 - 196, 1982.

51466 - SCHLICHTER, D. : Epidermal nutrition of the alcyonarian *Heteroxenia fuscescens* (EHRB.): Absorption of dissolved organic material and lost endogenous photosynthates. - Oecologia *53* : 40 - 49, 1982.

51467 - SCHLODDER, E., GRÄBER, P., WITT, H.T. : Mechanism of phosphorylation in chloroplasts. - In : BARBER, J. (ed.) : Electron Transport and Photophosphorylation. Pp. 105 - 175. Elsevier Biomedical Press, Amsterdam - New York - Oxford 1982.

51468 - SCHLODDER, E., RÖGNER, M., WITT, H.T. : ATP synthesis in chloroplasts induced by a transmembrane electric potential difference as a function of the proton concentration. - FEBS Lett. *138*: 13 - 18, 1982.

51469 - SCHMIDT, A. : Assimilation of sulfur. - In : SCHIFF, J.A., LYMAN, H. (ed.): On the Origins of Chloroplasts. Pp. 179 - 197. Elsevier/North-Holland, New York - Amsterdam - Oxford 1982. [Chloroplast.]

51470 - SCHMIDT, A., ERDLE, I., KÖST, H.-P. : Changes of C-phycocyanin in *Synechococcus* 6301 in relation to growth on various sulfur compounds. - Z. Naturforsch. *37 C* : 870 - 876, 1982.

*51471 - SCHMIDT, G.W., BARTLETT, S.G., GROSSMAN, A.R., CASHMORE, A.R., CHUA, N.-H.: Biosynthetic pathways of two polypeptide subunits of the light-harvesting chlorophyll *a/b* protein complex. - J. Cell Biol. *91* : 468 - 478, 1981.

51472 - SCHMIDT, K. : Carotenoids of photosynthetic bacteria. - In : ZABORSKY, O.R. (ed.) : CRC Handbook of Biosolar Resources. Vol.I/1. Pp. 69 - 72. CRC Press, Boca Raton 1982.

51473 - SCHMITT, M.R., EDWARDS, G.E. : Isolation and purification of intact peroxisomes from green leaf tissue. - Plant Physiol. *70* : 1213 - 1217, 1982.

51474 - SCHMITZ, K., KÜHN, R. : Fine structure, distribution and frequency of plasmodesmata and pits in the cortex of *Laminaria hyperborea* and *L. saccharina*. - Planta *154* : 385 - 392, 1982. [Photosynthates.]

51475 - SCHMUTZ, D., BRUNOLD, C. : Regulation of sulfate assimilation in plants XIII. Assimilatory sulfate reduction during ontogenesis of primary leaves of *Phaseolus vulgaris* L. - Plant Physiol. *70* : 524 - 527, 1982.

51476 - SCHNABL, H., ELBERT, C., KRÄMER, G. : The regulation of the starch-malate balances during volume changes of guard cell protoplasts. - J. exp. Bot. *33* : 996 - 1003, 1982.

51477 - SCHNEIDER, K. : Response of microalgae to CO_2, HCO_3^-, O_2 and pH. - In : ZABORSKY, O.R. (ed.) : CRC Handbook of Biosolar Resources. Vol.I/2. Pp. 125 - 133. CRC Press, Boca Raton 1982. [Ps.]

51478 - SCHNEPF, E. : Bau der Zelle (Prokaryoten, Eukaryoten). - In : HOPPE, W., LOHMANN, W., MARKL, H., ZIEGLER, H. (ed.) : Biophysik. 2nd Ed. Pp. 1 - 21. Springer-Verlag, Berlin - Heidelberg - New York 1982. [Chloroplast.]

*51479 - SCHOCH, S., SCHEER, H., SCHIFF, J.A., RÜDIGER, W., SIEGELMAN, H.W. : Pyropheophytin *a* accompanies pheophytin *a* in darkened light grown cells of *Euglena*. - Z. Naturforsch. *36 C* : 827 - 833, 1981.

51480 - SCHOLZ, B., BALLSCHMITER, K. : Chromatographic separation and analytical characterization of bacteriochlorophylls ap, agg and b. - J. Chromatogr. 252 : 269 - 282, 1982.

51481 - SCHÖNHERR, J. : Resistance of plant surfaces to water loss: transport properties of cutin, suberin and associated lipids. - In : LANGE, O.L., NOBEL, P.S., OSMOND, C.B., ZIEGLER, H. (ed.) : Physiological Plant Ecology II. Water Relations and Carbon Assimilation. Pp. 153 - 179. Springer-Verlag, Berlin - Heidelberg - New York 1982. [Resistances.]

51482 - SCHOPER, J.B., JOHNSON, R.R., LAMBERT, R.J. : Maize yield response to increased assimilate supply. - Crop Sci. 22 : 1184 - 1189, 1982.

51483 - SCHOTTE, J. : Einfluß exogener Faktoren auf Teilprozesse des photosynthetischen Elektronentransportes unter Berücksichtigung der Membranstruktur. - In : HOFFMANN, P., HIEKE, B. (ed.) : Photosynthese: Regulation und Evolution. Pp. 160 - 162. Humboldt-Universität, Berlin 1982.

51484 - SCHRADER, G.C., HORNER, R., SMITH, G.F. : An improved chamber for in $situ$ measurement of primary productivity by sea ice algae. - Can. J. Fish. aquat. Sci. 39 : 522 - 524, 1982.

51485 - SCHRAMM, R.W. : Carbon assimilation and flux connected with nitrogen fixation in legumes. - Isr. J. Bot. 31 : 131 - 139, 1982.

51486 - SCHRAUDOLF, H. : Effects of metronidazole on growth, chloroplast structure and differentiation in gametophytes of $Anemia$ $phyllitidis$ L.SW. - Protoplasma 113 : 144 - 149, 1982.

51487 - SCHREIBER, U., VALLE-TASCON, D. del : ATP synthesis with single turnover flashes in spinach chloroplasts: In $situ$ monitoring with the firefly luciferase method. - FEBS Lett. 150 : 32 - 37, 1982.

51488 - SCHREIBER, U., PFISTER, K. : Kinetic analysis of the light-induced chlorophyll fluorescence rise curve in the presence of dichlorophenyldimethylurea. Dependence of the slow-rise component on the degree of chloroplast intactness. - Biochim. biophys. Acta 680 : 60 - 68, 1982.

51489 - SCHREIBER, U., RIENITS, K.G. : ATP-induced absorbance changes around 515 nm following light-activation of the latent ATP-hydrolase in intact chloroplasts. - FEBS Lett 141 : 287 - 291, 1982.

51490 - SCHREIBER, U., RIENITS, K.G. : Complementarity of ATP-induced and light-induced absorbance changes around 515 nm. - Biochim. biophys. Acta 682 : 115 - 123, 1982.

*51491 - SCHUBER, M., KLUGE, M. : In $situ$ studies on crassulacean acid metabolism in $Sedum$ $acre$ L. and $Sedum$ $mite$ GIL. - Oecologia 50 : 82 - 87, 1981.

51492 - SCHUBERT, K.R. : Symbiotic nitrogen fixation in nodulated leguminous and nonleguminous angiosperms. - In : ZABORSKY, O.R. (ed.) : CRC Handbook of Biosolar Resources. Vol.I/1. Pp. 265 - 282. CRC Press, Boca Raton 1982. [Ps.]

51493 - SCHULER, F., BRANDT, P., WIEßNER, W. : Isolierung von PS II-Partikeln mit in $vivo$ Eigenschaften aus $Euglena$ $gracilis$, Stamm Z. - Z. Naturforsch. 37 C : 256 - 259, 1982.

51494 - SCHULTZ, G., SOLL, J., FIEDLER, E. : The biosynthesis of isoprenoid compound in the chloroplast from the compartmental view. - In : WINTERMANS, J.F.G.M., KUIPER, P.J.C. (ed.) : Biochemistry and Metabolism of Plant Lipids. Pp. 501 - 506. Elsevier Biomedical Press, Amsterdam 1982.

*51495 - SCHULZE, E.-D. : Carbon gain and wood production in trees of deciduous beech ($Fagus$ $silvatica$) and trees of evergreen spruce ($Picea$ $excelsa$). - Mitt. forst. Bundesversuchsanstalt (Wien) 142 : 105 - 123, 1981. [Ps.]

51496 - SCHULZE, E.-D. : Plant life forms and their carbon, water and nutrient re-
lations. - In : LANGE, O.L., NOBEL, P.S., OSMOND, C.B., ZIEGLER, H. (ed.) :
Physiological Plant Ecology II. Water Relations and Carbon Assimilation.
Pp. 615 - 676. Springer-Verlag, Berlin - Heidelberg - New York 1982. [Ps.]

51497 - SCHULZE, E.-D., HALL, A.E. : Stomatal responses, water loss and CO_2 assi-
milation rates of plants in contrasting environments. - In : LANGE, O.L.,
NOBEL, P.S., OSMOND, C.B., ZIEGLER, H. (ed.) : Physiological Plant Ecology
II. Water Relations and Carbon Assimilation. Pp. 181 - 230. Springer-Verlag,
Berlin - Heidelberg - New York 1982.

51498 - SCHULZE, E.-D., HALL, A.E., LANGE, O.L., WALZ, H. : A portable steady-state
porometer for measuring the carbon dioxide and water vapour exchanges of
leaves under natural conditions. - Oecologia 53 : 141 - 145, 1982.

51499 - SCHÜRMANN, P., ORTIZ, W. : Photosynthetic activity of isolated chloroplasts
from Euglena gracilis. - Planta 154 : 70 - 75, 1982.

51500 - SCHWARZ, H.P. : Effects of nuclear gene mutations on the structure and
function of plastids in pea. I. Physiological and spectroscopical studies.
- Egypt. J. Genet. Cytol. 11 : 69 - 79, 1982.

51501 - SCHWARZ, H.P., KLOPPSTECH, K. : Effects of nuclear gene mutations on the
structure and function of plastids in pea. The light-harvesting chlorophyll a/b protein. - Planta 155 : 116 - 123, 1982.

51502 - SCHWARZENBACH, F.H., HEGETSCHWEILER, T. : Wood as biomass for energy: Results of a problem analysis. - Experientia 38 : 22 - 27, 1982. [Production.]

51503 - SCHWEIGER, H.-G. : Interrelationship between chloroplasts and the nucleo-
-cytosol compartment in Acetabularia. - In : PARTHIER, B., BOULTER, D. (ed.):
Nucleic Acids and Proteins in Plants II. Structure, Biochemistry and Physiology of Nucleic Acids. Pp. 645 - 662. Springer-Verlag, Berlin - Heidelberg - New York 1982.

51504 - SCHWEIGER, H.-G. : The mechanism of the circadian rhythm of photosynthesis.
- In : HÉLÈNE, C., CHARLIER, M., MONTENAY-GARESTIER, T., LAUSTRIAT, G.(ed.):
Trends in Photobiology. Pp. 437 - 450. Plenum Press, New York - London
1982.

51505 - SCOTT, J.M., MARLOW, J.A. : A microcalorimeter with a range of 0.1-1.0 ca-
lories. - Limnol. Oceanogr. 27 : 585 - 590, 1982.

51506 - SCOTT, N.S., CAIN, P., POSSINGHAM, J.V. : Plastid DNA levels in albino and
green leaves of the "albostrians" mutant of Hordeum vulgare. - Z. Pflanzenphysiol. 108 : 187 - 191, 1982.

51507 - SCOTT, N.S., POSSINGHAM, J.V. : Leaf development. - In : SMITH, H., GRIER-
SON, D. (ed.) : The Molecular Biology of Plant Development. Botanical Mono-
graphs. Vol.18. Pp. 223 - 255. Blackwell Sci. Publ., Oxford - Edinburgh -
Boston - Melbourne 1982.   [Chloroplast, RuBPC.]

51508 - SCOTT, R.A., HAHN, J.E., DONIACH, S., FREEMAN, H.C., HODGSON, K.O. : Polarized X-ray absorption spectra of oriented plastocyanin single crystals.
Investigation of methionine-copper coordination. - J. amer. chem. Soc. 104 :
5364 - 5369, 1982.

51509 - SCOUFFLAIRE, C., LANNOYE, R., BARBER, J. : Chlorophyll fluorescence and thylakoid stacking changes: electrostatic screening versus charge neutralizat-
ion. - Photobiochem. Photobiophys. 4 : 249 - 256, 1982.

51510 - SECOR, J., FORD, D.M., SHIBLES, R. : Ontogenetic changes in ribulose-1,5-
-bisphosphate carboxylase-oxygenase activity in soybean leaves. - Plant
Sci. Lett. 27 : 147 - 154, 1982.

51511 - SECOR, J., McCARTY, D.R., SHIBLES, R., GREEN, D.E. : Variability and selection for leaf photosynthesis in advanced generations of soybeans. - Crop Sci.
22 : 255 - 259, 1982.

*51512 - SEDLETSKII, V.A. : Peredvizhenie i raspredelenie assimilyatov v vinogradno?
loze pod vliyaniem kol'tsevaniya plodovykh strelok. [Transport and distri-

bution of assimilates in grapevine under the influence of ringing of the fruit scapes.] - In : Vinogradarstvo na Yuge Ukrainy. Pp. 26 - 33. Odesskiĭ Sel'skokhozyaĭstvennyĭ Institut, Odessa 1980. [In R.]

51513 - SEELY, G.R., RUTKOSKI, A.M., KUSUMOTO, Y., SENTHILATHIPAN, V., SHAW, E.R.: Association of chlorophyll with amides on plasticized polyethylene particles - I. N,N-dimethylmyristamide. - Photochem. Photobiol. 36 : 633 - 640, 1982.

51514 - SEELY, G.R., SENTHILATHIPAN, V., KUSUMOTO, Y. : Photosynthesis model systems: interaction of chlorophyll with N-pyridylmyristamides on plasticized particles. - Annu. Rep. C.F.Kettering Res. Lab. 1982 : 40 - 41, 1982.

51515 - SEEMANN, J.R., BERRY, J.A. : Interspecific differences in the kinetic properties of RuBP carboxylase protein. - Carnegie Inst. Washington Year Book 81 : 78 - 83, 1982.

51516 - SEENI, S., GNANAM, A. : Growth of photoheterotrophic cells of peanut (*Arachis hypogaea* L.) in still nutrient medium. - Plant Physiol. 70 : 815 - 822, 1982. [Ps, Chl.]

51517 - SEENI, S., GNANAM, A. : Carbon assimilation in photoheterotrophic cells of peanut (*Arachis hypogaea* L.) grown in still nutrient medium. - Plant Physiol. 70 : 823 - 826, 1982.

51518 - SEIBERT, M., JANZEN, A.F., KENDALL-TOBIAS, M. : Light-induced electron transport across semiconductor electrode/reaction-center film/electrolyte interface. - Photochem. Photobiol. 35 : 193 - 200, 1982. [Ps.]

51519 - SEIBERT, M., KENDALL-TOBAIS, M.W. : Photoelectrochemical properties of electrodes coated with photoactive-membrane vesicles isolated from photosynthetic bacteria. - Biochim. biophys. Acta 681 : 504 - 511, 1982.

51520 - SEIFERT, B. : Die Größe der photosynthetischen Einheit als Verhältnis zwischen Gesamtchlorophyll und P 700 im Verlauf der Ergrünung etiolierter Weizenprimärblätter (*Triticum aestivum* L. cv. Hatri). - In : HOFFMANN, P., HIEKE, B. (ed.) : Photosynthese: Regulation und Evolution. (Colloquia Pflanzenphysiologie Nr.5.) Pp. 154 - 156. Humboldt-Universität, Berlin 1982.

51521 - SELAK, M.A., WHITMARSH, J. : Kinetics of the electrogenic step and cytochrome b_6 and f redox changes in chloroplasts. Evidence for a Q cycle. - FEBS Lett. 150 : 286 - 292, 1982.

51522 - SELLNER, K.G., LYONS, L., PERRY, E.S., HEIMARK, D.B. : Assessing physiological stress in *Thalassiosira fluviatilis (Bacillariophyta)* and *Dunaliella tertiolecta (Chlorophyta)* with DCMU-enhanced fluorescence. - J. Phycol. 18 : 142 - 148, 1982.

51523 - SELTZER, S., LIN, M. : Isolation and purification of retinals from purple-membranes from mass spectral analysis. - In : COLOWICK, S.P., KAPLAN, N.O. (ed.) : Methods in Enzymology. Vol.88. Pp. 542 - 546. Academic Press, New York - London - Paris - San Diego - San Francisco - São Paulo - Sydney - Tokyo - Toronto 1982.

51524 - SEMAN, I. : Počty chloroplastov v prieduchoch cukrovej repy rôznych genómových stupňov. [Number of chloroplasts in stomata of sugar beet of different genome grades.] - Sborník ÚVTIZ - Genet. Šlecht. 18 : 169 - 174, 1982. [In Slovak, ab : E, G, R.]

51525 - SEMBDNER, G., KLOSE, C. : Phytohormone in Chloroplasten. - In : HOFFMANN, P., HIEKE, B. (ed.) : Photosynthese: Regulation und Evolution. (Colloquia Pflanzenphysiologie Nr.5.) Pp. 98 - 114. Humboldt-Universität, Berlin 1982.

51526 - SEMENENKO, V.E. : Mekhanizmy éndogennoĭ regulyatsii fotosinteza i adaptivnye svoĭstva khloroplasta. [Mechanisms of endogenous regulation of photosynthesis and adaptive properties of chloroplast.] - In : Fiziologiya Fotosinteza. Pp. 164 - 187. Nauka, Moskva 1982. [In R.]

51527 - SEMIKHATOVA, O.A. : Rol' issledovaniĭ dykhaniya v razvitii teorii fotosintetícheskoĭ produktivnosti rasteniĭ. [Role of studies on respiration in the development of theory of photosynthetic plant productivity.] - Bot. Zh. 67 : 1025 - 1035, 1982. [In R, ab : E.]

51528 - SEMIKHATOVA, O.A., ZALENSKIĬ, O.V. : Sopryazhennost' protsessov fotosinteza
         i dykhaniya. [Coupling of photosynthesis and respiration processes.] - In :
         Fiziologiya Fotosinteza. Pp. 130 - 145. Nauka, Moskva 1982. [In R.]

51529 - SEMYCHAEVS'KYĬ, V.D. : Vykorystannya deyakykh algorytmiv avtomatychnoï kla-
         syfikatsiï dlya otsinky taksonomichnoï podibnosti vydiv *Cyanophyta* za bio-
         khimichnymy oznakamy. [Use of certain algorhythms of automatic classifi-
         cation for estimation of the taxonomic affinity of *Cyanophyta* by biochemi-
         cal characters.] - Ukr. bot. Zh. *39* (4) : 96 - 102, 113, 1982. [Bil; in
         Ukr., ab : E, R.]

51530 - SEN, A., WILLIAMS, W.P., BRAIN, A.P.R., QUINN, P.J. : Bilayer and non-linear
         transformations in aqueous dispersions of mixed *sn*-3-galactosyldiacylglyce-
         rols isolated from chloroplasts. A freeze-fracture study. - Biochim. bio-
         phys. Acta *685* : 297 - 306, 1982.

B51531 - SEN, D.N. : Environment and Plant Life in Indian Desert. - Geobios Inter-
         national, Jodhpur 1982. [Ps, Chl.]

51532 - SEN, N.K., PATRA, H.K., SARANGI, C.S., MISHRA, D. : Kinetin effect on some
         enzyme activities during senescence of excised rice leaves as subjected
         to water stress. - Biochem. Physiol. Pflanzen *177* : 577 - 584, 1982. [Chl.]

51533 - SENGER, H. : Efficiency of incident light utilization and quantum require-
         ment in microalgae. - In : ZABORSKY, O.R. (ed.) : CRC Handbook of Biosolar
         Resources. Vol.I/1. Pp. 55 - 58. CRC Press, Boca Raton 1982.

51534 - SENGER, H. : Response of microalgae to light intensity and light quality.
         - In : ZABORSKY, O.R. (ed.) : CRC Handbook of Biosolar Resources. Vol.I/2.
         Pp. 121 - 123. CRC Press, Boca Raton 1982. [Ps.]

51535 - SENGER, H. : Culture conditions of photosynthetic microorganisms. - In :
         ZABORSKY, O.R. (ed.) : CRC Handbook of Biosolar Resources. Vol.I/2. Pp.
         153 - 155. CRC Press, Boca Raton 1982.

51536 - SENGER, H. : The effect of blue light on plants and microorganisms. - Photo-
         chem. Photobiol. *35* : 911 - 920, 1982.

51537 - SENGUPTA, T., MUKHERJI, S. : Colchicine induction of protochlorophyllide
         formation & greening in primary leaves & differential effect of embryonic
         axis on chlorophyll production in cotyledons of etiolated mungbean (*Phaseo-
         lus aureus* L.) seedlings. - Indian J. exp. Biol. *20* : 537 - 540, 1982.

51538 - SENSER, M., BECK, E. : Frost resistance in spruce [*Picea abies* (L.) KARST.]:
         IV. The lipid composition of frost resistant and frost sensitive spruce
         chloroplasts. - Z. Pflanzenphysiol. *105* : 241 - 253, 1982.

51539 - SENTSOVA, O.Yu., ROMANOVA, N.I., MINEEVA, L.A., GUSEV, M.V. : Fotosintez
         obligatno atsidofil'noĭ odnokletochnoĭ vodorosli *Cyanidium caldarium* :
         zavisimost' ot pH i formy neorganicheskogo ugleroda. [Photosynthesis of
         obligately acidophilic unicellular alga *Cyanidium caldarium* : dependence
         on pH and the form of inorganic carbon.] - Dokl. Akad. Nauk SSSR *266* :
         762 - 765, 1982. [In R.]

51540 - SEO, S.W., OTA, Y. : [Role of the hull in the ripening of rice plant.II.
         Changes in photosynthesis and respiration of hull during ripening.] - Jap.
         J. Crop Sci. *51* : 105 - 109, 1982. [In Jap., ab : E.]

51541 - SEO, S.W., OTA, Y. : [Role of the hull in the ripening of rice plant. VI.
         Translocation of carbohydrates into fertile and sterile grains.] - Jap. J.
         Crop Sci. *51* : 570 - 576, 1982. [In Jap., ab : E.]

51542 - SEPP, E., WANNER, G., EDER, ¹ , KÖST, H.-P. : Immunochemical characterizat-
         ion of *Porphyridium cruentum* B-phycoerythrin: proof of cross-reaction bet-
         ween chromophore-free apoprotein and holoprotein-specific antibodies. - Z.
         Naturforsch. *37 C* : 1146 - 1156, 1982.

51543 - SERGEEVA, E.A. : Sposob polucheniya kontura lista rasteniĭ. [A method
         producing a contour of plant leaf.] - Bot. Zh. *67* : 234 - 235, 1982. [In
         R.]

51544 - SERRANO, A., RIVAS, J. : Purification of ferredoxin-NADP$^+$ oxidoreductase
from cyanobacteria by affinity chromatography on 2',5'-ADP-Sepharose 4B. -
Anal. Biochem. *126* : 109 - 115, 1982.

51545 - SERRANO, A., RIVAS, J., LOSADA, M. : Changes in fluorescence spectra by
nitrate and nitrite in a blue-green alga. - Photobiochem. Photobiophys.
4 : 257 - 264, 1982.

*51546 - ŠESTÁK, Z. : Současné světové trendy ve výzkumu fotosyntézy. [Present world
trends in photosynthesis research.] - In : Využitie Poznatkov o Fotosyntéze
v Produkčnom Procese Rastlín. Pp. 20 - 30. Vysoká Škola Pol'nohospodárska,
Nitra 1981. [In Czech.]

51547 - ŠESTÁK, Z. : Thin layer chromatography of chlorophylls - 2. - Photosynthetica
16 : 568 - 617, 1982.

*51548 - ŠESTÁK, Z., ČATSKÝ, J. : Bibliography of reviews and methods of photosynthe-
sis - 50, 51. - Photosynthetica *15* : 264 - 292, 612 - 639, 1981.

51549 - ŠESTÁK, Z., ČATSKÝ, J. : Bibliography of reviews and methods of photosynthesis - 52, 53. - Photosynthetica *16* : 285 - 315, 618 - 640, 1982.

B51550 - ŠESTÁK, Z., ČATSKÝ, J. (ed.) : Photosynthesis Bibliography. Vol.10/1, 10/2
- 1979. - Dr. W. Junk Publishers, The Hague 1982.

*51551 - SETHURAJ, M.R. : Yield components in *Hevea brasiliensis* - theoretical con-
siderations. - Plant Cell Environm. *4* : 81 - 83, 1981.

51552 - SETIF, P., QUAEGEBEUR, J.-P. : Primary processes in photosystem I. Identification and decay kinetics of the P-700 triplet state. - Biochim. biophys.
Acta *681* : 345 - 353, 1982.

51553 - ŠETLÍKOVÁ, E., ZACHLEDER, V., ŠETLÍK, I. : Effect of cell function inhibitors on photochemical activity of *Scenedesmus* in the course of its cell
cycle. - In : AKOYUNOGLOU, G., EVANGELOPOULOS, A.E., GEORGATSOS, J., PALAIO-
LOGOS, G., TRAKATELLIS, A., TSIGANOS, C.P. (ed.) : Cell Function and Differentiation, Part B. Biogenesis of Energy Transducing Membranes and Membrane
and Protein Energetics. Pp. 223 - 246. Alan R. Liss Inc., New York 1982.

51554 - SEVERINA, I.I. : Nystatin-induced increase in photocurrent in the system
"bacteriorhodopsin proteoliposome/bilayer planar membrane". - Biochim. biophys. Acta *681* : 311 - 317, 1982.

51555 - SHAHAK, Y. : Activation and deactivation of H$^+$-ATPase in intact chloroplasts.
- Plant Physiol. *70* : 87 - 91, 1982.

51556 - SHAHAK, Y. : The role of Mg^{2+} in the light activation process of the H$^+$-ATP-
ase in intact chloroplasts. - FEBS Lett. *145* : 223 - 229, 1982.

51557 - SHAHAK, Y., ADMON, A., AVRON, M. : Transmembrane electrical potential formation by chloroplast ATPase complex (CF$_1$-CF$_0$) proteoliposomes. - FEBS Lett.
150 : 27 - 31, 1982.

*51558 - SHAKHOV, A.A. : Fotoénergetika produktivnosti rasteniǐ. [Photoenergetics
of plant productivity.] - In : Preobrazovanie Solnechnoǐ Énergii. Pp. 158 -
- 170. Akademiya Nauk SSSR, Chernogolovka 1981. [In R.]

51559 - SHAKHOV, Y.A., NYRÉN, P., BALTSCHEFFSKY, M. : Reconstitution of highly purified proton-translocating pyrophosphatase from *Rhodospirillum rubrum*. -
FEBS Lett. *146* : 177 - 180, 1982. [Chromatophore.]

51560 - SHAMALA, T.R., DRAWERT, F., LEUPOLD, G. : Studies on *Scenedesmus acutus*
growth. I.Effect of autotrophic and mixotrophic conditions on the growth of
Scenedesmus acutus. - Biotechnol. Bioeng. *24* : 1287 - 1299, 1982. [Ps.]

51561 - SHAMALA, T.R., DRAWERT, F., LEUPOLD, G. : Studies on *Scenedesmus acutus*
growth. II. Effect of autotrophic and mixotrophic growth on the amino acid
and the carbohydrate composition of *Scenedesmus acutus*. - Biotechnol.Bioeng.
24 : 1301 - 1317, 1982. [Photosynthates.]

51562 - SHANK, C.V., GREENE, B.I. : Subpicosecond ultrafast laser technique - application and design. - In : ALFANO, R.R. (ed.) : Biological Events Probed

by Ultrafast Laser Spectroscopy. Pp. 417 - 427. Academic Press, New York - London - Paris - San Diego - San Francisco - São Paulo - Sydney - Tokyo - Toronto 1982. [Bacteriorhodopsin.]

51563 - SHANNON, J.C. : A search for rate-limiting enzymes that control crop production. - Iowa State J. Res. *56* : 307 - 322, 1982.

51564 - SHARENKOVA, Kh.A., SEMENENKO, V.E. : Kharakteristika fotosinteticheskogo vydeleniya kisloroda *Spirulina plantensis* (GOM) GEITL. [Characteristics of photosynthetic oxygen evolution in *Spirulina platensis* (GOM) GEITL.] - Fiziol. Rast. *29* : 572 - 577, 1982. [In R, ab : E.]

51565 - SHARIFI, M.R., NILSEN, E.T., RUNDEL, P.W. : Biomass and net primary production of *Prosopis glandulosa (Fabaceae)* in the Sonoran desert of California. - Amer. J. Bot. *69* : 760 - 767, 1982.

51566 - SHARKEY, T.D., BADGER, M.R. : Effects of water stress on photosynthetic electron transport, photophosphorylation, and metabolite levels of *Xanthium strumarium* mesophyll cells. - Planta *156* : 199 - 206, 1982.

51567 - SHARKEY, T.D., IMAI, K., FARQUHAR, G.D., COWAN, I.R. : A direct confirmation of the standard method of estimating intercellular partial pressure of CO_2. - Plant Physiol. *69* : 657 - 659, 1982.

51568 - SHARKEY, T.D., STEVENSON, G.F., PATON, D.M. : Effect of G, a growth regulator from *Eucalyptus grandis*, on photosynthesis. - Plant Physiol. *69* : 935 - - 938, 1982.

51569 - SHARMA, A.K., SINGH, B.B., SINGH, S.P. : Relationships among net assimilation rate, leaf area index and yield in soybean (*Glycine max* (L.) MERRILL) genotypes. - Photosynthetica *16* : 115 - 118, 1982.

51570 - SHARMA, M.L. : Aspects of salinity and water relations of Australian chenopods. - In : SEN, D.N., RAJPUROHIT, K.S. (ed.) : Contributions to the Ecology of Halophytes. Pp. 155 - 172. Dr. W. Junk Publ., The Hague - Boston - London 1982. [Ps.]

51571 - SHARMA, P., BERGMAN, B., HÄLLBOM, L., HOFSTEN, A.von : Ultrastructural changes in *Nostoc* of *Peltigera canina* in presence of SO_2. - New Phytol. *92* : 573 - 579, 1982.

51572 - SHATILOV, V.R., SOF'IN, A.V., ZABRODINA, T.M., KRETOVICH, W.L. : The role of chloroplast and cytoplasm in the NADP-glutamate dehydrogenase and glutamine synthetase synthesis in *Ankistrodesmus* cells. - Mol. cell. Biochem. *49* : 157 - 159, 1982.

51573 - SHAŬCHUK, S.M., GAPONENKA, V.I., ZHĚBRAKOVA, I.V. : Dasledavanne stanu khlarafilu i intênsiŭnastsi fotasintêzu ŭ zalatsistaĭ i zyalěnalistaĭ form klěnu yasenyalistaga. [Study of the state of chlorophyll and photosynthetic rate in aurea and green forms of *Acer negundo* L.] - Vestsi Akad. Navuk belarus. SSR, Ser. biyal. Navuk *1982* (2) : 26 - 29, 123, 1982. [In Belorus., ab : E, R.]

51574 - SHAVYALUKHA, V.S., CHAĬKA, M.Ts., LAMAN, M.A., GRYB, S.I. : Fiziyalagichnyya dasledavanni ŭ suvyazi z prablemami selektsyi zernevykh kul'tur u Belarusi. [Physiological studies on problems of selection of cereals in Belorussia.] - Vestsi Akad. Navuk belarus. SSR, Ser. biyal. Navuk *1982* (6) : 34 - 40, 133, 1982. [Chl; in Belorus., ab : E, R.]

51575 - SHAW, R.H., PEREIRA, A.R. : Aerodynamic roughness of a plant canopy : a numerical experiment. - Agr. Meteorol. *26* : 51 - 65, 1982.

*51576 - SHCHERBAKOV, V.Ya., KALUS, Yu.A. : Sopryazhennost' mezhdu ploshchad'yu list'-ev i produktivnost'yu rasteniĭ grechikhi i prosa. [Interrelation between leaf area and productivity of buckwheat and millet plants.] - Nauch.-tekh. Byull. vsesoyuz. selekts.-genet. Inst. *25* : 49 - 53, 1975. [In R.]

51577 - SHEEN, S.J., LOWE, R.H., BURTON, H.R. : Leaf proteins and chemical constituents in tobacco chlorophyll genotypes. - Beitr. Tabakforsch. int. *11* : 170 - 179, 1982.

51578 - SHEFFER, M., AVRON, M. : An unusual sensitivity to salt of ferredoxin-dependent photoreactions in *Dunaliella*. - Plant Sci. Lett. *25* : 241 - 246, 1982.

*51579 - SHEPANSKI, J.F., ANDERSON, R.W.,Jr. : Chlorophyll-*a* excited singlet state
absorption measured in the picosecond time regime. - Chem. Phys. Lett. *78* :
165 - 173, 1981.

51580 - SHEPARD, R.B. : Primary productivity and phytoplankton distribution in a
small Illinois (U.S.A.) lake. - Int. Rev. ges. Hydrobiol. *67* : 555 - 565,
1982.

51581 - SHERMAN, P.A., WIMMER, M.J. : Two types of kinetic regulation of the acti-
vated ATPase in the chloroplast photophosphorylation system. - J. biol.
Chem. *257* : 7012 - 7017, 1982.

51582 - SHERMAN, W.V. : Time-resolved fluorpmetry of bacteriorhodopsin. - Photochem.
Photobiol. *36* : 463 - 469, 1982.

51583 - SHEVYAKOVA, N.I. : Soleustoĭchivost' plastomnykh khlorofil'nykh mutantov
podsolnechnika. [Salt tolerance in chlorophyll plastid mutants of sunflower.]
- Fiziol. Rast. *29* : 317 - 324, 1982. [In R, ab : E.]

51584 - SHIBA, H., KAWASUMI, T., IGARASHI, Y., KODAMA, T., MINODA, Y. : The defici-
ent carbohydrate metabolic pathways and the incomplete tricarboxylic acid
cycle in an obligately autotrophic hydrogen-oxidizing bacterium. - Agr. biol.
Chem. (Tokyo) *46* : 2341 - 2345, 1982.

51585 - SHIBA, T., SIMIDU, U. : *Erythrobacter longus* gen.nov., sp.nov., an aerobic
bacterium which contains bacteriochlorophyll *a*. - Int. J. syst. Bacteriol.
32 : 211 - 217, 1982.

51586 - SHIBATA, H., FUJIWARA, T., SAWA, Y., OCHIAI, H. : Chloroplast development
in 4-thiouridine-cultured radish seedlings VI. Anabolic pathway of 4-thio-
uridine. - Plant Cell Physiol. *23* : 365 - 374, 1982. [Chl.]

51587 - SHIBATA, Y., ARUGA, Y. : Variations in chlorophyll *a* concentration and photosynthetic activity of phytoplankton in Tokyo Bay. - Mer *20* : 75 - 92,
1982.

51588 - SHIEH, Y.-J., KU, M.S.B., BLACK, C.C.,Jr. : Photosynthetic metabolism of
aspartate in mesophyll and bundle sheath cells isolated from *Digitaria sanguinalis* (L.) SCOP., a NADP$^+$-malic enzyme-C_4 plant. - Plant Physiol. *69* :
776 - 780, 1982.

51589 - SHIH, L.M., KAUR-SAWHNEY, R., FUHRER, J., SAMANTA, S., GALSTON, A.W. :
Effects of exogenous 1,3-diaminopropane and spermidine on senescence of oat
leaves. I. Inhibition of protease activity, ethylene production, and chlorophyll loss as related to polyamine content. - Plant Physiol. *70* : 1592 -
- 1596, 1982.

51590 - SHILO, M. : Diversity of photosynthetic prokaryotes. - In : SCHIFF, J.A.,
LYMAN, H. (ed.) : On the Origins of Chloroplasts. Pp. 9 - 26. Elsevier/
/North-Holland, New York - Amsterdam - Oxford 1982.

51591 - SHIMABUKU, M., KUDO, M., TAMAKI, K., MIYAGI, Y. : [Studies on dry matter
production of sugarcane in Okinawa. 3. Optimum LAI in relation to dry matter
production in sugarcane varieties.] - Jap. J. trop. Agr. *26* : 193 - 197,
1982. [In Jap., ab : E.]

51592 - SKIMAZAKI, K., GOTOW, K., KONDO, N. : Photosynthetic properties of guard
cell protoplasts from *Vicia faba* L. - Plant Cell Physiol. *23* : 871 - 879,
1982.

*51593 - SHIMIZU, T., MORI, M. : [Carotenoids in Russian comfrey and kale.] - Sagami
Joshi Daigaku Kiyo *43* : 235 - 244, 1980. [In Jap.]

*51594 - SHIMIZU, T., MORI, M., IWASAKI, M. : [Study of carotenoids.] - Sagami Joshi
Daigaku Kiyo *42* : 25 - 31, 1979. [In Jap.]

51595 - SHIMOKAWA, K. : Hydrophobic chromatographic purification of ethylene-enhan-
ced chlorophyllase from *Citrus unshiu* fruits. - Phytochemistry *21* : 543 -
- 545, 1982.

51596 - SHIMSHI, D., MAYORAL, M.L., ATSMON, D. : Responses to water stress in wheat and related wild species. - Crop Sci. *22* : 123 - 128, 1982. [Ps.]

*51597 - SHIMURA, S., YAMAGUCHI, Y., ARUGA, Y., FUJITA, Y., ICHIMURA, S. : Extra-cellular release of photosynthetic products by a pelagic blue-green alga, *Trichodesmium thiebautii*. - J. oceanogr. Soc. Jap. *34* : 181 - 187, 1978.

51598 - SHIN, I.V., FOGEL', V.R., KOTEL'NIKOV, A.I., LIKHTENSHTEĬN, G.I. : Issledo-vanie struktury i vnutrimolekulyarnoĭ dinamiki fotoaktivnogo pigment-belko-vogo kompleksa êozin-kazein. [Study of the structure and intramolecular dynamics of the eosin-kasein photoactive pigment-protein complex.] - Biofi-zika *27* : 5 - 9, 1982. [Ps model; in R, ab : E.]

51599 - SHIN, I.V., KOTEL'NIKOV, A.I., LIKHTENSHTEĬN, G.I. : Kinetika reaktsii foto-perenosa êlektrona v komplekse êozin-kazein. [Reaction kinetics of electron phototransfer in the eosin-casein complex.] - Biofizika *27* : 208 - 211, 1982. [Photorespiration model; in R, ab : E.]

51600 - SHIN, M. : Ferredoxin-NADP reductase. - In : ZABORSKY, O.R. (ed.) : CRC Handbook of Biosolar Resources. Vol.I/1. Pp. 115 - 119. CRC Press, Boca Raton 1982.

51601 - SHIN, M., SAKIHAMA, N., SUGIMOTO, N., MIYAZAWA, T., OSHINO, R. : Immobilized ferredoxin-NADP$^+$ reductase: preparation and properties. - J. Biochem. (Tokyo) *91* : 953 - 957, 1982.

51602 - SHINKAREV, V.P., KONONENKO, A.A., RUBIN, A.B. : Analiz termodinamicheskikh kharakteristik vzaimodeĭstviya vysokopotentsial'nogo tsitokhroma s dimerom bakteriokhlorofilla v reaktsionnykh tsentrakh khromatoforov *Ectothiorhodo-spira shaposhnikovii*. [Analysis of thermodynamic parameters of interaction between high-potential cytochrome and dimer of bacteriochlorophyll in the reaction centres of *Ectothiorhodospira shaposhnikovii* chromatophores.] - Biofizika *27* : 832 - 836, 1982. [In R, ab : E.]

51603 - SHINKAREV, V.P., RUBIN, A.B. : Simmetriya protsessa perenosa êlektronov na donornoĭ i aktseptornoĭ storonakh fotosinteticheskogo reaktsionnogo tsentra. [Symmetry of electron transfer on donor and acceptor sides of the photosynthetic reaction centre.] - Biofizika *27* : 722 - 724, 1982. [In R, ab : E.]

51604 - SHINOHARA, K., SAKURAI, H. : Light-induced ATP formation from acetyl phos-phate and ADP by broken spinach chloroplasts. - Plant Cell Physiol. *23* : 59 - 66, 1982.

51605 - SHINOZAKI, K., SASAKI, Y., SAKIHAMA, T., KAMIKUBO, T. : Coordinate light--induction of two mRNAs, encoded in nuclei and chloroplasts, of ribulose 1,5-bisphosphate carboxylase/oxygenase. - FEBS Lett. *144* : 73 - 76, 1982.

51606 - SHINOZAKI, K., SUGIURA, M. : The nucleotide sequence of the tobacco chloro-plast gene for the large subunit of ribulose-1,5-bisphosphate carboxylase//oxygenase. - Gene *20* : 91 - 102, 1982.

51607 - SHIOI, Y., SASA, T. : Separation of protochlorophylls esterified with dif-ferent alcohols from inner seed coats of three *Cucurbitaceae*. - Plant Cell Physiol. *23* : 1315 - 1321, 1982.

51608 - SHIPMAN, L.L. : Electronic structure and function of chlorophylls and their pheophytins. - In : GOVINDJEE (ed.) : Photosynthesis. Vol.1. Pp. 275 - 291. Academic Press, New York - London - Paris - San Diego - San Francisco - São Paulo - Sydney - Tokyo - Toronto 1982.

51609 - SHIPMAN, L.L. : Binding sites associated with inhibition of photosystem II. - In : MORELAND, D.E., St.JOHN, J.B. , HESS, F.D. (ed.) : Biochemical Res-ponses Induced by Herbicides. Pp. 23 - 35. American Chemical Society, Washington 1982.

*51610 - SHIRATORI, K., MORIKAWA, M., TAKASAKI, T., TOTSUKA, T. : [Preliminary investigation in net photosynthesis rate of kidney bean leaves injured by SO_2 fumigation.] - Kokuritsu Kogai Kenkyusho Tokubetsu Kenkyu Seika Hokoku [Rep. spec. Res. Project, nat. Inst. environm. Stud.] *2* : 113 - 118, 1978. [In Jap., ab : E.]

*51611 - SHLYK, A.A. : Current concept of organization of chlorophyll biosynthesis. -
         In : MAZLIAK, P., BENVENISTE, P., COSTES, C., DOUCE, R. (ed.) : Biogenesis
         and Function of Plant Lipids. Pp. 311 - 320. Elsevier/North-Holland Biomedi-
         cal Press, Amsterdam - New York - Oxford 1980.

 51612 - SHLYK, A.A. : Biyagenez fotasintètychnaga aparatu. [Biogenesis of the photo-
         synthetic apparatus.] - Vestsi Akad. Navuk belarus. SSR, Ser. biyal. Navuk
         1982 (6) : 21 - 28, 132, 1982. [In Belorus., ab : E, R.]

 51613 - SHLYK, A.A., VLASENOK, L.I., GRONSKAYA, N.I. : Raspredelenie novykh molekul
         khlorofilla v strukture khloroplastov pri vvedenii ^{14}C-δ-aminolevulinovoĭ
         kisloty v rasteniya. [Distribution of new chlorophyll molecules into the
         chloroplast structure of plants supplied with ^{14}C-δ-aminolevulinic acid.] -
         Fiziol. Rast. 29 : 720 - 727, 1982. [In R, ab : E.]

 51614 - SHMELEVA, V.L., IVANOV, B.N., RED'KO, T.P. : Transport èlektronov i foto-
         fosforilirovanie, svyazannye s fotovosstanovleniem kisloroda khloroplastami
         gorokha, vyrashchennogo pri razlichnoĭ osveshchennosti. [Electron transport
         and photophosphorylation coupled with oxygen photoreduction by chloroplasts
         of pea grown upon different illuminance.] - Biokhimiya 47 : 1104 - 1107,
         1982. [In R, ab : E.]

 51615 - SHOCHAT, S., OWENS, G.C., HUBERT, P., OHAD, I. : The dichlorophenyldimethyl-
         urea-binding site in thylakoids of Chlamydomonas reinhardii. Role of photo-
         system II reaction center and phosphorylation of the 32-35 kilodalton poly-
         peptide in the formation of the high-affinity binding site. - Biochim. bio-
         phys. Acta 681 : 21 - 31, 1982.

 51616 - SHOL'TS, K.F., REZNIK, G.I., MOSOLOVA, I.M., KOTEL'NIKOVA, A.V. : Sravnitel'-
         noe izuchenie deĭstviya nekotorykh induktorov pronitsaemosti na mitokhondrii
         i khloroplasty. [Comparative study of the effects of some perminductors on
         mitochondria and chloroplasts.] - Biokhimiya 47 : 447 - 454, 1982. [In R,
         ab : E.]

 51617 - SHOME, A., BHADURI, P.N. : Response of excised embryos of rice (Oryza sa-
         tiva L.) to X-rays. - Theor. appl. Genet. 61 : 135 - 139, 1982. [Chl.]

*51618 - SHOMER-ILAN, A., SAMISH, Y.B., KIPNIS, T., ELMER, D., WAISEL, Y. : Effects
         of salinity, N-nutrition and humidity on photosynthesis and protein meta-
         bolism of Chloris gayana KUNTH. - Plant Soil 53 : 477 - 486, 1979.

 51619 - SHOUSE, P., JURY, W.A., STOLZY, L.H., DASBERG, S. : Field measurement and
         modelling of cowpea water use and yield under stressed and well-watered
         growth conditions. - Hilgardia 50(6) : 1 - 25, 1982. [Growth analysis.]

 51620 - SHOWELL, M.S., FONG, F.K. : Elementary reconstitution of the water split-
         ting light reaction in photosynthesis. 3. Photooxidative properties of
         chlorophyll dihydrate on metal as catalyst for water photolysis. - J. amer.
         chem. Soc. 104 : 2773 - 2781, 1982.

 51621 - SHOWELL, M.S., FONG, F.K. : Molecular mechanisms of hydrogen generation
         in the chlorophyll water splitting light reaction. - In : Hydrogen Energy
         Progress IV. Vol.2. Pp. 715 - 724. 1982.

 51622 - SHUBIN, L.M., EROKHINA, L.G., PROSKURYAKOV, I.I., KRASNOVSKIĬ, A.A. :
         Obrazovanie svobodnykh radikalov pri fotovosstanovlenii allofikotsianina
         B. [Formation of free radicals in photoreduction of allophycocyanin B.] -
         Biofizika 27 : 724 - 726, 1982. [In R, ab : E.]

*51623 - SHULIENE, R., YANKYAVICHYUS, K. : Pervichnaya produktsiya. [Primary pro-
         duction.] - In : Gidrobiologicheskie Issledovaniya Ozer Dusya, Galstas,
         Shlavantas,Obyaliya. Pp. 7 - 22. 1977. [In R.]

*51624 - SHULIENE, R.I., MARCHYULENENE, D.P., YANKAVICHYUTE, G.Yu. : Ėkologiya fito-
         planktona v zalive Kurshyu-Mares. [Phytoplankton ecology in the lagoon Kur-
         šiŭ Marios.] - Gidrobiol. Zh. 15 (5) : 53 - 58, 1979. [In R, ab : E.]

*51625 - SHUMILOVA, A.A., FEDOSEENKO, A.A., STEPANOVA, A.M. : Vzaimodeĭstvie foto-
         sinteza i reduktsii nitratov v list'yakh rasteniĭ C_3 i C_4 tipa. [Interrelat-
         ion of photosynthesis and reduction of nitrates in leaves of plants of C_3
         and C_4 type.] - In : Fotosintez, Dykhanie i Organicheskie Kisloty. Pp. 26 -
         - 33. Izd. Voronezh. Universiteta, Voronezh 1980. [In R.]

51626 - **SHUSHANASHVILI, V.I.** : Lineĭnye razmery ribulozobisfosfatkarboksilazy i ko-
lichestvo fermenta v kletkakh ẻvgleny. [Linear dimensions and amount of
ribulose 1,5-bisphosphate carboxylase in *Euglena* cells.]- Fiziol. Rast.
29 : 1195 - 1202, 1982. [In R, ab : E.]

51627 - **SHUTILOVA, N.I., DEMIDOVA, L.N., KADOSHNIKOVA, I.G., KLIMOV, V.V., ZAKRZHEV-
SKIĬ, D.A.** : Issledovanie éffektivnosti khromatograficheskogo razdeleniya
pigment-belkovolipidnykh kompleksov reaktsionnykh tsentrov fotosistemy 1
i fotosistemy 2 khloroplastov gorokha na DÉAÉ-tsellyuloze.[The efficiency
of chromatographic separation of the pigment-protein-lipid complexes of the
reaction centres of photosystems 1 and 2 from pea chloroplasts on DEAE-cel-
lulose.] - Biokhimiya *47* : 317 - 322, 1982. [In R, ab : E.]

51628 - **SHUTILOVA, N.I., FALUDI-DÁNIEL, Á., KLIMOV, V.V.** : A rapid procedure for
isolating the photosystem II reaction centers in a highly enriched form.
The use of a chlorophyll *b*-less mutant of barley. - FEBS Lett. *138* : 255 -
- 257, 1982.

51629 - **SHUVALOV, V.A., KRASNOVSKY, A.A.** : Photochemical electron transfer in reaction centers of photosynthesis. - Sov. sci. Rev., Sect. D (Biol. Rev.) *3* :
1 - 24, 1982.

*51630 - **SHUVALOV, V.A., PARSON, W.W.** : Energies and kinetics of radical pairs involving bacteriochlorophyll and bacteriopheophytin in bacterial reaction centers. - Proc. nat. Acad. Sci. USA *78* : 957 - 961, 1981.

51631 - **SICHER, R.C.** : Reversible light-activation of ribulose bisphosphate carbo-
xylase/oxygenase in isolated barley protoplasts and chloroplasts. - Plant
Physiol. *70* : 366 - 369, 1982.

51632 - **SICHER, R.C., HARRIS, W.G., KREMER, D.F., CHATTERTON, N.J.** : Effects of
shortened day length upon translocation and starch accumulation by maize,
wheat, and pangola grass leaves. - Can. J. Bot. *60* : 1304 - 1309, 1982.[Ps.]

51633 - **SIEBERT, F., MÄNTELE, W., KREUTZ, W.** : Evidence for the protonation of two
internal carboxylic groups during the photocycle of bacteriorhodopsin. Investigation by kinetic infrared spectroscopy. - FEBS Lett. *141* : 82 - 87,
1982.

51634 - **SIEFERMANN-HARMS, D., NINNEMANN, H.** : Pigment organization in the light-
-harvesting chlorophyll-*a/b* protein complex of lettuce chloroplasts. Evidence obtained from protection by the chlorophylls against proton attack
and from excitation energy transfer. - Photochem. Photobiol. *35* : 719 -
- 731, 1982.

51635 - **SIEFERMANN-HARMS, D., ROSS, J.W., KANESHIRO, K.H., YAMAMOTO, H.Y.** : Reconstitution of monogalactosyldiacylglycerol of energy transfer from light-harvesting chlorophyll *a/b*-protein complex to the photosystems in Triton X-100-so-
lubilized thylakoids. - FEBS Lett. *149* : 191 - 196, 1982.

51636 - **SIEGELMAN, H.W., KYCIA, J.H.** : Molecular morphology of cyanobacterial phycobilisomes. - Plant Physiol. *70* : 887 - 897, 1982.

51637 - **SIGETI, Z., IMASHEVA, É.S., YAGLOVA, L.G.** : Fotoindutsirovannye izmeneniya
raznosti ẻlektricheskikh potentsialov membran rastitel'nykh kletok pri deĭstvii veshchestv tipa benzonitrila. [Light-induced changes of membrane electrical potentials in plant cells as affected by benzonitrile derivatives.] -
Fiziol. Rast. *29* : 345 - 349, 1982. [Ps; in R, ab : E.]

51638 - **SIGRIST, H., ALLEGRINI, P.R., KEMPF, C., SCHNIPPERING, C., ZAHLER, P.** :
5-Isothiocyanato-1-naphthalene azide and *p*-azidophenylisothiocyanate. Synthesis and application in hydrophobic heterobifunctional photoactive cross-
-linking of membrane proteins. - Europe. J. Biochem. *125* : 197 - 201, 1982.
[Bacteriorhodopsin.]

51639 - **SIGRIST, H., ZAHLER, P.** : Heterobifunctional cross-linking of bacteriorho-
dopsin by hydrophobic azidoarylisothiocyanates. - In : COLOWICK, S.P.,
KAPLAN, N.O. (ed.) : Methods in Enzymology. Vol.88. Pp. 207 - 213. Academic
Press, New York - London - Paris - San Diego - San Francisco - São Paulo -
Sydney - Tokyo - Toronto 1982.

51640 - SIKES, C.S., WHEELER, A.P. : Carbonic anhydrase and carbon fixation in cocco-
lithophorids. - J. Phycol. *18* : 423 - 426, 1982.

51641 - SILAEVA, A.M., TKACHUK, E.S. : Ul'trastrukturnaya organizatsiya khloroplastov list'ev pshenitsy pri razlichnoĭ vodoobespechennosti. [Ultrastructural
organization of wheat leaves chloroplasts under different water supply.] -
Fiziol. Biokhim. kul't. Rast. *14* : 148 - 152, 1982. [In R, ab : E.]

51642 - SILCOCK, R.G., WILSON, D. : The effects of leaf orientation on photosynthesis, transpiration and diffusive conductance of leaves of contrasting *Festuca* species. - New Phytol. *90* : 27 - 36, 1982.

51643 - SILVA, P.C. : Taxonomic classification of algae. - In : ZABORSKY, O.R.(ed.):
CRC Handbook of Biosolar Resources. Vol.I/1. Pp. 523 - 564. CRC Press, Boca
Raton 1982. [Chl, Car, Bil.]

51644 - SILVA, P.C. : Sources and collections of algae (including cyanobacteria). -
In : ZABORSKY, O.R. (ed.) : CRC Handbook of Biosolar Resources. Vol.I/1.
P. 593. CRC Press, Boca Raton 1982.

*51645 - SIMINEL, V.D., BABITSKIĬ, A.F. : K prirode geterozisa. [Nature of heterosis.]
- In : Povyshenie Éffektivnosti Geterozisa pri Selektsii Polevykh Kul'tur.
Pp. 5 - 14. Kishinev. Sel'skokhoz. Institut Im. M.V.Frunze, Kishinev 1981.
[Photophosphorylation.]

51646 - ŠIMON, J. : Tvorba biomasy a hlavní růstové charakteristiky ozimé a jarní
pšenice pri různěm vodním režimu, výsevku a hnojení dusíkatým hnojivem.
[Biomass formation and main growth characteristics in winter and spring
wheats at different water regime, sowing date and nitrogen fertilizing.] -
Rostl. Výroba (Praha) *28* : 389 - 397, 1982. [Growth analysis; in Czech,
ab : E, G, R.]

51647 - SIMON, P., DIETER, P., BONZON, M., GREPPIN, H., MARMÉ, D. : Calmodulin-de-
pendent and independent NAD kinase activities from cytoplasmic and chloroplastic fractions of spinach (*Spinacia oleracea* L.). - Plant Cell Rep. *1* :
119 - 122, 1982.

51648 - SIMPSON, D.J. : Effect of preparation temperature on freeze-fracture
ultrastructure of chloroplast thylakoid membranes. - Micron *13* : 323 - 324,
1982.

51649 - SIMSEK, M., DasSARMA, S., RajBHANDARY, U.L., KHORANA, H.G. : A transposable
element from *Halobacterium halobium* which inactivates the bacteriorhodopsin
gene. - Proc. nat. Acad. Sci. USA *79* : 7268 - 7272, 1982.

51650 - SINCLAIR, J., ARNASON, T. : The effects of alpha terthienyl on photosynthesis. - Can. J. Bot. *60* : 2565 - 2569, 1982.

51651 - SINCLAIR, J., COUSINEAU, C. : The deactivation of photosystem II in *Chlorella* as a second-order process. - Biochim. biophys. Acta *680* : 258 - 264,
1982.

51652 - SINCLAIR, T.R., ALLEN, L.H..Jr. : Carbon dioxide and water vapour exchange
of leaves on field-grown citrus trees. - J. exp. Bot. *33* : 1166 - 1175,
1982.

51653 - SINCLAIR, T.R., KNOERR, K.R. : Distribution of photosynthetically active
radiation in the canopy of a loblolly pine plantation. - J. appl. Ecol.
19 : 183 - 191, 1982.

51654 - SINESHCHĔKOV, V.A., GOSTIMSKIĬ, S.A., BELYAEVA, O.B. : Issledovanie kompleksa khlorofill *a*-karotin i migratsii énergii s karotinoidov v mutantakh
gorokha s ponizhennym soderzhaniem pigmentov. [Chlorophyll *a*-carotene complex and energy migration from carotenoids in pea mutants with a reduced
pigment content.] - Biol. Nauki *1982* (7) : 29 - 38, 1982. [In R.]

51655 - SINGH, A.K. : Some recent developments in the photochemistry and chemical
modification of bacteriorhodopsin. - J. sci. ind. Res. *41* : 665 - 673,
1982.

*51656 - SINGH, C., BHAN, A.K., KAUL, B.L. : Variation and correlation studies in
two species of *Datura*. - Herba hung. *17* : 25 - 30, 1978. [Chl, Car.]

51657 - SINGH, D.P., TURNER, N.C., RAWSON, H.M. : Effects of radiation, temperature
and humidity on photosynthesis, transpiration and water use efficiency of
oilseed rape (*Brassica campestris* L.). - Biol. Plant. *24* : 130 - 135, 1982.

51658 - SINGH, J.S., CHATURVEDI, O.P. : Photosynthetic pigments on plant bearing
surfaces in the Himalayas. - Photosynthetica *16* : 101 - 114, 1982.

51659 - SINGHVI, N.R., SHARMA, K.D. : Antagonistic effect of humic acid with di-
kegulac-sodium on growth and chlorophyll biosynthesis. - Sci. Cult. *48* :
366 - 367, 1982.

*51660 - SINHA, S.S.N. : Radiation studies in *Lens culinaris:* Ultrastructure of
chloroplast in the leaves of the normal and mutated plants. - Cytologia *44* :
13 - 19, 1979.

51661 - SIONIT, N., HELLMERS, H., STRAIN, B.R. : Interaction of atmospheric CO_2
enrichment and irradiance on plant growth. - Agron. J. *74* : 721 - 725, 1982.
[Growth analysis.]

B51662 - SIRENKO, L.A., KUREĬSHEVICH, A.V. : Opredelenie Soderzhaniya Khlorofilla
v Planktone Presnykh Vodoemov. [Determination of Chlorophyll Content in
Plankton of Freshwater Reservoirs.] - Naukova Dumka, Kiev 1982. [In R.]

51663 - SIRONVAL, C. : Application of the theory of the relations between pigments
to the description of energy transfers at the early stages of thylakoid
membrane greening. - In : AKOYUNOGLOU, G., EVANGELOPOULOS, A.E., GEORGATSOS,
J., PALAIOLOGOS, G., TRAKATELLIS, A., TSIGANOS, C.P. (ed.) : Cell Function
and Differentiation, Part B. Biogenesis of Energy Transducing Membranes and
Membrane and Protein Energetics. Pp. 53 - 65. Alan R. Liss Inc., New York
1982.

51664 - SKJOLDAL, H.R. : Vertical and small-scale horizontal distribution of chlorophyll a and ATP in subtropical beach sand. - Sarsia *67* : 79 - 83, 1982.

51665 - SKULACHEV, V.P. : The localized $\Delta\bar{\mu}H^+$ problem. The possible role of the local electric field in ATP synthesis. - FEBS Lett. *146* : 1 - 4, 1982.

51666 - SLATER, P.N., JACKSON, R.D. : Atmospheric effects on radiation reflected
from soil and vegetation as measured by orbital sensors using various
scanning directions. - Appl. Optics *21* : 3923 - 3931, 1982.

51667 - SLAVOVA, Ĭ., VAKLINOVA, S., FEDINA, I. : Vliyanie na ploidnostta i khrani-
telnata sreda v"rkhu p"pkuvaneto i intensivnostta na fotosintezata pri
zakharnoto tsveklo *in vitro*. [Effect of ploidy level and nutrient medium
on sugar beet budding and photosynthetic rate *in vitro*.] - Fiziol. Rast.
(Sofia) *8* (4) : 3 - 7, 1982. [In Bulg., ab : E, R.]

51668 - SLAWYK, G., COLLOS, Y. : ^{13}C and ^{15}N uptake by marine phytoplankton: II.
Results from a tropical area (Guinea Dome). - Rapp. Proc.-verb. Réun. Cons.
int. Explor. Mer *180* : 209 - 213, 1982.

51669 - SLOVACEK, R.E. : Intact chloroplast electron flow. Effects of ribose 5-phos-
phate. - Biochim. biophys. Acta *680* : 361 - 365, 1982.

51670 - SLOVACEK, R.E., VAUGHN, S. : Chloroplast sulfhydryl groups and the light
activation of fructose-1,6-bisphosphatase. - Plant Physiol. *70* : 978 - 981,
1982.

51671 - SLOVIN, J.P., TOBIN, E.M. : Synthesis and turnover of the light-harvesting
chlorophyll *a/b*-protein in *Lemna gibba* grown with intermittent red light:
possible translational control. - Planta *154* : 465 - 472, 1982.

51672 - SLUKA, Z.A. : Ob izmenchivosti soderzhaniya khlorofilla i ploshchadi list'-
ev u mkhov v zavisimosti ot ėkologicheskikh usloviĭ. [Relation between
chlorophyll content and leaf areas in mosses in dependence on ecological
factors.] - Vest. mosk. gos. Univ., Biol. *1982* (3) : 24 - 29, 1982. [In R,
ab : E.]

51673 - SMAKMAN, G., HOFSTRA, R. (J.J.) : Energy metabolism of *Plantago lanceolata*, as affected by change in root temperature. - Physiol. Plant. *56* : 33 - 37, 1982. [Ps.]

51674 - SMILLIE, R.M., NOTT, R. : Salt tolerance in crop plants monitored by chlorophyll fluorescence *in vivo*. - Plant Physiol. *70* : 1049 - 1054, 1982.

51675 - SMIRNOV, G.F., CHERNOZUBOV, A.M., VAĬPAN, T.N. : Transpiratsiya pshenitsy i yachmenya v usloviyakh iskusstvennogo klimata. [Transpiration in wheat and barley under conditions of controlled environment.] - Fiziol. Rast. *29* : 744 - 751, 1982. [Resistances; in R, ab : E.]

51676 - SMITH, A.G., RUBERY, P.H. : Investigation of the mechanism of action of a chlorosis-inducing toxin produced by *Pseudomonas phaseolicola*. - Plant Physiol. *70* : 932 - 938, 1982.

51677 - SMITH, A.M., WOOLHOUSE, H.W., JONES, D.A. : Photosynthetic carbon metabolism of the cool-temperate C_4 grass *Spartina anglica* HUBB. - Planta *156* : 441 - 448, 1982.

51678 - SMITH, B.N. : General characteristics of terrestrial plants (agronomic and forests) - C_3, C_4, and crassulacean acid metabolism plants. - In : ZABORSKY, O.R. (ed.) : CRC Handbook of Biosolar Resources. Vol.I/2. Pp. 99 - 103. CRC Press, Boca Raton 1982.

51679 - SMITH, C.M., FORK, D.C., SATOH, K. : The mechanism in red algae of tolerance of photosynthesis to high salt and air drying. - Carnegie Inst. Washington Year Book *81* : 65 - 68, 1982.

51680 - SMITH, D.F. : Observation and quantitative analysis of curvilinear regions of time-varying oxygen concentrations with an oxygen electrode and a minicomputer. - J. exp. mar. Biol. Ecol. *64* : 117 - 124, 1982.

51681 - SMITH, D.F., HORNER, S.M.J. : Laboratory and field measurements of aquatic productivity made by a minicomputer employing a dual oxygen electrode system. - Mar. Biol. *72* : 53 - 60, 1982.

51682 - SMITH, G.D., MUALLEM, A., HALL, D.O. : Hydrogenase-catalyzed photoproduction of hydrogen by photosystem I of the thermophilic blue-green algae *Mastigocladus laminosus* and *Phormidium laminosum*. - Photobiochem. Photobiophys. *4* : 307 - 319, 1982.

51683 - SMITH, J.A.C., MARIGO, G., LÜTTGE, U., BALL, E. : Adenine-nucleotide levels during crassulacean acid metabolism and the energetics of malate accumulation in *Kalanchoë tubiflora*. - Plant Sci. Lett. *26* : 13 - 21, 1982.

51684 - SMITH, K.M., GOFF, D.A., FAJER, J., BARKIGIA, K.M. : Chirality and structures of bacteriochlorophylls *d*. - J. amer. chem. Soc. *104* : 3747 - 3749, 1982.

51685 - SMITH, R.E.H. : The estimation of phytoplankton production and excretion by carbon-14. - Mar. Biol. Lett. *3* : 325 - 334, 1982.

51686 - SMITH, V.H. : The nitrogen and phosphorus dependence of algal biomass in lakes: An empirical and theoretical analysis. - Limnol. Oceanogr. *27* : 1101 - 1112, 1982. [Chl.]

51687 - SMITH, W.O.,Jr. : The relative importance of chlorophyll, dissolved and particulate material, and seawater to the vertical extinction of light. - Estuar. coast. Shelf Sci. *15* : 459 - 465, 1982.

51688 - SNEDAKER, S.C., BROWN, M.S. : Primary productivity in mangroves. - In : ZABORSKY, O.R. (ed.) : CRC Handbook of Biosolar Resources. Vol.I/2. Pp. 477 - 485. CRC Press, Boca Raton 1982.

*51689 - SNELGAR, W.P., BROWN, D.H., GREEN, T.G.A. : A provisional survey of the interaction between net photosynthetic rate, respiratory rate, and thallus water content in some New Zealand cryptogams. - New Zeal. J. Bot. *18* : 247 - 256, 1980.

51690 - SOKOLOV, Z.N., MATORIN, D.N., AKSENOV, S.I., BULYCHEV, A.A., VENEDIKTOV,
P.S. : Vliyanie izotopnogo zameshcheniya H_2O na D_2O na stabil'nost' i funktsional'nuyu aktivnost' fotosinteticheskikh membran zelenykh rasteniĭ. [In-
fluence of isotopic substitution of H_2O to D_2O upon stability and functional
activity of photosynthetic membranes of green plants.] - Vest. mosk. Univ.,
Biol. 1982 (3) : 51 - 57, 1982. [In R, ab : E.]

51691 - SOKOLOV, Z.N., MATORIN, D.N., VENEDIKTOV, P.S., GOL'TSEV, V.N. : Vliyanie
pH i zameshcheniya H_2O na D_2O na vydelenie kisloroda khloroplastami. [Effect
of pH and substitution of H_2O for D_2O on oxygen liberation by chloroplasts.]
- Biofizika 27 : 333 - 334, 1982. [In R, ab : E.]

51692 - SOLDATINI, G.F., ANTONIELLI, M., VENANZI, G., LUPATTELLI, M. : A comparison
of the metabolism of the ear and accompanying tissues in Zea mays L. I.
$^{14}CO_2$ assimilation and photorespiration. - Z. Pflanzenphysiol. 108 : 1 - 8,
1982.

51693 - SOLL, J., ROUGHAN, G. : Acyl-acyl carrier protein pool sizes during steady-
-state fatty acid synthesis by isolated spinach chloroplasts. - FEBS Lett.
146 : 189 - 192, 1982.

*51694 - SOLTANI, A., BRIENS, M., GOAS, M. : Sur le métabolisme glucidique d'Hedysa-
rum coronarium L. cultivé en présence de chlorure de sodium. - Compt. rend.
Acad. Sci. Paris, Sér. III 293 : 297 - 300, 1981. [Photosynthates.]

51695 - SOMERVILLE, C.R., OGREN, W.L. : Genetic modification of photorespiration. -
Trends biochem. Sci. 7 : 171 - 174, 1982.

51696 - SOMERVILLE, C.R., OGREN, W.L. : Isolation of photorespiration mutants in
Arabidopsis thaliana. - In : EDELMAN, M., HALLICK, R.B., CHUA, N.-H. (ed.):
Methods in Chloroplast Molecular Biology. Pp. 129 - 138. Elsevier Biomedical
Press, Amsterdam - New York - Oxford 1982.

51697 - SOMERVILLE, C.R., PORTIS, A.R.,Jr., OGREN, W.L. : A mutant of Arabidopsis
thaliana which lacks activation of RuBP carboxylase in vivo. - Plant Physiol.
70 : 381 - 387, 1982.

51698 - SOMMER, C., WINKLER, S. : Reaktionen im Gaswechsel von Fontinalis antipyre-
tica HEDW. nach experimentellen Belastungen mit Schwermetallverbindungen. -
Arch. Hydrobiol. 93 : 503 - 524, 1982.

51699 - SOMMER, U. : Die Periodizität des Phytoplanktons in Bodensee - ein Beispiel
für Sukzession. - Praxis Naturwiss., Biol. 31 : 323 - 329, 1982. [Producti-
vity.]

*51700 - SONG, P.-S. : Electronic spectroscopy of photobiological receptors. - Can.
J. Spectrosc. 26 : 59 - 72, 1981. [Chl, Car.]

51701 - SOSIŃSKA, A., MALESZEWSKI, S. : Effect of high oxygen concentration on photosynthesis in rape plants pretreated in low temperature. - Z. Pflanzenphysiol. 108 : 397 - 399, 1982.

51702 - SOUZA MACHADO, V., BANDEEN, J.D. : Genetic analysis of chloroplast atrazine resistance in Brassica campestris - cytoplasmic inheritance. - Weed Sci.
30 : 281 - 285, 1982.

51703 - SOUZA MACHADO, V., DITTO, C. : Tomato chloroplast photochemical sensitivity
to metribuzin. - Sci. Hort. 17 : 9 - 13, 1982.

51704 - SOWELL, J.B., KOUTNIK, D.L., LANSING, A.J. : Cuticular transpiration of
whitebark pine (Pinus albicaulis) within a Sierra Nevadan timberline eco-
tone, U.S.A. - Arctic alpine Res. 14 : 95 - 103, 1982. [Resistances.]

51705 - SPALDING, M.H., OGREN, W.L. : Photosynthesis is required for induction
of the CO_2-concentrating system in Chlamydomonas reinhardii. - FEBS Lett.
145 : 41 - 44, 1982.

51706 - SPALINK, J.-D., APPLEBURY, M.L., SPERLING, W., REYNOLDS, A.H., RENTZEPIS,
P.M. : Picosecond studies of bacteriorhodopsin intermediates from 11-cis
rhodopsin and 9-cis rhodopsin. - In : EISENTHAL, K.B., HOCHSTRASSER, R.M.,
KAISER, W., LAUBEREAU, A. (ed.) : Picosecond Phenomena. III. Pp. 307 - 309.
Springer-Verlag, Berlin - Heidelberg - New York 1982.

51707 - ŠPÁNIK, F. : Využívanie fotosynteticky aktívneho žiarenia hlavnými poľno-
         hospodárskymi plodinami vo vytypovaných okresoch Slovenska. [Utilization
         of photosynthetically-active radiation by main agricultural crops in selec-
         ted districts in Slovakia.] - Rost. Výroba (Praha) 28 : 765 - 772, 1982.
         [In Slovak, ab : E, G, R.]

51708 - SPARACE, S.A., MUDD, J.B. : Studies on chloroplast lipid metabolism: Stimu-
         lation of phosphatidylglycerol biosynthesis and analysis of the radioactive
         lipid. - In : WINTERMANS, J.F.G.M., KUIPER, P.J.C. (ed.) : Biochemistry and
         Metabolism of Plant Lipids. Pp. 111 - 119. Elsevier Biomedical Press, Amster-
         dam - New York - Oxford 1982.

51709 - SPARACE, S.A., MUDD, J.B. : Phosphatidylglycerol synthesis in spinach chlo-
         roplasts: characterization of the newly synthesized molecule. - Plant Phy-
         siol. 70 : 1260 - 1264, 1982.

*51710 - SPENCER-PHILLIPS, P.N.T., GAY, J.L. : Electron microscope autoradiography
         of ^{14}C photosynthate distribution at the haustorium-host interface in pow-
         dery mildew of Pisum sativum. - Protoplasma 103 : 131 - 154, 1980.

51711 - SPERANZA, M.L., ZAPPONI, M.C., IADAROLA, P. : Conformation and kinetic pro-
         perties of photosynthetic glyceraldehyde-3-phosphate dehydrogenase "in vi-
         vo". - Ital. J. Biochem. 31 : 22 - 27, 1982.

51712 - SPILLER, S.C., CASTELFRANCO, A.M., CASTELFRANCO, P.A. : Effects of iron
         and oxygen on chlorophyll biosynthesis. I. In vivo observations on iron
         and oxygen-deficient plants. - Plant Physiol. 69 : 107 - 111, 1982.

51713 - SPIRESCU, I., POLESCU, L., ATANASIU, L. : L'influence de la kinétine sur la
         croissance de l'algue Spirulina platensis (GOM) GEITL. - Rev. roum. Biol.-
         Biol.vég. 27 : 117 - 120, 1982. [Ps, Chl.]

51714 - SPITZNER, C., STARK, N. : Productivity of western larch and subsoil perco-
         lation rates on poor andic soils. - Soil Sci. 134 : 395 - 400, 1982.

51715 - SPORTELLI, L., MARTINO, G., CANNISTRARO, S. : Photobiological conversion
         of solar energy. Part I. Bacteriorhodopsin proton pump activity in reconsti-
         tuted liposomes. - Bioelectrochem. Bioenerg. 9 : 197 - 206, 1982. J. electro-
         anal. Chem. 141 : 197 - 206, 1982.

51716 - SPREITZER, R.J., JORDAN, D.B., OGREN, W.L. : Biochemical and genetic analy-
         sis of an RuBP carboxylase/oxygenase-deficient mutant and revertants of
         Chlamydomonas reinhardii. - FEBS Lett. 148 : 117 - 121, 1982.

51717 - SPREITZER, R.J., METS, L. : An assessment of arsenate selection as a method
         for obtaining nonphotosynthetic mutants of Chlamydomonas. - Genetics 100 :
         417 - 425, 1982. [Ps, Chl.]

*51718 - SPREITZER, R.J., METS, L.J. : Non-mendelian mutation affecting ribulose-
         -1,5-bisphosphate carboxylase structure and activity. - Nature 285 : 114 -
         - 115, 1980.

51719 - SPUDICH, E.N., SPUDICH, J.L. : Measurement of light-regulated phosphopro-
         teins of Halobacterium halobium. - In : COLOWICK, S.P., KAPLAN, N.O. (ed.):
         Methods in Enzymology. Vol.88. Pp. 213 - 216. Academic Press, New York -
         London - Paris - San Diego - San Francisco - São Paulo - Sydney - Tokyo -
         Toronto 1982.

*51720 - SQUIRE, G.R., CALLANDER, B.A. : Tea plantations. - In : KOZLOWSKI, T.T.
         (ed.) : Water Deficits and Plants Growth. Vol.VI. Woody Plant Communities.
         Pp. 471 - 510. Academic Press, New York - London - Toronto - Sydney - San
         Francisco 1981. [Ps.]

51721 - SRINIVASA RAO, N.K., SINGH, S.P. : Note on the contribution of different
         photosynthetic sites to grain yield in sorghum. - Indian J. agr. Sci. 52 :
         543 - 544, 1982.

51722 - SRINIVASA RAO, N.K., SINGH, S.P. : Defoliation and compensatory mechanism
         in sorghum. - Indian J. agr. Sci. 52 : 748 - 750, 1982. [Ps.]

51723 - SRIVASTAVA, R.A.K., MATHUR, S.N., DEV CHAUDHURY, M.N. : Physiological aspects of different levels of nitrogen utilization in *Camellia sinensis* L. with respect to yield & quality of made teas. - Indian J. exp. Biol. *20* : 152 - 155, 1982. [Chl.]

51724 - SRIVASTAVA, R.B., PARODA, R.S., LUTHRA, O.P., GOYAL, K.C. : Genetic architecture of yield and components of yield in durum wheat. - Indian J. agr. Sci. *52* : 58 -   , 1982.

51725 - STACHOWIAK, H., LEHMANN, D., HENDRICH, W. : On positron behaviour in thylakoids. - In : COLEMAN, P.G., SHARMA, S.C., DIANA, L.M. (ed.) : Positron Annihilation. Pp. 925 - 927. North-Holland Publ. Co., Amsterdam 1982.

51726 - STADNICHUK, I.N., LUKASHEV, E.P. : Struktura spektrov pogloshcheniya i fluorestsentsii i molekulyarnaya organizatsiya bakteriokhlorofilla v reaktsionnykh tsentrakh *Rhodopseudomonas sphaeroides*. [Absorption and fluorescence spectra and molecular organization of bacteriochlorophyll in reaction centres of *Rhodopseudomonas sphaeroides*.] - Mol. Biol. (Moskva) *16* : 991 - 997, 1982. [In R, ab : E.]

51727 - STADNIČUK, I. : Pigment-Protein-Komplexe in Pro- und Eukaryoten. - In : HOFFMANN, P., HIEKE, B. (ed.) : Photosynthese : Regulation und Evolution. (Colloquia Pflanzenphysiologie Nr.5.) Pp. 144 - 147. Humboldt-Universität, Berlin 1982.

51728 - STAEHELIN, L.A. : Freeze-fracture and freeze-etch electron microscopy of chloroplast membranes. - In : EDELMAN, M., HALLICK, R.B., CHUA, N.-H. (ed.): Methods in Chloroplast Molecular Biology. Pp. 821 - 833. Elsevier Biomedical Press, Amsterdam - New York - Oxford 1982.

51729 - STAMATOFF, J., EISENBERGER, P., BLASIE, J.K., PACHENCE, J.M., TAVORMINA, A., ERECINSKA, M., DUTTON, P.L., BROWN, G. : The location of redox centers in biological membranes determined by resonance X-ray diffraction. I.Observation of the resonance effect. - Biochim. biophys. Acta *679* : 177 - 187, 1982. [Photosynthetic reaction centre-cytochrome *c* membranes.]

51730 - STAMATOFF, J., LOZIER, R.H., GRUNER, S. : X-ray diffraction studies of light interactions with bacteriorhodopsin. - In : COLOWICK, S.P., KAPLAN, N.O. (ed.) : Methods in Enzymology. Vol.88. Pp. 282 - 286. Academic Press, New York - London - Paris - San Diego - San Francisco - São Paulo - Sydney - Tokyo - Toronto 1982.

51731 - STAMP, P. : Development of maize seedlings in dependence of temperature and mineral nutrition. - Z. Acker- Pflanzenbau *151* : 294 - 301, 1982. [Chl.]

51732 - STAMP, P. : Relations of pigment contents and activities of photosynthetic enzymes with the fatty acid composition of membrane lipids in leaves of maize seedlings depending on genotypes and changing temperatures. - Angew. Bot. *56* : 191 - 199, 1982.

51733 - STAMP, P., RAVE, G., DIEPENBROCK, W., HERZOG, H., KRIPPGANS, O., GEISLER,G.: Comparative analysis of dry matter accumulation in caryopses and seeds of aestivum wheats, durum wheat, rye, triticale, rape seed and pea by means of a growth function. - Z. Acker- Pflanzenbau *151* : 224 - 234, 1982.

51734 - STANCHER, B., ZONTA, F. : High-performance liquid chromatographic determination of carotene and vitamin A and its geometric isomers in foods. - J. Chromatogr. *238* : 217 - 225, 1982.

51735 - STANEK, R. : Wpływ zróżnicowanej temperatury nocy na wzrost młodych roślin fasoli (*Phaseolus vulgaris* L.). [Influence of differentiated night temperature on the growth of young bean plants (*Pheseolus vulgaris* L.)] - Acta agrobot. *34* : 219 - 230, 1982. [Ps; in Pol., ab : E.]

51736 - STANEV, V.P., PANDEV, S., KUDREV, T. : Influence of macroelement deficiency on photosynthesis and respiration in sunflower. - Dokl. bolg. Akad. Nauk *35* : 217 - 220, 1982.

51737 - STANEV, V.P., VELICHKOV, D.K., PANDEV, S.D. : Influence of S, Ca and Mg deficiency on the temperature and light dependence of photosynthesis in the pepper plant. - Dokl. bolg. Akad. Nauk *35* : 1271 - 1273, 1982.

51738 - STARZECKI, W., MYDLARZ, J. : Teoretyczne aspekty dokarmiania CO_2 roślin szklarniowych. [Theoretical aspects of CO_2 enrichment on greenhouse vegetable production.] - Zesz. nauk. Akad. roln. Krakow. *171* (Ogrodnictwo 9) : 5 - 23, 1982. [Ps; in Pol., ab : E, R.]

*51739 - STAUB, E. : Two chromatographic separation methods for chlorophyll analyses: TLC and its transformation to modern HPLC. - Arch. Hydrobiol. Beih. (Ergebn. Limnol.) *14* : 79 - 80, 1980.

*51740 - STEELE, J.H., HENDERSON, E.W. : Spatial patterns in North Sea plankton. - Deep-Sea Res. *26 A* : 955 - 963, 1979. [Chl.]

*51741 - STEEMANN NIELSEN, E. : The carbon-14 technique for measuring organic production by plankton algae. A report on the present knowledge. - Folia limnol. scand. *17* (Danish Limnology. Reviews and Perspectives) : 45 - 48, 1977.

51742 - STEENBERGEN, C.L.M. : Contribution of photosynthetic sulphur bacteria to primary production in Lake Vechten. - Hydrobiologia *95* : 59 - 64, 1982.

51743 - STEENBERGEN, C.L.M., KORTHALS, H.J. : Distribution of phototrophic microorganisms in the anaerobic and microaerophilic strata of Lake Vechten (The Netherlands). Pigment analysis and role in primary production. - Limnol. Oceanogr. *27* : 883 - 895, 1982.

51744 - STEENKAMP, J., DE VILLIERS, O.T., TERBLANCHE, J.H. : Translocation of sorbitol and other photosynthates in Golden Delicious apple shoots. - S. Afr. J. Sci. *78* : 335 - 336, 1982.

51745 - STEER, B.T. : The effect of growth temperature on dry weight and carbohydrate content of onion bulbs (*Allium cepa* L. cv. Creamgold) bulbs. - Aust. J. agr. Res. *33* : 559 - 563, 1982.

51746 - STEINBACK, K.E., MULLET, J.E., ARNTZEN, C.J. : Fractionation of thylakoid membrane protein complexes by sucrose density-gradient centrifugation. - In : EDELMAN, M., HALLICK, R.B., CHUA, N.-H. (ed.) : Methods in Chloroplast Molecular Biology. Pp. 863 - 872. Elsevier Biomedical Press, Amsterdam - New York - Oxford 1982.

51747 - STEINBACK, K.E., PFISTER, K., ARNTZEN, C.J. : Identification of the receptor site for triazine herbicides in chloroplast thylakoid membranes. - In : MORELAND, D.E., ST.JOHN, J.B., HESS, F.D. (ed.) : Biochemical Responses Induced by Herbicides. Pp. 37 - 55. Amer. chem. Soc., Washington 1982.

*51748 - STEINER, R., SCHÄFER, W., BLOS, I., WIESCHHOFF, H., SCHEER, H. : Δ2,10-phytadienol as esterifying alcohol of bacteriochlorophyll *b* from *Ectothiorhodospira halochloris*. - Z. Naturforsch. *36 C* : 417 - 420, 1981.

51749 - STEINER, R., WIESCHHOFF, H., SCHEER, H. : High-performance liquid chromatography of bacteriochlorophyll *b* and its derivatives as an aid for structure analysis. - J. Chromatogr. *242* : 127 - 134, 1982.

51750 - STEITZ, T.A., GOLDMAN, A., ENGELMAN, D.M. : Quantitative application of the helical hairpin hypothesis to membrane proteins. - Biophys. J. *37* : 124 - - 125, 1982. [Bacteriorhodopsin.]

51751 - STEMLER, A. : The functional role of bicarbonate in photosynthetic light reaction II. - In : GOVINDJEE (ed.) : Photosynthesis. Vol.II. Pp. 513 - 539. Academic Press, New York - London - Paris - San Diego - San Francisco - São Paulo - Sydney - Tokyo - Toronto 1982.

*51752 - STEPANOVA, A.M., NIKIFOROVA, L.F. : Élektroforeticheskiĭ analiz Ca^{2+}-ATFazy iz sopryagayushchego faktora khloroplastov gorokha i kukuruzy. [Electrophoretic analysis of Ca^{2+}-ATPase from the conjugating factor of pea and maize chloroplasts.] - In : Fotosintez, Dykhanie i Organicheskie Kisloty. Pp. 10 - 14. Izd. Voronezh. Universiteta, Voronezh 1980. [In R.]

51753 - STEPHEN, F.M., WALLIS, G.W., COLVIN, R.J., YOUNG, J.F., WARREN, L.O. : Pine tree growth and yield : Influence of species, plant spacing, vegetation, and pine tip moth control. - Arkansas Farm Res. *31* (2) : 10, 1982.

51754 - STERN, R.D., DENNETT, M.D., DALE, I.C. : Analysing daily rainfall measurements to give agronomically useful results. II. A modelling approach. -
Exp. Agr. *18* : 237 - 253, 1982. [Dry-matter production.]

51755 - STEWART, A.C. : The effect of cations on DCMU-insensitive electron transport in trypsin-treated spinach chloroplasts. - FEBS Lett. *139*: 279 - 282,
1982.

51756 - STEWART, A.C. : The effects of high concentrations of salts on photosynthetic electron transport in spinach (*Spinacia oleracea*) chloroplasts. - Biochem. J. *204* : 705 - 712, 1982.

51757 - STEWART, A.J., WETZEL, R.G. : Influence of dissolved humic materials on
carbon assimilation and alkaline phosphatase activity in natural algal-
-bacterial assemblages. - Freshwater Biol. *12* : 369 - 380, 1982.

51758 - STEWART, A.J., WETZEL, R.G. : Phytoplankton contribution to alkaline phosphatase activity. - Arch. Hydrobiol. *93* : 265 - 271, 1982. [Ps.]

51759 - STEWART, W.D.P., HAWKESFORD, M., ROWELL, P., CODD, G.A. : H_2 production
from sunlight and H_2O - the role of the uptake hydrogenase in N_2-fixing
cyanobacteria. - In : HALL, D.O., PALZ, W.(ed.):Photochemical,Photoelectro-
chemical and Photobiological Processes. Vol.1. Pp. 165 - 169. D. Reidel
Publ. Co., Dordrecht - Boston - London 1982.

51760 - STIDHAM, M.A., HEATH, R.L. : pH effects on light dependence of the stoichiometry of proton influx and efflux to electron transport in spinach chloroplasts. - Photosynthesis Res. *3* : 335 - 346, 1982.

51761 - STIDHAM, M.A., URIBE, E.G., WILLIAMS, G.J.III : Temperature dependence of
photosynthesis in *Agropyron smithii* RYDB. II. Contribution from electron
transport and photophosphorylation. - Plant Physiol. *69* : 929 - 934, 1982.

51762 - STIEGLER, G.L., MATTHEWS, H.M., BINGHAM, S.E., HALLICK, R.B. : The gene
for the large subunit of ribulose-1,5-bisphosphate carboxylase in *Euglena
gracilis* chloroplast DNA: location, polarity, cloning, and evidence for an
intervening sequence. - Nucl. Acids Res. *10* : 3427 - 3444, 1982.

51763 - STIGTER, C.J., JIWAJI, N.T., MAKONDA, M.M. : A calibration plate to deter-
mine the performance of infrared thermometers in field use. - Agr. Meteorol.
26 : 279 - 283, 1982. [Leaf temperature.]

51764 - STIGTER, C.J., MAKONDA, M.M., JIWAJI, N.T. : Improved field use of a simple
infrared thermometer. - Acta bot. neerl. *31* : 379 - 389, 1982.

51765 - STIGTER, C.J., MUSABILHA, V.M.M. : The conservative ratio of photosynthetically active to total radiation in the tropics. - J. appl. Ecol. *19* :
853 - 858, 1982.

51766 - STILES, J.I. : Restriction endonuclease cleavage map of the maize chloroplast genome. - In : SHERIDAN, W.F. (ed.) : Maize for Biological Research.
Pp. 275 - 276. Plant Mol. Biol. Association, Charlottesville 1982.

*51767 - STILLWELL, W., TIEN, H.T. : The requirement of a sealed membrane in oxygen
evolution. - Photobiochem. Photobiophys. *2* : 159 - 165, 1981.

51768 - STITT, M., LILLEY, R.McC., HELDT, H.W. : Adenine nucleotide levels in the
cytosol, chloroplasts, and mitochondria of wheat leaf protoplasts. - Plant
Physiol. *70* : 971 - 977, 1982.

51769 - STITT, M., MIESKES, G., SOLING, H.-D., HELDT, H.W. : On a possible role of
fructose 2,6-bisphosphate in regulating photosynthetic metabolism in leaves.
- FEBS Lett. *145* : 217 - 222, 1982.

51770 - ST.JOHN, J.B. : Effect of herbicides on the lipid composition of plant membranes. - In : MORELAND, D.E., ST.JOHN, J.B., HESS, F.D. (ed.) : Biochemical
Responses Induced by Herbicides. Pp. 97 - 109. Amer. Chem. Society, Washington 1982. [Ps, Chl, chloroplast.]

51771 - STOCKING, C.R., FRANCESCHI, V.R. : Some properties of the chloroplast envelope as revealed by electrophoretic mobility studies of intect chloroplasts.
- Plant Physiol. *70* : 1255 - 1259, 1982.

51772 - STODDART, J.L., THOMAS, H. : Leaf senescence. - In : BOULTER, D., PARTHIER,
        B. (ed.) : Nucleic Acids and Proteins in Plants I. Structure, Biochemistry
        and Physiology of Proteins. Pp. 592 - 636. Springer-Verlag, Berlin - Heidel-
        berg - New York 1982. [Ps.]

51773 - STOECKENIUS, W., BOGOMOLNI, R.A. : Bacteriorhodopsin and related pigments
        of halobacteria. - Annu. Rev. Biochem. *51* : 587 - 616, 1982.

51774 - STOEV, K., SLAVTCHEVA, T. : La photosynthèse nette chez la Vigne (*V. vini-
        fera* L.) et les facteurs écologiques. - Connais. Vigne Vin *16* : 171 - 185,
        1982.

51775 - STOYANOVA, Ĭ., PETROV, P.D. : Prouchvaniya v"rkhu produktivnostta i razmno-
        zhavaneto na roditelskite linii pri khibridniya sl"nchogled. [Studies on the
        productivity and the reproduction of sunflower hybrid parental lines.] -
        Rasteniev. Nauki *19* (3) : 46 - 52, 1982. [In Bulg., ab : E, R.]

51776 - STOYLOV, S.P., TODOROV, G., ZHIVKOV, A. : Dynamics of the electric charge
        of purple membranes as model membranes. - Stud. biophys. *90* : 59 - 60,
        1982.

51777 - STRAIGHT, R.C., SPIKES, J.D. : Effects of hydrostatic pressure on Hill re-
        action of isolated chloroplast preparations. - Physiol. Chem. Phys. *14* :
        109 - 110, 1982.

51778 - STREIT, L., FELLER, U. : Inactivation of N-assimilating enzymes and pro-
        teolytic activities in wheat leaf extracts: Effect of pyridine nucleotides
        and of adenylates. - Experientia *38* : 1176 - 1180, 1982. [Light-dependent
        nitrogen fixation.]

51779 - STREIT, L., FELLER, U. : Changing activities of nitrogen-assimilating enzy-
        mes during growth and senescence of dwarf beans (*Phaseolus vulgaris* L.). -
        Z. Pflanzenphysiol. *108* : 273 - 281, 1982. [Chl, photorespiration.]

51780 - STRIEGL, M. : Faktory ovlivňující tvorbu biomasy u jarních obilnin. [The
        factors affecting biomass formation in spring cereals.] - Rost. Výroba
        (Praha) *28* : 363 - 370, 1982. [In Czech, ab : E, G, R.]

51781 - STROTMANN, H., BRENDEL, K., BOOS, K.S., SCHLIMME, E. : Energy transfer
        inhibition in photosynthesis by anthraquinone dyes. - FEBS Lett. *145* : 11 -
        - 15, 1982.

51782 - STRUIK, P.C., DEINUM, B. : Effect of light intensity after flowering on the
        productivity and quality of silage maize. - Neth. J. agr. Sci. *30* : 297 -
        - 316, 1982. [Dry-matter production.]

51783 - STUMPF, D.K., JENSEN, R.G. : Photosynthetic CO_2 fixation at air levels of
        CO_2 by isolated spinach chloroplasts. - Plant Physiol. *69* : 1263 - 1267,
        1982.

51784 - STUPISHINA, Ye., CHEREZOV, S., GUSEV, N. : The properties of water and hy-
        drogen deuterium exchange under structural reconstitutions of chloroplast
        proteins. - Stud. biophys. *91* : 69 - 70, 1982.

51785 - STÜRZL, E., SCHERER, S., BÖGER, P. : Reconstitution of electron transport
        by cytochrome c-553 in a cell-free system of *Nostoc muscorum*. - Photosynthe-
        sis Res. *3* : 191 - 201, 1982.

51786 - SUBA, J., KISZELY-VÁMOSI, A., LÉGRÁDY, Gy., ORBÁN, S. : Examination of the
        photosynthetic fixation $^{14}CO_2$ on bryophyte and lichen species. - Acta
        bot. Acad. Sci. hung. *28* : 181 - 191, 1982.

51787 - SUCOFF, E. : Water relations of the aspens. - Agr. Exp. Sta. Univ. Minnesota
        tech. Bull. *338* : 1 - 27, 1982. [Resistances.]

B51788 - SUD'INA, E.G., LOZOVAYA, G.I. : Osnovy Évolyutsionnoĭ Biokhimii Rasteniĭ.
        [Fundamentals of Evolution Biochemistry of Plants.] - Naukova Dumka, Kiev
        1982. [Chl, Car; in R, ab : E.]

51789 - SUD'ĬNA, O.G., LOS', S.I. : Evolyutsiĭnyĭ pidkhid do vyvchennya biokhimiĭ
        syn'ozelenykh vodorosteĭ. [Evolutionary approach to the study of bioche-
        mistry of blue-green algae.] - Ukr. bot. Zh. *39* (4) : 65 - 71, 112, 1982.
        [In Ukr., ab : E, R.]

51790 - SUE, J.M., KNOWLES, J.R. : Ribulose-1,5-bisphosphate carboxylase: Fate of
the tritium label in [3-^{3}H]ribulose 1,5-bisphosphate during the enzyme catalyzed reaction. - Biochemistry 21 : 5404 - 5410, 1982.

51791 - SUE, J.M., KNOWLES, J.R. : Ribulose-1,5-bisphosphate carboxylase: Primary
deuterium kinetic isotope effect using [3-^{2}H]ribulose 1,5-bisphosphate. -
Biochemistry 21 : 5410 - 5414, 1982.

51792 - SUGIHARA, T., BLOUT, E.R., WALLACE, B.A. : Hydrophobic oligopeptides in solution and in phospholipid vesicles: synthetic fragments of bacteriorhodopsin. - Biochemistry 21 : 3444 - 3452, 1982.

51793 - SUGIMOTO, T., MIYAZAKI, J., KOKUBO, T., TANIMOTO, S., OKANO, M., MATSUMOTO,
M. : Light-driven electron transport through an asymmetric photosynthetic
liquid membrane. - J. chem. Soc.,chem. Commun. 1982 : 186 - 188, 1982.

51794 - SUGIYAMA, T. : PEP carboxylase from plants and photosynthetic bacteria. -
In : ZABORSKY, O.R. (ed.) : CRC Handbook of Biosolar Resources. Vol.I/1.
Pp. 207 - 210. CRC Press, Boca Raton 1982.

51795 - SUGIYAMA, T. : Photorespiration and various types of CO_2 fixation. b. C_4-
photosynthesis. - Recent Progr. nat. Sci. Jap. 7 : 56 - 58, 1982.

51796 - SUÏSALU, A.P. : Primenenie metoda FDMR dlya izucheniya khlorofillopodobnykh
molekul. [Application of FDMR technique for studying chlorophyll-like molecules.] - Eesti NSV Tead. Akad. Toim., Füüs., Mat. 31 : 150 - 154, 1982.
[In R, ab : E, Est.]

51797 - SUMPER, M. : The brown membrane of Halobacterium halobium: The biosynthetic
precursor of the purple membrane. - In : COLOWICK, S.P., KAPLAN, N.O. (ed.):
Methods in Enzymology. Vol.88. Pp. 391 - 395. Academic Press, New York -
London - Paris - San Diego - San Francisco - São Paulo - Sydney - Tokyo -
Toronto 1982.

51798 - SUN Bingrong, WANG Weiguang, ZHAO Xianduan : [Transient fluorescence, transient electrochromic change and photophosphorylation.] - Shengwu Huaxue Yu
Shengwu Wuli Jinzhan 45 : 32 - 34, 1982. [In Chin.]

51799 - SUNDBOM, E., ÖQUIST, G. : Temperature-induced changes of variable fluorescen-
ce-yield in intact leaves. - Plant Cell Physiol. 23 : 1161 - 1167, 1982.

51800 - SUNDBOM, E., STRAND, M., HÄLLGREN, J.E. : Temperature-induced fluorescence
changes. A screening method for frost tolerance of potato (Solanum sp.). -
Plant Physiol. 70 : 1299 - 1302, 1982.

51801 - SÜSS, K.-H. : Purification of stromal and membrane-bound ferredoxin-NADP$^+$-
-reductase. - In : EDELMAN, M., HALLICK, R.B., CHUA, N.-H. (ed.) : Methods
in Chloroplast Molecular Biology. Pp. 957 - 971. Elsevier Biomedical Press,
Amsterdam - New York - Oxford 1982.

51802 - SÜSS, K.H. : Topology and association of the subunits of chloroplast and
etioplast ATPase complex (CF_1-CF_0). A lactoperoxidase-catalyzed iodination
and limited proteolysis study. - Biochem. Physiol. Pflanzen 177 : 143 - 155,
1982.

51803 - SÜSS, K.-H., SCHMIDT, O. : Evidence for an α_3, β_3, γ, δ, I, II, ϵ, III$_5$
subunit stoichiometry of chloroplast ATP synthetase complex (CF_1-CF_0). -
FEBS Lett. 144 : 213 - 218, 1982.

51804 - SUTTON, M.R., ROSEN, D., FEHER, G., STEINER, L.A. : Amino-terminal sequence of the L, M, and H subunits of reaction centers from the photosynthetic
bacterium Rhodopseudomonas sphaeroides R-26. - Biochemistry 21 : 3842 -
- 3849, 1982.

51805 - SUWANKETNIKOM, R., HATZIOS, K.K., PENNER, D., BELL, D. : The site of electron transport inhibition by bentazon (3-isopropyl-1N-2,1-3-benzothiadiazin-
-(4)3N-one 2,2-dioxide) in isolated chloroplasts. - Can. J. Bot. 60 : 409 -
- 412, 1982.

51806 - SUZUKI, S., MURABAYASHI, M., MATSUNO, T. : [Effects of ozone on photosynthetic electron transport in spinach (III).] - Bull. Inst. environm. Sci. Technol. Yokohama nat. Univ. *8* : 81 - 87, 1982. [In Jap., ab : E.]

51807 - SWAMY, G.S., PILLAY, D.T.N. : Characterization of *Glycine max* cytoplasmic,
chloroplastic and mitochondrial tRNAs and synthetases for phenylalanine,
tryptophan and tyrosine. - Plant Sci. Lett. *25* : 73 - 84, 1982.

51808 - SWANK, J.C., BELOW, F.E., LAMBERT, R.J., HAGEMAN, R.H. : Interaction of
carbon and nitrogen metabolism in the productivity of maize. - Plant Physiol.
70 : 1185 - 1190, 1982. [Ps.]

51809 - SWARTHOFF, T., GAST, P., AMESZ, J., BUISMAN, H.P. : Photoaccumulation of
reduced primary electron acceptors of photosystem I of photosynthesis. -
FEBS Lett. *146* : 129 - 132, 1982.

51810 - SWARTHOFF, T., KRAMER, H.J.M., AMESZ, J. : Thin-layer chromatography of
pigments of the green photosynthetic bacterium *Prosthecochloris aestuarii*. -
Biochim. biophys. Acta *681* : 354 - 358, 1982.

51811 - SWEENEY, B.M., BERRY, J.A., FORK, D.C. : Chlorophyll fluorescence in *Gony-*
aulax polyedra as a function of temperature. - Carnegie Inst. Washington
Year Book *81* : 70 - 72, 1982.

51812 - SWEENEY, B.M., SATOH, K., FORK, D.C. : A comparison of the fluorescence and
delayed light emission in *Gonyaulax polyedra* in day and night phases of the
circadian rhythm in constant light. - Carnegie Inst. Washington Year Book
81 : 68 - 70, 1982.

51813 - SWENBERG, C.E. : Fluorescence decay kinetics and bimolecular processes in
photosynthetic membranes. - In : ALFANO, R.R. (ed.) : Biological Events
Probed by Ultrafast Laser Spectroscopy. Pp. 193 - 214. Academic Press, New
York - London - Paris - San Diego - San Francisco - São Paulo - Sydney -
Tokyo - Toronto 1982.

51814 - SWIECKI, T.J., ENDRESS, A.G., TAYLOR, O.C. : Histological effects of aqueous acids and gaseous hydrogen chloride on bean leaves. - Amer. J. Bot.
69 : 141 - 149, 1982. [Chloroplast.]

51815 - SWIETLIK, D., FAUST, M., KORCAK, R.F. : Effect of mineral nutrient sprays
on photosynthesis and stomatal opening of water stressed and unstressed
apple seedlings I. Complete nutrient sprays. - J. amer. Soc. hort. Sci.
107 : 563 - 567, 1982.

51816 - SWIETLIK, D., KORCAK, R.F., FAUST, M. : Effect of mineral nutrient sprays
on photosynthesis and stomatal opening of water-stressed and unstressed
apple seedlings II. Potassium sulfate sprays. - J. amer. Soc. hort. Sci.
107 : 568 - 572, 1982.

51817 - SWIETLIK, D., KORCAK, R.F., FAUST, M. : Physiological and nutritive effects
of K-pretreatment and KCl sprays on water-stressed and unstressed apple
seedlings. - J. amer. Soc. hort. Sci. *107* : 669 - 673, 1982. [Stomatal resistance.]

51818 - SWIFT, I.E., MILBORROW, B.V., JEFFREY, S.W. : Formation of neoxanthin, di-
adinoxanthin and peridinin from [^{14}C]-zeaxanthin by a cell-free system from
Ampidinium carterae. - Phytochemistry *21* : 2859 - 2864, 1982.

51819 - SYARGEĬCHYK, A.A. : Stan anatama-marfalagichnaĭ struktury listsyaŭ agurka
i kukuruzy ŭ suvyazi z nedakhopam boru. [Effect of boron deficiency on the
anatomical and morphological structure of cucumber and maize leaves.] -
Vestsi Akad. Navuk belarus. SSR, Ser. biyal. Navuk *1982* (2) : 30 - 34, 123,
1982. [Chloroplast; in Beloruss., ab : E, R.]

51820 - **SYARCHEÏCHYK, S.A.** : Stan plastydnykh pigmentaŭ listsyaŭ drěvavykh raslin u sfery pramyslovaga zabrudzhvannya pavetra CS_2, H_2S, SO_2. [State of plastid pigments of arboreal plants in the sphere of industrial pollution by CS_2, H_2S, SO_2.] - Vestsi Akad. Navuk belarus. SSR, Ser. biyal. Navuk *1982* (3) : 17 - 21, 121 - 122, 1982. [In Belorus., ab : E, R.]

51821 - **SÝKORA, M., DUBOVSKÝ, J.** : Frekvencia chlorofylových mutácií jarného jačmeňa po opakovanom ovplyvnení chemickými mutagénmi v M_6/M_2 generácii. [Frequency of chlorophyll mutations of spring barley after repeated treatment with chemical mutagens in M_6/M_2 generations.] - Pol'nohospodárstvo *28* : 111 - - 118, 1982. [In Slovak, ab : E, R.]

51822 - **SYVERTSEN, J.P.** : Minimum leaf water potential and stomatal closure in citrus leaves of different ages. - Ann. Bot. *49* : 827 - 834, 1982. [Resistances.]

51823 - **SYVERTSEN, J.P., LEVY, Y.** : Diurnal changes in citrus leaf thickness, leaf water potential and leaf to air temperature difference. - J. exp. Bot. *33* : 783 - 789, 1982. [Growth.]

51824 - **SZALAY, L., LEHOCZKI, E., HERCZEG, T., LASKAY, G.** : Photosynthetic properties of chemically modified membranes of algae and higher plants. - Zagad. Biofiz. współcz. *7* : 59 - 71, 1982.

51825 - **SZALAY, L., TOMBÁCZ, E., VÁRKONYI, Z., FALUDI-DÁNIEL, Á.** : Detergent effects on an albumin-chlorophyll complex model of photosynthetic protein-pigment complexes. - Acta phys. Acad. Sci. hung. *53* : 225 - 235, 1982.

51826 - **SZANIAWSKI, R.K., KIEŁKIEWICZ, M.** : Maintenance and growth respiration in shoots and roots of sunflower plants grown at different root temperatures. - Physiol. Plant. *54* : 500 - 504, 1982.

51827 - **SZAREK, S.R., SMITH, S.D., RYAN, R.D.** : Moisture stress effects on biomass partitioning in two Sonoran desert annuals. - Amer. Midl. Natur. *108* : 338 - - 345, 1982. [Growth analysis.]

51828 - **SZEMES, I., KÁDÁR, I., LÁSZTITY, B.** : Az őszi rozs ásványitápanyag-felvételének vizsgálata szabadföldi tartamkísérletben I. Szárazanyag-felhalmozódás, N-, P-, K-, Ca-, Mg-felvétel. [Investigations on the nutrient uptake of winter rye in a long-term field experiment I. Accumulation of dry matter and macronutrient (N, P, K, Ca, Mg) uptake.] - Agrokémia Talajtan *31* : 5 - 16, 1982. [In Hung., ab : E, G, R.]

51829 - **SZIGETI, Z., KAPLANOVÁ, M.** : Studies on chlorophyll photooxidation enhanced by benzonitriles *in vivo* and *in vitro*. - Photobiochem. Photobiophys. *4* : 299 - 305, 1982.

51830 - **SZIGETI, Z., SÁRVÁRI, É.** : Benzonitrilekkel kezelt kloroplasztiszok klorofill-protein komplexeinek vizsgálata. [Chlorophyll-protein complexes of chloroplasts treated with benzonitriles.] - Bot. Közl. *69* : 191 - 196, 1982. [In Hung., ab : E.]

51831 - **SZIGETI, Z., TÓTH, E., PALESS, G.** : Mode of action of photosynthesis inhibiting 4-hydroxybenzonitriles containing nitro group. - Photosynthesis Res. *3* : 347 - 356, 1982.

51832 - **SZUJKÓ-LACZA, J.** : Developmental morphology of *Armeniaca vulgaris* LAM. *(Rosaceae)*. - Acta bot. Acad. Sci. hung. *28* : 199 - 239, 1982.

51833 - **SZWARCBAUM (SHAVIV), I.** : Influence of leaf morphology and optical properties on leaf temperature and survival in three mediterranean shrubs. - Plant Sci. Lett. *26* : 47 - 56, 1982.

51834 - **TABBADA, R.A., FLORES, M.A.A.** : Influence of soil water stress on vegetative and reproductive growth of *Phaseolus vulgaris* cv. "White Baguio". - Kalikasan (Philipp. J. Biol.) *11* : 266 - 272, 1982. [Chl.]

51835 - TADROS, M.H., ZUBER, H., DREWS, G. : The polypeptide components from light-
       -harvesting pigment protein complex II (B800-850) of *Rhodopseudomonas sphae-
       roides*. Solubilization, purification and sequence studies. - Europe.J. Bio-
       chem. *127* : 315 - 318, 1982.

51836 - TAKABE, T., DEBENEDETTI, E., JAGENDORF, A.T. : Inhibition of chloroplast
       coupling factor by naphthylglyoxal. - Biochim. biophys. Acta *682* : 11 - 20,
       1982.

51837 - TAKAGI, N., INOUE, J. : [Seasonal changes in berry growth and photosynthetic
       rate of leaves of 'Muscat of Alexandria' grapes.] - J. Jap. Soc. hort. Sci.
       51 : 286 - 292, 1982. [In Jap., ab : E.]

51838 - TAKAGI, S., TAKEDA, K., KAMEYAMA, K., TAKAGI, T. : Visible circular dichro-
       ism of lutein acquired on dispersion in an aqueous solution in the presen-
       ce of a limited amount of sodium dodecyl sulfate and a dramatic change of
       the CD spectrum with concentration of the surfactant. - Agr. biol.Chem. *46* :
       2035 - 2040, 1982.

51839 - TAKAGI, S., TAKEDA, K., SHIROISHI, M. : Aggregation, configuration and par-
       ticle size of lutein dispersed by sodium dodecyl sulfate in various salt
       concentrations. - Agr. biol. Chem. *46* : 2217 - 2222, 1982.

51840 - TAKAGI, S., TAKEDA, K., TAKAGI, T. : Effect of addition of lipids on the
       novel optical activity of the complex between lutein and ovalbumin. - Agr.
       biol.Chem. *46* : 399 - 404, 1982.

51841 - TAKAHAMA, U. : Suppression of carotenoid photobleaching by kaempferol in
       isolated chloroplasts. - Plant Cell Physiol. *23* : 859 - 864, 1982.

51842 - TAKAHASHI, M., ASADA, K. : Dependence of oxygen affinity for Mehler react-
       ion on photochemical activity of chloroplast thylakoids. - Plant Cell Phy-
       siol. *23* : 1457 - 1461, 1982.

51843 - TAKAHASHI, M., FUKUZAWA, N. : A mechanism of "red-tide" formation. II. Ef-
       fect of selective nutrient stimulation on the growth of different phyto-
       plankton species in natural water. - Mar. Biol. *70* : 267 - 273, 1982.
       [Chl.]

51844 - TAKAHASHI, M., ICHIMURA, S. : Phytoplankton productivity in the Pacific Oce-
       an. - In : ZABORSKY, O.R. (ed.) : CRC Handbook of Biosolar Resources. Vol.
       I/2. Pp. 401 - 406. CRC Press, Boca Raton 1982.

51845 - TAKAHASHI, M., KOIKE, I., ISEKI, K., BIENFANG, P.K., HATTORI, A. : Phyto-
       plankton species' responses to nutrient changes in experimental enclosures
       and coastal waters. - In : GRICE, G.D., REEVE, M.R. (ed.) : Marine Mesocosms:
       Biological and Chemical Research in Experimental Ecosystems. Pp. 333 - 340.
       Springer-Verlag, New York - Heidelberg - Berlin 1982. [Ps.]

51846 - TAKAMI, S., KUMASHIRO, T. : Response of crop photosynthesis of rice to cli-
       matic conditions as affected by the canopy architecture. - Jap.agr.Res.quart.
       15 : 227 - 230, 1982.

51847 - TAKAMI, S., RAWSON, H.M., TURNER, N.C. : Leaf expansion of four sunflower
       (*Helianthus annuus* L.) cultivars in relation to water deficits. II. Diurnal
       patterns during stress and recovery. - Plant Cell Environm. *5* : 279 - 286,
       1982.

51848 - TAKAMIYA, K., DOI, M., OKIMATSU, H. : Isolation and purification of a ubi-
       quinone-cytochrome b-c_1 complex from a photosynthetic bacterium, *Rhodopseu-
       domonas sphaeroides*. - Plant Cell Physiol. *23* : 987 - 997, 1982.

51849 - TAKANO, M., TAKAHASHI, M., ASADA, K. : Reduction of photosystem I reaction
       center, P-700, by plastocyanin in stroma thylakoids from spinach: Lateral
       diffusion of plastocyanin. - Arch. Biochem. Biophys. *218* : 369 - 375,
       1982.

51850 - TAKEDA, H., HIROKAWA, T. : Studies on the cell wall of *Chlorella* III. In-
       corporation of photosynthetically fixed carbon into cell walls of synchro-
       nously growing cells of *Chlorella ellipsoidea*. - Plant Cell Physiol. *23*:
       1033 - 1040, 1982.

51851 - TAKEMOTO, B.K., NOBLE, R.D. : The effects of short-term SO_2 fumigation on photosynthesis and respiration in soybean *Glycine max*. - Environm. Pollut., Ser.A *28* : 67 - 74, 1982.

51852 - TAKEMOTO, J.Y., PETERS, J., DREWS, G. : Crosslinking of photosynthetic membrane polypeptides of *Rhodopseudomonas capsulata*. - FEBS Lett. *142* : 227 - - 230, 1982.

*51853 - TAKIO, S., TAKAOKI, T. : Light-induced quenching of atebrin fluorescence in relation to H^+ uptake and light-scattering change in isolated spinach chloroplasts. - Photobiochem. Photobiophys. *2* : 217 - 225, 1981.

51854 - TAKITA, T., KAWAKAMI, J. : Varietal difference of leaf photosynthetic ability in rice. - Annu. Rep. Div. Genet. nat. Inst. agr. Sci. Jap. *1982* : 34 - - 35, 1982.

51855 - TAKRURI, I.A.H., GILROY, J., BOULTER, D. : Amino acid sequence of ferredoxin from *Arctium lappa*. - Phytochemistry *21* : 325 - 327, 1982.

51856 - TALANOV, A.V., BEZDENEZHNYKH, V.A., KHILAKOV, N.I. : Ustanovka dlya issledovaniya gazoobmena intaktnykh rasteniĬ. [Device for studying gas exchange of intact plants.] - In : Vliyanie Faktorov VneshneĬ Sredy i Fiziologicheski Aktivnykh Veshchestv na Termorezistentnost' i Produktivnost' RasteniĬ. Pp. 142 - 150, 159. Karel'. Filial Akademii Nauk SSSR, Institut Biologii, Petrozavodsk 1982. [In R.]

*51857 - TALARICO, L., KOSOVEL, V. : Qualitative interpretation of the absorption spectra of total biliproteic extracts from some *Rhodophyta (Florideophyceae)*. - Boll. Soc. adr. Sci. *64* : 85 - 97, 1980.

51858 - TALARICO, L., KOSOVEL, V. : Osservazioni preliminari sulla risposta dei pigmenti fotosintetici di *Gracilaria verrucosa* (HUDS.) PAPENFUSS all'inquinamento di liquami urbani ed industriali (*Gigartinales, Florideophyceae*). [Preliminary observations on the response of photosynthetic pigments from *Gracilaria verrucosa* (HUDS.) PAPENFUSS (*Gigartinales, Florideophyceae*) to municipal and industrial waste waters.] - Naturalista sicil., S.IV *6*(Suppl.) *1* : 71 - 80, 1982. [In Ital., ab : E.]

51859 - TALARICO, L., KOSOVEL, V. : Spectral analysis of the biliproteins from *Gracilaria verrucosa* (HUDS.) PAPENFUSS (*Gigartinales, Florideophyceae*). - Photosynthetica *16* : 184 - 190, 1982.

51860 - TALLING, J.F. : Utilization of solar radiation by phytoplankton. - In : HÉLÈNE, C., CHARLIER, M., MONTENAY-GARESTIER, T., LAUSTRIAT, G. (ed.) : Trends in Photobiology. Pp. 619 - 631. Plenum Press, New York - London 1982.

51861 - TAMKIVI, R.P. : Issledovanie gomoperenosa ênergii s pomoshch'yu spektral'no selektivnoĬ kinetiki nizkotemperaturnoĬ fluorestsentsii khlorofillov. [Energy homotransfer *via* spectrally selective kinetics of low-temperature fluorescence of chlorophylls.] - Eesti NSV Tead. Akad. Toim., Füüs., Mat. *31* : 187 - 191, 1982. [In R, ab : E, Est.]

51862 - TAMMINEN, T. : Effects of ammonium effluents on planktonic primary production and decomposition in a coastal brackish water environment. II. Interrelations between abiotic and biotic components of the planktonic ecosystem. - Neth. J. Sea Res. *15* : 349 - 361, 1982. [Chl.]

51863 - TAMMINEN, T. : Effects of ammonium effluents on planktonic primary production and decomposition in a coastal brackish water environment. I. Nutrient balance of the water body and effluent tests. - Neth. J. Sea Res. *16* : 455 - 464, 1982.

51864 - TAN, B.H., HALLORAN, G.M. : Variation and correlations of proline accumulation in spring wheat cultivars. - Crop Sci. *22* : 459 - 463, 1982. [Chl.]

51865 - TAN, C.S., BUTTERY, B.R. : Response of stomatal conductance, transpiration, photosynthesis, and leaf water potential in peach seedlings to different watering regimes. - HortScience *17* : 222 - 223, 1982.

51866 - TAN, C.S., BUTTERY, B.R. : The effects of soil moisture stress to various fractions of the root system on transpiration, photosynthesis, and internal water relations of peach seedlings. - J. amer. Soc. hort. Sci. *107* : 845 - - 849, 1982.

51867 - TANABE, T. : [Studies of the effect of root-cutting treatment on growth and yield in direct sowing culture of paddy rice. II. Changes in the physiological characteristics induced by root-cutting treatment and its relationship to growth behaviour.] - Jap. J. Crop Sci. *51* : 316 - 324, 1982. [Chl; in Jap., ab : E.]

51868 - TANABE, Y., SANO, M., KAWASHIMA, N. : Changes in free amino acids in white and green tissues of variegated tobacco leaves during water stress. - Plant Cell Physiol. *23* : 1229 - 1235, 1982. [Chl.]

51869 - TANAKA, A., TSUJI, H. : Calcium-induced formation of chlorophyll *b* and light-harvesting chlorophyll *a/b*-protein complex in cucumber cotyledons in the dark. - Biochim. biophys. Acta *680* : 265 - 270, 1982.

51870 - TANAKA, I. : [Effective use of solar energy in agriculture [2] Plant species specificity for photosynthesis and productivity.] - Nogyo Oyobi Engei *57* : 896 - 900, 1982. [In Jap.]

51871 - TANAKA, K., KAKUNO, T., YAMASHITA, J., HORIO, T. : Purification and properties of chlorophyllase from greened rye seedlings. - J. Biochem. (Tokyo) *92* : 1763 - 1773, 1982.

51872 - TANAKA, K., KONDO, N., SUGAHARA, K. : Accumulation of hydrogen peroxide in chloroplasts of SO_2-fumigated spinach leaves. - Plant Cell Physiol. *23* : 999 - 1007, 1982.

51873 - TANAKA, K., MITSUHASHI, H., KONDO, N., SUGAHARA, K. : Further evidence for inactivation of fructose-1,6-bisphosphatase at the beginning of SO_2 fumigation. Increase in fructose-1,6-bisphosphate and decrease in fructose-6-phosphate in SO_2-fumigated spinach leaves. - Plant Cell Physiol. *23* : 1467 - -1470, 1982.

51874 - TANAKA, K., OTSUBO, T., KONDO, N. : Participation of hydrogen peroxide in the inactivation of Calvin-cycle SH enzymes in SO_2-fumigated spinach leaves. - Plant Cell Physiol. *23* : 1009 - 1018, 1982.

51875 - TANAKA, O., NASU, Y., YANASE, D., TAKIMOTO, A., KUGIMOTO, M. : pH dependence of the copper effect on flowering, growth and chlorophyll content in *Lemna paucicostata* 6746. - Plant Cell Physiol. *23* : 1479 - 1482, 1982.

51876 - TANCRÈDE, P., MUNGER, G., LEBLANC, R.M. : Excess free energies of interaction of chlorophyll *a* with monogalactosyldiacylglycerol and phytol. A mixed monolayer study. - Biochim. biophys. Acta *689* : 45 - 54, 1982.

51877 - TANDY, N.E., DILLEY, R.A., HERMODSON, M.A., BHATNAGAR, D. : Evidence for an interaction between protons released in chloroplast photosystem II water oxidation and the 8000 M_r hydrophobic subunit of the energy-coupling complex. - J. biol. Chem. *257* : 4301 - 4307, 1982.

51878 - TANFORD, C. : Simple model for the chemical potential change of a transported ion in active transport. - Proc. nat. Acad. Sci. USA *79* : 2882 - 2884, 1982. [Bacteriorhodopsin.]

51879 - TANG Chong-qin, ZHANG Qi-de, ZUO Bao-yu, LOU Shi-qing, LIN Shi-qing, KUANG Ting-yun : [The structure and function of chloroplast membranes X. Effects of magnesium- and potassium-ions on the light-induced pH changes of chloroplast suspensions.] - Acta Phytophysiol. sin. *8* : 163 - 172, 1982. [In Chin., ab : E.]

51880 - TANG, J., NORRIS, J.R. : Theoretical calculations of kinetics of the radical pair P^F state in bacterial photosynthesis. - Chem. Phys. Lett. *92* : 136 - 140, 1982.

51881 - TANG, Z.C., KOZLOWSKI, T.T. : Physiological, morphological, and growth responses of *Platanus occidentalis* seedlings to flooding. - Plant Soil *66* : 243 - 255, 1982. [Stomatal resistance, growth analysis.]

51882 - TANG, Z.C., KOZLOWSKI, T.T. : Some physiological and growth responses of *Betula papyrifera* seedlings to flooding. - Physiol. Plant. *55* : 415 - - 420, 1982. [Stomatal resistance, growth analysis.]

51883 - TANG, Z.C., KOZLOWSKI, T.T. : Some physiological and morphological responses of *Quercus macrocarpa* seedlings to flooding. - Can. J. Forest Res. *12* : 196 - - 202, 1982. [Stomatal resistance.]

*51884 - TANIYAMA, T. : [Crops and environmental pollution [4] Air pollution by sulphur oxides and physiology of damaged agricultural products.] - Nogyo Oyobi Engei *53* : 1421 - 1426, 1978. [Ps; in Jap.]

*51885 - TANIYAMA, T. : [Crops and environmental pollution [16] Photosynthesis, transpiration and organogenesis of crops in relation to photochemical oxidants.] - Nogyo Oyobi Engei *54* : 1551 - 1555, 1979. [In Jap.]

*51886 - TANIYAMA, T. : Studies on injuroius effects of air pollutants on crop plants XVI Effects of acid rain on apparent photosynthesis and grain yield of wheat, barley and rice plants. - Rep. environm. Sci. Mie Univ. *6* : 87 - 101, 1981.

51887 - TANIYAMA, T., MIZUNO, T. : [Studies on injurious effects of air pollutants on crop plants. XVII Interrelation between air pollutants and yield of rice plants in the circumferential area of Ise Bay.] - Rep. environm. Sci. Mie Univ. *7* : 101 - 114, 1982. [In Jap., ab : E.]

51888 - TANNO, J.A., WEBSTER, T.R. : Variegation in *Selaginella martensii* f. *albovariegata*. II. Plastid structure in mature leaves. - Can. J. Bot. *60* : 2384 - - 2393, 1982. [Chloroplast.]

51889 - TANTAWY, M.M., GRIMME, L.H. : The bleaching induced by the phenylpyridazinone herbicide *Metflurazon* in *Chlorella* is a metabolic and reversible process. - Pestic. Biochem. Physiol. *18* : 304 - 314, 1982.

51890 - TANTON, T.W. : Environmental factors affecting the yield of tea (*Camellia sinensis*). I. Effects of air temperature. - Exp. Agr. *18* : 47 - 52, 1982. [Growth analysis.]

51891 - TANTON, T.W. : Environmental factors affecting the yield of tea (*Camellia sinensis*). II. Effects of soil temperature, day length, and dry air. - Exp. Agr. *18* : 53 - 63, 1982. [Growth analysis.]

51892 - TAPIE, P., HAWORTH, P., HERVO, G., BRETON, J. : Orientation of the pigments in the thylakoid membrane and in the isolated chlorophyll-protein complexes of higher plants III. A quantitative comparison of the low-temperature linear dichroism spectra of thylakoids and isolated pigment-protein complexes. - Biochim. biophys. Acta *682* : 339 - 344, 1982.

51893 - TARCHEVSKIĬ, I.A. : Mekhanism vliyaniya zasukhi na fotosinteticheskoe usvoenie CO_2. [Mechanism of drought effects on photosynthetic CO_2 uptake.] - In : Fiziologiya Fotosinteza. Pp. 118 - 129. Nauka, Moskva 1982. [In R.]

51894 - TATARINTSEV, N.P., MAKAROV, A.D. : Issledovanie osobennosteĭ reaktsii fotofosforilirovaniya s pomoshch'yu εADP - fluorestsentnogo analoga ADP. [Study of photophosphorylation, using εADP as a fluorescent analogue of ADP.] - Biokhimiya *47* : 1928 - 1931, 1982. [In R, ab : E.]

51895 - TAYLOR, H.M., MASON, W.K., BENNIE, A.T.P., ROWSE, H.R. : Responses of soybeans to two row spacings and two soil water levels. I. An analysis of biomass accumulation, canopy development, solar rediation interception and components of seed yield. - Field Crop Res. *5* : 1 - 14, 1982.

51896 - TAYLOR, S.E., TERRY, N., HUSTON, R.P. : Limiting factors in photosynthesis III. Effects of iron nutrition on the activities of three regulatory enzymes of photosynthetic carbon metabolism. - Plant Physiol. *70* : 1541 - 1543, 1982.

B51897 - TEARE, I.D., PEET, M.M. (ed.) : Crop-Water Relations. - John Wiley & Sons, New York - Chichester - Brisbane - Toronto - Singapore 1982. [Ps.]

51898 - TEARE, I.D., SIONIT, N., KRAMER, P.J. : Changes in water status during water stress at different stages of development in wheat. - Physiol. Plant. *55* : 296 - 300, 1982. [Stomatal resistance.]

*51899 - TECHY, F., AGHION, J. : Photoreduction and bound cytochrome *c* by dissolved
         ascorbate in suspensions of vesicles made of lecithin, cardiolipin, chloro-
         phyll *a* and cytochrome *c*. - Photobiochem. Photobiophys. *2* : 347 - 353,
         1981.

 51900 - TEDDERS, W.L., WOOD, B.W., SNOW, J.W. : Effects of feeding by *Monelliopsis
         nigropunctata, Monellia caryella,* and *Melanocallis caryaefoliae* on growth
         of pecan seedlings in the greenhouse. - J. econ. Entomol. *75* : 287 - 291,
         1982. [Chl.]

*51901 - TEERI, J.A., OVERTON, J. : Chloroplast ultrastructure in two Crassulacean
         species and an F_1 hybrid with differing biomass $\delta^{13}C$ values. - Plant Cell
         Environm. *4* : 427 - 431, 1981.

 51902 - TEH, K.H., SWANSON, C.A. : Sulfur dioxide inhibition of translocation in bean
         plants. - Plant Physiol. *69* : 88 - 92, 1982. [Ps.]

 51903 - TEIXEIRA, C. : A influência das variações nictemeral e sazonal sobre as cur-
         vas de luz-fotossíntese. [Influence of diurnal and seasonal variations on
         light curves of photosynthesis.] - Bol. Inst. oceanogr. (São Paulo) *31* :
         55 - 67, 1982. [In Port., ab : E.]

*51904 - TEIXEIRA, C., VIEIRA, A.A.H. : Nutrient experiment using *Phaeodactylum tri-
         cornutum* as an assay organism. - Bol. Inst. oceanogr. (São Paulo) *25* : 29 -
         - 42, 1976. [Ps, Chl.]

 51905 - TENHUNEN, J.D. : The diurnal course of leaf gas exchange of the C_4 species
         Amaranthus retroflexus under field conditions in a "cool" climate: compari-
         son with the C_3 species *Glycine max* and *Chenopodium album*. - Oecologia *53* :
         310 - 316, 1982.

 51906 - TENHUNEN, J.D., LANGE, O.L., JAHNER, D. : The control by atmospheric factors
         and water stress of midday stomatal closure in *Arbutus unedo* growing in a
         natural macchia. - Oecologia *55* : 165 - 169, 1982. [Stomatal resistance.]

 51907 - TENHUNEN, J.D., TENHUNEN, L.C., ZIEGLER, H., STICHLER, W., LANGE, O.L. :
         Variation in carbon isotope ratios of *Sempervivoideae* species from different
         habitats of Teneriffe in the spring. - Oecologia *55* : 217 - 224, 1982.

*51908 - TERASAWA, T., ASANO, H., HIROSE, S. : [Ecological studies on environmental
         adaptation in weeds. I. The effect of density on growth and seed productive
         structure of large crabgrass and common purslane.] - Weed Res. *25* : 10 - 16,
         1980. [In Jap., ab : E.]

*51909 - TERASAWA, T., ASANO, H., HIROSE, S. : [Ecological studies on environmental
         adaptation in weeds. 3. The effect of soil moisture on growth and seed pro-
         ductive structure of large crabgrass and common purslane.] - Weed Res. *26* :
         14 - 18, 1981. [Dry-matter distribution; in Jap., ab : E.]

*51910 - TERASAWA, T., ASANO, H., HIROSE, S. : [Ecological studies on environmental
         adaptation in weeds. 4. The effect of shading on growth and seed productive
         structure of large crabgrass and common purslane.] - Weed Res. *26* : 19 - 23,
         1981. [In Jap., ab : E.]

 51911 - TERPSTRA, W. : Studies on chlorophyllase. The mechanism of the action of
         lecithin liposomes on enzyme activity and the function of the carbohydrate
         moiety of the enzyme. - Biochim. biophys. Acta *681* : 233 - 241, 1982.

 51912 - TERPUGOV, E.L., CHEKULAEVA, L.N., LAZAREV, Yu.A. : Issledovanie vliyaniya
         degidratatsii na bakterial'nyĭ rodopsin metodom rezonansnoĭ spektroskopii
         kombinatsionnogo rasseyaniya sveta. [Study of dehydration effects on bacte-
         riorhodopsin by resonance Raman spectroscopy.] - Mol. Biol. (Moskva) *16* :
         814 - 820, 1982. [In R, ab : E.]

 51913 - TERRY, K.L. : Nitrate uptake and assimilation in *Thalassiosira weissflogii*
         and *Phaeodactylum tricornutum:* interactions with photosynthesis and with
         the uptake of other ions. - Mar. Biol. *69* : 21 - 30, 1982.

 51914 - TERRY, K.L., CAPERON, J. : Phytoplankton assimilation of carbon, nitrogen,
         and phosphorus in response to enrichments with deep-ocean water. - Deep-Sea
         Res. *A 29* : 1251 - 1258, 1982.

51915 - TERRY, N., LOW, G. : Leaf chlorophyll content and its relation to the intra-cellular localization of iron. - J. Plant Nutr. 5 : 301 - 310, 1982.

51916 - TEZUKA, Y. : Seasonal variation of dominant phytoplankton, chlorophyll and nutrient levels in nearshore waters of the south basin of Lake Biwa. - Jap. J. Limnol. 43 : 215 - 220, 1982.

51917 - THEBUD, R., SANTARIUS, K.A. : Effects of high-temperature stress on various biomembranes of leaf cells in situ and in vitro. - Plant Physiol. 70 : 200 - - 205, 1982. [Ps.]

51918 - THEG, S.M., HOMANN, P.H. : Light-, pH- and uncoupler-dependent association of chloride with chloroplast thylakoids. - Biochim. biophys. Acta 679 : 221 - 234, 1982.

51919 - THEG, S.M., JOHNSON, J.D., HOMANN, P.H. : Proton efflux from thylakoids induced in darkness and its effect on photosystem II. - FEBS Lett. 145 : 25 - - 29, 1982.

51920 - THIAGARAJAH, M.R., HUNT, L.A. : Effects of temperature on leaf growth in corn (Zea mays). - Can J. Bot. 60 : 1647 - 1652, 1982.

51921 - THOMAS, D.A., ANDRÉ, M. : The response of oxygen and carbon dioxide exchanges and root activity to short term water stress in soybean. - J. exp. Bot. 33 : 393 - 405, 1982.

51922 - THOMAS, D.R., JALIL, M.N.H., COOKE, R.J., YONG, B.C.S., ARIFFIN, A., McNEIL, P.H., WOOD, C. : The synthesis of palmitoylcarnitine by etio-chloroplasts of greening barley leaves. - Planta 154 : 60 - 65, 1982.

51923 - THOMAS, H. : Leaf senescence in a non-yellowing mutant of Festuca pratensis I. Chloroplast membrane polypeptides. - Planta 154 : 212 - 218, 1982.

51924 - THOMAS, H. : Leaf senescence in a non-yellowing mutant of Festuca pratensis II. Proteolytic degradation of thylakoid and stroma polypeptides. - Planta 154 : 219 - 223, 1982.

51925 - THOMASSET, B., THOMASSET, T., BARBOTIN, J.-N., VEJUX, A. : Photoacoustic spectroscopy of active immobilized chloroplast membranes. - Appl. Optics 21 : 124 - 126, 1982.

51926 - THOMASSET, B., THOMASSET, T., VEJUX, A., JEANFILS, J., BARBOTIN, J.-N., THOMAS, D. : Immobilized thylakoids in a cross-linked albumin matrix.Effects of cations studied by electron microscopy, fluorescence emission, photo-acoustic spectroscopy, and kinetic measurements. - Plant Physiol. 70 : 714 - - 722, 1982.

51927 - THOMPSON, S.L., SCHNEIDER, S.H. : Carbon dioxide and climate: has a signal been observed yet? - Nature 295 : 645 - 646, 1982.

51928 - THORHAUG, A. : Primary productivity of seagrasses. - In : ZABORSKY, O.R. (ed.) : CRC Handbook of Biosolar Resources. - Vol.I/2. Pp. 471 - 475. CRC Press, Boca Raton 1982.

51929 - THORNE, G.N. : Distribution between parts of the main shoot and the tillers of photosynthate produced before and after anthesis in the top three leaves of main shoots of Hobbit and Maris Huntsman winter wheat. - Ann. appl. Biol. 101 : 553 - 559, 1982.

51930 - THORNE, J.H. : Temperature and oxygen effects on ^{14}C-photosynthate unloading and accumulation in developing soybean seeds. - Plant Physiol. 69 : 48 - 53, 1982.

51931 - THURNAUER, M.C., RUTHERFORD, A.W., NORRIS, J.R. : The effect of ambient redox potential on the transient electron spin echo signals observed in chloroplasts and photosynthetic algae. - Biochim. biophys. Acta 682 : 332 - - 338, 1982.

51932 - TIBONI, O., CIFERRI, O. : A rapid procedure for the purification of elongation factor Tu(EF-Tu/chl) from spinach chloroplasts. - FEBS Lett. 146 : 197 - 200, 1982.

51933 - TICHÁ, I. : Photosynthetic characteristics during ontogenesis of leaves. 7.
Stomata density and sizes. - Photosynthetica *16* : 375 - 471, 1982.

51934 - TICHÁ, I., ČATSKÝ, J. : Regulation des photosynthetischen CO_2-Transportes
durch Außenluft-CO_2-Konzentration im Laufe der Blattontogenese. - In :
HOFFMANN, P., HIEKE, B. (ed.) : Photosynthese: Regulation und Evolution. Pp.
244 - 246. Humboldt-Universität, Berlin 1982.

51935 - TICHÁ, I., PEISKER, M., ČATSKÝ, J. : Dunkelatmung, apparente Quantenausbeute
und CO_2-Kompensationskonzentration bei Primärblättern von *Phaseolus vulgaris*
L. - In : HOFFMANN, P., HIEKE, B. (ed.) : Photosynthese: Regulation und
Evolution. Pp. 214 - 217. Humboldt-Universität, Berlin 1982.

51936 - TICHÁ, I., POSPÍŠILOVÁ, J., ČATSKÝ, J. : Einfluss von Wasserstress auf die
CO_2-Abhängigkeit der Photosyntheserate bei Bohnenpflanzen. - In : UNGER, K.,
SCHUH, J. (ed.) : Umwelt-Stress. Pp. 217 - 220. Martin-Luther-Universität,
Halle 1982.

51937 - TIEDE, D.M., MUELLER, P., DUTTON, P.L. : Spectrometric and voltage clamp
characterization of monolayers of bacterial photosynthetic reaction centers.
- Biochim. biophys. Acta *681* : 191 - 201, 1982.

51938 - TIEMANN, R., WITT, H.T. : Salt dependence of the electrical potential at the
photosynthetic membrane in steady-state light and its structural consequen-
ce. - Biochim. biophys. Acta *681*: 202 - 211, 1982.

51939 - TIEN, H.T. : Light-induced redox reactions in pigmented BLM. - Bioelectro-
chem. Bioenerg. *9* : 559 - 570, 1982. J. electroanal. Chem. *141* : 559 - 570,
1982.

51940 - TIESZEN, L.L. : Biomass accumulation and primary production. - In : COOMBS,
J., HALL, D.O. (ed.) : Techniques in Bioproductivity and Photosynthesis.
Pp. 16 - 20. Pergamon Press, Oxford - New York - Toronto - Sydney - Paris -
Frankfurt 1982.

51941 - TIKHOMIROV, A.A., SID'KO, F.Ya. : Sostoyanie pigmentnogo apparata i formirovanie struktury tsenozov redisa v svyazi s ikh produktivnost'yu pri razlich-
noǐ intensivnosti i spektre izlucheniya. [State of pigment apparatus and
structure of radish cenoses in relation to their productivity at various
light spectrum and intensity.] - Fiziol. Rast. *29* : 457 - 464, 1982. [in R,
ab : E.]

51942 - TIKHOMIROV, A.A., SID'KO, F.Ya.: Photosynthesis and structure of radish and
wheat canopies as affected by radiation of different energy and spectral
composition. - Photosynthetica *16* : 191 - 195, 1982.

51943 - TIKHONOV, A.N., TIMOSHIN, A.A., RUUGE, É.K., BLYUMENFEL'D, L.A. : Poverkh-
nostnyǐ potentsial tilakoidnoǐ membrany, fotoindutsirovannoe pogloshchenie
protonov i fotofosforilirovanie v khloroplastakh. [Surface potential of
the thylakoid membrane, photoinduced proton absorption, and photophosphorylation in chloroplasts.] - Dokl. Akad. Nauk SSSR *266*: 730 - 733, 1982.
[In R.]

*51944 - TIMÁR, J., BIHARI, F. : Possibilities of control of atrazine-resistant
Amaranthus retroflexus L. with herbicides of the Budapesti vegyimüvek (Budapest Chemical Works) on the basis of their effects on the Hill-reaction:.
In : Proceedings 21st Hung. Annual Meeting Biochem. Pp. 69 - 70. Veszprém
1981.

51945 - TIMKO, M.P., ALHADEFF, M., SCHIFF, J.A. : Developmental changes in thylakoid
membranes during plastid morphogenesis in *Euglena gracilis*. - In : AKOYUNO-
GLOU, G., EVANGELOPOULOS, A.E., GEORGATSOS, J., PALAIOLOGOS, G., TRAKATELLIS,
A., TSIGANOS, C.P. (ed.) : Cell Function and Differentiation, Part B. Bio-
genesis of Energy Transducing Membranes and Membrane and Protein Energetics.
Pp. 201 - 221. Alan F. Liss Inc., New York 1982.

51946 - TINGEY, D.T., TAYLOR, G.E.,Jr. : Variation in plant responses to ozone: a
conceptual model of physiological events. - In : UNSWORTH, M.H., ORMROD,
D.P. (ed.) : Effects of Gaseous Air Pollution in Agriculture and Horticulture. Pp. 113 - 138. Butterworth Scientific, London - Boston - Sydney -
Wellington - Durban - Toronto 1982. [Ps.]

51947 - TINGEY, D.T., THUTT, G.L., GUMPERTZ, M.L., HOGSETT, W.E. : Plant water status influences ozone sensitivity of bean plants. - Agr. Environm. 7 : 243 -
- 254, 1982. [Stomatal resistance.]

51948 - TISHCHENKO, N.N., MAGOMEDOV, I.M., DZHUMANOVA, É. : Fiziologo-biokhimiches-
kaya kharakteristika pshenitsy, vyrashchennoǐ na razlichnom azotnom fone.
[A physiological and biochemical characteristics of wheat grown under different nitrogen supply.] - Tr. prikl. Bot. Genet. Selek. 72 (2) : 109 - 114,
1982. [In R, ab : E.]

51949 - TITLYANOVA, A.A., NURMEDOV, S.S. : Produktsionno-destruktsionnye protsessy
i balans rastitel'nogo veshchestva v pustynnoǐ ékosisteme Zapadnoǐ Turkme-
nii. [Production and destruction processes and balance of plant matter in
a desert ecosystem of Western Turkmenia.] - Ékologiya 1982 (3) : 31 - 37,
1982. [In R.]

51950 - TITUS, J.E., STONE, W.H. : Photosynthetic response of two submerged macrophytes to dissolved inorganic carbon concentration and pH. - Limnol. Oceanogr. 27 : 151 - 160, 1982.

51951 - TIWARI, S.C. : Estimation of nitrogen content in the elevated grasslands
of Pauri hills, Garhwal Himalayas. 1. Standing state in soil-vegetations
ingredients. - Indian J. Ecol. 9 : 210 - 217, 1982. [Ps standing crop.]

51952 - TKACHENKO, F.P., KOVAL', V.T., POGREBNYAK, I.I. : Pigmentnyǐ sklad vydiv
rodu Cladophora KÜTZ. [Pigment composition of various species of the genus
Cladophora KÜTZ.] - Ukr. bot. Zh. 39 (4) : 72 - 75, 112, 1982. [In Ukr.,
ab : E,R.]

51953 - TKACHUK, R., KUZINA, F.D. : Chlorophyll analysis of whole rapeseed ker-
nels by near infrared reflectance. - Can. J. Plant Sci. 62 : 875 - 884,
1982.

51954 - TOBIESSEN, P. : Dark opening of stomata in successional trees. - Oecologia
52: 356 - 359, 1982. [Resistances.]

51955 - TODOROV, G., SOKEROV, S., STOYLOV, S.P.S. : Interfacial electric polariza-
bility of purple membranes in solution. - Biophys. J. 40 : 1 - 5, 1982.
[Bacteriorhodopsin.]

51956 - TOFT, N.L., PEARCY, R.W. : Gas exchange characteristics and temperature relations of two desert annuals: A comparison of a winter-active and summer-
-active species. - Oecologia 55 : 170 - 177, 1982.

51957 - TOKUNAGA, F., IWASA, T. : The photoreaction cycle of bacteriorhodopsin:
Low-temperature spectrophotometry. - In : COLOWICK, S.P., KAPLAN, N.O. (ed.):
Methods in Enzymology. Vol.88. Pp. 163 - 167. Academic Press, New York -
London - Paris - San Diego - San Francisco - São Paulo - Sydney - Tokyo -
Toronto 1982.

51958 - TOLBERT, N.E. : Leaf peroxisomes. - Ann. New York Acad. Sci. 386 [KINDL,
H., LAZAROW, P.B. (ed.) : Peroxisomes and Glyoxysomes] : 254 - 268, 1982.

51959 - TOLLENAAR, M., DAYNARD, T.B. : Effect of source-sink ratio on dry matter
accumulation and leaf senescence of maize. - Can. J. Plant Sci. 62 : 855 -
- 860, 1982.

51960 - TOMBATS, É., VARKONI, Z., SALAI, L. : Iskusstvennyǐ khlorofill-belkovyǐ
kompleks (prigotovlenie, spektral'nye svoǐstva, sravnenie s pigment-protei-
novym kompleksom in vivo). [Artificial chlorophyll-protein complex (preparation, spectral properties, comparison with a pigment-protein complex in
vivo).] - Zh. prikl. Spektrosk. 36 (1) : 64 - 72, 1982. [In R, ab : E.]

B51961 - TOMBESI, L. : Elementi di Climatologia Agraria e Valutazione della Produtti-
vità Ambientale. [Elements of Agricultural Climatology and Assessment of
Environmental Production.] - Istituto Sperimentale per la Nutrizione delle
Piante, Roma 1982. [Ps; in Ital.]

51962 - TOOMING, Kh.G. : Optimal'naya fotosintetlcheskaya deyatel'nost' posevov
       pri tsenotlcheskom vzaimodeĭstvii rastenlĭ. [Optimum photosynthesis in
       crops under cenotic interaction of plants.] - Fiziol. Rast. *29* : 964 - 971,
       1982. [In R, ab : E.]

51963 - TORIU, S., WATANABE, T. : [Effects of foliar-applied herbicides on photo-
       synthesis and growth of naked barley.] - Weed Res. *27* : 191 - 197, 1982.
       [In Jap., ab : E.]

51964 - TORNABENE, T.G. : Microorganisms as hydrocarbon producers. - Experientia
       38 : 43 - 46, 1982. [Ps.]

51965 - TORSVIK, T., DUNDAS, I. : The classification of halobacteria. - In : COLO-
       WICK, S.P., KAPLAN, N.O. (ed.) : Methods in Enzymology. Vol.88. Pp. 360 -
       - 368. Academic Press, New York - London - Paris - San Diego - San Francis-
       co - São Paulo - Sydney - Tokyo - Toronto 1982. [Bacteriorhodopsin.]

*51966 - TÓTH, L.G. : Anwendung von Dialysiersäckchen bei der Bestimmung der Pro-
       duktion des Bakterio- und Phytoplanktons. - In : III International Hydro-
       biol. Symposium. Pp. 405 - 412. Bratislava 1981.

51967 - TOYOSHIMA, Y., FUKUTAKA, E. : A protein essential for recovering oxygen
       evolution in cholate-treated chloroplasts. - FEBS Lett. *150* : 223 - 227,
       1982.

51968 - TREBST, A. : Sulfur assimilation in higher plants and algae. - In : ZABOR-
       SKY, O.R. (ed.) : CRC Handbook of Biosolar Resources. Vol.I/1. Pp. 289 -
       - 298. CRC Press, Boca Raton 1982. [Ps.]

51969 - TREHARNE, K.J. : Hormonal control of photosynthesis and assimilate distri-
       bution. - In : McLAREN, J.S. (ed.) : Chemical Manipulation of Crop Growth
       and Development. Pp. 55 - 66. Butterworth Scientific, London - Boston -
       Durban - Singapore - Sydney - Toronto - Wellington 1982.

51970 - TREMPE, M.R., OHGI, K., GLITZ, D.G. : Ribosome structure. Localization of
       7-methylguanosine in the small subunits of *Escherichia coli* and chloroplast
       ribosomes by immunoelectron microscopy. - J. biol. Chem. *257* : 9822 - 9829,
       1982.

51971 - TRENCH, R.K. : Physiology, biochemistry, and ultrastructure of cyanellae.
       - In : ROUND, F.E., CHAPMAN, D.J. (ed.) : Progress in Phycological Research.
       Vol.1. Pp. 257 - 288. Elsevier Biomedical Press, Amsterdam 1982. [Ps.]

51972 - TRENCH, R.K. : Cyanelles. - In : SCHIFF, J.A., LYMAN, H. (ed.) : On the
       Origins of Chloroplasts. Pp. 55 - 76. Elsevier/North-Holland, New York -
       Amsterdam - Oxford 1982. [Ps, Chl, Car, Bil.]

51973 - TRIFONOVA, I.S., UL'YANOVA, D.S., CHEBOTAREV, E.N. : Pervichnaya produktsi-
       ya, soderzhanie khlorofilla i organicheskoe veshchestvo sestona v Onezhskom
       ozere letom 1977 g. [Primary production, chlorophyll content and organic
       matter of sestom in the Onega Lake in summer 1977.] - Gidrobiol. Zh. *18*(5):
       106 - 109, 1982. [In R.]

51974 - TRIOLO, L., CERVIGNI, T., GIACOMELLI, M. : Photosynthetic characteristics
       of a yellow green mutant of durum wheat. - Agrochimica *26* : 192 - 203,
       1982.

51975 - TRIPATHY, B.C., SUBBALAKSHMI, B., MOHANTY, P. : Problems and possibilities
       in controlling oxygen inhibition of photosynthesis. - Proc. Indian nat. Sci.
       Acad. *B 48* : 271 - 305, 1982.

*51976 - TRIPODI, G. : ER-plastidial membrane relationships in *Ochrosphaera neapoli-
       tana* SCHUSSNIG (*Haptophyta*). - In : SCHWEMMLER, W., SCHENK, H.E.A. (ed.) :
       Endocytobiology, Endosymbiosis and Cell Biology. Vol.I. Pp. 617 - 622. Wal-
       ter de Gruyter & Co., Berlin - New York 1980. [Chloroplast.]

*51977 - TRIPODI, G. : A convoluted membranous structure associated to fibrils in the
       mitochondrial and plastidial matrix during sporogenesis in red algae. - In :
       SCHWEMMLER, W., SCHENK, H.E.A. (ed.) : Endocytobiology, Endosymbiosis and
       Cell Biology. Vol.I. Pp. 817 - 823. Walter de Gruyter & Co., Berlin - New
       York 1980. [Chloroplast.]

*51978 - TRIPODI, G., DE MASI, F. : The post-fertilization stages of red algae: the fine structure of the fusion cell of *Erythrocystis*. - J. submicr. Cytol.*9* : 389 - 401, 1977. [Chloroplast.]

51979 - TRIPODI, G., SANTISI, S. : A study on the cell covering of *Symbiodinium*, a symbiote of the octocoral *Eunicella*. - J. submicr. Cytol. *14* : 613 - 620, 1982. [Chloroplast.]

51980 - TRISSL, H.-W., KUNZE, U., JUNGE, W. : Extremely fast photoelectric signals from suspensions of broken chloroplasts and of isolated chromatophores. - Biochim. biophys. Acta *682* : 364 - 377, 1982.

51981 - TROCINE, R.P., RICE, J.D., WELLS, G.N. : Photosynthetic response of seagrasses to ultraviolet-A radiation and the influence to visible light intensity. - Plant Physiol. *69* : 341 - 344, 1982.

51982 - TROENG, E., LINDER, S. : Gas exchange in a 20-year-old stand of Scots pine. I. Net photosynthesis of current and one-year-old shoots within and between seasons. - Physiol. Plant. *54* : 7 - 14, 1982.

51983 - TROENG, E., LINDER, S. : Gas exchange in a 20-year-old stand of Scots pine. II. Variation in net photosynthesis and transpiration within and between trees. - Physiol. Plant. *54* : 15 - 23, 1982.

51984 - TROSPER, T.L., FRANK, H.A., NORRIS, J.R., THURNAUER, M.C. : Magnetophotoselection studies on *Rhodopseudomonas viridis* reaction centers. - Biochim. biophys. Acta *679*: 44 - 50, 1982.

51985 - TRÜPER, H.G. : Sulfur assimilation in photosynthetic bacteria. - In : ZABORSKY, O.R. (ed.) : CRC Handbook of Biosolar Resources. Vol.I/1. Pp. 283 - - 287. CRC Press, Boca Raton 1982. [Ps.]

51986 - TRÜPER, H.G., FISCHER, U. : Anaerobic oxidation of sulphur compounds as electron donors for bacterial photosynthesis. - Phil. Trans. roy. Soc. London, Ser.B *298* : 529 - 542, 1982.

51987 - TSEL'NIKER, Yu.L., OSIPOVA, O.P., NIKOLAEVA, M.K. : Fiziologicheskie aspekty adaptatsii list'ev k usloviyam osveshcheniya. [Physiological aspects of leaves adaptation to irradiance.] - In : Fiziologiya Fotosinteza. Pp. 187 - - 203. Nauka, Moskva 1982. [In R.]

51988 - TSUCHIYA, M., OGO, T. : [Analytical studies on the process of growth and production of mat rush (*Juncus decipiens* NAKAI) I. Effects of tip cutting on tillering, stem elongation and dry matter production.] - Jap. J. Crop Sci. *51* : 126 - 131, 1982. [Growth analysis; in Jap., ab : E.]

51989 - TSUDA, M. : Methods for extraction of pigment chromophore. - In : COLOWICK, S.P., KAPLAN, N.O. (ed.) : Methods in Enzymology. Vol.88. Pp. 552 - 561. Academic Press, New York - London - Paris - San Diego - San Francisco - São Paulo - Sydney - Tokyo - Toronto 1982. [Bacteriorhodopsin.]

51990 - TSUDA, M. : Effect of pressure on visual pigment and purple membrane. - In : COLOWICK, S.P., KAPLAN, N.O. (ed.) : Methods in Enzymology. Vol.88. Pp. 714 - - 722. Academic Press, New York - London - Paris - San Diego - San Francisco - São Paulo - Sydney - Tokyo - Toronto 1982.

*51991 - TSUDA, M., GLACCUM, M., NELSON, B., EBREY, T.G. : Light isomerizes the chromophore of bacteriorhodopsin. - Nature *287* : 351 - 353, 1980.

51992 - TSUDA, M., HAZEMOTO, N., KONDO, M., KAMO, N., KOBATAKE, Y., TERAYAMA, Y. : Two photocycles in *Halobacterium halobium* that lacks bacteriorhodopsin. - Biochem. biophys. Res. Commun. *108* : 970 - 976, 1982.

51993 - TSUJIKAWA, I. : [Magnetic field effects on chemical and biological systems.]- Nippon Butsuri Gakkaishi *37* : 856 - 858, 1982. [In Jap.]

51994 - TSUKIDA, K., SAIKI, K., TAKII, T., KOYAMA, Y. : Separation and determination of *cis/trans*-β-carotenes by high-performance liquid chromatography. - J. Chromatogr. *245* : 359 - 364, 1982.

51995 - TSUKIHARA, T., KATSUBE, Y., HASE, T., WADA, K., MATSUBARA, H. : Evolutio-
         nary relationship between [2Fe-2S]ferredoxin and an ancestral ferredoxin. -
         In : KIMURA, M. (ed.) : Molecular Evolution, Protein Polymorphism and the
         Neutral Theory. Pp. 299 - 312. Japan Scientific Societies Press, Tokyo
         1982. Springer-Verlag, Berlin 1982.

51996 - TSUKIHARA, T., KOBAYASHI, M., NAKAMURA, M., KATSUBE, Y., FUKUYAMA, K.,
         HASE, T., WADA, K., MATSUBARA, H. : Structure-function relationship of
         [2Fe-2S]ferredoxins and design of a model molecule. - BioSystems 15 : 243 -
         - 257, 1982.

51997 - TSUZUKI, M., MIYACHI, S., WINTER, K., EDWARDS, G.E. : Localization of car-
         bonic anhydrase in Crassulacean acid metabolism plants. - Plant Sci. Lett.
         24 : 211 - 218, 1982.

51998 - TSVIRKO, M.P., STEL'MAKH, G.F. : Lazernaya fluorestsentnaya spektroskopiya
         verkhnikh vozbuzhdennykh sostoyaniĭ khlorofillopodobnykh molekul. [Laser
         fluorescence spectroscopy of upper excited states of chlorophyll-like mole-
         cules.] - Eesti NSV Tead. Akad. Toim., Füüs., Mat. 31 : 124 - 128, 1982.
         [In R, ab : E, Est.]

51999 - TU, S.-I., HUTCHINSON, H., CAVANAUGH, J.R. : Interaction between Gramicidin-
         -A and bacteriorhodopsin in reconstituted purple membrane. - Biochem. bio-
         phys. Res. Commun. 106 : 23 - 29, 1982.

52000 - TUBA, Z. : A relatív fényintenzitás, a fotoszintetikus pigmentarányok és
         a specifikus levélterület közötti összefüggések egy többszintű erdőtársu-
         lásban. [Correlation between relative irradiance, photosynthetic pigment
         ratios and specific leaf area in a multilayer community.] - Acta Acad. paed.
         nyíregyháziensis 9/F : 55 - 67, 1982. [In Hung.]

52001 - TUNDISI, J.G., MATSUMURA TUNDISI, T. : Estudios limnológicos no sistema de
         lagos do médio Rio Doce, Minas Gerais, Brasil. [Limnological studies in the
         River Doce Valley Lake System, Minas Gerais, Brasil.] - An. II Seminário
         region. Ecologia 1982 : 133 - 258, 1982. [Chl; in Port., ab : E.]

52002 - TUNDO, P., KURIHARA, K., KIPPENBERGER, D.J., POLITI, M., FENDLER, J.H. :
         Chemisch unsymmetrische, polymerisierte Tensid-Vesikeln: Herstellung und
         mögliche Verwendung bei der künstlichen Photosynthese. - Angew. Chem. 94 :
         73 - 74, 1982.

52003 - TURNER, F.T., McCAULEY, G.N. : Rice. - In : TEARE, I.D., PEET, M.M. (ed.):
         Crop-Water Relations. Pp. 307 - 350. John Wiley & Sons, New York - Chiches-
         ter - Brisbane - Toronto - Singapore 1982. [Ps.]

52004 - TURNER, N.C., BURCH, G.J. : The role of water in plants. - In : TEARE, I.D.,
         PEET, M.M. (ed.) : Crop-Water Relations. Pp. 73 - 126. John Wiley & Sons,
         New York - Chichester - Brisbane - Toronto - Singapore 1982. [Ps.]

52005 - TUSKAN, G.A., DE LA CRUZ, A.A. : Solar input and energy storage in a five-
         -year-old American sycamore plantation. - Forest Ecol. Manag. 4 : 191 - 198,
         1982.

52006 - TYANKOVA, L., TRIFONOV, A., KUSMANOVA, R. : A spectroscopic (ATR) approach
         to the behaviour of proline- and sucrose-treated biomembranes towards dehyd-
         ration and rewatering. - Biochem. Physiol. Pflanzen 177 : 509 - 514, 1982.
         [Chloroplast.]

52007 - TYMMS, M.J., SCOTT, N.S., POSSINGHAM, J.V. : Chloroplast and nuclear DNA
         content of cultured spinach leaf discs. - J. exp. Bot. 33 : 831 - 837,
         1982.

52008 - TYSZKIEWICZ, E., BOTTIN, H., ROUX, E. : Redox potential-dependent ATP synthe-
         sis in darkness in spinach chloroplasts. - Bioelectrochem. Bioenerg. 9 :
         157 - 166, 1982. J. electroanal. Chem. 141 : 157 - 166, 1982.

52009 - UCHIDA, N., WADA, Y., MURATA, Y. : [Studies on the changes in the photosyn-
thetic activity of a crop leaf during its development and senescence. II.
Effect of nitrogen deficiency on the changes in the senescing leaf of rice.]
- Jap. J. Crop Sci. *51* : 577 - 583, 1982. [In Jap., ab : E.]

*52010 - UCHIMIYA, H., CHEN, K., WILDMAN, S.G. : Polypeptide composition of fraction
I protein as an aid in the study of plant evolution. - In : REDEI, G.P.
(ed.) : Stadler Genetics Symposium. Vol.9. Pp. 83 - 89. University of Missou-
ri, Columbia 1977.

52011 - UDALOVA, G.V., YURINA, N.G., ODINTSOVA, M.S. : Ribosomy khloroplastov: bel-
kovyĭ sostav sub"edinits. [Ribosomes of chloroplasts: protein composition
of subunits.] - Dokl. Akad. Nauk SSSR *263* : 1492 - 1497, 1982. [In R.]

52012 - UEHARA, K., SHIBATA, K., NAKAMURA, H., TANAKA, M. : Aggregation of chlorophyll *a* in aqueous dimethyl sulfoxide. - Chem. Lett. *1982*: 1445 - 1448,
1982.

52013 - UENO, T., MAEKAWA, A., SUZUKI, T. : [Effect of lipoxygenase on the determination of carotene in vegetables.]- Bitamin *56* (2) : 83 - 89, 1982. [In Jap.,
ab : E.]

*52014 - UHRIG, H. : Regeneration of protoplasts of dihaploid potato plants bleached
by a herbicide (SAN 6706). - Mol. gen. Genet. *181* : 403 - 405, 1981. [Chl.]

52015 - ULMANN, L. : Vliv termínu sklizně na produkci biomasy ovsa. [Effect of harvesting term on oat biomass production.] - Rostl. Výroba (Praha) *28* : 69 -
- 74, 1982. [In Czech, ab : E, G, R.]

52016 - UMOESSIEN, S.N., FORWARD, D.F. : Effect of gibberellic acid on the distribution of products of photosynthesis in sunflower. - Ann. Bot. *50* : 465 -
- 472, 1982.

52017 - UNGER, K. : Zur Modellierung der Reaktionsnorm von Kulturpflanzen und deren
Bedeutung für die Prüfung von Stress-Reaktionen. - In : UNGER, K., SCHUH,J.
(ed.) : Umwelt-Stress. Pp. 190 - 199. Martin-Luther-Universität, Halle 1982.
[Dry-matter production.]

B52018 - UNGER, K., SHUH, J. (ed.) : Umwelt-Stress. - Martin-Luther-Universität,
Halle 1982. [Ps.]

52019 - UNO, S. : Distribution and standing stock of chlorophyll *a* in the Antarctic
Ocean, from December 1980 to January 1981. - Mem. nat. Inst. polar Res.,
spec. Issue *23* : 20 - 27, 1982.

52020 - UNSWORTH, M.H. : Exposure to gaseous pollutants and uptake by plants. - In :
UNSWORTH, M.H., ORMROD, D.P. (ed.) : Effects of Gaseous Air Pollution in
Agriculture and Horticulture. Pp. 43 - 63. Butterworth Scientific, London -
Boston - Sydney - Wellington - Durban - Toronto 1982. [Resistances.]

B52021 - UNSWORTH, M.H., ORMROD, D.P. (ed.) : Effects of Gaseous Air Pollution in
Agriculture and Horticulture. - Butterworth Scientific, London - Boston -
Sydney - Wellington - Durban - Toronto 1982. [Ps.]

52022 - URIBE, E.G., STARK, B. : Inhibition of photosynthetic energy conversion by
cupric ion. Evidence for Cu^{2+}-coupling factor 1 interaction. - Plant Physiol. *69* : 1040 - 1045, 1982.

52023 - URSINO, D.J., HUNTER, D.M., LAING, R.D., KEIGHLEY, J.L.S. : Nitrate modification of photosynthesis and photoassimilate export in young nodulated
soybean plants. - Can. J. Bot. *60* : 2665 - 2670, 1982.

52024 - USPENSKAYA, N.Ya., ALEKSANDROV, A.Yu., NOVAKOVA, A.A., KUZ'MIN, R.N., KONO-
NENKO, A.A., RUBIN, A.B. : Issledovanie membrannykh ferredoksinov *Rhodopseudomonas sphaeroides* metodom mèssbauèrovskoĭ spektroskopii. [Mössbauer
study of ferredoxins in membranes of *Rhodopseudomonas sphaeroides*.] - Mol.
Biol. *16* : 830 - 836, 1982. [In R, ab : E.]

52025 - USUDA, H., EDWARDS, G.E. : Influence of varying CO_2 and orthophosphate
concentrations on rates of photosynthesis, and synthesis of glycolate and
dihydroxyacetone phosphate by wheat chloroplasts. - Plant Physiol. *69* :
469 - 473, 1982.

52026 - USUKURA, J., YAMADA, E. : Freeze-substitution and freeze-etching method for
studying the ultrastructure of photoreceptive membrane. - In : COLOWICK,
S.P., KAPLAN, N.O. (ed.) : Methods in Enzymology. Vol.88. Pp. 118 - 123.
Academic Press, New York - London - Paris - San Diego - San Francisco -
São Paulo - Sydney - Tokyo - Toronto 1982. [Bacteriorhodopsin.]

*52027 - UTKIN, A.I., DYLIS, N.V., SOLNTSEVA, O.N. : Pervichnaya produktivnost' i
vertikal'naya biogeotsenoticheskaya struktura 83-letnego bereznyaka volo-
sistoosokovogo. [Primary productivity and vertical biogeocoenotic structure
of a 83-year-old stand of *Betuletum pilosae caricosum*.] - Byul. moskov.
Obshch. Ispyt. Prirody, Otd. biol. *85* : 100 - 117, 1980. [In R, ab : E.]

52028 - UTLEY, J.H.P. : Electrochemical reactions applied to polyenes and caroteno-
ids. - In : BRITTON, G., GOODWIN, T.W. (ed.) : Carotenoid Chemistry and
Biochemistry. Pp. 97 - 105. Pergamon Press, Oxford - New York - Toronto -
Sydney - Paris - Frankfurt 1982.

52029 - UTSUNOMIYA, N., YAMADA, H., KATAOKA, I., TOMANA, T. : [The effect of fruit
temperature on the maturation of satsuma mandarin (*Citrus unshiu* MARC.)
fruits.] - J. jap. Soc. hort. Sci. *51* : 135 - 141, 1982. [Chl; in Jap., ab :
E.]

52030 - UZO, J.O. : Inheritance of "Nsukka Yellow" aroma, ascorbic acid and carotene
in *Capsicum annuum* L. - Crop Res. *22* : 77 - 83, 1982.

52031 - VACEK, K., VALENT, O., ŠKŮTA, A. : Photopotential of chlorophyll *a* containing lecithin BLM formed with and without carotene. - Gen. Physiol. Biophys.
2 : 135 - 139, 1982.

52032 - VAKLINOVA, S.G., GOUSHTINA, L.M., LAZOVA, G.N. : Carboanhydrase activity
in chloroplasts and chloroplast fragments. - Dokl. bolg. Akad. Nauk *35* :
1721 - 1724, 1982.

52033 - VALADON, L.R.G., MUMMERY, R.S. : The effect of light and of 2-(4-chlorophe-
nylthio)-triethylamine hydrochloride on chlorophyll and carotenoid contents
of mung bean seedlings. - Ann. Bot. *49* : 247 - 256, 1982.

52034 - VALANNE, N., ARO, E.-M. : Incorporation of 5-aminolevulinic acid and turnover rate of seven chlorophyll-protein complexes in the moss *Ceratodon purpureus*. - Photobiochem. Photobiophys. *4* : 53 - 61, 1982.

52035 - VALANNE, N., ARO, E.-M., NIEMI, H. : Photosynthetic apparatus of *Ceratodon
purpureus*. - J. Hattori bot. Lab. *53* : 171 - 179, 1982.

52036 - VALANNE, N., ARO, E.-M., RINTAMÄKI, E. : Leaf and chloroplast structure of
two aquatic *Ranunculus* species. - Aquat. Bot. *12* : 13 - 22, 1982.

52037 - VALKIRS, G.E., FEHER, G. : Topography of reaction center subunits in the
membrane of the photosynthetic bacterium, *Rhodopseudomonas sphaeroides*. -
J. Cell Biol. *95* : 179 - 188, 1982.

52038 - VALKUNAS, L.L., GAÏZHAUSKAS, E., KUDZHMAUSKAS, Sh.P. : Teoreticheskoe issledovanie absorbtsionnykh izmeneniĭ v reaktsionnykh tsentrakh fotosinteziruyushchikh bakteriĭ. [Theoretical investigation of absorption changes in the
reaction centres of photosynthetic bacteria.] - Eesti NSV Tead. Akad. Toim.,
Füüs., Mat. *31* : 215 - 218, 1982. [In R, ab : E, Est.]

52039 - VALLE, E.M., CARRILLO, N., VALLEJOS, R.H. : Functional sulfhydryl groups
of ferredoxin-NADP⁺oxidoreductase. - Biochim. biophys. Acta *681* : 412 - 418,
1982.

*52040 - VAN, T.K., HALLER, W., BOWES, G. : Some aspects of the competitive biology
of *Hydrilla*. - Proc. EWRS Symp. aquatic Weeds *5* : 117 - 126, 1978. [Ps.]

52041 - VAN ASSCHE, C.J., CARLES, P.M. : Photosystem II inhibiting chemicals. Mole-
cular interaction between inhibitors and a common target. - In : MORELAND,
D.E., ST.JOHN, J.B., HESS, F.D. (ed.) : Biochemical Responses Induced by
Herbicides. Pp. 1 - 21. Amer. Chem. Society, Washington 1982.

*52042 - VAN BENNEKOM, A.J., GIESKES, W.W.C., TIJSSEN, S.B. : Eutrophication of Dutch coastal waters. - Proc. roy. Soc. London B *189* : 359 - 374, 1975. [Primary production.]

*52043 - VAN BEST, J.A., MATHIS, P. : Apparatus for the measurement of small absorption change kinetics at 820 nm in the nanosecond range after a ruby laser flash. - Rev. sci. Instrum. *49* : 1332 - 1335, 1978. [Ps.]

52044 - VAN DEN AVYLE, M.J., ALLARD, D.W., DREIER, T.M., CLARK, W.J. : Effects of diel phytoplankton migrations on chlorophyll *a* vertical profiles in a central Texas pond. - Texas J. Sci. *34* : 69 - 78, 1982.

52045 - VAN DEN BERG, G., BRANDSE, M., TIPKER, J. : Effects of substituted 2-phenyl-amino-1,4,5,6-tetrahydropyrimidines on ATP formation in isolated spinach chloroplasts. - Z. Naturforsch. *37 C* : 651 - 657, 1982.

52046 - VAN DEN BERG, G., TIPKER, J. : Quantitative structure-activity relationships of the inhibition of photosynthetic electron flow by substituted diphenyl ethers. - Pestic. Sci. *13* : 29 - 38, 1982.

52047 - VANDEN DRIESSCHE, T., LANNOYE, R. : Time-dependent effects of auxin and anti-auxin on photosynthesis in *Acetabularia*. - Int. J. Chronobiol. *8* : 97 - 104, 1982.

*52048 - VAN DER CAMMEN, J.C.J.M., GOEDHEER, J.C. : Fluorescence lifetime spectra of greening etiolated bean leaves. - Photobiochem. Photobiophys. *3* : 159 - - 165, 1981.

52049 - VAN DER CAMMEN, J.C.J.M., GOEDHEER, J.C. : Fluorescence lifetime spectra measured after illumination of etiolated bean leaves at low temperatures. - Photobiochem. Photobiophys. *4* : 145 - 152, 1982.

52050 - VAN DER MERWE, N.J. : Carbon isotopes, photosynthesis, and archaeology. - Amer. Sci. *70* : 596 - 606, 1982.

52051 - VAN DER TOORN, J., MOOK, J.H. : The influence of environmental factors and management on stands of *Phragmites australis* 1. Effects of burning, frost and insect damage on shoot density and shoot size. - J. appl. Ecol. *19* : 477 - 499, 1982. [Growth analysis.]

52052 - VanderZEE, D., KENNEDY, R.A. : Plastid development in seedlings of *Echinochloa crus-galli* var. *oryzicola* under anoxic germination conditions. - Planta *55* : 1 - 7, 1982.

52053 - VAN DIJCK, P.W.M., VAN DAM, K. : Bacteriorhodopsin in phospholipid vesicles. - In : COLOWICK, S.P., KAPLAN, N.O. (ed.) : Methods in Enzymology. Vol.88. Pp. 17 - 25. Academic Press, New York - London - Paris - San Diego - San Francisco - São Paulo - Sydney - Tokyo - Toronto 1982.

52054 - VAN EPPS, G.A., BARKER, J.R., McKELL, C.M. : Energy biomass from large rangeland shrubs of the intermountain United States. - J. Range Manage. *35* : 22 - 25, 1982.

52055 - VAN GORKOM, H.J., THIELEN, A.P.G.M. : Primary and associated reactions in photosystem II. - In : HÉLÈNE, C., CHARLIER, M., MONTENAY-GARESTIER, T., LAUSTRIAT, G. (ed.) : Trends in Photobiology. Pp. 579 - 586. Plenum Press, New York - London 1982.

52056 - VAN GORKOM, H.J., THIELEN, A.P.G.M., GORREN, A.C.F. : The secondary electron acceptor of photosystem II. - In : TRUMPOVER, B.L. (ed.) : Function of Quinones in Energy Conserving Systems. Pp. 213 - 225. Academic Press, New York 1982.

52057 - VAN GRONDELLE, R., DUYSENS, L.N.M. : Photoreaction 2 and the oxygen evolving system. - In : HALL, D.O., PALZ, W. (ed.) : Photochemical, Photoelectrochemical and Photobiological Processes. Vol.1. Pp. 147 - 149. D.Reidel Publ.Co., Dordrecht - Boston - London 1982.

52058 - VAN GRONDELLE, R., DUYSENS, L.N.M. : Primary light absorption processes and thermodynamics in photosynthesis. - In : ZABORSKY, O.R. (ed.) : CRC Handbook of Biosolar Resources. Vol.I/1. Pp. 11 - 36. CRC Press, Boca Raton 1982.

52059 - VAN GRONDELLE, R., KRAMER, H.J.M., RIJGERSBERG, C.P. : Energy transfer in
the B800-850-carotenoid light-harvesting complex of various mutants of *Rhodopseudomonas sphaeroides* and of *Rhodopseudomonas capsulata*.-Biochim.biophys.
Acta *682* : 208 - 215, 1982.

52060 - VAN HASSELT, P.R., WASSEN, M.J. : Combined effects of salt (NaCl) and air
pollution (SO$_2$) stress on *Cucumis* plants. - In : UNSWORTH, M.H., ORMROD,
D.P. (ed.) : Effects of Gaseous Air Pollution in Agriculture and Horticulture. Pp. 499 - 500. Butterworth Scientific, London - Boston - Sydney - Wellington - Durban - Toronto 1982. [Ps.]

52061 - VAN RENSEN, J.J.S. : Molecular mechanisms of herbicide action near photosystem II. - Physiol.Plant. *54* : 515 - 521, 1982.

*52062 - VAN STEVENINCK, M.E., VAN STEVENINCK, R.F.M. : Plastids with densely stain-
ing thylakoid contents in *Nymphoides indica*. II. Characterization of stain-
able substance. - Protoplasma *103* : 343 - 360, 1980.

52063 - VAN STRATEN, G., HERODEK, S. : Estimation of algal growth parameters from
vertical primary production profiles. - Ecol. Model. *15* : 287 - 311, 1982.

52064 - VAPAAVUORI, E., NURMI, A. : Chlorophyll-protein complexes in *Salix* sp.
"aquatica gigantea" under strong and weak light II. Effect of water stress
on the chlorophyll-protein complexes and chloroplast ultrastructure. - Plant
Cell Physiol. *23* : 791 - 801, 1982.

52065 - VAPAAVUORI, E.M., VALANNE, N.K.S. : Activities of ribulose-1,5-bisphosphate
carboxylase-oxygenase in *Salix* sp. during water stress. - Photosynthetica
16 : 1 - 6, 1982.

52066 - VARKEY, P.J., NADAKAVUKAREN, M.J. : Influence of leaf differentiation on
the developmental pathway of *Coleus* chloroplasts. - New Phytol. *92* : 273 -
- 278, 1982.

52067 - VARLET-GRANCHER, C., BONHOMME, R., CHARTIER, M., ARTIS, P. : Efficience de
la conversion de l'énergie solaire par un couvert végétal. - Acta oecol.-
Oecol.Plant. *3* : 3 - 26, 1982.

52068 - VASANDER, H. : Plant biomass and production in virgin, drained and fertili-
zed sites in a raised bog in southern Finland. - Ann. bot. fenn. *19* : 103 -
- 125, 1982.

52069 - VASHAKMADZE, G.Sh., KRENDELEVA, T.E., KUKARSKIKH, G.P., KHRAMOVA, G.A.,
RUBIN, A.B. : O rekonstruktsii funktsii sopryagayushchego kompleksa khloroplastov na fosfolipidnykh vezikulakh. [Reconstitution of the coupling chloroplast complex function on phospholipid vesicles.] - Biokhimiya *47* : 1556 -
- 1562, 1982. [In R, ab : E.]

52070 - VASS, I. : Connection of thermoluminescence and phase transitions in chloroplasts. - In : 4 Conference Lumin. Szeged, Conf. Dig. Pp. 147 - 151.
Szeged 1982.

52071 - VASS, I., DEMETER, S. : Classification of Photosystem II inhibitors by thermodynamic characterization of the thermoluminescence of inhibitor-treated
chloroplasts. - Biochim. biophys. Acta *682* : 496 - 499, 1982.

*52072 - VASSEUR, P., JOUANY, J.-M., FERARD, J.-F., TOUSSAINT, B. : Interêt du dosage
de l'A.T.P. en tant que critère d'écotoxicité algue chez les algues. - In :
LECLERC, H., DIVE, D. (ed.) : Les Tests de Toxicité Algue en Milieu Aqua-
tique. (Acute Aquatic Ecotoxicological Tests.) Pp. 207 - 226. Institut National de la Santé et de la Recherche Medicale, Paris 1981.

*52073 - VATER, J., GAUDSZUN, T., SCHARNOW, H., SALNIKOW, J. : Competition of pyri-
doxal 5'-phosphate with ribulose 1,5-bisphosphate and effector sugar phospha-
tes at the reaction centers of the spinach ribulose 1,5-bisphosphate carbo-
xylase/oxygenase. - Z. Naturforsch. *35 C* : 416 - 422, 1980.

52074 - VEBLEN, T.T. : Growth patterns of *Chusquea* bamboos in the understory of
Chilean *Nothofagus* forests and their influences in forest dynamics. - Bull.
Torrey bot. Club *109* : 474 - 487, 1982. [Dry-matter accumulation.]

52075 - VECCHI, M., ENGLERT, G., MAYER, H. : Chromatographische Trennung und Iden-
tifizierung diastereomerer Carotinoide mit grossem räumlichen Abstand der
chiralen Zentren. - Helv. chim. Acta 65 : 1050 - 1058, 1982.

52076 - VECHER, A.S., MASNYĬ, M.N., GAPONENKO, V.I., BALEVA, E.F. : Aktivnost' ri-
bulozodifosfatkarboksilazy na rannykh stadiyakh prorastaniya semyan rzhi.
[Ribulose bisphosphate carboxylase activity at the early stages of rye seed
germination.] - Dokl. Akad. Nauk belorus. SSR 26 : 369 - 371, 384, 1982.
[In R, ab : E.]

52077 - VEERANJANEYULU, K., DAS, V.S.R. : Photoacoustic spectroscopy - leaf absorption spectra. - J. exp. Bot. 33 : 515 - 519, 1982. [Chl.]

52078 - VEIERSKOV, B., ANDERSEN, A.S., STUMMANN, B.M., HENNINGSEN, K.W. : Dynamics
of extractable carbohydrates in Pisum sativum. II. Carbohydrate content
and photosynthesis of pea cuttings in relation to irradiance and stock
plant temperature and genotype. - Physiol. Plant. 55 : 174 - 178, 1982.

52079 - VEJSADOVÁ, H., LAŠTŮVKA, Z. : The importance of FeEDTA for reversibility
of phosphate-induced chlorosis in maize (Zea mays L.). - Biol. Plant. 24 :
401 - 406, 1982.

52080 - VENDELAND, J.S., BRUCK, D.K., SINCLAIR, T.R. : Differential starch accumulation in the leaf mesophyll layers of soybean. - Crop Sci. 22 : 1251 -
- 1252, 1982. [Photosynthates.]

52081 - VENDELAND, J.S., SINCLAIR, T.R., SPAETH, S.C., CORTES, P.M. : Assumptions
of plastochron index: evaluation with soya bean under field drought conditions. - Ann. Bot. 50 : 673 - 680, 1982.

52082 - VENEDIKTOV, P.S., KRENDELEVA, T.E., RUBIN, A.B. : Pervichnye protsessy fotosinteza i fiziologicheskoe sostoyanie rastitel'nogo organizma. [Primary
processes of photosynthesis and physiological state of the plant organism.]
- In : Fiziologiya Fotosinteza. Pp. 55 - 76. Nauka, Moskva 1982. [In R.]

52083 - VENKATARAMANA, S., DAS, V.S.R. : Distribution of nitrogen assimilating enzymes in relation to photosynthesis in certain C_4 grasses. - Z. Pflanzenphysiol. 105 : 289 - 296, 1982.

52084 - VENTUROLI, G., MELANDRI, B.A. : The localized coupling of bacterial photophosphorylation. Effect of antimycin A and N,N-dicyclohexylcarbodiimide in
chromatophores from Rhodopseudomonas sphaeroides, Ga, studied by single
turnover event analysis. - Biochim. biophys. Acta 680 : 8 - 16, 1982.

52085 - VERDUIN, J. : Components contributing to light extinction in natural waters:
Method of isolation. - Arch. Hydrobiol. 93 : 303 - 312, 1982. [Chl.]

52086 - VERETENNIKOV, A.V. : Fotosintez seyantsev sosny i eli v teplitsakh s poli-
ètilenovym ukrytiem. [Photosynthesis of pine and fire seedlings in green-
houses covered by polyethylene foil.] - Les. Zh. 1982 (4) : 20 - 23, 1982.
[In R.]

52087 - VERETENNIKOV, A.V. : Vliyanie drenazha na protsess fotosinteza drevesnykh
rasteniĭ. [Effect of drainage on photosynthesis of woody plants.] - In :
Regulyatsiya Fiziologicheskikh Protsessov Rasteniĭ. Pp. 41 - 47. Izdatel'-
stvo Voronezhskogo Universiteta, Voronezh 1982. [In R.]

52088 - VERETENNIKOV, A.V., CHECHUEVA, T.A. : Sutochnyĭ khod fotosinteza seyantsev
Pinus sylvestris i Picea abies (Pinaceae) v usloviyakh Severa. [The diurnal
course of photosynthesis in the seedlings of Pinus sylvestris and Picea
abies (Pinaceae) under conditions of the North.] - Bot. Zh. 67 : 1521 -
- 1523, 1982. [In R.]

52089 - VERMAAS, W.F.J., GOVINDJEE : Bicarbonate effects on chlorophyll a fluorescence transients in the presence and the absence of diuron. - Biochim. biophys. Acta 680 : 202 - 209, 1982.

52090 - **VERMAAS, W.F.J., GOVINDJEE** : Bicarbonate or carbon dioxide as a requirement
for efficient electron transport on the acceptor side of photosystem II. -
In : GOVINDJEE (ed.) : Photosynthesis. Vol.II. Pp. 541 - 558. Academic Press,
New York - London - Paris - San Diego - San Francisco - São Paulo - Sydney -
Tokyo - Toronto 1982.

52091 - **VERMAAS, W.F.J., VAN RENSEN, J.J.S., GOVINDJEE** : The interaction between
bicarbonate and the herbicide ioxynil in the thylakoid membrane and the effects of amino acid modification on bicarbonate action. - Biochim. biophys.
Acta *681* : 242 - 247, 1982.

52092 - **VERMÉGLIO, A.** : Electron transfer between primary and secondary electron
acceptors in chromatophores and reaction centers of photosynthetic bacteria.
- In : TRUMPOWER, B.L. (ed.) : Function of Quinones in Energy Conserving
Systems. Pp. 169 - 180. Academic Press, New York 1982.

52093 - **VERMEGLIO, A., CARRIER, J.-M.** : Modulation de la respiration par la lumière
chez la bactérie photosynthétique *Rhodopseudomonas sphaeroïdes* : périodici-
té de deux en fonction du rang de l'éclair. - Compt. rend. Acad. Sci. Paris,
Sér.III *295* : 147 - 150, 1982.

52094 - **VERMEGLIO, A., PAILLOTIN, G.** : Structure of *Rhodopseudomonas viridis* reaction centers. Absorption and photoselection at low temperature. - Biochim.
biophys. Acta *681* : 32 - 40, 1982.

52095 - **VERNET, T., FLECK, J., DURR, A., FRITSCH, C., PINCK, M., HIRTH, L.** : Expression of the gene coding for the small subunit of ribulosebisphosphate carboxylase during differentiation of tobacco plant protoplasts. - Europe. J.
Biochem. *126* : 489 - 494, 1982.

52096 - **VERNON, L.P., CARDON, S.** : Direct spectrophotometric measurement of photosystem I and photosystem II activities of photosynthetic membrane preparations from *Cyanophora paradoxa*, *Phormidium laminosum*, and spinach. - Plant
Physiol. *70* : 442 - 445, 1982.

52097 - **VERNOTTE, C., BRIANTAIS, J.-M., MAISON-PETERI, B.** : Analysis of the kinetics
of the cation-induced increase in photosystem II fluorescence in isolated
thylakoids. - Biochim. biophys. Acta *681* : 11 - 14, 1982.

52098 - **VIALE, A.M., ANDREO, C.S., VALLEJOS, R.H.** : The interaction of phenylglyoxal
with soluble and membrane-bound chloroplast coupling factor 1. - Biochim.
biophys. Acta *682* : 135 - 144, 1982.

*52099 - **VICHERKOVÁ, M., SOUČKOVÁ, L.** : Changes of growth analysis values in the
wheat and sunflower plant stands of different density. - Scr. Fac. Sci. nat.
Univ. purkynianae brun., Biol. *6* (2/3) : 69 - 82, 1976.

52100 - **VICTOR, D.M.** : Variation of ^{14}C-labeled photosynthate recovery from roots
and rooting media of warm season grasses. - Crop Sci. *22* : 362 - 366,
1982.

52101 - **VIDAL, J., GODBILLON, G., GADAL, P.** : Estimation of *Sorghum* leaf phosphoenolpyruvate carboxylase protein using an immunoadsorbent column. - Phytochemistry *21* : 2829 - 2830, 1982.

52102 - **VIETINGHOFF, U.** : Berücksichtigung von Umwelt-Stress in einem mathematischen
Modell für ein aquatisches Ökosystem. - In : UNGER, K., SCHUH, J. (ed.) :
Umwelt-Stress. Pp. 345 - 353. Martin-Luther-Universität, Halle 1982.

52103 - **VINCENT, W.F., VINCENT, C.L.** : Factors controlling phytoplankton production in Lake Vanda (77°S). - Can. J. Fish. aquat. Sci. *39* : 1602 - 1609,
1982. [Chl.]

52104 - **VINCENZINI, M., MATERASSI, R., TREDICI, M.R., FLORENZANO, G.** : Hydrogen
production by immobilized cells - I. Light dependent dissimilation of organic substances by *Rhodopseudomonas palustris*. - Int. J. Hydrogen Energy
7 : 231 - 236, 1982.

52105 - **VINCENZINI, M., MATERASSI, R., TREDICI, M.R., FLORENZANO, G.** : Hydrogen
production by immobilized cells - II. H_2-photoproduction and waste-water
treatment by agar-entrapped cells of *Rhodopseudomonas palustris* and *Rhodospirillum molischianum*. - Int. J. Hydrogen Energy *7* : 725 - 728, 1982.

52106 - VINES, H.M., ARMITAGE, A.M., CHEN, S.-S., TU, Z.-P., BLACK, C.C.,Jr. :
A transient burst of CO_2 from geranium leaves during illumination at various light intensities as a measure of photorespiration. - Plant Physiol. *70* : 629 - 631, 1982.

52107 - VINKLER, C., KORENSTEIN, R. : Characterization of external electric field-driven ATP synthesis in chloroplasts. - Proc. nat. Acad. Sci. USA *79* : 3183 - 3187, 1982.

52108 - VINKLER, C., KORENSTEIN, R., FARKAS, D.L. : External electric field-driven ATP synthesis in chloroplast: a slow, ATP synthase-dependent reaction. - FEBS Lett. *145* : 235 - 240, 1982.

*52109 - VINOGRADOV, B.V. : Distantsionnaya indikatsiya v ěkologicheskoǐ botanike. [Remote sensing in ecological botany.] - Zh. obshch. Biol. *37* : 47 - 55, 1976. [In R, ab : E.]

52110 - VINOGRADOV, B.V. : Distantsionnoe izmerenie fitomassy. [Remote measurement of phytomass.] - Issled. Zemli Kosmosa *1982* (5) : 36 - 45, 1982. [In R, ab : E.]

52111 - VIRO, M., KLOPPSTECH, K. : Expression of genes for plastid membrane proteins in barley under intermittent light conditions. - Planta *154* : 18 - 23, 1982.

*52112 - VIRZO DE SANTO, A., DE LUCA, P., ALFANI, A. : CAM metabolism, transpiration and moisture uptake among ten species of atmospheric Tillandsias. - G. bot. ital. *109* : 309 - 310, 1975.

*52113 - VIRZO DE SANTO, A., DE LUCA, P., ALFANI, A. : Adattamenti fisiologici alla vita "aerea" nel genere *Tillandsia*. [Physiological adaptations to "aerial" life in the genus *Tillandsia*.] - G. bot. ital. *111* : 195 - 210, 1977. [CAM; in Ital., ab : E.]

52114 - VISSER, C.M. : Evolutionary roots of catalysis by nicotinamine and flavins in C-H oxidoreductases and in photosynthesis. - Origins Life *12* : 165 - 179, 1982.

52115 - VÍTEK, L. : Dynamika tvorby nadzemní biomasy travního porostu při stupňovaných dávkách dusíku a při vyšší frekvenci sklizní. [Dynamics of the aboveground biomass production in grassland at gradated nitrogen rates and higher frequency of cutting.] - Rost. Výroba (Praha) *28* : 263 - 271, 1982. [In Czech, ab : E, G, R.]

*52116 - VIVEKANANDAN, M. : Biochemical composition of plastidic and amitrole-induced aplastidic leaf cells of *Canna edulis* KER. - Isr. J. Bot. *30* : 33 - 39, 1981.

*52117 - VIVEKANANDAN, M. : Respiratory metabolism of amitrole-induced aplastidic mosophyll cells of *Canna edulis* KER. - Indian J. Plant Physiol. *24* : 206 - - 211, 1981. [Chloroplast.]

52118 - VLADIMIROVA, M.G., MARKELOVA, A.G., KASATKINA, T.I. : Svoǐstva fotosinteticheskogo apparata lishennogo kletochnoǐ stenki mutanta *Chlamydomonas reinhardii* CW-15. [Properties of photosynthetic apparatus in a cell-wall-free mutant of *Chlamydomonas reinhardii* CW-15.] - Fiziol. Rast. *29* : 80 - 87, 1982. [In R, ab : E.]

52119 - VLADIMIROVA, M.G., MARKELOVA, A.G., SEMENENKO, V.E. : Vyyavlenie lokalizatsii ribulozobisfosfatkarboksilazy v pirenoidakh odnokletochnykh vodoroslei tsitoimmunofluorestsentnym metodom. [Identification of ribulose bisphosphate carboxylase localization in the pyrenoids of unicellular algae by cyto-immunofluorescent method.] - Fiziol. Rast. *29* : 941 - 950, 1982. [In R, ab : E.]

52120 - VLK, J., RŮŽIČKOVÁ, M., SEDLÁČKOVÁ, M. : Diference ve fotosyntetické intenzitě tuřínu. [Differences in photosynthetic rate of turnip.] - Sborník ÚVTIZ - Genet. Šlecht. (Praha) *18* : 77 - 79, 1982. [In Czech, ab : E, G, R.]

52121 - VOEVODIN, A.V., KONDRATENKO, V.I., KOROLEV, A.M. : Primenenie zamedlennoĭ
fluorestsentsii dlya izucheniya vliyaniya gerbitsidov na fotosinteticheskiĭ
transport élektronov i fotofosforilirovanie. [Use of delayed fluorescence
to study the effect of herbicides on photosynthetic electron transport and
photophosphorylation.] - Byul. vsesoyuz. nauch.-issled. Inst. Zashchity
Rast. *52* : 61 - 67, 1982. [In R, ab : E.]

52122 - VOGELMANN, T.C., LARSON, P.R., DICKSON, R.E. : Translocation pathways in
the petioles and stem between source and sink leaves of *Populus deltoides*
BARTR. ex MARSH. - Planta *156* : 345 - 358, 1982.

52123 - VOGT, K.A., GRIER, C.C., MEIER, C.E., EDMONDS, R.L. : Mycorrhizal role in
net primary production and nutrient cycling in *Abies amabilis* ecosystems
in western Washington. - Ecology *63* : 370 - 380, 1982.

52124 - VOLENEC, J.J., NELSON, C.J. : Diurnal leaf elongation of contrasting tall
fescue genotypes. - Crop Sci. *22* : 531 - 535, 1982. [Stomatal resistance.]

52125 - VOLKOVA, E.B., ZOLOTUKHIN, I.G., LISOVSKIĬ, G.M., ROKHLIN, G.N., SID'KO,
F.Ya., TIKHOMIROV, A.A., FEDOROV, V.V. : Fotobiologicheskaya éffektivnost'
nekotorykh istochnikov sveta dlya svetokul'tury. [Photobiological effecti-
vity of some light sources used in horticulture.] - Svetotekhnika *1982* (9):
1 - 3, 1982. [PhAR and production; in R, ab : E.]

52126 - VOLKOVA, N.V., LOZINSKIĬ, M.O., MUSHKETIK, L.S., VASILENOK, L.I., KANIVETS,
N.P., YASNIKOV, A.A. : O vzaimodeĭstvii étoniya s fotofosforiliruyushcheĭ
sistemoĭ khloroplastov. [Interaction of ethonium with the photophosphoryla-
ting system of chloroplasts.] - Dokl. Akad. Nauk ukr. SSR, Ser. B *1982* (2):
37 - 39, 1982. [In R.]

52127 - VOLODARSKIĬ, N.I., BYSTRYKH, E.E. : Ispol'zovanie pokazateleĭ pervichnykh
reaktsiĭ fotosinteza dlya diagnostiki vysokoĭ produktivnosti yarovoĭ pshenitsy. [Use of indexes of primary photosynthetic reactions for predicting
high productivity in spring wheat.] - Dokl. vsesoyuz. Akad. sel'skokhoz.
Nauk im. V.I.Lenina *1982* (12) : 4 - 6, 1982. [In R.]

52128 - VOLOVIK, O.I., VASILENOK, L.I., VOLKOVA, N.V., MUSHKETIK, L.S., KANIVETS,
N.P., YASNIKOV, A.A. : O roli atsetolfosfata v fotofosforilirovanii khloroplastov gorokha. [Role of acetolphosphate in photophosphorylation of pea
chloroplasts.] - Fiziol. Biokhim. kul't. Rast. *14* : 23 - 28, 102, 1982.
[In R, ab : E.]

52129 - VONK, C.R., RIBÔT, S.A. : Assimilate distribution and the role of abscisic
acid and zeatin in relation to flower-bud blasting, induced by lack of
light in *Iris* cv. Ideal. - Plant Growth Regulation *1* : 93 - 105, 1982.

52130 - VONSHAK, A., MASKE, H. : Algae: growth techniques and biomass production. -
In : COOMBS, J., HALL, D.O. (ed.) : Techniques in Bioproductivity and Photo-
synthesis. Pp. 66 - 77. Pergamon Press, Oxford - New York - Toronto - Sydney -
Paris - Frankfurt 1982.

52131 - VOROB'EVA, T.N., LUKASHEV, E.P., RIZNICHENKO, G.Yu. : Issledovanie funktsional'noĭ organizatsii aktseptornogo uchastka élektron-transportnoĭ tsepi v
reaktsionnykh tsentrakh fotosinteziruyushchikh bakteriĭ *Rhodopseudomonas
sphaeroides*. [Functional organization of acceptor part of electron-transport chain in reaction centres of photosynthetic bacteria *Rhodopseudomonas
sphaeroides*.] - Biol. Nauki *1982* (7) : 44 - 51, 1982. [In R.]

52132 - VOŠKERUŠA, J. : Dynamika nárůstu a kvalita nadzemní biomasy máku. [The
increment dynamics and quality of above-ground biomass in poppy.] - Rostl.
Výroba (Praha) *28* : 841 - 850, 1982. [In Czech, ab : E, G, R.]

*52133 - VOSKOBOĬNIKOV, G.M. : Élektronno-mikroskopicheskoe issledovanie kletok *Ahn-
feltia tobuchiensis* iz razlichnykh chasteĭ talloma. [Electron microscopic
study of the cells of *Ahnfeltia tobuchiensis* from different parts of the
thallome.] - In : Biologiya Anfel'tsii. Pp. 21 - 27. Dal'nevostochnyĭ
Nauchnyĭ Tsentr Akademii Nauk SSSR. Institut Biologii Morya, Vladivostok
1980. [Chloroplast; in R, ab : E.]

52134 - **VOSKRESENSKAYA, N.P.** : Regulyatornaya rol' sinego sveta v fotosinteze. [Regulatory role of blue light in photosynthesis.] - In : Fiziologiya Fotosinteza. Pp. 203 - 220. Nauka, Moskva 1982. [In R.]

52135 - **VOSKRESENSKAYA, N.P., DROZDOVA, I.S., MOSKALENKO, A.A., CHETVERIKOV, A.G., TSEL'NIKER, Yu.L.** : Perestroĭki fotosinteticheskogo apparata pod vliyaniem dlitel'nogo deĭstviya krasnogo i sinego sveta. [Rearrangements in photosynthetic apparatus in response to long-term exposure of plants to red and blue light.] - Fiziol. Rast. *29* : 447 - 456, 1982. [In R, ab : E.]

52136 - **VOZNESENSKIĬ, V.L., GLAGOLEVA, T.A., ZUBKOVA, E.K., MAMUSHINA, N.S., FILIPPOVA, L.A., CHULANOVSKAYA, M.V.** : Metabolizm ^{14}C pri dlitel'nom vyrashchivanii khlorelly v prisutstvii $^{14}CO_2$. [Metabolism of ^{14}C in *Chlorella* during prolonged cultivation in the presence of $^{14}CO_2$.] - Fiziol. Rast. *29* : 564 - - 571, 1982. [In R, ab : E.]

52137 - **VOZNESENSKIĬ, V.L., SHCHERBATYUK, A.S.** : Graduirovanie uglekislotnykh gazoanalizatorov. [Calibration of CO_2 infra-red gas analysers.] - Fiziol. Biokhim. kul't. Rast. *14* : 600 - 606, 1982. [In R.]

52138 - **VREŠTIAK, P.** : Zhodnotenie vybraných taxónov drevín podľa veľkosti fotosynteticky aktívneho povrchu. [Evaluation of some taxa of woody plants according to photosynthetically active surface.]-Acta Univ. Agr., Fac. agron. (Brno) *A28* (3/4) : 495 - 498, 1980. [In Slovak.]

52139 - **VU, C.V., ALLEN, L.H., Jr., GARRARD, L.A.:** Effects of supplemental UV-B radiation on primary photosynthetic carboxylating enzymes and soluble proteins in leaves of C_3 and C_4 crop plants. - Physiol. Plant. *55* : 11 - 16, 1982.

52140 - **VU, C.V., ALLEN, L.H., Jr., GARRARD, L.A.** : Effects of UV-B radiation (280- -320 nm) on photosynthetic constituents and processes in expanding leaves of soybean (*Glycine max* (L.) MERR.). - Environm. exp. Bot. *22* : 465 - 473, 1982.

52141 - **VUČINIČ, Ž., NEŠIČ, G., RADENOVIČ, Č.** : Delayed fluorescence as an *in situ* probe of fluidity changes in maize photosynthetic membranes. - Period. Biol. *84* : 223 - 226, 1982.

52142 - **VYSHKVARTSEV, D.I., NGUEN TAK AN, KONOVALOVA, G.V., KHARLAMENKO, V.I.** : Faktory opredelyayushchie produktivnost' bukhty Nyafu Yuzhno-Kitaĭskogo Morya. [Factors determining productivity of the Nhaphu Bay, South China Sea.] - Biol. Morya *1982* (6) : 17 - 23, 1982. [Ps; in R, ab : E.]

52143 - **WADA, H., INUBUSHI, K., TAKAI, Y.** : [Easily decomposable organic matter in paddy soils (III) Relationship between chlorophyll-type compounds and mineralisable nitrogen.] - Nippon Doyo Hiryogaku Zasshi [Jap. J. Soil Sci. Plant Nutr.] *53* : 380 - 384, 1982. [In Jap., ab : E.]

*52144 - **WAGENER, K.** : Methane production by mariculture on land. - In : CHARTIER,P., PALZ, W. (ed.) : Energy from Biomass. Vol.1. Pp. 64 - 69. D.Reidel Publ. Co., Dordrecht - Boston - London 1981.

52145 - **WAGENER, K.** : Fuel gas production by mariculture on land. - In : PALZ, W., GRASSI, G. (ed.) : Energy from Biomass. Vol.2. Pp. 166 - 175. D.Reidel Publ. Co., Dordrecht - Boston - London 1982.

52146 - **WAGHMODE, A.P., JOSHI, G.V.** : Chemical composition of leaves of halophytes & sediments in estuarine habitat. - Indian J. mar. Sci. *11* : 104 - 106, 1982. [Chl.]

52147 - **WAGHMODE, A.P., JOSHI, G.V.** : Photosynthetic and photorespiratory enzymes and metabolism of ^{14}C-substrates in isolated leaf cells of the C_4 species of *Aeluropus lagopoides* L. - Photosynthetica *16* : 17 - 21, 1982.

52148 - **WAGNER, G.** : Secondary ion movements in *Halobacterium halobium*. - In : COLOWICK, S.P., KAPLAN, N.O. (ed.) : Methods in Enzymology. Vol.88. Pp. 344 - - 349. Academic Press, New York - London - Paris - San Diego - San Francisco - São Paulo - Sydney - Tokyo - Toronto 1982.

52149 - **WAGNER, G.J.** : Compartmentation in plant cells: the role of the vacuole. - In : CREASY, L.L., HRAZDINA, G. (ed.) : Cellular and Subcellular Localization in Plant Metabolism. Pp. 1 - 45. Plenum Press, New York - London 1982. [Chloroplast, photosynthates.]

52150 - **WAGNER, R., CARRILLO, N., JUNGE, W., VALLEJOS, R.H.** : On the conformation of reconstituted ferredoxin:NADP⁺ oxidoreductase in the thylakoid membrane. Studies *via* triplet lifetime and rotational diffusion with eosin isothiocyanate as label. - Biochim. biophys. Acta *680* : 317 - 330, 1982.

52151 - **WAGNER, R., JUNGE, W.** : Coupling factor for photophosphorylation labeled with eosin isothiocyanate: activity, size, and shape in solution. - Biochemistry *21* : 1890 - 1899, 1982.

52152 - **WAKAMATSU, K.** : The light oxidation and the dark-reduction of plastcyanin in subchloroplast particles. - Seikatsu Kagaku [Science human Life] *13* (2/3) : 67 - 78, 1982.

52153 - **WALBOT, V., HOISINGTON, D.A.** : Isolation of mesophyll and bundle sheath chloroplasts from maize. - In : EDELMAN, M., HALLICK, R.B., CHUA, N.-H. (ed.) : Methods in Chloroplast Molecular Biology. Pp. 211 - 219. Elsevier Biomedical Press, Amsterdam - New York - Oxford 1982.

52154 - **WALDREN, R.P.** : Corn. - In : TEARE,I.D.,PEET,M.M. (ed.) : Crop-Water Relations. Pp. 187 - 212. John Wiley & Sons, New York - Chichester - Brisbane - Toronto - Singapore 1982. [Ps.]

52155 - **WALKER, R.R., TÖRÖKFALVY, E., DOWNTON, W.J.S.** : Photosynthetic responses of the citrus varieties Rangpur lime and Etrog citron to salt treatment. - Aust. J. Plant Physiol. *9* : 783 - 790, 1982.

52156 - **WALKER, T.** : Use of a Secchi disc to measure attenuation of underwater light for photosynthesis. - J. appl. Ecol. *19* : 539 - 544, 1982.

52157 - **WALLACE, A.** : Effect of nitrogen fertilizer and nodulation on lime-induced chlorosis in soybeans. - J. Plant Nutr. *5* : 363 - 368, 1982.

52158 - **WALLACE, A., SAMMAN, Y.S., WALLACE, G.A.** : Correction of lime-induced chlorosis in soybeans in a glasshouse with sulfur and an acidifying iron compound. - J. Plant Nutr. *5* : 949 - 953, 1982.

52159 - **WALLACE, B.A.** : Comparison of bacteriorhodopsin and rhodopsin molecular structure. - In : COLLOWICK, S.P., KAPLAN, N.O. (ed.) : Methods in Enzymology. Vol.88. Pp. 447 - 462. Academic Press, New York - London - Paris - San Diego - San Francisco - São Paulo - Sydney - Tokyo - Toronto 1982.

52160 - **WALLACE, B.A., HENDERSON, R.** : Location of the carboxyl terminus of bacteriorhodopsin in purple membrane. - Biophys. J. *39* : 233 - 239, 1982.

52161 - **WALLACE, D.C.** : Structure and evolution of organelle genomes. - Microbiol. Rev. *46* : 208 - 240, 1982.

*52162 - **WALLERSTEIN, I., OGREN, R., MONSELISE, S.P.** : Rapid and slow translocation of ¹⁴C-sucrose and ¹⁴C-assimilates in *Citrus* and *Phaseolus* with special reference to ringing effect. - J. hort. Sci. *53* : 203 - 208, 1978.

52163 - **WALLSGROVE, R.M., LEA, P.J., MIFLIN, B.J.** : The development of NAD/P/H-dependent and ferredoxin-dependent glutamate synthase in greening barley and pea leaves. - Planta *154* : 473 - 476, 1982.

52164 - **WALOSZCZYK, K.** : Einfluß von Temperatur und Bodenwassergehalt auf Aufgang und Herbstwachstum von Winterweizen (Klimakammerversuch). - In : UNGER, K., SCHUH, J. (ed.) : Umwelt-Stress. Pp. 253 - 256. Martin-Luther-Universität, Halle 1982.

52165 - **WALSH, P., LEGENDRE, L.** : Effets des fluctuations rapides de la lumière sur la photosynthèse du phytoplancton. - J. Plankton Res. *4* : 313 - 327, 1982.

52166 - **WALTER, G.** : Lichtregulation der Chlorophyllbildung bei Pro- und Eukaryoten. - In : HOFFMANN, P., HIEKE, B. (ed.) : Photosynthese: Regulation und Evolution. (Colloquia Pflanzenphysiologie Nr.5.) Pp. 125 - 134. Humboldt-Universität, Berlin 1982.

52167 - WALTER, G., HOFFMANN, P. : Zur Lichtstabilisierung der Photosynthesepigmen-
te höherer Pflanzen. - In : UNGER, K., SCHUH, J. (ed.) : Umwelt-Stress. Pp.
225 - 229. Martin-Luther-Universität, Halle 1982.

52168 - WALTERS, D.R., AYRES, P.G. : Translocation of ^{14}C-labelled photoassimilates
to roots in barley: effects of mildew on partitioning in roots and the mi-
totic index. - Plant Pathol. 31 : 307 - 313, 1982.

52169 - WALTON, D.W.H. : Instruments for measuring biological microclimates for
terrestrial habitats in polar and high alpine regions: A review. - Arctic
alp. Res. 14 : 275 - 286, 1982. [Also radiation.]

52170 - WALTON, N.J. : Glyoxylate decarboxylation during glycollate oxidation by pea
leaf extracts: significance of glyoxylate and extract concentrations. -
Planta 155 : 218 - 224, 1982.

52171 - WANG, C.-B., TIEN, H.T., LOPEZ, J.R., LIU, Q.-Y., JOSHI, N.B., HU, Q.-Y. :
Photoelectrochemical properties of bilayer lipid membranes containing co-
valently linked porphyrin-quinone and other complexes. - Photobiochem. Photobiophys. 4 : 177 - 184, 1982.

52172 - WANG, C.Y., CHENG, S.H., KAO, C.H. : Senescence of rice leaves. VII. Proline accumulation in senescing excised leaves. - Plant Physiol. 69 : 1348 -
- 1349, 1982. [Chl.]

52173 - WANG Guo-qiang, XU Ya-nan, LI Shu-jun : [Effects of chemical modification
by acetic anhydride on chloroplast coupling factor.] - Acta Phytophysiol.
sin. 8 : 223 - 230, 1982. [In Chin., ab : E.]

52174 - WANG Wan-li, LIN Zhi-ping, ZHANG Xiu-ying, WU Ya-hua : [On the effect of
soil drought during the period from the end of flowering to ripening on
the grain filling and matter translocation in the wheat plant.] - Acta
Phytophysiol. sin. 8 : 67 - 80, 1982. [Ps; in Chin., ab : E.]

52175 - WANG, Y.-Ch., HANADA, K. : [Effects of defoliation on growth of main stem
and primary tillers in rice seedlings.] - Jap. J. Crop Sci. 51 : 455 - 461,
1982. [Photosynthates; in Jap., ab : E.]

52176 - WANG, Y.Ch., HANADA, K. : [Translocation of ^{14}C-assimilate among main stem
and tillers in rice plants.] - Jap. J. Crop Sci. 51 : 483 - 491, 1982. [In
Jap., ab : E.]

52177 - WANG Yu-qin, LIN Zhen-wu, WU Shao-bo, ZHAO Yu-ju, TANG Yu-wei : [Enhance-
ment of chloroplast development by 6-benzylaminopurine in etiolated wheat
leaves. I. Effect of 6-BA on chlorophyll formation.] - Acta Phytophysiol.
sin. 8 : 45 - 52, 1982. [In Chin., ab : E.]

52178 - WANNER, G., VIGIL, E.L., THEIMER, R.R. : Ontogeny of microbodies (glyoxysomes) in cotyledons of dark-grown watermelon (Citrullus vulgaris SCHRAD)
seedlings. Ultrastructural evidence. - Planta 156 : 314 - 325, 1982.

52179 - WARDLAW, I.F. : Assimilate movement in Lolium and Sorghum leaves. III. Carbon dioxide concentration effects on the metabolism and translocation of
photosynthate. - Aust. J. Plant Physiol. 9 : 705 - 713, 1982.

52180 - WAREMBOURG, F.R. : Répartition et devenir des assimilats dans les écosystè-
mes prairiaux. - Acta oecol.-Oecol.gen. 3 : 75 - 90, 1982.

52181 - WAREMBOURG, F.R., MONTANGE, D., BARDIN, R. : The simultaneous use of ^{14}CO$_2$
and ^{15}N$_2$ labelling techniques to study the carbon and nitrogen economy of
legumes grown under natural conditions. - Physiol. Plant. 56 : 46 - 55,
1982.

52182 - WARING, R.H. : Land of the giant conifers. - Natur. Hist. 91(10) : 55 -
- 62, 1982. [Ps.]

52183 - WARING, R.H., SCHROEDER, P.E., OREN, R. : Application of the pipe model theory to predict canopy leaf area. - Can. J. Forest Res. 12 : 556 - 560,
1982.

*52184 - WARSHEL, A. : Interpretation of resonance Raman spectra of biological mole-
          cules. - Annu. Rev. Biophys. *6* : 273 - 300, 1977. [Chl.]

 52185 - WASIELEWSKI, M.R. : Synthetic approaches to photoreaction center structure
          and function. - In : FONG, F.K. (ed.) : Light Reaction Path of Photosynthe-
          sis. Pp. 234 - 276. Springer-Verlag, Berlin - Heidelberg - New York 1982.

 52186 - WASIELEWSKI, M.R., SMITH, U.H., NORRIS, J.R. : ESR study of the primary
          electron donor in highly ^{13}C-enriched *Chlorobium limicola* f. *thiosulfato-
          philum*. - FEBS Lett. *149* : 138 - 140, 1982.

 52187 - WASMUND, N. : Probleme der ^{14}C-Methode. - Wiss. Z. Wilhelm-Pieck-Univ.
          Rostock, naturwiss. Reihe *31* (6) : 37 - 41, 1982.

 52188 - WASMUND, N., KOWALCZEWSKI, A. : Production and distribution of benthic
          microalgae in the littoral sediments of Mikołajskie Lake. - Ekol. pol. *30* :
          287 - 301, 1982. [Ps, Chl.]

 52189 - WASTERNACK, C. : Nucleotidinterkonversion, Nukleotidkompartimentierung
          und intrazellulärer Nukleotidtransport unter besonderer Berücksichtigung
          der Adenylate. - In : HOFFMANN, P., HIEKE, B. (ed.) : Photosynthese: Regu-
          lation und Evolution. (Colloquia Pflanzenphysiologie Nr.5.) Pp. 61 - 78.
          Humboldt-Universität, Berlin 1982.

 52190 - WASTERNACK, C., WEISSER, J. : Potentiation by dihydrothymine of 5-fluoro-
          uracil inhibition during chloroplast development and RNA synthesis in *Eu-
          glena gracilis*. - Biochem. Physiol. Pflanzen *177* : 757 - 768, 1982.

 52191 - WATANABE, A., PRICE, C.A. : Translation of mRNAs for subunits of chloroplast
          coupling factor 1 in spinach. - Proc. nat. Acad. Sci. USA *79* : 6304 - 6308,
          1982.

 52192 - WATANABE, K., NAKAJIMA, Y. : Vertical distribution of chlorophyll *a* along
          45° E in the Southern Ocean, 1981. - Mem. nat. Inst. polar Res. spec. Issue
          23 : 73 - 86, 1982.

 52193 - WAVARE, R.A., MOHANTY, P. : Aluminium stimulation of photoelectron trans-
          port in spheroplasts of cyanobacterium *Synechococcus cedrorum*. - Photobio-
          chem. Photobiophys. *3* : 327 - 335, 1982.

*52194 - WAX, E., LOCKAU, W. : Stoichiometric photophosphorylation in thylakoids from
          the blue-green alga, *Anabaena variabilis*. - Z. Naturforsch. *35C* : 98 - 105,
          1980.

 52195 - WEBB, J.A. : Partial purification of galactinol synthase from leaves of
          Cucurbita pepo. - Can. J. Bot. *60* : 1054 - 1059, 1982. [Photosynthates.]

 52196 - WEBER, H.J., BOGOMOLNI, R.A. : The isolation of *Halobacterium* mutant strains
          with defects in pigment synthesis. - In : COLOWICK, S.P., KAPLAN, N.O. (ed.):
          Methods in Enzymology. Vol.88. Pp. 379 - 390. Academic Press, New York -
          London - Paris - San Diego - San Francisco - São Paulo - Sydney - Tokyo -
          Toronto 1982. [Bacteriorhodopsin.]

 52197 - WEBER, H.J., SARMA, S., LEIGHTON, T. : The *Halobacterium* group: microbial
          methods. - In : COLOWICK, S.P., KAPLAN, N.O. (ed.) : Methods in Enzymology.
          Vol.88. Pp. 369 - 373. Academic Press, New York - London - Paris - San Diego
          - San Francisco - São Paulo - Sydney - Tokyo - Toronto 1982. [Bacteriorho-
          dopsin.]

 52198 - WEEGE, K.-H., HELLER, R., SCHRÖDER, R., LUX, H. : Optimisierung der Bestan-
          desstruktur der Zuckerrübe zur effektiven Nutzung von Umwelt- und Intensi-
          vierungsfaktoren. - In : HOFFMANN, P., HIEKE, B. (ed.) : Photosynthese :
          Regulation und Evolution. (Colloquia Pflanzenphysiologie Nr.5.) Pp. 229 -
          - 236. Humboldt-Universität, Berlin 1982.

 52199 - WEI Chin, TANG Xiao-yi, LI You-ze : [Effect of ethanol on photosynthesis
          of chloroplast.] - Acta Phytophysiol. sin. *8* : 149 - 155, 1982. [In Chin.,
          ab : E.]

52200 - WEIDINGER, G., PFEIFER, F., GOEBEL, W. : Plasmids in halobacteria: Restriction maps. - In : COLOWICK, S.P., KAPLAN, N.O. (ed.) : Methods in Enzymology.
Vol.88. Pp. 374 - 379. Academic Press, New York - London - Paris - San Diego - San Francisco - São Paulo - Sydney - Tokyo - Toronto 1982. [Bacterio-
rhodopsin.]

52201 - WEIDNER, M., KÜPPERS, U. : Metabolic conversion of ^{14}C-aspartate, ^{14}C-mala-
te and ^{14}C-mannitol by tissue disks of *Laminaria hyperborea:* Role of phosphoenolpyruvate carboxykinase. - Z. Pflanzenphysiol. *108* : 353 - 364, 1982.

52202 - WEIL, J.H., MUBUMBILA, M., KUNTZ, M., KELLER, M., STEINMETZ, A., CROUSE, E.J.,
BURKARD, G., GUILLEMAUT, P., SELDEN, R., McINTOSH, L., BOGORAD, L., LÖFFEL-
HARDT, W., MUCKE, H., BOHNERT, H.J. : Gene mapping studies and sequence determination on chloroplast transfer RNAs from various photosynthetic organisms. - In : AKOYUNOGLOU, G., EVANGELOPOULOS, A.E., GEORGATSOS, J., PALA-
IOLOGOS, G., TRAKATELLIS, A., TSIGANOS, C.P. (ed.) : Cell Function and Differentiation, Part B. Biogenesis of Energy Transducing Membranes and Membrane and Protein Energetics. Pp. 321 - 331. Alan R. Liss Inc., New York
1982.

52203 - WEINER, J., GRODZIŃSKI, W., GÓRECKI, A., PERZANOWSKI, K. : Standing crop
and above-ground production of vegetation in arid Mongolian steppe with
Caragana. - Pol. ecol. Stud. *8* : 23 - 39, 1982.

52204 - WEINSTEIN, L.H., ALSCHER-HERMAN, R. : Physiological responses of plants to
fluorine. - In : UNSWORTH, M.H., ORMROD, D.P. (ed.) : Effects of Gaseous
Air Pollution in Agriculture and Horticulture. Pp. 139 - 167. Butterworth
Scientific, London - Boston - Sydney - Wellington - Durban - Toronto 1982.
[Ps.]

52205 - WEIS, E. : Influence of light on the heat sensitivity of the photosynthetic
apparatus in isolated spinach chloroplasts. - Plant Physiol. *70* : 1530 -
- 1534, 1982.

52206 - WEIS, E. : The influence of metal cations and pH on the heat sensitivity
of photosynthetic oxygen evolution and chlorophyll fluorescence in spinach
chloroplasts. - Planta *154* : 41 - 47, 1982.

52207 - WEISS, A. : An experimental study of net radiation, its components and
prediction. - Agron. J. *74* : 871 - 874, 1982.

*52208 - WEISS, C. : Electronic absorption spectra of chlorophylls. - In : DOLPHIN,
D. (ed.) : The Porphyrins. Vol.III, Part A. Pp. 211 - 223. Academic Press,
New York - London 1978.

52209 - WELLBURN, A.R. : Bioenergetic and ultrastructural changes associated with
chloroplast development. - Int. Rev. Cytol. *80* : 133 - 191, 1982.

52210 - WELLBURN, A.R. : Effects of SO_2 and NO_2 on metabolic function. - In :
UNSWORTH, M.H., ORMROD, D.P. (ed.) : Effects of Gaseous Air Pollution in
Agriculture and Horticulture. Pp. 169 - 187. Butterworth Scientific, Lon-
don - Boston - Sydney - Wellington - Durban - Toronto 1982. [Ps.]

52211 - WELLBURN, A.R., ROBINSON, D.C., WELLBURN, F.A.M. : Chloroplast development
in low light-growth barley seedlings. - Planta *154* : 259 - 265, 1982.

52212 - WELLS, G.N., NACHTWEY, D.S. : The effects of ultraviolet irradiation on
photosynthesis by *Ruppia maritima* L. (widgeon grass). - In : CALKINS, J.
(ed.) : The Role of Solar Ultraviolet Radiation in Marine Ecosystems. Pp.
555 - 562. Plenum Publ. Co., London - New York 1982.

52213 - WELLS, R., SCHULZE, L.L., ASHLEY, D.A., BOERMA, H.R., BROWN, R.H. : Cultivar differences in canopy apparent photosynthesis and their relationship
to seed yield in soybeans. - Crop Sci. *22* : 886 - 890, 1982.

52214 - WENG, J.-H., TAKEDA, T., AGATA, W., HAKOYAMA, S. : [Studies on dry matter
and grain production of rice plants. I. Influence of the reserved carbohydrate until heading stage and the assimilation products during the ripening period on grain production.] - Jap. J. Crop Sci. *51* : 500 - 509,
1982. [In Jap., ab : E.]

52215 - WENG, J.-H., TAKEDA, T., AGATA, W., HAKOYAMA, S. : [Studies on dry matter
         and grain production of rice plants. II. Varietal differences in dry matter
         productivity before the heading stage.] - Jap. J. Crop Sci. *51* : 510 - 518,
         1982. [In Jap., ab : E.]

52216 - WENG, J.-H., TAKEDA, T., AGATA, W., HAKOYAMA, S. : [Studies on dry matter
         and grain production of rice plants. III. Analysis of dry matter producti-
         vity before heading stage.] - Jap. J. Crop Sci. *51* : 519 - 528, 1982. [In
         Jap., ab : E.]

52217 - WERNER, I., WEISE, G. : Biomass production of submersed macrophytes in a
         selected stretch of the river Zschopau (south GDR) with special regard
         to orthophosphate incorporation. - Int. Rev. ges. Hydrobiol. *67* : 45 - 62,
         1982.

*52218 - WESTHOFF, P., ZIMMERMANN, K., BOEGE, F., ZETSCHE, K. : Regulation of the
         synthesis of ribulose-1,5-bisphosphate carboxylase and its subunits in the
         flagellate *Chlorogonium elongatum*. II. Coordinated synthesis of the large
         and small subunits. - Z. Naturforsch. *36 C* : 942 - 950, 1981.

52219 - WESTOBY, M. : Frequency distributions of plant size during competitive
         growth of stands: the operation of distribution-modifying functions. - Ann.
         Bot. *50* : 733 - 735, 1982.

52220 - WETTSTEIN, D. von : Rapporteur's summary: Origin and evolution of plastid
         proteins. - In : SCHIFF, J.A., LYMAN, H. (ed.) : On the Origins of Chloro-
         plasts. Pp. 263 - 273. Elsevier/North-Holland, New York - Amsterdam - Ox-
         ford 1982.

52221 - WETTSTEIN, D. von, MØLLER, B.L., HØYER-HANSEN, G., SIMPSON, D. : Mutants in
         the analysis of the photosynthetic membrane polypeptides. - In : SCHIFF,
         J.A., LYMAN, H. (ed.) : On the Origins of Chloroplasts. Pp. 243 - 255.
         Elsevier/North-Holland, New York - Amsterdam - Oxford 1982.

52222 - WETTSTEIN, D. von, MØLLER, B.L., KOENIG, F., HENRY, L., HØYER-HANSEN, G. :
         Polypeptides associated with photosystem II. - In : HALL, D.O., PALZ, W.
         (ed.) : Photochemical, Photoelectrochemical and Photobiological Processes.
         Vol.1. Pp. 118 - 124. D.Reidel Publ.Co., Dordrecht - Boston - London
         1982.

52223 - WETZSTEIN, H.Y., SOMMER, H.E. : Leaf anatomy of tissue-cultured *Liquidambar
         styraciflua* (*Hamamelidaceae*) during acclimatization. - Amer. J. Bot. *69* :
         1579 - 1586, 1982. [Chloroplast.]

52224 - WEYERS, J.D.B., PATERSON, N.W., FITZSIMONS, P.J., DUDLEY, J.M. : Metabolic
         inhibitors block ABA-induced stomatal closure. - J. exp. Bot. *33* : 1270 -
         - 1278, 1982.

52225 - WHATLEY, J.M. : Ultrastructure of plastid inheritance: green algae to angio-
         sperms. - Biol. Rev. *57* : 527 - 569, 1982.

52226 - WHATLEY, J.M., HAWES, C.R., HORNE, J.C., KERR, J.D.A. : The establishment
         of the plastid thylakoid system. - New Phytol. *90* : 619 - 629, 1982.

52227 - WHEELER, W.N. : Response of macroalgae to light quality, light intensity,
         temperature, CO_2, HCO_3^-, O_2, mineral nutrients, and pH. - In :ZABORSKY, O.R.
         (ed.) : CRC Handbook of Biosolar Resources. Vol.I/2. Pp. 157 - 184. CRC
         Press, Boca Raton 1982. [Ps.]

52228 - WHITFIELD, D.W.A., MEHLENBACHER, L.A., LABINE, C. : Solar radiation atte-
         nuation in a hillside jack pine forest. - Can. J. Bot. *60* : 1913 - 1922,
         1982. [Leaf area index.]

52229 - WHITMARSH, J., BOWYER, J.R., CROFTS, A.R. : Modification of the apparent
         redox reaction between cytochrome *f* and the Rieske non-sulfur protein. -
         Biochim. biophys. Acta *682* : 404 - 412, 1982.

52230 - WIDMER, E., SOUKUP, M., ZELL, R., BROGER, E., LOHRI, B., MARBET, R., LUKÁČ,
         T. : Technische Verfahren zur Synthese von Carotinoiden und verwandten Ver-
         bindungen aus 6-Oxo-isophoron. VI. Synthese von Rhodoxanthin und (*3RS,3'RS*)-

-Zeaxanthin; Zugänge zur C_{15}-Ringkomponente über 3-Oxo-jonon-Derivate. - Helv. chim. Acta *65* : 944 - 957, 1982.

52231 - WIDMER, E., ZELL, R., GRASS, H., MARBET, R. : Technische Verfahren zur Synthese von Carotinoiden und verwandten Verbindungen aus 6-Oxo-isophoron. VII. Synthese von Rhodoxanthin und (3*RS*, 3'*RS*)-Zeaxanthin aus der C_{15}-Ringkomponente. - Helv. chim. Acta *65* : 958 - 967, 1982.

*52232 - WIĘCKOWSKI, S. : On the green and red complexes derived from spinach thylakoid membranes by the action of Triton X-100 in low salt medium. - Pol. ecol. Stud. *7* : 331 - 334, 1981.

*52233 - WIĘCKOWSKI, S. : Activities of photosystem reaction centers at various phases of greening of etiolated bean seedlings. - Pol. ecol. Stud. *7* : 335 - - 339, 1981.

52234 - WIEMKEN, V., BACHOFEN, R. : Probing the topology of proteins in the chromatophore membrane of *Rhodospirillum rubrum* G-9 with proteinase K. - Biochim. biophys. Acta *681* : 72 - 76, 1982.

52235 - WIEN, H.C. : Dry matter production, leaf area development, and light interception of cowpea lines with broad and narrow leaflet shape. - Crop Sci. *22* : 733 - 737, 1982.

52236 - WIENCKE, C. : Effect of osmotic stress on thylakoid fine structure in *Porphyra umbilicalis*. - Protoplasma *111* : 215 - 220, 1982.

*52237 - WIERZBICKI, B. : Intermittent light as physiological and ecological factor of photosynthesis of pine and spruce. - Acta Physiol. Plant. *2* : 69 - 80, 1980.

*52238 - WIERZBICKI, B. : The use of intermittent illumination in studies on photosynthesis of pine and spruce of various physiological condition. - Acta Physiol. Plant. *2* : 81 - 91, 1980.

52239 - WIESE, M.V. : Crop management by comprehensive appraisal of yield determining variables. - Annu. Rev. Phytopathol. *20* : 419 - 432, 1982.

52240 - WIESNER, B., LEUPOLD, D., VOIGT, B. : Gekoppelte nichtlineare Absorptionsmessungen in der langwelligen und in der Soret-Bande des Chlorphyll-*a in vivo* und *in vitro*. - In : HOFFMANN, P., HIEKE, B. (ed.) : Photosynthese: Regulation und Evolution. Pp. 118 - 120. Humboldt-Universität, Berlin 1982.

52241 - WIESNER, B., LEUPOLD, D., VOIGT, B., MATORIN, D.N. : Intensitätsabhängigkeit der Fluoreszenzquantenausbeute von PS I- und PS II-Defektmutanten sowie DCMU-behandelten Blättern von *Pisum sativum* bei Anregung mit Nanosekundenimpulsen. - In : HOFFMANN, P., HIEKE, B. (ed.) : Photosynthese: Regulation und Evolution. Pp. 121 - 123. Humboldt-Universität, Berlin 1982.

52242 - WILCOX, H.A. : The ocean as a supplier of food and energy. - Experientia *38* : 31 - 35, 1982. [Production.]

52243 - WILDMAN, S.G. : Further aspects of fraction-1 protein evolution. - In : SCHIFF, J.A., LYMAN, H. (ed.) : On the Origins of Chloroplasts. Pp. 229 - - 242. Elsevier/North-Holland, New York - Amsterdam - Oxford 1982.

52244 - WILHELM, C., EISENBEIS, G., WILD, A., ZAHN, R. : *Nanochlorum eucaryotum* : A very reduced coccoid species of marine chlorophyceae. - Z. Naturforsch. *37 C* : 107 - 114, 1982. [Chl, chloroplast.]

52245 - WILHELM, C., WILD, A. : Growth and photosynthesis of *Nanochloris eucaryotum*, a new and extremely small eucaryotic green alga. - Z. Naturforsch. *37 C* : 115 - 119, 1982.

52246 - WILLATT, S.T., OLSSON, K.A. : Root distribution and water uptake by irrigated soybeans on a duplex soil. - Aust. J. Soil Res. *20* : 139 - 146, 1982. [Growth analysis.]

52247 - WILLENBRINK, J. : Physiology of carbon fixation in benthic marine algae. - In : SRIVASTAVA, L.M. (ed.) : Synthetic and Degradative Processes in Marine Macrophytes. Pp. 7 - 22. W.de Gruyter, Berlin 1982.

52248 - WILLERT, D.J. von, ELLER, B.M., BRINCKMANN, E., BAASCH, R. : CO_2 gas exchange and transpiration of *Welwitschia mirabilis* HOOK. fil. in the central Namib desert. - Oecologia *55* : 21 - 29, 1982.

52249 - WILLIAMS, B.A., GURNER, P.J., AUSTIN, R.B. : A new infra-red analyser and portable photosynthesis meter. - Photosynthesis Res. *3* : 141 - 151, 1982.

52250 - WILLIAMS, L.E., DEJONG, T.M., PHILLIPS, D.A. : Effect of changes in shoot carbon-exchange rate on soybean root nodule activity. - Plant Physiol. *69* : 432 - 436, 1982.

52251 - WILLIAMS, P.J.LeB., JENKINSON, N.W. : A transportable microprocessor-controlled precise Winkler titration suitable for field station and shipboard use. - Limnol. Oceanogr. *27* : 576 - 584, 1982.

52252 - WILLIAMS, S.L., McROY, C.P. : Seagrass productivity: The effect of light on carbon uptake. - Aquat. Bot. *12* : 321 - 344, 1982.

52253 - WILLIAMS, W.P., SEN, A., QUINN, P.J. : Protein lipid interactions in the photosynthetic membrane. - Biochem. Soc. Trans. *10* : 335 - 338, 1982.

52254 - WILSON, D. : Overcoming physiological limitations to production from herbage. - Rep. Welsh Plant Breeding Sta. Aberystwyth *1981* : 202 - 215, 1982.

52255 - WILSON, D. : Response to selection for dark respiration rate of mature leaves in *Lolium perenne* and its effects on growth of young plants and simulated swards. - Ann. Bot. *49* : 303 - 312, 1982. [Growth analysis.]

52256 - WILSON, D., JONES, J.G. : Effect of selection for dark respiration rate of mature leaves on crop yields of *Lolium perenne* cv. S23. - Ann. Bot. *49* : 313 - 320, 1982.

52257 - WINK, M., HARTMANN, T. : Localization of the enzymes of quinolizidine alkaloid biosynthesis in leaf chloroplasts of *Lupinus polyphyllus*. - Plant Physiol. *70* : 74 - 77, 1982.

52258 - WINKLER, F.J., KEXEL, H., KRANZ, C., SCHMIDT, H.-L. : Parameters affecting the $^{13}CO_2/^{12}CO_2$ isotope discrimination of the ribulose-1,5-bisphosphate carboxylase reaction. - In : SCHMIDT, H.-L., FÖRSTEL, H., HEINZINGER, K. (ed.) : Stable Isotopes. Pp. 83 - 89. Elsevier Scientific Publishing Company, Amsterdam 1982.

52259 - WINNER, W.E., KOCH, G.W. : Water relations and SO_2 resistance of mosses. - J. Hattori bot. Lab. *1982* (52) : 431 - 440, 1982. [Ps.]

52260 - WINNER, W.E., KOCH, G.W., MOONEY, H.A. : Ecology of SO_2 resistance. IV. Predicting metabolic responses of fumigated shrubs and trees. - Oecologia *52* : 16 - 21, 1982. [Ps.]

52261 - WINTER, E. : Salt tolerance of *Trifolium alexandrinum* L. III. Effect of salt on ultrastructure of phloem and xylem transfer cells in petioles and leaves. - Aust. J. Plant Physiol. *9* : 239 - 250, 1982. [Chloroplast.]

52262 - WINTER, K. : Properties of phosphoenolpyruvate carboxylase in rapidly prepared, desalted leaf extracts of the Crassulacean acid metabolism plant *Mesembryanthemum crystallinum* L. - Planta *154* : 298 - 308, 1982.

52263 - WINTER, K., FOSTER, J.G., EDWARDS, G.E., HOLTUM, J.A.M. : Intracellular localization of enzymes of carbon metabolism in *Mesembryanthemum crystallinum* exhibiting C_3 photosynthetic characteristics or performing Crassulacean acid metabolism. - Plant Physiol. *69* : 300 - 307, 1982.

52264 - WINTER, K., FOSTER, J.G., SCHMITT, M.R., EDWARDS, G.E. : Activity and quantity of ribulose bisphosphate carboxylase- and phosphoenolpyruvate carboxylase-protein in two Crassulacean acid metabolism plants in relation to leaf age, nitrogen nutrition, and point in time during a day/night cycle. - Planta *154* : 309 - 317, 1982.

52265 - WINTER, K., HOLTUM, J.A.M., EDWARDS, G.E., O'LEARY, M.H. : Effect of low relative humidity on $\delta^{13}C$ value in two C_3 grasses and in *Panicum milioides*, a C_3-C_4 intermediate species. - J. exp. Bot. *33* : 88 - 91, 1982.

52266 - WINTER, K., TENHUNEN, J.D. : Light-stimulated burst of carbon dioxide uptake following nocturnal acidification in the Crassulacean acid metabolism plant *Kalanchoë daigremontiana*. - Plant Physiol. *70* : 1718 - 1722, 1982.

52267 - WINTER, K., USUDA, H., TSUZUKI, M., SCHMITT, M., EDWARDS, G.E., THOMAS, R.J., EVERT, R.F. : Influence of nitrate and ammonia on photosynthetic characteristics and leaf anatomy of *Moricandia arvensis*. - Plant Physiol. *70* : 616 - - 625, 1982.

52268 - WIRTZ, W., STITT, M., HELDT, H.W. : Light activation of Calvin cycle enzymes as measured in pea leaves. - FEBS Lett. *142* : 223 - 226, 1982.

*52269 - WITT, H.T. : On the molecular machine of photosynthesis: overview and some recent results. - In : SKULACHEV, V.P., HINKLE, P.C. (ed.) : Chemiosmotic Proton Circuits in Biological Membranes in Honor of Peter Mitchell. Pp. 221 - - 244. Addison-Wesley, Reading, Mass. 1981.

*52270 - WITT, U., KOSKE, P.H., KUHLMAN, D., LENZ, J., NELLEN, W. : Production of *Nannochloris* spec. *(Chlorophyceae)* in large-scale outdoor tanks and its use as a food organism in marine aquaculture. - Aquaculture *23* : 171 - 181, 1981. [Chl.]

52271 - WITTENBACH, V.A. : Effect of pod removal on leaf senescence in soybeans. - Plant Physiol. *70* : 1544 - 1548, 1982.

52272 - WITTENBACH, V.A., LIN, W., HEBERT, R.R. : Vacuolar localization of proteases and degradation of chloroplasts in mesophyll protoplasts from senescing primary wheat leaves. - Plant Physiol. *69* : 98 - 102, 1982.

52273 - WITTWER, S.H. : Solar energy and agriculture. - Experientia *38* : 10 - 13, 1982. [Ps.]

52274 - WŁOCH, E., WIĘCKOWSKI, S. : Karotenoidy aparatu fotosyntetycznego. [Carotenoids of the photosynthetic apparatus.] - Postępy Biochem. *28* : 331 - - 352, 1982. [In Pol.]

52275 - WODZICKI, T.J., RAKOWSKI, K., STARCK, Z., PORANDOWSKI, J., ZAJĄCZKOWSKI,S.: Apical control of xylem formation in the pine stem. I. Auxin effects and distribution of assimilates. - Acta Soc. Bot. Pol. *51* : 187 - 201, 1982.

52276 - WOJCIECHOWSKI, A. : Syndrom C_4 na tle rozwoju ontogenetycznego. [C_4 syndrome on the ontogenetic background.] - Wiadom. bot. *26* : 29 - 50, 1982. [In Pol.]

52277 - WOLEDGE, J., DENNIS, W.D. : The effect of temperature on photosynthesis of ryegrass and white clover leaves. - Ann. Bot. *50* : 25 - 35, 1982.

52278 - WOLLMAN, F.A., BENNOUN, P. : A new chlorophyll-protein complex related to photosystem I in *C. reinhardtii*. - In : HALL, D.O., PALZ, W. (ed.) : Photochemical, Photoelectrochemical and Photobiological Processes. Vol.1. Pp. 156 - 159. D.Reidel Publ.Co., Dordrecht - Boston - London 1982.

52279 - WOLLMAN, F.-A., BENNOUN, P. : A new chlorophyll-protein complex related to photosystem I in *Chlamydomonas reinhardii*. - Biochim. biophys. Acta *680* : 352 - 360, 1982.

52280 - WOLOSIUK, R.A., HERTIG, C.M., NISHIZAWA, A.N., BUCHANAN, B.B. : Enzyme regulation in C_4 photosynthesis. Role of Ca^{2+} in thioredoxin-linked activation of sedoheptulose bisphosphate from corn leaves. - FEBS Lett. *140* : 31 - - 35, 1982.

52281 - WONG, D. : Primary processes of oxygen-evolving photosynthesis. - In : ALFANO, R.R. (ed.) : Biological Events Probed by Ultrafast Laser Spectroscopy. Pp. 3 - 25. Academic Press, New York - London - Paris - San Diego - San Francisco - São Paulo - Sydney - Tokyo - Toronto 1982.

52282 - WOO, K.C., MOROT-GAUDRY, J.F., SUMMONS, R.E., OSMOND, C.B. : Evidence for the glutamine synthetase/glutamate synthase pathway during the photorespiratory nitrogen cycle in spinach leaves. - Plant Physiol. *70* : 1514 - 1517, 1982.

52283 - **WOO, K.C., OSMOND, C.B.** : Stimulation of ammonia and 2-oxoglutarate-dependent O_2 evolution in isolated chloroplasts by dicarboxylates and the role of the chloroplast in photorespiratory nitrogen recycling. - Plant Physiol. *69* : 591 - 596, 1982.

52284 - **WOODHOUSE, R.M., NOBEL, P.S.** : Stipe anatomy, water potentials, and xylem conductances in seven species of ferns (Filicopsida). - Amer. J. Bot. *69* : 135 - 140, 1982. [Stomatal resistance.]

52285 - **WOODROW, I.E.** : Sedoheptulose-1,7-bisphosphatase from wheat chloroplasts. - In : COLOWICK, S.P., KAPLAN, N.O. (ed.) : Methods in Enzymology. Vol.90. Pp. 392 - 396. Academic Press, New York - London - Paris - San Diego - San Francisco - São Paulo - Sydney - Tokyo - Toronto 1982.

52286 - **WOODS, J.D., ONKEN, R.** : Diurnal variation and primary production in the ocean - preliminary results of a Lagrangian ensemble model. - J. Plankton Res. *4* : 735 - 756, 1982.

52287 - **WOOLHOUSE, H.W.** : Leaf senescence. - In : SMITH, H., GRIERSON, D. (ed.) : The Molecular Biology of Plant Development. Pp. 256 - 281. Blackwells, Oxford 1982. [Ps.]

*52288 - **WOŹNY, A.** : Wpływ cytokinin na rozwój plastydów w splątku *Ceratodon purpureus* BRID. i w liścieniach ogórka (*Cucumis sativus*). [Effect of cytokinins on plastid development in protonema of *Ceratodon purpureus* BRID. and in cotyledons of *Cucumis sativus*.] - Uniw. im. Adama Mickiewicza Poznaniu, Ser. Biol. *10* : 1 - 37, 1978. [Chloroplast; in Pol., ab : E.]

52289 - **WRAIGHT, C.A.** : Current attitudes in photosynthesis research. - In : GOVINDJEE (ed.) : Photosynthesis. Vol. 1. Pp. 17 - 61. Academic Press, New York - London - Paris - San Diego - San Francisco - São Paulo - Sydney - Tokyo - Toronto 1982.

52290 - **WRAIGHT, C.A.** : The involvement of stable semiquinones in the two-electron gates of plant and bacterial photosystems. - In : TRUMPOVER, B.L. (ed.) : Function of Quinones in Energy Conserving Systems. Pp. 181 - 197. Academic Press, New York - London - Paris - San Diego - San Francisco - São Paulo - Sydney - Tokyo - Toronto 1982.

52291 - **WRIGHT, L.A.,Jr., MURPHY, T.M.** : Short-wave ultraviolet light closes leaf stomata. - Amer. J. Bot. *69* : 1196 - 1199, 1982. [Stomatal resistance.]

52292 - **WU Guang-yao, DENG Yue-fen, LU Chong-en, PU Zong-shi, XU Shi-rong** : [On the change of photosynthetic characteristics in *Setaria italica*.] - Acta Phytophysiol. sin. *8* : 111 - 116, 1982. [In Chin., ab : E.]

52293 - **WU Guo-liang, ZHONG Ze-pu, BAI Ke-zhi, WANG Fa-zhu, CUI Cheng** : [The effects of light quality on the growth and development of *Anabaena azollae*.] - Acta bot. sin. *24* : 46 - 53, 1982. [Chl; in Chin., ab : E.]

52294 - **WU Min-xian, ZHA Jing-juan, CHEN Jing-zhi, SHI Jiao-nai** : [Studies on plant phosphoenolpyruvate carboxylase VI. Stabilizing effect of glucose-6-phosphate and glycine to sorghum leaf PEP carboxylase.] - Acta Phytophysiol. sin. *8* : 9 - 16, 1982. [In Chin., ab : E.]

52295 - **WU Min-xian, ZHA Jing-juan, TANG Xiao-yi, SHI Jiao-nai** : [Studies on plant phosphoenolpyruvate carboxylase. VIII. Light-induced formation of PEP-carboxylase in C_4 plants.] - Acta Phytophysiol. sin. *8* : 101 - 110, 1982. [In Chin., ab : E.]

*52296 - **WU Xiang-yu** :[Photosynthetic carbon metabolism in plants.] - Hua Hsueh Tung Pao *1* : 55 - 62, 1979. [In Chin.]

52297 - **WULFF, R.D., STRAIN, B.R.** : Effects of CO_2 enrichment on growth and photosynthesis in *Desmodium paniculatum*. - Can. J. Bot. *60* : 1084 - 1091, 1982.

52298 - **WUN, C.-K., LITSKY, W.** : Rapid determination of algal chlorophyll by gas-liquid chromatography. - Environm. Sci. Technol. *16* : 335 - 338, 1982.

52299 - WYDRZYNSKI, T.J. : Oxygen evolution in photosynthesis. - In : GOVINDJEE (ed.) : Photosynthesis. Vol.1. Pp. 469 - 506. Academic Press, New York - London - Paris - San Diego - San Francisco - São Paulo - Sydney - Tokyo - Toronto 1982.

52300 - WYSE, R.E., SAFTNER, R.A. : Reduction in sink-mobilizing ability following periods of high carbon flux. - Plant Physiol. *69* : 226 - 228, 1982. [Photosynthates.]

52301 - XIA Shu-fang, ZHANG Zhen-qing, YU Xin-jian : [The diurnal variation in carbohydrate contents and in photosynthate export in *Zea mays* leaves.] - Acta Phytophysiol. sin. *8* : 141 - 148, 1982. [In Chin., ab : E.]

52302 - XU Da-quan, SHEN Yun-gang : [Exploring the relationship between the photosynthate level and the operation of photosynthetic apparatus.] - Acta Phytophysiol. sin. *8* : 173 - 186, 1982. [In Chin., ab : E.]

52303 - YAANISO, R.V. : Model' prevrashcheniya tsentrov pri vyzhiganii provala v spektrakh khlorofilla. [A model of site interconversions at hole-burning in the spectra of chlorophyll.] - Eesti NSV Tead. Akad. Toim., Füüs., Mat. *31* : 161 - 165, 1982. [In R, ab : E, Est.]

52304 - YAKUBOVA, M.M., YULDASHEV, Kh., KHAMIDOV, B.M. : Fotosintez i metabolizm ugleroda C^{14} u khlopchatnika v svyazi s yavleniem geterozisa. [Photosynthesis and metabolism of ^{14}C in cotton in relation to the heterosis phenomenon.] - Dokl. Akad. Nauk tadzh. SSR *25* (2) : 112 - 116, 1982. [In R, ab : Tajik.]

52305 - YAKUSHĚNKA, I.K. : Sezonnaya dynamika kol'kastsi khlarafilu i karatynoidaŭ u listsyakh topalyaŭ. [Seasonal dynamics of chlorophyll and carotenoid amounts in *Populus* leaves.] - Vestsi Akad. Navuk belarus. SSR, Ser. biyal. Navuk *1982*(3) : 21 - 25, 122, 1982. [In Belorus., ab : E, R.]

52306 - YAMADA, M., ISHIGE, T., OHKAWA, Y. : Dry matter production in maize resulting from crosses between Japanese flint and U.S. dent. - Annu. Rep. Div. Genet. nat. Inst. agr. Sci. Jap. *1982* : 36 - 37, 1982.

52307 - YAMADA, Y., SATO, F., WATANABE, K. : Photosynthetic carbon metabolism in cultured photoautotrophic cells. - In : FUJIWARA, A. (ed.): Plant Tissue Culture 1982. Pp. 249 - 250. Maruzen, Tokyo 1982.

52308 - YAMAGUCHI, Y., SHIBATA, Y. : Standing stock and distribution of phytoplankton chlorophyll in the Southern Ocean south of Australia. - Trans. Tokyo Univ. Fish. *1982* (5) : 111 - 128, 1982.

52309 - YAMAMOTO, T., WATANABE, S. : [The effects of leaf water stress and leaf burn on photosynthesis of 'Bartlett' pear trees.] - J. jap. Soc. hort. Sci. *51* : 19 - 28, 1982. [In Jap., ab : E.]

52310 - YAMAMOTO, Y., UEDA, T., SHINKAI, H., NISHIMURA, M. : Preparation of O$_2$-evolving photosystem-II subchloroplasts from spinach. - Biochim. biophys. Acta *679* : 347 - 350, 1982.

52311 - YAMANAKA, G., LUNDELL, D.J., GLAZER, A.N. : Molecular architecture of a light-harvesting antenna. Isolation and characterization of phycobilisome subassembly particles. - J. biol. Chem. *257* : 4077 - 4086, 1982.

52312 - YAMASHITA, J., TAKAMIYA, K., YAMANAKA, T. : Bacterial electron transport. - Rec. Progr. nat. Sci. Jap. *7* : 33 - 39, 1982.

52313 - YAMASHITA, K. : Effects of expansion of the π-electron system on photocurrent quantum yields for porphyrin photocells: magnesium and zinc tetrabenzporphyrin senzitizers. - Chem. Lett. *1982* : 1085 - 1088, 1982.

*52314 - YAMASHITA, T. : ⌐Some topics on the Tris-washing treatment on chloroplasts.]
- Tanpakushitsu Kakusan Koso, Bessatsu 21 : 82 - 95, 1979. [In Jap.]

52315 - YAMASHITA, T. : Effects of uncouplers on photo-reactivation of Tris(pH 8.8)
and 2,6-dichlorophenol indophenol-treated chloroplasts. - Plant Cell Physiol. 23 : 833 - 841, 1982.

52316 - YAMAUCHI, M., YAMADA, Y. : Photorespiratory CO_2 release from L-[U-^{14}C] seri-
ne in tomato leaves. - Soil Sci. Plant Nutr. 28 : 99 - 107, 1982.

52317 - YANAGIMOTO, M., SAITOH, H. : Evaluation tests of a large spiral blue green
alga, Oscillatoria sp., for biomass production. - J. Fermentation Technol.
60 : 305 - 310, 1982.

*52318 - YANKAVICHYUTE, G.Yu., YANKYAVICHYUS, K.K., SHULIENE, R.Yu. : Rol' fitoplank-
tona v samoochishchenii vody severnoĭ chasti zaliva Kurshyu-Marēs. [Role
of phytoplankton in autopurification of water in the northern part of lagoon
of Kuršiy Marios.] - Liet. TSR Mokslu Akad. darb., Ser. C 1980 [3(91)] :
11 - 17, 1980. [Primary production; in R, ab : Lithu.]

52319 - YANYUSHIN, M.F. : Vydelenie vodoroda i gidrogenaznaya aktivnost' v sin-
khronnoĭ kul'ture Chlamydomonas reinhardii v svyazi s anaerobnoĭ degrada-
tsieĭ krakhmala. [Hydrogen evolution and hydrogenase activity in a synchronous culture of Chlamydomonas reinhardii with respect to anaerobic degradation of starch.]- Fiziol. Rast. 29 : 121 - 126, 1982. [In R, ab : E.]

52320 - YANYUSHIN, M.F. : Aktivatsiya gidrogenazy i fotovydelenie vodoroda v sin-
khronnoĭ kul'ture Chlamydomonas reinhardii pri anaèrobnoĭ adaptatsii na
svetu. [Activation of hydrogenase and light-induced hydrogen evolution in
synchronous culture of Chlamydomonas reinhardii during anaerobic adaptation
in the light.] - Fiziol. Rast. 29 : 1126 - 1133, 1982. [In R, ab : E.]

B52321 - YASNIKOV, A.A. : Organicheskie Katalizatory Kofermenty i Fermenty. [Organic
Catalysers, Coenzymes and Enzymes.] - Naukova Dumka, Kiev 1982. [Ps; in R.]

52322 - YASNIKOV, A.A., ALIEV, D.A., MURADOV, A.Z., VOLKOVA, N.V., KANIVETS, N.P.,
VASILENOK, L.I., MUSHKETIK, L.S., BORISEVICH, A.N., PEL'KIS, P.S. : Rol'
pirofosfata v fotofosforilirovanii v khloroplastakh rasteniĭ. [Role of
pyrophosphate in photophosphorylation in chloroplasts of plants.] - Dokl.
Akad. Nauk SSSR 264 : 1508 - 1510, 1982. [In R.]

52323 - YE Ji-yu, TANG Chong-qin, QIAN Lu-ping : [Inhibition of unsaturated fatty
acids on PS-II electron transport in chloroplast and derepression by divalent cations.] - Acta Phytophysiol.sin. 8 : 267 - 272, 1982. [In Chin.,
ab : E.]

52324 - YENTSCH, C.S., PHINNEY, D.A. : The use of attenuation of light by particulate matter for the estimate of phytoplankton chlorophyll with reference
to the Coastal Zone Color Scanner. - J. Plankton Res. 4 : 93 - 102, 1982.

52325 - YIN Hong-chang, SHEN Yung-kang : [Recent advances in photosynthesis research.] - Tzu Jan Tsa Chih 2 (4) : 202 - 207, 1979. [In Chin.]

52326 - YOCUM, C.F. : Purification of ferredoxin and plastocyanin. - In : EDELMAN,
M., HALLICK, R.B., CHUA, N.-H. (ed.) : Methods in Chloroplast Molecular
Biology. Pp. 973 - 981. Elsevier Biomedical Press, Amsterdam - New York -
Oxford 1982.

*52327 - YOCUM, C.F., YERKES, C.T., BLANKENSHIP, R.E., SHARP, R.R., BABCOCK, G.T. :
Stoichiometry, inhibitor sensitivity, and organization of manganese associated with photosynthetic oxygen evolution. - Proc. nat. Acad. Sci. USA
78 : 7507 - 7511, 1981.

*52328 - YODER, J.A., ATKINSON, L.P., LEE, T.N., KIM, H.H., McCLAIN, C.R. : Role of
Gulf Stream frontal eddies in forming phytoplankton patches on the outer
southeastern shelf. - Limnol. Oceanogr. 26 : 1103 - 1110, 1981. [Chl.]

52329 - YOKOTA, A., HAGA, S., KITAOKA, S. : Glycolate excretion by air-grown Euglena gracilis z. - Photosynthesis Res. 3 : 363 - 367, 1982.

52330 - **YOKOTA, A., KITAOKA, S.** : Synthesis, excretion, and metabolism of glycolate under highly photorespiratory conditions in *Euglena gracilis* Z. - Plant Physiol. *70* : 760 - 764, 1982.

52331 - **YOKOYAMA, H., HSU, W.J., POLING, S., HAYMAN, E.** : Bioregulation of pigment biosynthesis by onium compounds. - In : MORELAND, D.E., ST.JOHN, J.B., HESS, F.D. (ed.) : Biochemical Responses Induced by Herbicides. Pp. 153 - - 173. Amer. chem. Soc., Washington 1982.

52332 - **YOKOYAMA, H., HSU, W.J., POLING, S.M., HAYMAN, E.** : Chemical regulation of carotenoid biosynthesis. - In : BRITTON, G., GOODWIN, T.W. (ed.) : Carotenoid Chemistry and Biochemistry. Pp. 371 - 385. Pergamon Press, Oxford - New York - Toronto - Sydney - Paris - Frankfurt 1982.

52333 - **YOSHIDA, K., KONISHI, M.** : Inhibitory effect of continuous far-red preirradiation on chlorophyll accumulation of *Pharbitis nil* cotyledons. - Photochem. Photobiol. *36* : 241 - 244, 1982.

52334 - **YOUNG, D.R., SMITH, W.K.** : Simulation studies of the influence of understory location on transpiration and photosynthesis of *Arnica cordifolia* on clear days. - Ecology *63* : 1761 - 1770, 1982.

52335 - **YOUNG, E., HAND, J.M., WIEST, S.C.** : Osmotic adjustment and stomatal conductance in peach seedlings under severe water stress. - HortScience *17* : 791 - - 793, 1982.

52336 - **YOUNIS, H.M., MORJANA, N.A.** : Resolution of the chloroform-released CF_1 into ATPase complexes differing in their ATPase activity. Identification of one subunit as a putative natural inhibitor of the ATPase. - FEBS Lett. *140* : 317 - 319, 1982.

52337 - **YOUNIS, H.M., MORJANA, N.A.** : Isolation and properties of a natural inhibitor of the chloroplast adenosine triphosphatase. - FEBS Lett. *140* : 320 - - 324, 1982.

52338 - **YU, M.H., GLAZER, A.N.** : Cyanobacterial phycobilisomes. Role of the linker polypeptides in the assembly of phycocyanin. - J. biol.Chem. *257* : 3429 - - 3433, 1982.

52339 - **YU, M.-H., MILLER, G.W.** : Formation of delta-aminolevulinic acid in etiolated and iron-stressed barley. - J. Plant Nutr. *5* : 1259 - 1271, 1982.

52340 - **YU, P.-L., DREWS, G.** : Polyadenylated messenger RNA isolated from cells of *Rhodopseudomonas capsulata* induced to synthesize the photosynthetic apparatus. - FEMS Microbiol. Lett. *14* : 233 - 236, 1982.

52341 - **YUEN, M.J., SHIPMAN, L.L., KATZ, J.J., HINDMAN, J.C.** : Energy transfer in self-assembled chlorophyll *a* systems. - Photochem. Photobiol. *36* : 211 - - 222, 1982.

*52342 - **YUSHKOV, V.I., YUSHKOV, P.I.** : Nekotorye zakonomernosti radial'nogo peredvizheniya ^{14}C-assimilyatov v osevykh organakh drevesnykh rasteniĭ. [Some regularities of radial transport of ^{14}C-assimilates in stem organs of woody plants.] - In : Populyatsionnye i Biogeotsenologicheskie Issledovaniya v Gornykh Temnokhvoĭnykh Lesakh Srednego Urala. Pp. 105 - 112. Ural'. Gos. Universitet, Sverdlovsk 1979. [In R.]

52343 - **YUSHKOV, V.I., YUSHKOV, P.I.** : Ottok iz fotosinteziruyushchikh organov i raspredelenie ^{14}C-assimilyatov u molodykh derev'ev sosny obyknovennoĭ i berezy pushistoĭ. [^{14}C-assimilate efflux from photosynthesizing organs and their distribution in young trees of *Pinus silvestris* and *Betula pubescens*.] - In : Biogeotsenologicheskie Issledovaniya na Urale. Pp. 78 - 98, II. Ural'. Gos. Universitet, Sverdlovsk 1982. [In R.]

52344 - **ZABUGA, V.F., SHCHERBATYUK, A.S.** : Ėkologiya fotosinteza sosny obyknovennoĭ lesostepnogo Predbaĭkal'ya. [Ecology of photosynthesis of *Pinus silvestris* in the forest-steppe pre-Baikal region.] - Ėkologiya *1982* (5) : 76 - 78, 1982. [In R.]

52345 - ZAKHAROVA, N.I., SHUBIN, V.V., BUKHOV, N.G., KARAPETYAN, N.V. : Vliyanie vtorichnoĭ struktury na skorost' temnovogo vosstanovleniya P700$^+$ askorbatom kompleksa fotosistemy I. [Effect of secondary structure on the rate of P700$^+$ dark restoration with ascorbate.] - Biofizika 27 : 572 - 577, 1982. [In R, ab : E.]

*52346 - ZAKHAR'YANTS, I.L., ALEKSEEVA, L.N. : Ēkologo-fiziologicheskoe izuchenie fotosinteza i dykhaniya kak faktorov produktivnosti u mnogoletnikh rasteniĭ Yugo-Zapadnogo Kyzylkuma.[Ecophysiological study of photosynthesis and respiration as factors of productivity in perennial plants of Southwestern Kyzylkum.] - In : Fiziologiya i Biokhimiya Dikorastushchikh Kormovykh Rasteniĭ Uzbekistana. Pp. 15 - 26. Fan, Tashkent 1975. [In R.]

52347 - ZAMSKI, E., UMIEL, N. : Streptomycin resistance in tobacco : IV. Effects of the drug on the ultrastructure of plastids and mitochondria in cotyledons of germinating seeds. - Z. Pflanzenphysiol. 105 : 143 - 148, 1982.

52348 - ZANETTI, G., CIDARIA, D., CURTI, B. : Preparation of apoprotein from spinach ferredoxin-NADP$^+$ reductase. Studies on the resolution process and characterization of the FAD reconstituted holoenzyme. - Europe. J. Biochem. 126 : 453 - 458, 1982.

52349 - ZANNONI, D. : ATP synthesis coupled to light-dependent non-cyclic electron flow in chromatophores of Rhodopseudomonas capsulata. - Biochim. biophys. Acta 680 : 1 - 7, 1982.

52350 - ZAURALOV, O.A., ZHIDKIN, V.I. : Posledeĭstvie okhlazhdeniya na rost i fotosintez rasteniĭ prosa. [After-effect of cooling on millet growth and photosynthesis.] - Fiziol. Rast. 29 : 98 - 103, 1982. [In R, ab : E.]

52351 - ZEBROWER, M., LOACH, P.A. : Efficiency of light-driven metabolite transport in the photosynthetic bacterium Rhodospirillum rubrum. - J. Bacteriol. 150: 1322 - 1328, 1982.

52352 - ZEIGER, E., FIELD, C. : Photocontrol of the functional coupling between photosynthesis and stomatal conductance in the intact leaf. Blue light and PAR-dependent photosystems in guard cells. - Plant Physiol. 70 : 370 - 375, 1982.

52353 - ZEIGER, E., SCHWARTZ, A. : Longevity of guard cell chloroplasts in falling leaves: Implication for stomatal function and cellular aging. - Science 218 : 680 - 681, 1982.

52354 - ZEINALOV, Yu. : Existence of two different ways for oxygen evolution in photosynthesis and the photosynthetic unit concept. - Photosynthetica 16 : 27 - 35, 1982.

52355 - ZEĬNALOV, Yu., MASLENKOVA, L. : V"rkhu interpretatsiyata na spektralno-prekhodniya efekt. I. Vliyanie na intenziteta na svetlinata v"rkhu efekta na Blinks. [Interpretation of the chromatic transient effect. I. Influence of irradiance on the Blinks effect.] - Fiziol. Rast. (Sofiya) 8 (1) : 10 - - 18, 1982. [In Bulg., ab : E, R.]

52356 - ZEĬNALOV, Yu., MASLENKOVA, L. : V"rkhu interpretatsiyata na spektralno-prekhodniya efekt. II. Vliyanie na opticheskata pl"tnost v"rkhu efekta na Blinks. [Interpretation of the chromatic transient effect. II. Influence of absorbance on the Blinks effect.] - Fiziol. Rast. (Sofiya) 8 (3) : 3 - - 11, 1982. [In Bulg., ab : E, R.]

52357 - ZEITZSCHEL, B. : General description of phytoplankton productivity in marine and estuarine environments. - In : ZABORSKY, O.R. (ed.) : CRC Handbook of Biosolar Resources. Vol.I/2. Pp. 395 - 399. CRC Press, Boca Raton 1982.

52358 - ZEITZSCHEL, B. : Phytoplankton productivity in the Indian Ocean. - In : ZABORSKY, O.R. (ed.) : CRC Handbook of Biosolar Resources. Vol.I/2. Pp. 417 - 427. CRC Press, Boca Raton 1982.

52359 - ZELEŇÁKOVÁ, E., POLEK, B. : Effect of imissions with a prevailing SO_2 component on chlorophyll changes in apricot leaves. - Biológia (Bratislava) 37 : 901 - 907, 1982.

52360 - ZELENSKIĬ, M.I., SHITOVA, I.P., MOGILEVA, G.A. : Izuchenie fotosinteticheskikh kharakteristik genealogicheski svyazannykh sortov pshenitsy. [A study on photosynthetic characteristics of genealogically related wheat cultivars.] - Tr. prikl. Bot. Genet. Selek. 72 (2) : 18 - 26, 1982. [In R, ab : E.]

52361 - ZELITCH, I. : The close relationship between net photosynthesis and crop yield. - BioScience 32 : 796 - 802, 1982.

52362 - ZEN'KEVICH, É.I., GURINOVICH, G.P., SAGUN, E.I. : Perenos ėnergii ėlektronnogo vozbuzhdeniya s uchastiem molekul khlorofilla i ego proizvodnykh. [Electron excitation energy transfer with the participation of the molecules of chlorophyll and its derivatives.] - Eesti NSV Tead. Akad. Toim., Füüs. Mat. 31 : 180 - 186, 1982. [In R, ab : E, Est.]

52363 - ZHANG Qi-de, TANG Chong-qin, LI Shi-yi, LIN Shi-qing, LOU Shi-quing, KUANG Ting-yun : [Structure and function of chloroplast membrane. XI. The effects of linolenic acid on the structure and the absorption and fluorescence spectra of wheat chloroplast membranes as well as regulations by $MgCl_2$.] - Acta bot. sin. 24 : 326 - 333, 1982. [In Chin., ab : E.]

52364 - ZHANG Xiankong, LIU Mei, LIU Qifang, WANG Houle, LI Shanghao : Preliminary separation and characteristics of phycocyanin in blue-green Gymnodinium. - Kexue Tongbao 27 : 1000 - 1003, 1982.

52365 - ZHOU Pei-Zhen, LI Liang-Bi, ZHAI Xiao-Jing, ZHANG Zheng-Dong, MA Gui-Zhi : The relationship between cation-induced fluorescence and membrane stacking in isolated spinach chloroplasts. - Photosynthesis Res. 3 : 123 - 130, 1982.

52366 - ZIELINSKĬ, R.E., PRICE, C.A. : Preparation of cytochrome b_{559} from spinach. - In : EDELMAN, M., HALLICK, R.B., CHUA, N.-H. (ed.) : Methods in Chloroplast Molecular Biology. Pp. 933 - 944. Elsevier Biomedical Press, Amsterdam - New York - Oxford 1982.

*52367 - ZIELKE, H., FÖRSTEL, H. : Oxygen isotope fractionation during photosynthesis and respiration of a marine diatom. - In : KLEIN, E.R., KLEIN, P.D. (ed.) : Stable Isotopes: Proceedings of the Third International Conference. Pp. 205 - 214. Academic Press, New York - San Francisco - London 1979.

52368 - ZIKA, R.G., FINE, R.A. : The marine micronutrients of nitrogen, phosphorus, and silicon. - In : ZABORSKY, O.R. (ed.) : CRC Handbook of Biosolar Resources. Vol.I/2. Pp. 537 - 550. CRC Press, Boca Raton 1982.

52369 - ZILINSKAS, B.A. : Isolation and characterization of the central component of the phycobilisome core of Nostoc sp. - Plant Physiol. 70 : 1060 - 1065, 1982.

*52370 - ZILKAH, S., GRESSEL, J. : Correlations in phytotoxicity between white and green calli of Rumex obtusifolius, Nicotiana tabacum and Lycopersicon esculentum. - Pestic. Biochem. Physiol. 9 : 334 - 339, 1978. [Ps, Chl.]

*52371 - ZIMMER, B., PARADIES, H.H., WERZ, G. : On the quaternary structure of D-ribulose-1,5-bisphosphate carboxylase from Dasycladus clavaeformis ROTH.(AG.) in solution and in the crystalline state. - In : WOODCOCK, C.L.F. (ed.) : Progress in Acetabularia Research. Pp. 83 - 94. Academic Press, New York - San Francisco - London 1977.

52372 - ZIMMERMAN, J.K., LECHOWICZ, M.J. : Responses to moisture stress in male and female plants of Rumex acetosella L. (Polygonaceae). - Oecologia 53 : 305 - - 309, 1982.

52373 - ZIMMERMANN, U., KÜPPERS, G., SALHANI, N. : Electric field-induced release of chloroplasts from plant protoplasts. - Naturwissenschaften 69 : 451 - 452, 1982.

*52374 - ZLOKOLICA, M., GERIĆ, C., GERIĆ, I. : Fotohemijska aktivnost izolovanih hloroplasta kukuruza u zavisnosti od mineralne ishrane i intenziteta svetlosti. [Photochemical activity of maize chloroplasts in relation to the mineral nutrition and illuminance.] - Arh. poljopr. Nauke 42 : 443 - 456, 1981/4. [In Croat., ab : E.]

*52375 - ZOBEL, D.B. : Local variation in intergrading *Abies grandis* - *A. concolor* populations in the central Oregon Cascades. II. Stomatal reaction to moisture stress. - Bot. Gaz. *135* : 200 - 210, 1974. [Ps, stomatal resistance.]

52376 - ZONTA, F., STANCHER, B., BIELAWNY, J. : High-performance liquid chromatography of fat-soluble vitamins. Separation and identification of vitamins D_2 and D_3 and their isomers in food samples in the presence of vitamin A, vitamin E and carotene. - J. Chromatogr. *246* : 105 - 112, 1982.

52377 - ZUBOV, B.V., SULIKOV, N.A., CHERNAVSKAYA, N.M., CHERNAVSKIĬ, D.S., CHIZHOV, I.V. : Kinetika pervichnykh reaktsiĭ fototsikla bakteriorodopsina pri nizkikh temperaturakh. [Kinetics of primary stages of bacteriorhodopsin photocycle at low temperatures.] - Biofizika *27* : 357 - 361, 1982. [In R, ab : E.]

52378 - ZÜLLIG, H. : Untersuchungen über die Stratigraphie von Carotinoiden in geschichteten Sediment von 10 Schweizer Seen zur Erkundung früherer Phytoplankton-Entfaltungen. - Schweiz. Z. Hydrol. *44* : 1 - 98, 1982.

52379 - ZURAWSKI, G., BOTTOMLEY, W., WHITFELD, P.R. : Structures of the genes for the β and ε subunits of spinach chloroplast ATPase indicate a dicistronic mRNA and an overlapping translation stop/start signal. - Proc. nat. Acad. Sci. USA *79* : 6260 - 6264, 1982.

52380 - ZÜRRER, H. : Hydrogen production by photosynthetic bacteria. - Experientia *38* : 64 - 66, 1982.

*52381 - ZÜRRER, H., BACHOFEN, R. : Effect of L-methionine-DL-sulphoximine on the photoproduction of hydrogen by *Rhodospirillum rubrum*. - Experientia *36* : 1166 - 1167, 1980.

52382 - ZÜRRER, H., BACHOFEN, R. : Aspects of growth and hydrogen production of the photosynthetic *Rhodospirillum rubrum* in continuous culture. - Biomass *2* : 165 - 174, 1982.

*52383 - ZURZYCKI, J. : Metody izotopowe w badaniach pochodzenia tlenu fotosyntetycznego. [Isotopic methods in the investigation of the origin of photosynthetic oxygen.] - Zesz. nauk. Uniw. jagielloń. *492* (Prace Biol. mol. 5) : 111 - 123, 1978. [In Pol., ab : E.]

*52384 - ZURZYCKI, J. : Light-adaptation phenomena of the photosynthetic apparatus at the cellular level. - Pol. ecol. Stud. *7* : 365 - 376, 1981.

52385 - ZURZYCKI, J. : Komórka fotosyntetyczna jako modelowy układ fotorecepcji. [Photosynthetic cell as a model system for study on photoreception.] - Zesz. nauk. Uniw. jagielloń. *689* (Prace Biol. mol. 9) : 11 - 18, 1982. [In Pol., ab : E.]

52386 - ZVALINSKIĬ, V.I., LITVIN, F.F. : Analiz svetovoĭ krivoĭ fotosinteza pri nepreryvnom osveshchenii. [Analysis of photosynthesis light curve under continuous illumination.] - Biofizika *27* : 202 - 207, 1982. [In R, ab : E.]

52387 - ZVALINSKIĬ, V.I., LITVIN, F.F. : Svetovye krivye fotosinteza pri impul'snom osveshchenii. [Photosynthesis light curves under flash illumination.] - Biofizika *27* : 410 - 414, 1982. [In R, ab : E.]

52388 - ZWERLING, H., MEHLHORN, R., PACKER, L., MacELROY, R. : A computer technique for structural studies of bacteriorhodopsin. - In : COLOWICK, S.P., KAPLAN, N.O. (ed.) : Methods in Enzymology. Vol.88. Pp. 772 - 784. Academic Press, New York - London - Paris - San Diego - San Francisco - São Paulo - Sydney - Tokyo - Toronto 1982.

Authors' names are presented in the form in which they appear in the respective publication. The names from papers published in Cyrillic characters are transcri-
bed as shown on p. III of this volume. Alternative spellings and forms of the name
of the same author are usually cross-indexed. The numbers in *italics* refer to publication in which the respective author acts as editor.

A

AARESKJOLD, K. 48410
AARONSON, S. 48411
ABBOTT, M.R. 48412
ABBOTT, M.S. 49104
ABBOUD, M. 50732
ABDEL-BAR, M.Z. 49904
ABDULAEV, N.G. 48413, 50513, 51007-8
ABDULLAEV, A.A. 48414
ABDULLAEV, Kh.A. 48415
ABE, S. 49752
ABER, J.D. 48416
ABO, F. 50969
ABRAHAM, R.J. 48417
ABRAHAMSON, W.G. 48418
ABRAMCHIK, L.M. 48419-20
ÁBRÁNYI, A. 49253
ABROSOV, N.S. 48421-2
ACKER, S. 48423, 51114
ACKERSON, R.C. 48424
ADAMS, M.S. 50018
ADAMS, M.W. 48425
ADAMS, W.A. 48451
ADAMSON, H. 48426-7
ADLERCREUTZ, P. 48428-9
ADMON, A. 48430, 51557
ADYGEZALOV, V.F. 48431, 48467
AFLALO, C. 48432
AFRIA, B.S. 48433
AGAEV, M.G. 50547
AGALIDIS, I. 51268
AGATA, W. 48434-5, 52214-6
AGHION, J. 51899
AGUIRRE, R. 48436-7, 49023
AHL, P.L. 48438
AHMED, A.M. 48439
AHMED, J. 48440
AIBA, S. 48441, 50934
AIGLE, N. 49524
AKAO, S. 48442
AKAZAWA, T. 48443-4, 50894-6
ÅKERLUND, H.-E. 48445, 48501
AKHMEDOV, G.A. 48465
AKHMEDOV, Yu.D. 48414, 49962-3
AKIMOV, Yu.A. *49139*
AKINYEMIJU, O.A. 48446

AKOYUNOGLOU, A. 48526
AKOYUNOGLOU, G. *48426-7*, 48447, *48447,
   B48448, 48525,* 48526, *48680, 48777,
   48878,* 48921, 48943, *49209, 49274,
   49541, 49867, 49886, 49891, 49910,
   50032, 50181, 50394, 50590, 50808,
   51012, 51290, 51553, 51663, 51945,
   52202*
AKSENOV, S.I. 51690
ALABACK, P.B. 48449
ALAGARSWAMY, G. 49798
ALAM, S.M. 48450-1
ALBAN, D.H. 48452
ALBERGONI, F.G. 48453, 51261
ALBERTE, R.S. 49188, 49674, 50498
ALBERTSSON, P.-Å. 48454
ALBRACHT, S.P.J. 48455
ALBRECHT-ELLMER, K.J. 48455
ALEKSANDROV, A.Yu. 50912, 52024
ALEKSEEVA, L.N. 52346
ALEMDAG, I.S. 48456-9
ALEXANDRE, D.Y. 48460-2
ALFANI, A. 52112-3
ALFANO, A.J. 48463
ALFANO, R.R. *49295, 49532, 49866, 50093,
   50108, 50641, 50866,* 51074, *51074,
   51562, 51813, 52281*
ALFICH, R.A. 49326
ALHADEFF, M. 51945
ALI, S.M. 48464
ALIEV, D.A. 48465-7, 50780, 52322
ALINA, B.A. 48468
ALJARO, M.-E. 50300
ALKEMA, J. 48469
ALLAKHVERDIEV, S.I. 50177-8
ALLARD, D.W. 52044
ALLDREDGE, A.L. 48470-1
ALLEGRINI, P.R. 51638
ALLEN, L.H.,Jr. 51652, 52139-40
ALLEN, M.F. 51156
ALLEWELDT, G. 48472-3
ALM, P. 50988
ALMELA, L. 50855
ALMON, H. 48474

CHUNAEV, A.S. 49006-8, 50617
CHUNG, B. 49009
CHUNG, H.H. 49010
CHUPROVA, N.A. 50287
CHURCH, M.R. 49030
CHURCHILL, S.P. 50608
CHUSEL, M. 50031
CIBIK, S.J. 50605
CIDARIA, D. 52348
CIFERRI, O. 51932
CILENTO, G. 50846
CINTAS, A.M. 49595
CIOFFI, L. 50358
CIONE, J. 50532
CIPRIANO, F. 50185
CLAPP, R.E. 49011
CLARK, A.D. 50010
CLARK, A.J. 49975
CLARK, D.K. 49604
CLARK, N.A. 51328
CLARK, P.R. 49384
CLARK, W.C. *B49012*
CLARK, W.J. 52044
CLARKE, I.E. 49013-4
CLARKE, J.M. 49015
CLARKE, R.H. 49016-7, 49882
CLAYTON, R.K. 49018
CLIJSTERS, H. 48960
CLINE, K. 50114
CLOSS, G.L. 49019, 50908
CLOUGH, B.F. 49021, *49021*
COATES, D.B. 50791
COCHRANE, H. 49707
COCHRANE, M.P. 49022
COCKBURN, W. 48610
COCKING, E.C. 51055
COCQUEMPOT, M.F. 48437, 49023-4, 49997
CODD, G.A. 50341, 50826, 51759
CODDINGTON, J.M. 48724, 49025
CODISPOTI, L.A. 50730
COGDELL, R.J. 48654, 49026-8, 49137
COHEN, C.J. 49029
COHEN, R.R.H. 49030
COHEN, S.S. 49031-2
COHEN, Y. 50418
COHEN-BAZIRE, G. *50740*
COLASANTE, L.O. 51131
COLBEAU, A. 50024
COLBOW, K. 49033, *49033*
COLEMAN, J.R. 48723, 49034
COLEMAN, P.G. *51725*
COLIJN, C.M. 49035
COLLARD, F. 48847, 49997
COLLETT, B. 50017
COLLINS, C.D. 49036-7
COLLOS, Y. 51668
COLMAN, B. 48723, 49334
COLOWICK, S.P. *48413, 48438, 48524,*
      48585, 48593, 48630, 48644,
      48741, 48760, 48817, 48905, 48986,
      49046-7, 49082, 49089, 49183-5,
      49247, 49296, 49306, 49313, 49324,
                              (continued)

COLOWICK, S.P. (continued)
      49414, 49459, 49540, 49706, 49819,
      49871, 50061, 50127, 50134, 50157,
      50216, 50220-2, 50312, 50349, 50399,
      50420-2, 50460, 50503, 50579, 50603,
      50621, 50668, 50784, 50795, 50869,
      50924-5, 50959, 51003, 51039, 51328,
      51523, 51639, 51719, 51730, 51797,
      51957, 51965, 51989-90, 52026,
      52053, 52148, 52159, 52196-7, 52200,
      52285, 52388
COLVIN, R.J. 51753
COMINS, H.N. 49038
COMPÈRE, P. *51178-9*
CONE, R.A. 48438
CONN, E.E. 49103
CONNELI, C.R. 51442
CONNOLLY, J.S. 49039-40
CONNOR, D.J. 49186
CONOVER, C.A. 49041
CONRAD, J.M. 51118
CONRAD, P.L. 51118
CONSTABLE, G.A. 49042
CONUS, G. 50487
COOK, I.H. 50166
COOKE, J.R. 51233
COOKE, R.J. 51922
COOMBS, J. *48637, 48768,* 49043-4, *49043-*
      -4, B49045, 49694, *49694,* 49700,
      49828-9, 50389-90, 50485, 50509-10,
      51940, 52130
COON, M.J. *48545*
COOPER, A. 49046-7
COOPER, S.R. 50161
COPPING, L.G. 49818
CORK, D.J. 49048-9
CORMIER, M.J. 49991
CORNELIUS, M.J. 48722, 50137
CORNIC, G. 49050
CORNIDES, I. 51020
CORNILLON, P. 49051
CORNWELL, K.L. 50368
CORRE, B. 48978
CORREIA, O.C.A. 49052
CORTES, P.M. 52081
CORTIJO, M. 49053
COSPER, E. 49054-6
COSSINS, E.A. 49554
COSTA, B. 49057
COSTA, J.A. 51131
COSTA, S. 49058
COSTANZO, E. 49058
COSTE, B. 49059
COSTES, C. 48889, 49060, 49245, 49391,
      51611
COTTON, N.P.J. 49061
COTTON, T.M. 49062, 50775
COUDERT, A. 49391
      see COUDRET, A.
COUDRET, A. 49063-5, 49390
      see COUDERT, A.
COUGHLAN, S.J. 49066-7

K

KABAKI, N. 50041
KABSCH, U. 50042
KACPERSKA, A. 50043
KÁDÁR, I. 51828
KADISH, K.M. *49359*
KADOSHNIKOV, S.I. 48713
KADOSHNIKOVA, I.G. 51627
KAGEYAMA, M. *50251*
KAHN, J.S. 50044
KAIN, J.M. 50045
KAISER, G. 50046
KAISER, W.. *49315,* 50047-8, *51706*
KAISER, W.M. 50049
KAISER-JARRY, K. 48823
KAITALA, V. 50050
KAK, S.N. 50051
KAKHNOVICH, L.V. B50052
KAKUBARI, Y. 50630, 50632
KAKUNO, T. 50053-4, 51871
KALANDADZE, A.N. 50704
KALER, V.L. 50055
KALEZIĆ, R. 48956
KALIR, A. 50056-7
KALISKY, O. 50058
KALLIDUMBIL, V. 49421
KALOSAKAS, K. 48526
KALOSHIN, A.G. 49965
KALUS, Yu.A. 51576
KALYANASUNDARAM, K. 49130
KAMBAYASHI, M. 51419
KAMEYAMA, K. 51838
KAMIKUBO, T. 51605
KAMIŃSKA, Z. 50059
KAMIYA, A. 50060
KAMO, N. 49778, 50061, 51992
KAN, K. 50062
KANAYAMA, K. 51440
KANAZAWA, T. 50101
KANAZAWA, Y. 50063
KANDA, H. 50064
KANDA, M. 50286
KANDAUROVA, G.S. 49237
KANDYA, A.K. 50065
KANEKI, Y. 50979
KANEMASU, E.T. 48977, 50165, 51014
KANEMATSU, S. 48543
KANESHIRO, K.H. 51635
KANIUGA, Z. 50707
KANIVETS, N.P. 52126, 52128, 52322
KANNAN, S. 50066, 51225
KANNINEN, M. 49730, 50067, 51171
KANSTAD, S.O. 48882
KANTCHEVA, M.R. 50068-9
KAO, C.H. 52172
KAPCHINA, V. 49949
KAPITULA, A.M. 51221
KAPLAN, A. 50070, 50589
KAPLAN, D. 51089

KAPLAN, N.O. *48413, 48438, 48524,*
    48585, 48593, 48630, 48644, 48741,
    48760, 48817, 48905, 48986, 49046-
    -7, 49082, 49089, 49183-5, 49247,
    49256, 49296, 49306, 49313, 49324,
    49414, 49459, 49540, 49547, 49706,
    49819, 49825, 49871, 50061, 50127,
    50134, 50157, 50216, 50220-2,
    50312, 50349, 50399, 50420-2,
    50460, 50503, 50579, 50603, 50621,
    50668, 50784, 50795, 50869, 50874,
    50924-5, 50959, 51003, 51039,
    51268, 51328, 51523, 51639, 51719,
    51730, 51797, 51957, 51965, 51989-
    -90, 52026, 52053, 52148, 52159,
    52196-7, 52200, 52285, 52388
KAPLAN, S. 50071-2, *50072*
KAPLANOVÁ, M. 51829
KAPPEL, W.K. 51171
KAPPEN, L. 50844
KAPTEIN, R. 50883
KARADGE, B.A. 50073
KARAMANOS, A.J. 50074
KARAPETYAN, N.V. 49954, 50397-8,
    51220, 52345
    see KARAPETYAN, N.W.
KARAPETYAN, N.W. 50075
    see KARAPETYAN, N.V.
KARCZMARCZYK, S.J. 49200
KARG, V. 49575
KARIYA, K. 50076
KARLMAN, S.-G. 50077
KARMANOV, V.G. 50078
KARMARKAR, S.M. 50079
KARNAUKHOV, V.N. 50080
KARNEEVA, N.V. 50081
KAROLEWSKI, P. 50257
KARP, M. 50500
KARPILOVA, I.F. 48811, 50082
KARUNEN, P. 50083
KARVALY, B. 50084
KARVALY, B.E. 49654
KARVOVSKAYA, E.A. 49614
KASATKINA, T.I. 52118
KASEMIR, H. 48725, 50085
KASPAROVA, I.S. 48415
KASPRZYK, Z. 49988
KASTSYUKEVICH, G.S. 51245
KATAGIRI, S. 50086
KATAOKA, I. 52029
KATAOKA, M. 50087
KATAOKA, T. 50213
KATHJU, S. 49517
KATOH, S. 48518, 48544, 50088-9,
    50201
KATOH, T. 49958, 50299, 50919
KATSUBE, Y. 51995-6
KATSUMI, M. 50090
KATSURA, T. 50091
KATTAWAR, G.W. 50092
KATTNER, G. 48842

MAYORAL, M.L.   51596
MAZLIAK, P.   *51611*
MAZUR, B.J.   50645
MAZZOLINI, A.P.   50646
McCAIG, T.N.   49015
McCARTHY, J.J.   51340
McCARTHY, S.A.   50647
McCARTY, D.R.   51511
McCARTY, R.E.   48510, 49890, 50648,
    50769-71
McCAULEY, G.N.   52003
McCLAIN, C.R.   52328
McCONATHY, R.K.   50660
McCOWN, B.H.   48973
McCRACKEN, J.L.   50649
McCREE, K.J.   50650-3
McDANIEL, M.E.   50654
McDANIEL, R.D.   50655
McDONNEL, A.   50656
MCDONOUGH, T.J.,Jr.   50358
McEWAN, A.G.   50657
McFADDEN, B.A.   48686, 51197, 51382
McGANN, W.J.   49457
McINTOSH, A.R.   49841
McINTOSH, L.   50258, 52202
McINYRE, G.I.   50658
McKAY, W.   51401
McKELL, C.M.   52054
McKELLAR, H.N.,Jr.   50171
McLACHLAN, J.   48855
McLACHLAN, J.L.   50475
McLAREN, J.S.   *49387, 49818, 49857,*
    49980, 50137, 50149, *B50659, 51066,*
    51362, 51969
McLAUGHLIN, S.B.   50660
McLEAN, M.B.   50661
McMICHAEL, B.L.   48690, 50662
McMILLAN, C.   50663
McNEIL, P.H.   51922
McNEIL, P.L.   49858
McPHERSON, H.G.   48679
McROY, C.P.   52252
McWHA, J.A.   49696
McWILLIAM, J.R.   50664
MÉALLIER, P.   50665
MEDINA, E.   50666
MEDVEDEV, A.A.   50168
MEEK, I.T.   48581
MEESON, B.W.   50667
MEHLENBACHER, L.A.   52228
MEHLHORN, R.   52388
MEHLHORN, R.J.   50668
MEHTA, S.L.   51405
MEHTA, V.B.   50886
MEI Zhen-an   50669
MEIER, C.E.   52123
.MEIER, D.   50441-3, 50670
MEINDL, U.   50671
MEINHARDT, S.W.'   50672-3
MEINZER, F.C.   50674-6
MEISTER, A.   50677-8
MELACK, J.M.   50679

MELANDRI, B.A.   48938, 50680, 50994,
    52084
MELCHIONNA, M.   50681
MELCHIOR, G.L.   50764
MELETIOU-CHRISTOU, M.S.   49203
MELILLO, J.M.   48416
MELIS, A.   50682-3
MELLER, E.   49270, 49884, 50684
MELLINA, E.G.   48879
MEL'NIKOVA, L.M.   49600
MEMOTO, T.   50202
MENCZEL, L.   50685
MENDOZA, A.M.R.   49331
MENGEL, K.   50034, 50686, B50687, 51456
MENON, M.A.   51252
MENOUX, Y.   50688
MENYHÉRT, Z.   51211
MERBACH, W.   49467
MERCHANT, S.   50186
MERINO, J.   50689
MERLINI, L.   48629
MERRETT, M.J.   50690
MERRIEN, A.   48738
MERRITT, R.H.   50691
MERWE, N.J. van der
    see VAN DER MERWE, N.J.
MESHINEV, T.A.   49510
MESSIER, J.   50692
METEĬKO, T.Ya.   50693
METS, L.   49506, 50694
METS, L.J.   51717-8
METZ, J.G.   50695
METZGER, P.   48993
METZGER, U.   50195, 50696-7
METZIG, G.   50698
METZNER, H.   49407, 50075, 50699, 50998
MEUNIER, J.-C.   49392, 50734
MEYER, G.   50181
MEYER, H.-U.   50700
MEYER, M.   49316
MEYERS, S.P.   50701-2, 50750
MEZHUNTS, B.Kh.   50703
MGALOBLISHVILI, M.P.   50704
MICHAEL, D.   50862
MICHAEL, G.   50705-6
MICHALCZUK, L.   49293
MICHALSKI, W.P.   50707
MICHEL, H.   50708-10
MICHEL-BEYERLE, M.E.   51446
MICHELSON, A.M.   50711
MICHEL-WOLWERTZ, M.-P.   49541
MIESKES, G.   51769
MIFLIN, B.J.   50371, 52163
MIGINIAC, É.   48967
MIHALCIUC, V.   48718
MIKHAĬLOV, V.V.   49158
MIKHAĬLOVA, S.A.   48420
MIKHAĬLOVA, T.P.   50712
MIKHEEVA, S.A.   50713-4
MIKOLÁŠ, J.   49347
MIKULOVICH, T.P.   50715
MIKULSKA, E.   50716

NOZAWA, T. 49776
NOZDRINA, V.N. 49405
NUGENT, J.H.A. 49342-4, 50913-4
NUMATA, M. 50915
NURMEDOV, S.S. 51949
NURMI, A. 50916, 52064
N'Y, D.S. see DANG SUEN N'Y
NYGRÉN, M. 49730
NYRÉN, P. 51559

O

OBERLANDER, R.M. 50191
OBOLONSKIĬ, V.V. 50917-8, 51354
O'BRIEN, J. 51178
OBSHATKO, L.A. 50597
OCHIAI, H. 50919, 51440, 51586
ODERWALD, R.G. 50611
ODINTSOVA, M.S. 52011
ODUMANOVA-DUNAEVA, G.A. 50078
OELZE-KAROW, H. 50556, 50920-1
OESTERHELT, D. 49522, 50399-401, 50619,
    50710, 50922-5
OETTMEIER, W. 50926-31
OGAWA, S. 50932
OGAWA, T. 50933-5
OGBUEHI, S.N. 50936
OGDEN, J. 49474
OGINO, C. 50937
OGO, T. 51988
OGREN, R. 52162
OGREN, W.L. 50938, 51695-7, 51705,
    51716
OGURA, F. 49938
OHAD, I. 49544, 50939, 51011-2, 51615
OHGI, K. 51970
OH-HAMA, T. 49741, 50940-1
OHIRA, K. 50542
OHKAWA, Y. 52306
OHKI, K. 50942-3
OHMANN, E. 50944
OHMORI, M. 50945
OHNISHI, J. 50946
    see OHNISHI, J.-I.
OHNISHI, J.-I. 50947
    see OHNISHI, J.
OHNISHI, O. 50948
OHSAKA, A. 49920
OHSUGI, R. 50949
OHTA, H. 49926
OHYA, T. 50950
OHYAMA, K. 50951, 51433
OI, V.T. 50952
OIE, T. 50953
OIKAWA, M. 49490
OISHI, O. 49959
OKADA, M. 49959
OKAKA, S. 48543

OKAMURA, M.Y. 48900, 49157, 49305,
    50954-5
OKANO, M. 51793
OKAYAMA, S. 50956
OKAZAKI, H. 51226
O'KEEFE, D.P. 51177
O'KELLY, C.J. 50957
OKIMATSU, H. 51848
OKU, T. 50958
OKUBO, A. 48493
OLDFIELD, E. 50959
O'LEARY, J.W. 50783
O'LEARY, M.H. 49368, 50960-1, 52265
OLECH, K. 50962
OLEKSYN, J. 50257, 50491
OLIVEIRA, L. 50963
OLIVEIRA, S.A.de see DE OLIVEIRA, S.A.
OLIVER, G. 50537
OLIVER, R.P. 50964
OLLINGER, O. 51213
OLSEN, L.F. 49074, 50965-6
OLSEN, R.A. 50967
OLSON, N.H. 49463
OLSSON, K.A. 52246
OLUFAJO, O.O. 50968
OMASA, K. 50969
ONDOK, J.P. 50970-1
O'NEAL, S.W. 50972
O'NEIL, K.J. 50973
ONISHI, J.C. 50974
ONKEN, R. 52286
ONO, T. 50975
ONO, Y. 50976
ONOUE, Y. 50977-8
OOTA, T. 50979
OOTAKE, Y. 50980
OPARINA, L.A. 50981
OPRITOV, V.A. 50982
OPUTA, C.O. 49112
ÖQUIST, G. 49440, 50983-8, 51799
ORBÁN, S. 51786
OREN, R. 52183
ORLYUK, A.P. 50989
ORMOS, P. 50135
ORMROD, D.P. *48666, 48733,* 48734,
    49147, 49210, *49976, 50366, 50792,*
    50990-1, 50990, 51052, *51946,*
    52020, B52021, 52060, 52204, 52210
ORON, G. 50992
OROZCO, E.M.,Jr. 49703
ORR, G.L. 50993
ORSENIGO, M. 51242-4
ORSOLINO, R.S. 50063
ORT, D.R. 49624, 50606, 50994
ORTIZ, W. 51499
ORTOIDZE, T.V. 50628
OSAKI, M. 50995
OSBORN, J.F. 49384
OSBORNE, L.L. 50996
OSHIMA, T. *50251*
OSHINO, R. 51601
OSIPOV, A.V. 50517

Q

QASIM, S.Z. 49197-8
QIAN Lu-ping 52323
QUAEGEBEUR, J.-P. 51552
QUAGLIARIELLO, E. *49975*
QUAGLIETTA CHIARANDA, F. 51444
QUEIROZ, O. 48861-4, 51120, 51201-2
QUINN, P.J. 49615, 51530, 52253
QUINONES, F.A. 51203
QUINTANILHA, A. 48905
QUISENBERRY, J.E. 48690
QUIZENBERRY, J.E. 51204

R

RABE, R. 51205-6
RABINOWITCH, H.D. 51207
RABOTYAGOV, V.D. 50316
RACANELLI, T. 50061
RACCHI, M.L. 51242, 51244
RACKOVSKY, S. 51208
RADEMAKER, H. 51209
RADENOVIĆ, Č. 52141
RADHAKRISHNA, K. 51210
RADHAKRISHNAN, R. 48631, 49896
RADICE, M. 51261
RADICS, L. 51211
RADIN, J.W. 49922, 51212
RADMER, R. 51213
      see RADMER, R.J.
RADMER, R.J. 48659
      see RADMER, R.
RADZIEVSKAYA, M.G. 51192
RAFFERTY, C.N. 49185
RAGHAVENDRA, A.S. 51214
RAGIMOV, V.I. 48467
RAHAT, M. 48923
RAHIRE, M. 49259
RAI, H. 51215
RAINS, D.W. 49091, *49469, 50723, 50776, 51089*
RAISON, J.K. 48693, 51216
RAJBHANDARY, U.L. 51649
RAJENDRUDU, G. 49131-2, 51217-8
RAJORA, O.P. 51219
RAJPUROHIT, K.S. *50023, 50079, 50123*
RAK, E. 48576, 50233
RAKHIMBERDIEVA, M.G. 49954, 50397-8, 51220
RAKHTSEENKA, I.N. 51221
RAKOWSKI, K. 52275
RAM, H. 51222
RAMACHANDRA REDDY, A. 51223
RAMAMURTHY, S. 51224
RAMANI, S. 50066, 51225
RAMBAL, S. 49424
RAMBIER, M. 49068, 49661

RAMIREZ, F. 51226
RAMÍREZ, J.M. 49563-4, 49596-7
RAMIREZ-PONCE, M.P. 49563
RAMOS, C. 51227-8
RAMOS, J.L. 49669, 51229-31
RAMSHAW, J.A.M. 51232
RAMUS, J. 51319
RAND, L. 51137
RAND, R.H. 51233
RANDALL, D. 51061
RANFT, H. 50706
RANSON, D. 49869-70
RANTY, B. 51234-5
RAO, A.N. 51236
RAO, C.M. see MOHAN RAO, C.
RAO, K.K. 49692, 49695, 50473, 51237-9
RAO, K.S. 49136
RAO, L.V.M. 51240
RAO, M.B. see BABU RAO, M.
RAO, M.R.K. 49678
RAO, N.K.S. see SRINIVASA RAO, N.K.
RAO, P.S. 51241
RAPER, C.D.,Jr. 49101
RAPER, W.G.C. 49950
RAPP, M. 50408
RASCHKE, E. 50698
RASCHKE, R.L. 51265
RASCIO, N. 48939, 50592-3, 51242-4
RASKIN, V.I. 51245
RASMUSSEN, V.P. 49715
RAST, D.M. 49746
RASULOV, B.Kh. 51246-9
RATTIGAN, B.M. 49911
RAULIN, F. 51250
RAUNIO, R. 50500
RAUSCH, U. 48658
RAVANEL, P. 50534
RAVE, G. 51733
RAVEN, J.A. 48640, 49072, 51251
RAVINDRAN, P.N. 51252
RAWAT, M.S. 51219
RAWITZ, E. 48721
RAWSON, H.M. 49042, 51253-5, 51657, 51847
RAYNER, J.H. 51157
RAYNES, D.A. 51076
RAZJIVIN, A.P. 48779, 51256
      see RAZZHIVIN, A.P.
RAZUMOVA, N.A. 51257
RAZZHIVIN, A.P. 48778
      see RAZJIVIN, A.P.
REAL, F. 48825
REARDON, E.M. 51173
REBEIZ, C.A. 48660-1, 49121-2, 49271-3, 50647, 51258-60
REBELLO, A. De LUCA
      see DE LUCA REBELLO, A.
RECALCATI, L.M. 48453, 51261
REDDY, A.R. see RAMACHANDRA REDDY, A.
RÉDEI, E. 51262
RÉDEI, G.P. *52010*
REDENBAUGH, K. 48910

SINGH, C. 51656
SINGH, D.P. 51657
SINGH, J.S. 51658
SINGH, M.B. 49202
SINGH, R. 51222
SINGH, S.P. 51569, 51721-2
SINGH, Y.D. 51051
SINGHVI, N.R. 51659
SINHA, S.K. 50143, 51405
SINHA, S.S.N. 51660
SIONIT, N. 51661, 51898
SIRENKO, L.A. B51350, B51662
SIREVÅG, R. 50481
SIRONVAL, C. 49275, 51663
SITZMANN, E.V. 49019
SKJOLDAL, H.R. 51664
SKLAN, D. 51207
SKULACHEV, V.P. 48612, 49250-1, 51007,
    51665, 52269
ŠKŮTA, A. 52031
SLADKÝ, Z. 49985
SLATER, E.C. 48455, 49975
SLATER, J.H. 50772
SLATER, P.N. 51666
SLAVOVA, Ĭ. 51667
SLAVTCHEVA, T. 51774
SLAWYK, G. 50732, 51668
SLAYMAN, C.L. 49212, 49625
SLEEP, J.A. 49140
SLEMNEV, N.N. 48780
SLOVACEK, R.E. 51669-70
SLOVIN, J.P. 51671
SLUKA, Z.A. 51672
SMAKMAN, G. 51673
SMERAGE, G.H. 50015
SMETACEK, V. 51069
SMILLIE, R.M. 49812-5, 51674
SMIRNOV, G.F. 51675
SMITH, A.G. 51676
SMITH, A.M. 51677
SMITH, B.N. 50538, 50663, 51678
SMITH, C.M. 51679
SMITH, D.F. 51680-1
SMITH, D.W. 50611
SMITH, G.D. 51239, 51682
SMITH, G.F. 51484
SMITH, G.M. 49765
SMITH, H. 49102, 50766, 51507, 52287
SMITH, J.A.C. 51683
SMITH, J.P. 50161
SMITH, K.M. 48417, 48858, 49359, 49885,
    51684
SMITH, M.G. 51113
SMITH, R.E.H. 51685
SMITH, S.D. 51827
SMITH, S.M. 48653
SMITH, T.D. 50471
SMITH, U.H. 52186
SMITH, V.H. 51686
SMITH, W.K. 49535, 52334
SMITH, W.O.,Jr. 51687
SMITH, W.R. 50781

SMITHERS, J. 51124
SMYTH, D.A. 50236
SNEDAKER, S.C. 51688
SNEL, J.F.H. 51179-80
SNELGAR, W.P. 49638-9, 51689
SNOW, J.W. 51900
SNOZZI, M. 48569, 49127
SOF'IN, A.V. 51572
SOFROVÁ, D. 49838, 50536
SOKEROV, S. 51955
SOKOLOV, Z.N. 51690-1
SOLÁROVÁ, J. 51160
    see SOLAROVA, Ya.
SOLAROVA, Ya. 48975
    see SOLÁROVÁ, J.
SOLDATINI, G.F. 51692
SOLEL, Z. 50833
SOLING, H.-D. 51769
SOLL, H.-J. 50931
SOLL, J. 49397, 51494, 51693
SOLNTSEVA, O.N. 52027
SOLOMATKIN, V.P. 51321
SOLON, J. 50636
SOLONENKO, T.A. 50209
SOLOV'EV, E.V. 50078
SOLOVJEVA, N.A. 48516
SOLTANI, A. 51694
SOMERVILLE, C.R. 51695-7
SOMMER, C. 51698
SOMMER, H.E. 52223
SOMMER, U. 51699
SONG, P.-S. 51414, 51700
SOPORY, S.K. 51240
SOSIŃSKA, A. 51701
SOTTA, B. 48967
SOUČKOVÁ, L. 52099
SOUKUP, M. 52230
SOUZA, O.C. de see DE SOUZA, O.C.
SOUZA MACHADO, V. 51702-3
SOWELL, J.B. 51704
SPAETH, S.C. 52081
SPALDING, M.H. 51705
SPALINK, J.-D. 51706
ŠPÁNIK, F. 49500, 51707
SPARACE, S.A. 51708-9
SPENCER, C. 49450
SPENCER, N.R. 48948
SPENCER-PHILLIPS, P.N.T. 51710
SPENER, F. 51361
SPERANZA, M.L. 51711
SPERLING, W. 48645, 49185, 51465, 51706
SPIKES, J.D. 51777
SPILLER, S.C. 51712
SPIRESCU, I. 48551, 51713
SPIRO, T.G. 49285
SPITZNER, C. 51714
SPORTELLI, L. 51715
SPREITZER, R.J. 51716-8
SPUDICH, E.N. 51719
SPUDICH, J.L. 48759, 51719
SQUIRE, G.R. 51720
SQUIRE, R.O. 48553

This index contains a selection of primary items chosen according to their importance in photosynthesis research and to their relevance and occurrence. The word "Photosynthesis" is not regarded as a main theme, but partial processes, photosynthetic parameters and the factors affecting photosynthesis are listed. The processes and other characteristics are summarized into several themes when presented in combination with individual factors, *e.g.* canopy functioning, carbon fixation pathways, electron transport chain, chlorophyll, gas exchange, ecosystem and plant productivity (including photosynthate translocation and distribution), photorespiration, resistances to CO_2 and water vapour transfer, *etc.*

Several items from branches related to photosynthesis research were also chosen for convenience, *e.g.* dealing with respiration, plant growth and development, water relations, anatomy, bioclimatology, *etc.* These items contain only references to papers within the scope of this bibliography.

A more detailed information may be obtained by combining the items. Several very comprehensive items are therefore included.

A

Abscisic acid   see   Growth regulators ...

Absorbance in canopy   see   Canopy, radiation profile

Accumulation of dry matter   see   Biomass distribution...; Dry-matter production...;
        Ecosystem production ...

Achlorophyllous cells and organs, respiration in   see   Respiration of achlorophyllous
        tissues in light, light inhibition of respiration

Action spectra   see   Irradiance, spectral composition ...

Adenosine triphosphate   see   ATP

Aerodynamic methods, bioclimatological methods (sampling, measurement of wind, rain,
        dew, *etc.*)
        49102, 49126, 49194, 49604, 50092, 50698, 51123, 51666, 52109, 52169

Age of algae, leaf, plant   see   Ontogeny...; Canopy, leaf age; Position of leaf on
        plant ...

Agrotechnics and canopy functioning   50164, 51082

Agrotechnics and carotenoids   51096

Agrotechnics and chlorophyll   49499, 51083, 51096, 51147, 51574

Agrotechnics,and ecosystem and plant productivity
        48655, 49000, 49496-7, 49500, 49844, 49995, 50164, 50226, 50274, 51126,
        51306, 51409, 51411, 51512, 51895, 52099

Agrotechnics and electron transport chain   51079

Agrotechnics and gas exchange   48599, 48632, 48655, 51306, 51433, 52086-7

Agrotechnics and respiration   48599, 51306

Air-flow rate   see   Wind ...

Albedo, canopy   see   Canopy, radiation distribution

Algae and photosynthetic bacteria, cultivation (*cf.* also Algae mass cultures produc-
       tivity)
       48441, B48650, B48671, 48912, 49055, 49505, 49937, 49950, 51010, 51101,
       51345, 51535, 51560, 51644, 51843, 52130, 52144-5, 52197

Algae and secondary production of reservoirs
       48996, 49070, 49098, 49116, 49146, 49189, 49197-8, 49275, 49316, 49417,
       49445, 49468, 49487, 49647, 49879, 50180, 50355, 50409, 50741, 51010, 51123,
       B51350, 51374, 51466, 52270, 52318

Algae, blue-green , chromatophores in   see Chromatophore ...

Algae carotenoids   see   Xanthophyll of algae

Algae chlorophylls   see   Chlorophylls *c, d*

Algae, CO_2 and O_2 exchange   see Gas exchange in algae

Algae, depth distribution in reservoirs
       48412, 48484, 48492-3, 48629, 48641-3, 48728, 48769, 48842, 48932, 49070,
       49098-9, 49116, 49196, 49283, 49317, 49339, 49350-1, 49416, 49425, 49483,
       49487-8, 49518-9, 49629, 49729, 49735, 49777, 49784, 49799-800, 49874, 49889,
       49933, 49977, 50047, 50180, 50219, 50239-40, 50243, 50308, 50375, 50430,
       50544, 50601, 50679, 50727, 50730, 50751, 50864, 50957, 50992, 51060, 51124,
       51138, 51143, 51183-4, 51210, 51215, 51363, 51580, 51587, 51623, 51664,
       51687, 51740, 51742-3, 51758, 51844, 51860, 52001, 52019, 52044, 52063,
       52103, 52188, 52192, 52328, 52358

Algae in cosmonautics B48671

Algae in sediments 50022, 51286, 51664, 52146, 52378

Algae in sewage cleaning   48441, 48566, 48648, 48651, 49086, 49950, 51010, 51858

Algae life cycles see Ontogeny of algae ...

Algae mass cultures productivity (*cf.* also Algae and photosynthetic bacteria, culti-
       vation)
       48411, 48441, 48566, 48592, 48648, B48650, 48651, B48671, 48672, 48675,
       48678, 48706, 48786, 48937, 48981, 49275, 49667, 50045, 50385, 50613, 50690,
       50741, 50934, 51010, 51406, 51560, 52142, 52144-5, 52270, 52317

Algae photosynthesis and production
       48541, 48566-7, 48642, 48672, 48728, 48769-70, 48808, 48825, 48866, 48886,
       48937, 48981, 48996, 49149, 49189, 49195, 49197-8, 49258, 49269, 49288,
       49311, 49316-7, 49350-1, 49364, 49400, 49416-8, 49445, 49518, 49525, 49581,
       49591, 49630, 49679, 49728-9, 49735, 49768, 49824, 49876, 49968, 49977,
       50022, 50180, 50185, 50203, 50218-9, 50239, 50355, 50373, 50375, 50385,
       50428, 50430, 50457, 50467, 50566, 50600, 50679, 50730-2, 50772, 50864,
       50899, 50972, 50996, 51069, 51139, 51215, 51284, 51313, B51350, 51374,
       51387, 51395, 51438, 51450, 51459, 51560, 51580, 51587, 51623-4, 51668,
       51741, 51757-8, 51844, 51860, 51904, 51914, 51928, 51964, 51966, 51973,
       52085, 52103, 52142, 52165, 52188, 52251, 52286, 52318, 52358

Algae, primary productivity in reservoirs (*cf.* also Chlorophyll and production of
       algae and water reservoirs)
       49086, 49581, 50731, 50880, 51313, 51406, B52018, 52102, 52136, 52188

Algae, primary productivity, methods {*cf.* also O_2 determination (other than O_2 elec-
       trode); O_2 electrode}
       48484, 48493, 48866, 49070, 49160, 49468, 49494, 49504, 49552, 49604, 49630,
       49811, 49987, 50018, 50185, 50340, 50423, 50457, 50698, 50805, 50871, 50988,
       51044, 51105, 51161, 51265, 51323, 51580, 51699, 51741, 51966, 52063, 52072,
       52085, 52187, 52251, 52324

Algae synchronous cultures   see Algae and photosynthetic bacteria, cultivation;
        Ontogeny of algae ...

Altitude see Pressure, altitude ...

Amino acids see Proteins, amino acids, nucleic acids ...

δ-Aminolevulinic acid   see Chlorophyll biosynthesis ...

Amphistomatous leaf, gas exchange in (cf. also Leaf epidermis, stomata) 50169, 50783

Anaerobic atmosphere see N_2, anaerobic atmosphere ...

Antibiotics and carbon fixation pathways   48685, 49401, 49427, 51122, 52218, 52295

Antibiotics and carotenoids   49061, 49952, 49960

Antibiotics and chlorophyll   48582, 49094, 49209, 49263, 49430, 49544, 49559, 49952,
        51415, 51451

Antibiotics and chloroplast (chromatophore)
        49035, 49209, 49411, 49683, 49960, 50381, 50715, 50765, 50882, 51270, 51345,
        51415, 51451, 51559, 51879, 52288, 52347

Antibiotics and electron transport chain
        48430, 48474, 48582, 48688, 48795, 49078, 49544, 49593, 49624, 49722, 49739,
        49870, 49897, 49915, 49960-1, 50036, 50141, 50315, 50353, 50562, 50672,
        50704, 50770, 50865, 50975, 51177, 51276, 51321, 51429, 51467, 51490, 51559,
        51616, 51836, 51919, 51938, 51999, 52084, 52107, 52315

Antibiotics and gas exchange   48978, 49544, 50589, 50644, 50704, 51504, 52320

Antibiotics and photorespiration   49981

Antigens see Electron transport chain, serological analysis

Antitranspirants 48538, 48709, B48895, 49723, 50165, B50687, 51092, 51787, B51897

Architecture of canopy see Canopy ...

Architecture of thylakoid (photosynthetic membranes) see Electron transport chain
        localization in thylakoid

Assimilates see Photosynthates ...

Assimilation chamber
        48453, 48649, 48754, 48959, 49297, 49398, 49591, 49936, 50224, 50485, 50487,
        50630, 50887, 51228, 51253, 51395, 51441, 51484, 51698, 51786, 51856, 52020,
        52249

ATP   48430, 48432, 48484, 48569, 48594, 48602, 48680, 48702, 48748, 48860, 48865,
        48956, 49014, 49035, 49050, 49116, 49124, 49143, 49165, 49348, 49366, 49383,
        49428, 49507, 49551, 49569, 49593, 49624, 49626, 49649, 49709, 49714, 49782,
        49784, 49870, 49877, 49902, 49970, 49994, 50101, 50118, 50190, 50231-2,
        50235, 50280, 50319, 50391, 50411, 50449, 50633, 50648, 50732, 50765, 50786,
        50794, 50834, B50878, 50921, 50975, 50994, 51008, 51117, B51168, 51171,
        51190, 51259, 51263, 51275, 51301, B51350, 51356, 51440, 51467-8, 51487,
        51489-90, 51528, 51557, 51566, 51604, 51645, 51664-5, 51669, 51683,
        51743, 51768, 51781, 51803, 51836, 51893-4, 52008, 52022, 52069, 52084,
        52107-8, 52127, 52173, 52189, 52194, 52209, 52211, 52269, 52314, B52321,
        52336-7, 52349

ATP, methods   50101, 50711, 51487

ATPase, coupling factor 1
        48432, 48495-7, 48510, 48522, 48569, 48580, 48582, 48601, 48619, 48681,
        48709, 48748, 48762, 48787-8, 48795, 48860, 48955, 49061, 49066, 49104,
        49143, 49151, 49240, 49276, 49279, 49343, 49383, 49460, 49517, 49569 ,
        49590, 49625-6, 49683, 49793, 49830, 49842, 49887, 49897, 49902, 49961,
        49978, 50033, 50044, 50071, 50186, 50197, 50258, 50260, 50393, 50411, 50506,
        50568-9, 50648, 50701, 50728-9, 50768-71, 50799, 50814, 50858, 50860,
        B50878, 50893, 50982, 50994, 51047, 51117-8, B51168, 51467, 51487, 51490,
        51555-7, 51559, 51581, 51614, 51618, 51665, 51683, 51746, 51752, 51781,
        51802, 51836, 51877, 51894, 51918, 52022, 52069, 52084, 52098, 52107-8,
        52149, 52151, 52173, 52191, 52199, 52220-1, 52269, 52279, 52289, B52321,
        52336, 52349, 52379

ATPase, methods   48510, 49830, 49978, 50769, 50799, 51116, 51752, 52151

Autotrophy see Carbon metabolism types ...

B

Bacteria, photosynthatic see Photosynthetic bacteria ...

Bacteriochlorophylls
        48569, 48612, 48654, 48682, 48699, 48706, 48778-9, 48799, 48802, 48804,
        48831, 48859, 48900, 48991, 49016, 49019, 49027-8, 49040, 49062, 49137,
        49178, 49180-1, 49209, 49256, 49286, 49307, 49379, 49393-4, 49451, 49457,
        49481, 49533, 49583, 49596-7, 49721, 49764, 49774-6, 49789, 49835, 49848-9,
        49873, 49885, 49910, 49935, 49943, 49974, 50071-2, 50093, 50098, 50173,
        50189, 50223, 50245, 50247, 50415, 50495, 50524-6, 50539, 50832, 50866,
        50868, B50878, 50908-9, 50939, 50954, 50974, 51015, 51021, 51040-1, 51063,
        51100, B51168, 51183, 51208-9, 51260, 51388, 51452, 51454, 51480, 51585,
        51602, 51608, 51629-30, 51684, 51700, 51726, 51743, 51748-9, B51788, 51809-
        -10, 51835, 51848, 51871, 51937, 51984, 52038, 52058-9, 52084, 52092, 52094,
        52131, 52186, 52208, 52274, 52312, 52349

Bacteriorhodopsin see *Halobacterium* photosynthesis

Bacteriorhodopsin, methods
        48413, 48438, 48524, 48585, 48630, 48644, 48741, 48760, 48817, 48905, 48986,
        49082, 49089, 49183-5, 49247, 49296, 49306, 49313, 49324, 49414, 49459,
        49540, 49706, 49819, 49871, 50061, 50134, 50151, 50157, 50216, 50220-2,
        50312, 50349, 50399, 50420-2, 50460, 50503, 50579, 50603, 50621, 50668,
        50710, 50784, 50795, 50869, 50923-5, 50959, 51003, 51039, 51328, 51523,
        51633, 51638-9, 51706, 51719, 51730, 51797, 51957, 51965, 51989-90, 52026,
        52053, 52148, 52159, 52196-7, 52200, 52388

Bacteriorhodopsin, model   see Model of *Halobacterium* photosynthesis

Bibliographies of photosynthesis, biographies   51548-9, B51550

Biliproteins see also Phycocyanins; Phycoerythrins

Biliproteins absorption spectra *in vitro*
        B48650, 48654, 48821, 48867, 49106, 49681, 50289, 51174, 51264, 51276,
        52227, 52369

Biliproteins absorption spectra *in vivo*
        49166, 49213, 49224, 49565, 49573, 49577, 49865, 49958, 50109, 50159, 50289,
        51174, 51452, 51470, 51545, 51622, 51636, 51857, 51859, 51971, 52364

Biliproteins and production of algae and water reservoirs   B48650

Biliproteins biosynthesis, precursors   48858, 48867, 50943, 51260, 51460

Biliproteins chemical structure   48654, 48821, 49106, 49169, 49573, 49681, 50283, B51168, 51338, B51788

Biliproteins complexes *in vitro*   52338

Biliproteins complexes *in vivo*   48506, 49565-6, 49573, 49681, 49865, 50230, 50777

Biliproteins degradation   50283

Biliproteins determination, electrophoresis and other methods   49289, 49865, 51264, 51542, 52369

Biliproteins determination, spectral methods   49359, 49681, 50885, 51174, 51264, 52154

Biliproteins energetic states *in vitro*   49452

Biliproteins energetic states *in vivo*   48845-6, B49433, 49440, 49566, 49865, 50113, 50986, 51452

Biliproteins fluorescence *in vitro*   49452, 49681, 50464, 51174, 52369

Biliproteins fluorescence *in vivo*   48845-6, 49213, 49440, 49566, 49573, 49577, 49865, 49958, 50109, 50189, 50299, 50831, 51452, 51545, 51622, 51636, 51971, 52058, 52364

Biliproteins in model systems   48821, 49168, 49452

Biliproteins in mutants see Mutants, biliproteins in

Biliproteins in photosynthesis mechanism   48717, 48846, B49433, 49482, 50113, 50986, 51074, 51319, 51452

Bioclimatological methods see Aerodynamic methods ...

Biological clock see Diurnal changes ...

Biomass distribution and redistribution in plant
        48418, 48421-2, 48425, 48434, 48439, 48449, 48456-9, 48485, 48504, 48517,
        48523, 48548, 48556, 48564, 48573, 48579, 48605, 48614, 48624, 48626, 48711,
        48719-20, 48738, 48775, 48780, 48839, 48871, 48884, 48893, 48902, 48930,
        48948, 48957, 48966, 48977, 48999, 49009, 49021, 49042, 49087, 49113, 49144,
        49158, 49165, 49205, 49207, 49241, 49282, 49284, 49345, 49347, 49355, 49357,
        49384-5, 49415, 49425, 49431, 49467, 49471, 49493, 49510, 49536, 49549-50,
        B49561, 49578, 49592, 49599, 49602, 49617, 49619, 49638, 49645, 49668,
        49677, 49686, 49730, 49732-3, 49766, 49783, 49785, 49803, 49847, 49855,
        49861, 49955-6, 49979, 50002, 50013, 50017, 50019, 50034, 50037, 50063,
        50086, 50103, 50148-9, 50152, 50165-7, 50175, 50204, 50213, 50226, 50281,
        50288, 50302, 50327-8, 50367, 50384, 50396, 50405, 50408, 50446, 50463,
        50502, 50511, 50537, 50548, 50550, 50610-1, 50636, 50652, 50686, 50691,
        50712, 50718, 50731, 50753, 50757, 50789, 50791, 50796, 50806, 50812, 50841,
        50843, 50856, 50873, 50907, 50915, 50973, 50984-5, 50999, 51001, 51009,
        51014, 51033, 51049, 51057, 51075, 51086, 51094, 51131, 51157, 51182, 51217,
        51271, 51283, 51306, 51369-71, 51405, 51409, 51411, 51419, 51448, 51475,
        51482, 51496, 51502, B51531, 51540, 51551, 51565, 51646, 51661, 51782,
        51808, 51827, 51881-3, 51886-7, 51895, B51897, 51908-10 , 51951,
        51959, B51961, 52003, 52005, 52023, 52027, 52054, 52068, 52074, 52123,
        52140, 52198, 52203, 52219, 52235, 52242, 52246, 52250, 52265, 52297, 52343,
        52372

Biopotentials see Chloroplast and chromatophore biopotentials

Biosphere production see Ecosystem production ...

Blinks effect see Emerson effect, Blinks effect

Books on photosynthesis see General aspects ...

Boundary layer of air see Resistance, leaf boundary layer

Bundle sheaths see Carbon metabolism types...; Carbon fixation pathways, comparison
    in mesophyll and bundle sheath cells

C

$^{13}C/^{12}C$ ratio, $\delta^{13}C$
        48542, 48640, 49299, 49337, 49365, 49368, 49455, B49561, 49759, 49807,
        49965-7, 50370, 50374, 50609, 50663, 50744, 50940, 51071-2, 51289, 51447,
        51491, 51678, 51907, 52050, 52136, 52258, 52265, 52307

^{14}C, ^{11}C, ^{13}C see Carbon isotopes ...

C_3 pathway of carbon fixation
        48433, 48500, 48542, 48620-2, 48657, 48670, 48692-3, 48730, 48768, 48857,
        48930, 48945, 48999, 49020, 49038, 49043, 49050, 49065, 49131-2, 49150,
        49220, 49249, 49262, 49366, 49501, 49546, 49756, 49760, 49788, 49807,
        49860, 49863, 49967, 49999, 50011, 50043, 50273-4, 50382, 50390, 50538,
        50547, B50687, 50752, 50754, 50786, 50815, 50933, 50938, 50940, 50944,
        50999, 51004, 51054, 51072, 51081, 51098, 51137, 51152-3, 51167, B51168,
        51181, 51217, 51227, 51234, 51275, 51305, 51308, 51341, 51447, 51485,
        51494, 51517, 51625, 51678, 51794, 51870, 51905, 51907, 51928, 51997,
        52106, 52139, 52247-8, 52263, 52265, 52267, 52276, 52287, B52321, 52325,
        52361

C_4 pathway of carbon fixation
        48433, 48500, 48542, 48622, 48670, 48692-3, 48730, 48768, 48857, 48883,
        48919, 48930, 48945, 48999, 49038, 49132, 49249, 49262, 49299, 49366,
        49491, 49501, 49546, 49556, B49561, 49755-7, 49759-60, 49788, 49807, 49860,
        49967, 49999, 50011, 50043, 50123, 50126, 50183, 50279, 50382, 50390, 50547,
        B50687, 50752, 50754, 50761, 50786, 50815, 50836, 50933, 50938, 50940,
        50949, 50960-1, 50999, 51004, 51054, 51061, 51071-2, 51081, 51137, 51152-3,
        51167, B51168, 51181, 51217, 51227, 51234, 51289, 51360, 51447, 51485,
        51563, 51588, 51625, 51677-8, 51692, 51794-5, 51870, 51905, 51928, 51975,
        51997, 52083, 52106, 52139, 52267, 52276, 52280, 52292, 52295, B52321,
        52361

C_3, C_4, CAM pathways, comparison see Carbon metabolism types ...

Calibration of infre-red analyser see Infra-red analyser ...

Caloric values see Energy content ...

Calorimetry see Energy content ...

Calvin-Benson cycle see C_3 pathway of carbon fixation

CAM   48481-2, 48504, 48542, 48610, 48622, 48732, 48768, 48861-4, 48898, 49038,
        49446, 49475, 49579, 49637, 49756, 49863, 49899, 49967, 50011, 50115, 50126,
        50182, 50187, 50192-3, 50382, 50390, 50466, 50538, 50547, 50571, 50607,
        50609, 50666, B50687, 50761, 50797, 50891-2, 50938, 50960, 50999, 51081,
        51107, 51113, B51168, 51201-2, 51223, 51381, 51389, 51407-8, 51476, 51491,
        51678, 51683, 51870, 51901, 51907, 51997, 52112-3, 52262-3, 52266

Canopy, CO_2 profiles 48716

Carbonic anhydrase
        48542, 48945, 49628, 49944, 50210-1, 50461, 50911, 51640, 51705, 51997,
        52032

Carboxylation   see   Carbon fixation pathways ...

Carboxylation resistance   see   Resistance, carboxylation and excitation

Carotenes
        48568, 48626, 48654, 48674, 48727, 48824, B48841, 48855, 48899, 48964,
        49014, 49108, 49117, 49139, 49141, 49156, 49218, 49242, 49280, 49285, 49336,
        49340, 49426, 49458, 49552, 49570, 49655-7, 49664, 49904, 49952, 50009,
        50032, 50060, 50212, 50237, 50246, 50284, 50287, 50398, 50440-1, 50444,
        50448, 50478, 50505, 50520, 50525, 50536, 50556, 51033, 51056, 51148, 51172,
        51207, 51276, 51288, 51292, 51299, 51316, 51462, 51593-4, 51643, 51676,
        51734, B51788, 51889, 51974, 52000, 52013, 52030, 52033, 52116, 52130,
        52140, 52211, 52232, 52274, 52331-2, 52376, 52378

Carotenoids absorption spectra *in vitro*
        48672, 48674, 48727, 48964, 49141, 49245, 49261, 49336, 49753, 50505, 50667,
        51276, 51348, 51462, 51472, 51593, 51700, 51734, 51810, 51839, 52058, 52227,
        52378

Carotenoids absorption spectra *in vivo*
        48672, 48699, 49752-4, 49960, 50237, 50525, 50667, 51268, 51634, 51838,
        51841

Carotenoids, and production of algae and water reservoirs 48710, 48727, 49055,
        49796, 51406, 52378

Carotenoids biosynthesis, precursors
        48840, B48841, 48855, 48896-7, 48913-4, 49013-4, 49309, B49422, 49663,
        B49749, 49793, 50120, 50398, 50440, 50514, 50520, 50556, 50602, 50643,
        50717, 50746, 50803, 50946, 51056, 51099, 51260, 51292, 51348, 51399, 51451,
        51462, 51494, 51818, 52028, 52033, 52230-1, 52274, B52321, 52331-2, 52385

Carotenoids chemical structure
        48410, 48469, 48569, 48654, 48727, 48840, B48841, 48874, 49108, 49117,
        49141, 49340, 49849, 50237, 50643, 50717, 50759, 50803, 50901, 51099, 51348,
        51472, 51818, 51839, 51994, 52230-1, 52274, 52378

Carotenoids complexes *in vitro*   48989, 51432, 51840

Carotenpids complexes *in vivo*   48613, 48654, 49027, 49752-4, 49772, 49838, 50237,
        50444, 50623, 50667, 50759, 50997, 51654, 51820, 51835, 52059, 52232, 52274

Carotenoids degradation   48837, 49426, 49822, 50005, 50529, 50904, 51262, 51399,
        51676, 51820, 51841, 52013, 52378

Carotenoids determination see also Pigments determination, sampling and extraction

Carotenoids determination, column chromatography
        48410, 48598, 48899, 48964, 50448, 50505, 50803, 51018, 51424, 51593-4,
        51734, 51994, 52013, 52075, 52376

Carotenoids determination, electrophoresis and other methods   48654, 51316, 51835,
        52028

Carotenoids determination, paper chromatography, thin-layer chromatography
        48513, 48855, 48913, 49139, 49336, 49458, 49754, 50440, 51018, 51424, 51593-
        -4, 52378

Carotenoids determination, spectral methods 48855, 48964, 49141, 50440, 51424,
        51839, 52378

Carotenoids energetic states *in vitro*   48990, 50495, 50623, 51288, 51368, 51700,
    52362

Carotenoids energetic states *in vivo*   48713, 49456, 50524-5, 50623, 50759, 51654,
    52058

Carotenoids fluorescence *in vivo*   51700

Carotenoids in flowers   48469, 48840, 50162-3, 50439, 50717, 50803

Carotenoids in model systems   48989-90, 50445, 52031

Carotenoids in mutants see Mutants, carotenoids in

Carotenoids in photosynthesis mechanism
    48594, 48713, 48717, 48824, 49061, 49078, 49482, 49532, 49960, 49974-5,
    50005, 50009, 50225, 50237, 50403, 50444-5, 50524-5, 50618, 50620, 50623-4,
    50759, B50878, B51168, 51177, 51268, 51368, 51455, 51634, 51654, B51788,
    52058, 52084, 52274, B52321

Carotenoids in physiology of photosynthesis   50478, 51462, 51945

Carotenoids in seeds and fruits   48469, 48840, 48899, 48913-4, 48964, 49426, 49458,
    49655-7, 50163, 50439, 51207, 51348, 51424, 52029-30

Carotenoids precursors see Carotenoids biosynthesis, precursors

Chamber, assimilation see Assimilation chamber

Chemiosmotic hypothesis, proton transport in chloroplast
    48497, 48557, 48580, 48594, B48600, 48601, 48708, 48715, 48717, 48789-90,
    48832, 48938, 49074, 49078, 49124, 49138, 49143, 49551, 49622, 49624-5,
    49714, 49722, 49771, 49842, 49869-70, 49900, 49915, 49960-1, 49970, 49974,
    50035-6, 50070, 50231-2, 50261, 50275, 50280, 50391, 50411, 50471, 50506,
    50648, 50657, 50668, B50687, 50711, 50714, 50771, 50794, 50860, B50878,
    50893, 50953, 50994, 51008, B51168, 51177, 51250, 51275, 51278, 51321,
    51331, 51353, 51467-8, 51489, 51518, 51556-7, 51559, 51634, 51665, 51669,
    51760, 51798, 51806, 51836, 51878, 51918-9, 51938, 51943, 51999, 52022,
    52069, 52090, 52107, 52148-9, 52173, 52194, 52199, 52209, 52221-2, 52269,
    52289-90, 52312, 52315, B52321

Chlorobium chlorophylls see Bacteriochlorophylls

Chlorophyll absorption spectra *in vitro*
    48516, 48569, 48633, 48654, 48660-1, 48713, 48727, 48805, 48828, 48831,
    48833, 48849, 48872, 48903, 48962-3, 49011, 49033, 49135, 49235, 49336,
    49359, 49436, 49451, 49458, 49531, 49623, 49753, 49849, 49912, 49948, 50093,
    B50144, 50197, 50247, 50325, 50505, 50512, 50667, 50677-8, 50851, 50867,
    50978, 51013, 51036, 51065, 51095, 51135, B51168, 51264, 51276, 51339,
    51479-80, 51513-4, 51579, 51700, 51748, 51784, B51788, 51810, 51825, 51960,
    52012, 52058, 52064, 52118, 52185, 52208, 52227, 52240, 52244, 52341

Chlorophyll absorption spectra *in vivo*
    48466, 48506, 48569, 48583, 48613, 48654, 48656, B48671, 48699, 48713,
    48778-9, 48787, 48823, 48831, 48847, 48849-53, 48859, 48896, 48906, 48918,
    49006, 49024, 49027-8, 49033, 49122, 49134, 49137, 49180, 49224, 49256,
    49271, 49274, 49276, 49307, 49373-4, 49379, 49393, 49453, 49495, 49508-9,
    49531, 49541, 49559, 49564, 49596-7, 49614, 49635-6, 49648, 49650, 49753,
    49769, 49772-4, 49787, 49789, 49838, 49849, 49873, 49917, 49928, 49930,
    49943, 49945, 49954, 49959, 49971, 50027, 50080, 50098, 50111, 50113, B50144,
    50159, 50179, 50190, 50223, 50244-5, 50289, 50295, 50324, 50398, 50402,
    50424, 50427, 50433, 50441, 50455, 50498-9, 50525-6, 50539, 50590, 50616-7,
    50667, 50677-8, 50768, 50780, 50808, 50866, 50916, 50939, 50954, 51074,
                                                              (continued)

Chlorophyll absorption spectra *in vivo* (continued)
    51083, 51100, 51114, B51168, 51351-2, 51426, 51439, 51446, 51470, 51479,
    51493, 51500, 51537, 51585, 51630, 51634, 51726, B51788, 51869, 51925,
    51945, 51971, 52058-9, 52077, 52094, 52177, 52193, 52240, 52244, 52278-9,
    52310, 52345, 52363

Chlorophyll and its products determination see also Pigments determination, samp-
    ling and extraction

Chlorophyll and its products determination, column chromatography
    48682, 48696, 49575, 49584, 49753, 49953, 51161, 51607, 51739, 51749

Chlorophyll and its products determination, electrophoresis and other methods
    49171, 49753, 51426, 51746, 51835, 52034, 52298

Chlorophyll and its products determination, *in vivo*
    48831, 48881, 49123, 49171, 49371, 49453, 49464, 49533, 49688, 49959, 50026-
    -7, B50144, 50206, 50214, 50499, 50656, 50678, 50708, 50868, 51074, 51114,
    51426, 51449, 51493, 51746, 52077, 52341

Chlorophyll and its products determination, paper chromatography, thin-layer chro-
    matography 48513, 48633, 48661, 49120, 49123, 49321, 49458, 49986, 50173,
    50257, 50440, 50957, 51260, 51293, 51479-80, 51547, B51662, 51743, 51810

Chlorophyll and its products determination, spectral methods
    48463, 48588, 48645, 48654, 48656, 48849, 48906, 48921, 49039-40, 49120,
    49123, 49201, 49277, 49330, 49371, 49394, 49531, 49584, 49770, 49813, 50026-
    -7, B50144, 50215, 50255, 50316, 50358, 50440, 50525, 50596, 50639, 50677-8,
    50698, 50719, 50763, 50855, 50868, 50885, 50978, 51053, 51161, 51293, 51449,
    51562, B51662, 51734, 51953, 52324

Chlorophyll, and production of algae and water reservoirs
    48412, 48470-1, 48492-3, 48550, 48554, 48566, B48650, B48671, 48710, 48727-8,
    48808, 48825, 48842, 49001, 49054-6, 49059, 49070, 49098-9, 49116, 49146,
    49168, 49176, 49196-8,49268-9, 49316-7, 49338-9, 49350-1, 49416-7, 49425,
    49447-8, 49483, 49485-8, 49518-9, 49552, 49591, 49603-4, 49629-30, 49679,
    49728-9, 49735, 49777, 49784, 49795-6, 49799-800, 49802, 49874-6, 49879-80,
    49968, 49977, 49987, 50014, 50022, 50047, 50064, 50092, 50171, 50180,
    50202-3, 50218, 50239, 50255, 50264, 50308, 50322, 50340, 50342, 50369,
    50373, 50375, 50409, 50418, 50451, 50497, 50522, 50549, 50566, 50601, 50613,
    50679, 50722, 50730, 50732, 50751, 50805, 50864, 50934, 50988, 51010, 51013,
    51069, 51123-4, 51138-9, 51143, 51161, 51183-4, 51210, 51265, 51285-7,
    51319, 51387, 51406, 51459, 51522, 51587, B51662, 51664, 51686-7, 51740,
    51743, 51843, 51860, 51862, 51904, 51916, 51973, 52001, 52019, 52044, 52085,
    52103, 52156, 52188, 52192, 52245, 52251, 52270, 52308, 52324, 52328

Chlorophyll and production of higher plants 49556, 49599, 49610, 50105, 51373,
    51658

Chlorophyll biosynthesis, precursors
    48426-7, B48448, 48521, 48554, 48569, 48628, 48635, 48639, 48660-1, 48682,
    48858, 48897, 48903, 48921, 48941, 48943, 48982-3, 49079, 49083, 49121-2,
    49153, 49209, 49219, 49263, 49270-4, 49276, 49343, B49422, 49429-30, 49451,
    49453, 49476, 49515, 49541, 49648, 49664-5, 49670, 49738, 49793, 49812-3,
    49815, 49843, 49867, 49872, 49884-5, 49910, 49929-30, 49945, 49985, 50025-7,
    50085, 50094, 50120, 50138, 50173, 50247, 50320, 50419, 50440, 50453, 50514,
    50520, 50523, 50556, 50647, 50684, 50735-6, 50746, 50802, 50808, 50816,
    50831, 50920-1, 50941, 50946, 50963-4, 51080, 51155, B51168, 51240, 51243-5,
    51258-60, 51290, 51292, 51339, 51351-2, 51442, 51445, 51451, 51460, 51464,
    51494, 51537, 51607, 51611-3, 51659, 51663, 51671, 51712, 51770, B51788,
    51869, 51945, 52014, 52033-4, 52048-9, 52163, 52166-7, 52177, 52185, 52190,
    52208, 52295, 52333, 52339

Chlorophyll chemical structure
    48417, 48469, 48569, 48634-5, 48654, 48660-1, 48682, 48801, 48805, 48849,
    48872-3, 48963, 49011, 49016, 49033, 49272, 49321, 49848-9, 50088, 50093,
    50247, 50614, 50647, 50941, 50953, 51095, 51100, 51155, 51258-60, 51276,
    51331, 51479-80, 51608, 51684, 51700, 51748-9, B51788, 51792, 52058, 52185,
    52208, 52281

Chlorophyll complexes *in vitro*
    48805, 48828, 48872-3, 48963, 49016, 49201, B49433, 49439, 49452, 50247,
    50444, 50656, 50964, 51011, 51065, 51273, 51432, 51516, 51635, B51788,
    51825, 51876, 51899, 51960, 52012, 52185,

Cholorophyll complexes *in vivo*
    48419-20, 48423, 48466, 48494-7, 48501-2, 48525-6, 48535-6, 48545,
    48569, 48582-3, 48588, 48612-3, 48626, 48654, 48668, 48699, 48713, 48715,
    48773, 48778-9, 48787-8, 48796, 48823, 48831-2, 48834, 48846, 48850-3,
    48859, 48889, 48916, 48943, 49006, 49016-7, 49027-8, 49033, 49134, 49137,
    49171, 49179, 49182, 49209, 49213, 49235, 49240, 49556, 49271, 49374, 49379,
    49394, 49434-5, 49440, 49442, 49453, 49457, 49465, 49508, 49513, 49533,
    49543-4, 49564, 49570, 49596-7, 49614, 49634-6, 49642, 49648, 49676, 49682-
    -3, 49752-4, 49769-76, 49787, 49794, 49838, 49868, 49872, 49910, 49917,
    49928, 49931-2, 49935, 49945, 49959, 50025-8, 50071-2, 50088, 50093, 50179,
    50181, 50223, 50244, 50320-1, 50324-6, 50350, 50402, 50426-7, 50433,
    50441-2, 50444, 50488, 50498-9, 50515, 50526, 50599, 50641, 50649, 50656,
    50661, 50667, 50678, 50695, 50780, 50804, 50835, 50866, 50916, 50939, 50950,
    50953-4, 50964, 50986-7, 50997, 51011, 51019, 51041, 51063, 51114, B51168,
    51245, 51259-60, 51267, 51290, 51352, 51426, 51454, 51471, 51493, 51573,
    51602, 51611-2, 51628-30, 51634, 51654, 51663, 51671, 51679, 51726-7, 51746,
    51813, 51820, 51829-31, 51835, 51869, 51892, 51923, 52031, 52034-5, 52048-9,
    52059, 52064, 52077, 52084, 52092, 52111, 52135, 52220-2, 52232, 52245,
    52253, 52278-9, 52281, 52345

Chlorophyll degradation
    48417, 48470, 48479, 48483, 48489, 48506, 48530, 48545, 48559, 48569, 48633,
    48660-1, 48682, 48725, 48749, 48805, 48830, 48837, 48873, 48896, 48995,
    49001, 49016, 49019, 49040, 49094, 49120, 49135, 49174, 49179, 49196, 49198,
    49235, 49317, 49319, 49321, 49342, 49359, 49378, 49381, 49395, 49426, 49477,
    49508-9, 49518-9, 49526, 49539, 49575, 49584, 49623, 49665, 49675, 49711,
    49731, B49749, 49812-3, 49815, 49822, 49848, 49874, 49903, 49948, 49977,
    50062, 50093, 50107, 50110-1, 50154, 50168, 50178, 50247, 50250, 50257,
    50292-3, 50308, 50330, 50375, 50415, 50445, 50474, 50549, 50558, 50596,
    50612, 50616-7, 50625, 50654, 50661, 50722, 50832, 50866, 50898, 50910-1,
    50913, 50934, 50954, 50967, 51013, 51021, 51036, 51040-1, 51051,
    51095, 51123, 51135, 51156, 51185, 51199, 51208, 51210, 51262, 51265, 51285-
    -7, 51311, 51336, 51339, 51345, 51397, 51399, 51479-80, 51589, 51608, 51621,
    51627, 51630, 51634, 51676, 51684, 51796, 51810, 51820, 51829-30, 51871,
    51889, 51899, 51973, 52014, 52019, 52029, 52143, 52158, 52172, 52188, 52192,
    52208, 52313, 52362, 52384

Chlorophyll delayed light emission, luminescence *in vitro*   48559, 50188, 50247,
    50253, 51336, 52362

Chlorophyll delayed light emission, luminescence *in vivo*
    48557-8, 48680-1, 48810-1, 49177, 49191-2, 49420, 49453, 49502, 49583,
    49594, 49649, 49676, 49854, 49946-7, 50036, 50038-9, 50080, 50110, 50188,
    50217, 50245, 50253, 50260, 50316, 50359, 50363-4, 50397, 50595, 50627-8,
    50768, 50978, 51140, 51320, 51334, 51346, 51454, 51630, 51674, 51690, 51812,
    51824, 52041, 52057, 52070-1, 52082, 52121, 52141, 52152

Chlorophyll determination   see   Chlorophyll and its products determination ...

Chlorophyll energetic states *in vitro* (*cf.* also Chlorophyll in model systems)
    48463, 48805, 48873, 49011, 49016-7, 49096, 49300, 49359, 49435, 49438,
    49452, 49532, 49882, 49912, 50093, 50146, 50495-6, 51288, 51336, 51367-8,
    51579, 51608, 51700, 51876, 52185, 52303, 52362

                                                                    (continued)

Chlorophyll in photosynthesis mechanism (continued)
50197, 50223, 50245, 50250, 50282, 50321, 50325, 50350, 50364, 50403, 50426, 50433, 50435, 50444-5, 50499, 50617, 50641, 50649, 50661, 50835, B50878, 50913, 50954, 50986, 51015, 51019, 51074, 51102, 51114, B51168, 51177, 51276-7, 51319, 51368, 51414, 51426, 51452, 51471, 51493, 51611, 51620, 51629-30, 51634, 51746, 51813, 51980, 51984, 52037, 52057, 52084, 52089, 52092, 52111, 52152, 52222, 52274, 52310, 52345, 52349, 52354

Chlorophyll in physiology of photosynthesis
48514, 48656, 48679, 48960, 49447, 49496, 49514-5, 50105, B50144, 50147, 50183, 50215, 50296, 50329, 50429, 50455, 50469-70, 50478, 51004, 51059, 51222, 51462, 51573, 51896, 51945, 52009, 52370

Chlorophyll in seeds and fruits
48469, 48516, 48537, 48739, 49083, 49135, 49219, 49426, 49458, 49655-7, 49904, 49990, 50163, 50294, 50439, 51199, 51207, 51595, 51953, 52029

Chlorophyll luminescence see Chlorophyll delayed light emission ...

Chlorophyll, methods see Chlorophyll and its products determination ...

Chlorophyll number see Chlorophyll in physiology of photosynthesis

Chlorophyll precursors see Chlorophyll biosynthesis ...

Chlorophyll unit see Photosynthetic (chlorophyll) unit

Chlorophyllase
48682, 49272, 50062, 51198-9, 51339, 51595, B51788, 51871, 51911, 51972

Chlorophyllase and other enzymes of chlorophyll synthesis and degradation, methods
49827, 49930, 51595, 51871, 52154

Chlorophylls *a, b* content and their ratio
48414, 48427, 48433, 48439, 48447, 48489, 48495-6, 48515, 48526, 48528, 48535, 48551, 48561, 48568, 48572, 48583, 48591, 48612, 48626, 48633, 48636, 48654, 48656, 48662, 48668, 48674, 48679, B48701, 48710, 48768, 48809, 48811, 48849, 48853, 48892, 48897, 48916, 48918, 48958, 48960, 49006- -7, 49016, 49152, 49165, 49188, 49208, 49248, 49283, 49333, 49336, 49346, 49362, 49371, 49417, 49426, 49437, 49451, 49454, 49458, 49463, 49473, 49496, 49498-9, 49510, 49514-5, 49531, 49543, 49570, 49600, 49609, 49614, 49635-6, 49638, 49642, 49652, 49655, 49660, 49664-5, 49670, 49696, 49764, 49766, 49803-4, 49813, 49820, 49846, 49854, 49872, 49903, 49952, 49954, 49959, 49983-4, 49989, B50052, 50076, 50083, 50093, 50099, B50144, 50162-3, 50168, 50173, 50183, 50209, 50224, 50246, 50257, 50278, 50284, 50287, 50294, 50320, 50324, 50326, 50398, 50402, 50416, 50432, 50437, 50439-41, 50443-4, 50454-5, 50470, 50478, 50480, 50499, 50501, 50528, 50548, 50552, 50556, 50600, 50608, 50627, 50693, 50701-2, 50720-2, 50746, 50750, 50762, 50816, 50854-6, 50898, 50920-1, 50934, 50985, 51004, 51033, 51054, 51057, 51059, 51067, 51089, 51096, 51114, 51127-8, 51136, 51148, 51151-3, 51172, 51186, 51220, 51259-60, 51276, 51293, 51297, 51339, 51357, 51373, 51375, 51380, 51413, 51416, 51426, 51439, 51475, 51500, 51516, 51520, 51526, 51538, 51573, 51613, 51627, 51643, 51654, 51677, 51713, 51786, 51820, 51830, 51900, 51903, 51952, 51974, 51997, 52000, 52032, 52058, 52064, 52082, 52130, 52140, 52146, 52211, 52221, 52227, 52244-5, 52263-4, 52272, 52274, 52301, 52305, 52310, 52359-60, 52374

Chlorophylls *c, d*
48710, 48849, 48972, 49054, 49069, 49221, 49371, 49623, 50477, 40613, 50722, 50978, 51452, 51643, 51743, 51952, 52130, 52227, 52244

Chlorophylls, *Chlorobium* see Bacteriochlorophylls ...

Chloroplast see also Phycobilisome; Pyrenoid; Ribosome of chloroplast; Stroma of chloroplast; Thylakoid

                                                            (continued)

Chloroplast, isolated, gas exchange by (continued)
49900, 49972-3, 50111, 50248, 50391, 50424, 50567, 50700, 50721, 50773,
50927, 50983, 50985, 51070, 51301, 51443, 51499, 51755, 51783, 51967, 52025,
52206, 52222, 52283, 52302, 52315

Chloroplast isolation
48477, 48791, 48822, 48835, 48876, 48924, 49171, 49219, 49242, 49379-80,
49917, 50390, 50411, 50449, 50694, 50719, 50773, 50824, 51053, B51168, 51173,
51711, 52153

Chloroplast, localization of electron transport chain in thylakoid see Electron
transport chain localization in thylakoid

Chloroplast movements 49357, 49762-3, 50671, 52384

Chloroplast ontogeny see Chloroplast and chromatophore replication, ontogeny

Chloroplast outer membrane see Chloroplast envelope

Chloroplast proteins (and other photosynthetic proteins), methods
48419, 48423, 48536, 48615-6, 49003, 50309, 51115, 51273, 51315, 51344,
51746, 51803, 51932, 51970, 52037

Chloroplast ribosome see Ribosome of chloroplast

Chloroplast ultrastructure (*cf*. also Chloroplast envelope; Stroma of chloroplast;
Thylakoid, granum)
48447, B48448, 48514, 48526, 48535, 48547, 48561, 48569, 48603, 48626,
48685, 48690-1, 48790, 48820, 48892, 48919, 48928, 48939, 48965, 49002,
49035, 49079, 49100, 49136, 49165, 49167, 49199, 49212, 49223, 49234, 49242,
49263, 49276, 49289-90, 49322, 49372, 49413, 49426, 49454, 49463, 49489,
49505, 49543, B49561, 49587, 49595, 49614, 49641, 49682, 49745, 49755,
49812, 49858, 49886, 49939-40, 49958, B50052, 50071, 50082, 50098, 50174,
50187, 50200, 50284, 50303, 50325, 50380-1, 50386-8, 50394, 50402, 50406,
50414, 50426, 50439, 50440-1, 50443, 50472, 50504, 50521, 50528, 50536,
50589, 50592-3, 50640, 50670, 50681, 50716, 50720-1, 50760, 50778, 50802,
50807-8, 50818, 50833, 50856, 50886, 50932, 50939, 50946, 50949, 50963,
51000, 51022, 51047-8, 51148, B51168, 51173, 51192, 51200, 51242-4, 51269,
51281-2, 51297, 51300, 51302, 51320, 51329-30, 51361, 51379, 51381,
51385, 51394, 51407-8, 51413, 51421-3, 51451, 51478, 51486, 51500-1, 51507,
51526, 51538, 51648, 51654, 51660, 51728, B51788, 51814, 51824, 51831, 51879,
51888, 51901, 51923, 51926, 51971, 51976-9, 52026, 52035-7, 52052, 52062,
52064, 52066, 52209, 52211, 52221, 52223, 52225-6, 52236, 52244, 52267,
52272, 52276, 52288, 52297, 52302, 52347, 52363

Chromatophore in photosynthetic bacteria and blue-green algae (*cf*. also Chloroplast
and chromatophore...)
48569, 48576, 49027, 49057, 49127, 49213, 49223, 49379, 49393, 49943, 49958,
49996, 50071, 50098, 50303, 50866, 50886, 50939, 50974, B51168, 51294,
51300, 51365, 51571, 51971, 52037, 52312

Circular dichroism see Dichroisms ...

Clark electrode see O_2 electrode

Clock, biological see Diurnal changes ...

CO_2 and algae productivity 48566, 49175, 49950, 50727, 52136

Cold hardiness   see   Temperature, low ...

Combustion heat   see   Energy content ...

Compensation irradiance
        48541, 48945, 48958, 49188, 49357, 49448, 49845, 50007, 50011, 50482, 50889,
        51109-10, 52040, 52227

Compensation point, CO_2 see CO_2 compensation concentration

Compensation point, light   see Compensation irradiance

Competition in ecosystem   48683, 49733, 50301, 51295, 52219

Conductance for transfer of gases   see   Resistance ...

Contribution of individual organs to yield formation   see Biomass distribution and
        redistribution; Photosynthate translocation ...

Cosmic radiation   see   Ionizing radiation ...

Coupling factor 1   see   ATPase ...

Cover, vegetative   see   Canopy ...; Ecosystem ...

Crassulacean Acid Metabolism   see   CAM

Cultivar differences, canopy functioning   49005, 51082

Cultivar differences, carbon fixation pathways   48478, 48811, 48826, 49805, 49949,
        50143, 50782, 50847-8, 50850, 51731

Cultivar differences, carotenoids   49139, 49657, 50163, 51262

Cultivar differences, chlorophyll   48465, 48577, 48811, 48826, 48904, 48958, 49215,
        49378, 49498, 49570, 49652, 49657, 49666, 49731, 49804-5, 50066, 50076,
        50163, 50209, 50454-5, 50469-70, 50552, 50654, 50854, 51080, 51083, 51222,
        51236, 51262, 51373, 51574, 51717, 51731, 51821, 51953, 52360

Cultivar differences, chloroplast   49292, 50174, 51384-5, 51524

Cultivar differences, ecosystem and plant productivity
        48446, 48473, 48487, 48538, 48555, 48655, 48738, 48887, 48893, 48904, 49858,
        48966, 48977, 49029, 49077, 49112, 49119, 49144, 40205, 49239, 49241, 49335,
        49347, 49352, 49378, 49514, 49592, 49666, 49677, 49804, 49956, 50139, 50327,
        50407, 50438, 50446, 50450, 50470, 50532-3, 50552, 50753, 50782, 50789-90,
        50841, 50854, 50906, 51082, 51125, 51131, 51248, 51254, 51373, 51384, 51419,
        51576, 51591, 51721, 51775, 51808, 51854, 52127, 52132, 52213-5, 52235,
        52350, 52360

Cultivar differences, electron transport chain
        48465, 48904, 48980, 50530, 50552, 50847, 50850, 51079, 51153, 51236, 51373,
        51385, 51703, 52127, 52360

Cultivar differences, gas exchange
        48446, 48453, 48473, 48504, 48555, 48655, 48811, 48826, 48904, 48958-9,
        49000, 49029, 49077, 49118-9, 49215, 49241, 49322, 49501, 49511, 49514,
        49601, 49618, 49666, 49949, 50139, 50142-3, 50167, 50265-6, 50273, 50454,
        50469-70, 50475, 50801, 50827, 50847, 50854, 50903, 51030, 51145, 51188,
        51247-8, 51717, 51854, B51897, 52213, 52216, 52304, 52350

Cultivar differences, photorespiration   48826, 51249, 52216

Cultivar differences, resistances to CO_2 and water vapour transfer
        48475, 48958, 49077, 49210, 49241, 49618, 50165, 50642, 51030, 51247-8,
        B51897, 52154

Cultivar differences, respiration   49029, 49322, 50396, 50903, 51249

Cultivation of algae and photosynthetic bacteria see Algae and photosynthetic bac-
        teria, cultivation; Algae mass cultures productivity

Cuticular CO_2 and O_2 exchange   49932, 51004

Cuticular resistance see Resistance, cuticular

Cytochromes
        48436-7, 48494-8, 48544, 48569, 48597, 48625, 48676-7, 48684, 48717, 48740,
        48758, 48764, 48788-9, 48831, 48835, 48859, 48889, 48961, 49026, 49073-4,
        49078, 49092, 49226-7, 49240, 49276, 49285, 49409-10, 49481, 49634, 49648,
        49671, 49687, 49721, 49764, 49812, 49815, 49825, 49864, 49883, 49913-5,
        50054, 50104, 50118, 50196-7, 50201, 50268, 50271, 50334-7, 50471, 50560,
        50562-3, 50580, 50672-3 50719, 50859, 50913-4, 50931, 50954-5, 50965-6,
        50984, 51015, 51085, 51163, B51168, 51176-7, 51320-1, 51425, 51519-21,
        51602, 51669, 51729, 51785, 51848, 51899, 51972, 51986, 52220, 52222, 52229,
        52234, 52245, 52253, 52389, 52312, 52349

Cytochromes, methods   48889, 49226, 49828, 50196, 51848, 52184, 52366

D

Dark CO_2 fixation
        48540, 48702, 48800, 48856, 49202, 49568, 49916, 50666, 51202, 51517, 51692,
        52247

Decapitation see Defoliation, decapitation ...

Defoliation, decapitation, ear and root removal, effect on carbon fixation pathways
        49469

Defoliation, decapitation, ear and root removal, effect on chlorophyll   48995,
        49804, 49985, 51198, 51867

Defoliation, decapitation, ear and root removal, effect on chloroplast   49641

Defoliation, decapitation, ear and root removal, effect on ecosystem and plant
        productivity   48523, 48624, 48992, 49144, 49293, 49467, 49536-7, 49716,
        49719, 49732-3, 49804, 50015, 50153, 50226, 50843, 51125, 51131, 51409,
        51721, 51959, 52175

Defoliation, decapitation, ear and root removal, effect on electron transport chain
        48904

Defoliation, decapitation, ear and root removal, effect on gas exchange   48624,
        48992, 49389, 49469, 49719, 50015, 51193, 51383, 51722

Defoliation, decapitation, ear and root removal, effect on resistances to CO_2 and
        water vapour transfer   48624, 49469, 49719

Defoliation, decapitation, ear and root removal, effect on respiration   49010,
        50015, 51867

Desiccation of tissue see Water saturation deficit

Deuterium oxide, tritium oxide   48743-4, 49700, 49722, 50232, 50931, 50940, 51364-
        -5, 51437, 51454, 51690-1, 51790-1, 51841, 52007

Drought and chloroplast   51384-5, 52006

Drought and ecosystem and plant productivity   48464, 48473, B48998, 49142, 49281-2,
        49388, 49521, 49715, 49852, 50165, 50175, 50241, 50638, 50906-7, 51043,
        51091, 51204, B51531, 51532, 51720, 51787, B51897, 52003

Drought and electron transport chain   50986-7, 51385

Drought and gas exchange   48472-3, 48692, B48998, B49621, 49637, 49806, 49934,
        50165, 50193, 50241, 50266, 50662, 51043, 51497, B51897, 51921

Drought and photorespiration   51921

Drought and resistances to CO_2 and water vapour transfer   48627, B48998, 49038,
        49444, B49621, 50193, 50241, 51043, 51204, 51497, 51787

Drought and respiration   51921

Dry-matter production, gravimetric determination   50384

E

Ear removal see Defoliation, decapitation, ear and root removal ...

Ecosystem production, primary productivity (terrestrial) (*cf*. also Biomass...)
        48411, 48425, 48451-2, 48464, 48473, 48490, 48523, 48556, 48605, 48693,
        48719, 48721, 48729, 48731, 48747, 48771, 48806, 48839, 48843, 48887, 48901,
        48904, 48908-10, 48926, 48958, 48973, 49009, 49020-1, 49029, 49102, 49118,
        49125, 49144, 49186, 49193, 49228, 49231, 49241, 49287, 49297, 49320, 49329,
        49341, 49378, 49382, 49417, 49443, 49496, 49501, 49514, 49521, 49536-7,
        49549-50, 49555, 49599, 49613, 49651, 49678, 49686, 49691, 49693-4, 49701,
        49732, 49820, 49921, 49968, 49980, 50016, 50103, 50139, 50149, 50183, 50204,
        50288, 50357, 50384, 50408, 50458, 50463, B50476, 50536-7, 50552, 50574,
        50581, 50591, 50597, 50611, 50651, 50790-1, 50841, 50843, 50854, 50877,
        50888, 50906-7, 50989, 51014, 51049, 51106, 51153, 51181-2, 51196, 51203,
        51217, 51254, 51307, 51370-1, 51373, 51384, 51411-2, 51502, 51569, 51576,
        51688, 51714, 51753, 51775, 51780, 51808, 51828, 51908-10, 51949, B51961,
        52027, 52079, 52099, 52109, 52182, 52198, 52203, 52215, 52239, 52242, 52254,
        52256, 52306, 52346, 52360

Ecotypes, geographical types, and canopy functioning   51146

Ecotypes, geographical types, and carbon fixation pathways   49501

Ecotypes, geographical types, and carotenoids   49609, 50515, 50762, 51656, 51658

Ecotypes, geographical types, and chlorophyll   48755, 49609, 50515-6, 50762, 51061,
        51656, 51658

Ecotypes, geographical types, and ecosystem and plant productivity   48693, 48729,
        48839, 49021, 49297, 49686, 50103, 50999, 51014, 51061, 51091, 51656, 51688,
        52054

Ecotypes, geographical types, and electron transport chain   49501, 49524

Ecotypes, geographical types, and gas exchange   48632, 48693, 49297, 49501, 49747,
        50516, 50861, 51061, 51418, 51688, 51786, 51846, 52334

Ecotypes, geographical types, and resistances to CO_2 and water vapour transfer
        49747, 50861, 51704

Ecotypes, geographical types, and respiration   49747, 51688

Efficiency, photochemical (*cf.* also Irradiance and gas exchange, analysis of light
       curves)
       48572, 48599, 48632, 48958, 49150, 49366, 49718, 50511, 50754-5, 50889,
       50984-5, 51136, 51181, 51186, 51903, 51983, 52227, 52352

Electron paramagnetic resonance see EPR, NMR

Electron spin resonance see EPR, NMR

Electron transport chain activity
       48572, 48580, 48688, 48693, 48834, 48902, 49004, 49150, 49165, 49407, 49794,
       50276, 50321, 50536, 50721, 50773, 50927, 50946, 50983, 51102, 51132, 51578,
       51755-6, 51785, 51793, 51806, 52045, 52314

Electron transport chain activity, methods   50719

Electron transport chain components see Cytochromes; Ferredoxin...; Ferredoxin-NADP
       reductase; NADP...; O$_2$ evolution...; Photosystems...; Plastocyanin; Quinones

Electron transport chain components and carbon fixation pathways   52280

Electron transport chain, general aspects see General aspects on carbon fixation
       and electron transport chain, books

Electron transport chain localization in thylakoid
       48420, 48494-8, 48503, 48625, 48684, 48699, 48790, 48831, 48935,
       48961, 49033, 49073-4, 49092, 49212, 49531, 49589, 49671, 49687, 49769,
       49772-3, 49787, 49883, 50035-6, 50071, 50088, 50223, 50261, 50352, 50441,
       50560-2, 50628, 50793, 50818, 50866, B50878, 50916, 50926, 50930, 50954,
       50994, 51047, 51085, 51132, 51260, 51276, 51330, 51388, 51612, 51892, 51975,
       52141, 42149, 52221, 52234, 52253, 52269, 52289

Electron transport chain model see Model ...

Electron transport chain, serological analysis   51132, 51358, 52380

Emerson effect, Blinks effect   48920, 48976, 49825, 51330, 51390, 52355-6

Energization of thylakoid see Chemiosmotic hypothesis ...

Energy content in biomass 48418, 48618, 48719, 49766, 50065, B50476, 50537, 50587,
       50781, 51001, 51409, 51496, 52005, 52054, 52203, 52256, 52380

Energy content in biomass, methods   49047, 51505

Energy utilization, plant and ecosystem
       48487, 48655, 48731, 48756, 48909-10, 49000, 49058, 49060, 49125, 49228,
       49239, 49287, 49366, 49534, 49555, 49694, 49834, 49845, 50149, 50207, 50280,
       50511, 50574, 50581-3, 50586-7, 50600, 51181, 51707, 51964, 52005, 52273,
       52361

Enzymes and carbon fixation pathways   49944, 50723, 51031, 51975

Enzymes and carotenoids   49822

Enzymes and chlorophyll   48707, 48916, 49822, 49983, 52365

Enzymes and chloroplast (chromatophore)
       48543, 48576, 48619, 48929, 49093, 49111, 49153, 49232, 49244, 49558,
       49648, 49832, 49981, 49991, 50031, 50205, 50233, 50277, 50431, 50520, 50700,
       B51168, 51190, 51314, 51349, 51559, 51572, 51647, 51769, 51771, 51801,
       51807, 51874, 51896, 51922, 52083, 52257

Enzymes and electron transport chain
        48544, 48707, 49739, 50044, 50277, 50471, 50530, 50598, 50770, 50931, 51132,
        51150, 51176, 51190, 51334, 51489, 51581, 51647, 51803, 51986, 52039

Enzymes and gas exchange   50000-1

Enzymes and photorespiration   48589, 48919, 49893, 52283

Enzymes and photosynthates   49725, 51266

Enzymes of carbon fixation pathways other than RuBPC, PEPC, malic enzyme, malate
        dehydrogenase
        48477-8, 48500, 48542, 48569, 48572, 48621-2, 48663, 48693, 48918-9, 48949-
        -54, 48979, 49299, 49396, 49428, 49449, 49454, 49501, 49516, 49553, B49561,
        49568, 49648, 49659, 49710, 49734, 49756-7, 49765, 49782, 49816, 49863,
        49900, 49962-4, 49972, 50042-3, 50056, 50125-7, 50182, 50259, 50391-2,
        50431-2, 50449, 50480, 50557, 50626, 50734, 50819, 50828, 50836, 50999,
        51004, 51104, 51150, 51167, B51168, 51171, 51223, 51266, 51314, 51324, 51389,
        51475, 51503, 51517, 51584, 51670, 51677, 51711, 51769, 51790, 51873-4,
        51896, 51945, 51948, 52147, 52247, 52263, 52268, 52280, 52285

Enzymes of carbon fixation pathways other than RuBPC, PEPC, malic enzyme, malate
        dehydrogenase, methods
        48477, 48950, 48952, 49044, 50734, 50897, 51314, 51670, 52268, 52285

Enzymes of chlorophyll synthesis and degradation see Chlorophyll, enzymes...; Chlo-
        rophyllase

Enzymes of electron transport chain, methods   51132

Enzymes of glycollate cycle, methods   48658

Enzymes of photorespiration see Photorespiration enzymes; Enzymes of glycollate
        cycle, methods

Epidermis see Leaf epidermis

EPR, NMR (methods and results)
        48414, 48417, 48454-5, 48498, 48507, 48532-3, 48558, 48684, 48699, 48735,
        48743-4, 48773, 48784, 48793, 48799, 48801, 48810-1, 48836, B48841, 48859,
        48905, 48968-9, 49016-7, 49019, 49025-6, 49085, 49089, 49092, 49180-2,
        49304, 49325, 49342-3, 49359, 49436-7, 49450, 49456, 49526, 49567, 49586,
        49589, 49687, 49720, 49781, 49848-9, 49882, 50039, 50093, 50106-8, 50110-3,
        50179, 50223, 50247, 50415, 50471, 50559-60, 50562, 50603, 50624, 50627-8,
        50643, 50649, 50661, 50668, 50725, 50735-6, 50842, 50865, 50883, 50908-9,
        50913-4, 50941, 50954-5, 50959, 51041, 51065, 51155, 51216, 51347, 51446,
        51552, 51620-2, 51684, 51792, 51880, 51931, 51984, 52075, 52135, 52185-6,
        52233, 52299, 52327

Ethylene see Gases, organic ...

Evolution see Phylogeny ...

Excitation resistance see Resistance, carboxylation and excitation

Exhaust gases see Pollution of air ...

Exposure chamber see Assimilation chamber

Extension growth, leaf dimensions
        48425, 48473, 48475, 48480, 48485, 48523, 48548, 48556, 48605, 48624, 48711,
        48827, 48884, 48887, 48948, 48973, 48977, 48981, 49029, 49042, 49076, 49113,
        B49114, 49187, 49204, 49241, 49282, 49329, 49355, 49431, 49536, 49613,
                                                               (continued)

Extension growth, leaf dimensions (continued)
    49627, 49765, 49831, 49846, 49861, 50083, 50124, 50128, 50153, 50166, 50208,
    50224, 50266, 50284, 50302, 50327-8, 50367, 50405, 50452, 50502, 50511,
    50550, 50555, 50662, 50691, 50779, 50796, 50812, 50863, 50873, 50906-7,
    50980, 51043, 51075, 51133, 51191, 51203, 51255, 51271, 51306-7, 51373,
    51507, B51531, 51733, 51832, 51847, 51881-2, 51890-1, B51897, 51948, 51982-3,
    51988, 52016, B52018, 52081, 52124, 52164, 52182, 52235, 52246

Extraction of pigments see Pigments determination, sampling and extraction

Exudation of photosynthates see Photosynthate translocation ...

F

Fatty acids see Lipids, fatty acids ...

Ferredoxin, ferredoxin-NADP reductase, methods   49003, 51544, 51601, 51801, 52326,
    52348

Ferredoxin, flavoproteins, rubredoxin
    48436-7, 48500, 48532-3, 48542, 48544, 48569, 48622, 48694, 48763, 48767,
    48774, 48869, 48961, 49026, 49074, 49078, 49133, 49151, 49360, 49392, 49408,
    49567, 49671, 49692, 49698, 49742-3, 49750-1, 49841, 49848, 49919, 50071,
    50118, 50234-5, 50334, 50337, 50471, 50508, 50561, 50563, 50615, 50661,
    50842, 51047, 51150, 51263, 51301, 51321, 51343, 51375, 51578, 51614, 51621,
    B51788, 51855, 51972, 51995-6, 52024, 52039, 52135, 52150, 52152, 52163,
    52280, 52312

Ferredoxin-NADP reductase, pteridines
    48544, 48933-4, 49003, 49266, 49406, 49659, 49751, 49812, 49883, 50236,
    50337, 50615, 50626, 51047, 51263, 51425, 51600, 51614, 51972, 52039, 52150,
    52348

Flashes of light see Irradiance, flash ...

Flavoproteins see Ferredoxin ...

Flooding and chlorophyll   50383, 51679

Flooding and ecosystem and plant productivity   48839, B48998, 49284, 49847, 50242,
    50270, 50383, 50873, 51881-3

Flooding and gas exchange   48813, B48998, 50270

Flooding and resistances to CO_2 and water vapour transfer   48475, 48813, 48815,
    50383, 50873, 51881-3

Flooding and respiration   B48998, 50270

Fluorescence, methods   52043

Fluorine see Pollution of air ...

Foliage see Canopy ...

Fraction I protein see Ribulose 1,5-bisphosphate carboxylase

Frost hardiness see Temperature, low ...

Fungus diseases see Phytopathological effects ...

Fusicoccin see Growth regulators ...

G

Gas exchange, general aspects see General aspects on CO_2 exchange ...

Gas exchange in algae
        48428-9, 48441, 48506, 48541, 48551, 48565, 48571, 48640, 48647, 48649,
        48652, 48667, B48671, 48678, 48702, 48717, 48749, 48769, 48787, 48847,
        48854, 48911, 48978, 49030-1, 49036, 49055, 49070, 49075, 49140, 49148,
        49189, 49213, 49220-1, 49224, 49254, 49257, 49283, 49291, 49364, 49416-
        -7, 49423, 49447-8, 49468, 49472, 49483, 49503-4, 49544, 49562, 49591,
        49640, 49673, 49682, 49728-9, 49780, 49833, 49937, 49984, 50018, 50070,
        50176, 50180, 50195, 50218, 50239, 50259, 50339, 50353, 50416, 50418, 50425,
        50428-30, 50478, 50588-9, 50600, 50618, 50633, 50644, 50772, 50826, 50839,
        50864, 50887, 50934, 50972, 51109-12, 51139-40, 51183, 51189, 51251, 51284,
        51302, 51319, 51324, 51340, B51350, 51387, 51438, 51462, 51499, 51504, 51522,
        51526, 51534, 51539, 51560, 51564, 51587, 51597, 51624, 51640, 51650, 51680-
        -2, 51685, 51705, 51713, 51717, 51845, 51889, 51904, 51913, 51945, 51971,
        52082, 52118, 52156, 52165, 52188, 52190, 52227, 52245, 52247, 52356

Gas exchange in isolated chloroplasts see Chloroplast, isolated, gas exchange by

Gas exchange in isolated bacteria see Photosynthetic bacteria, gas exchange in

Gas exchange model see Model ...

Gas exchange of organs other than leaf   48827, 49022, 49587, 50406, 50604, 51517,
        51540, 51688, 51692, 51907, 51930

Gases, organic, and chlorophyll   51199

Gases, organic, and gas exchange   49878, 51024

Gases, organic, and resistances to CO_2 an water vapour transfer . 51024

Gasometric methods, generally   48915, 50331

Gasometric system, closed and semiclosed   48453, 48574, 48723, 49846, 50050, 50485,
        50490, 50887, 51094, 52249

Gasometric system, open   48453, 48467, 48716, 48959, 49331, 49398, 49466, 49700,
        50475, 50485, 50487, 50546, 50630, 50632, 50675, 51186, 51386, 51498, 51696,
        51738, 51856, 51963

General aspects on carbon fixation pathways and electron transport chain; books
        48422, B48448, B48560, B48600, B48650, B48671, B48701, B48841, 48843, B49012,
        B49045, B49081, B49114, B49419, B49422, B49433, 49547, B49561, B49620-1,
        B49691, 49712, B49749, B49853, B49908, B50052, B50144, B50346, 50614,
        B50659, B50687, 50721, B50878, 50944, 51046, B51168, 51330, B51350, B51463,
        B51531, B51662, B51788, B51897, B52018, B52021, 52296, B52321, 52325

General aspects on CO_2 exchange, photorespiration and productivity; books
        B48560, B48671, 48843, 48895, B48946, B48998, B49012, B49045,
        B49081, B49114, B49419, B49422, B49527, B49561, B49620-1, B49691, 49712,
        B49749, B49767, B49853, B49908, B49921, B50052, B50144, B50346, B50476,
        B50659, B50687, 50786, 50849, 50875-6, B50878, 50944, 51153, 51159, B51168,
        B51350, 51497,   B51531, 51558, B51897, B51961, B52018, B52021, B52321,
        52357

Genetics *cf.* also Mutagens ...; Mutants ...

Genetics, and canopy functioning 48617

Genetics,and ecosystem and plant productivity
        B48560, 48617, 48698, 48875, 48887, 48902, 49088, 49101, 49119, 49612,
        49666, 49678, 49992, 50582, 50585-6, 50702, 50750, 50889, 50989, 51125,
        51181, 51211, 51511, 51569, 51775, 52256, 52306

Genetics of carbon fixation pathways
        48623, 48653, 48761, 48766, 48788, 48877, 48960, 49153 -4, 49259, 49402,
        49556, 49659, 50142, 50229, 50258, 50290, 50540, 50702, 50750, 50847-8,
        50850, 51025, 51047, 51436, 51577, 51605-6, 51732, 51762, 52010, 52095,
        52304, 52379

Genetics of carotenoids 48964, 49006, 49333, 49822, 50602, 51298-9, 51577, 51732,
        52030, 52305

Genetics of chlorophyll
        48521, 48960, 49006-7, 49333, 49503, 49544, 49556, 49666, 49822, 50072,
        50501, 50701-2, 50750, 50848, 50884, 51136, 51187, 51252, 51297-8, 51511,
        51573, 51577, 51649, 51732, 52196, 52305

Genetics of chloroplast (chromatophore)   ,
        B48448, 48695, 48761, 48766, 48781, 48788, 48877, 48890, 48929, 48954,
        49153, 49559, 49703, 49868, 49886, 49996, 50121, 50150, 50309-10, 50521,
        50629, 50685, 50750, 50773, 51026, 51028, 51047-8, B51168, 51241, 51308-10,
        51435, 51460, 51501, 51506-7, 51524, 51766, 51971, 52161, 52202, 52225,
        52287, 52340, 52347

Genetics of electron transport chain
        48748, 48788, 48877, 48902, 48955, 49556, 49659, 49887, 50121, 50258, 50276,
        50701, 50850, 51025, 51308, 51435, 51645, 51702, 51804, 52191, 52379

Genetics of gas exchange
        48555-6, B48560, 48902, 48960, 49088, 49119, 49202, 49501, 49503, 49544,
        49556, 49612, 49666, 49672, 49678, 50142, 50265-6, 50274, 50413, 50550,
        50750, 50847-50, 50861, 50879, 50889, 51136, 51418, 51435-6, 51497, 51511,
        51667, 51851, B51897, 52120, 52304

Genetics of photorespiration   49556, 51606

Genetics of resistances to CO_2 and water vapour transfer
        48556, 48698, 49556, 49612, 50861, 51136, 51378, 51497, 51787, 52124

Genetics of respiration   49612, 50889, 51851, 52255

Glycollate metabolism see Photorespiration ...

Glyoxysome see Peroxisome ...

Granum see Thylakoid ...

Gravimetric determination of photosynthesis see Dry-matter production ...

Gross photosynthetic rate
        48435, 48723, 48839, 48882, 48973, 49382, 49417, 49469, 49513, 49768, 49957,
        50117, 50372, 50382, 50479, 50511, 50550, 50564-5, 50604, 50606, 50679,
        50692, 50776, 50861-2, 51042, 51189, 51307, 51688, 51905, 51921, 52277

Growth analysis, methods
        48440, 48452, 48490, 48637, 48687, B48946, 49384, 49461, B49908, 50063,
        50139, 50888, 51940, 52110

Growth analysis, net assimilation rate, leaf area ratio, relative growth rate
48434-5, 48473, 48538, B48560, 48655, 48666, 48693, 48698, 48729, 48737-8,
48857, 48871, 48875, 48887, 48893, B48946, 48948, 48957-8, 48973, 49009,
49187, 49190, 49229, 49239, 49241, 49345, 49352, 49388, 49415, 49424, 49496,
49500-1, 49549, 49556, 49571, 49574, 49612, 49677, 49705, 49707, 49716,
49719, 49766, 49783, 49810, 49834, 49845-6, 49852, B49908, 49909, 49956,
49995, 50004, 50011-2, 50074, 50100, 50128, 50149, 50166, 50183, 50204,
50207, 50286, 50301, 50338, 50429, 50446-7, 50459, 50470, 50511, 50532-3,
50548, 50581-2, 50585-7, 50610, 50651, 50653, 50689, 50757, 50796, 50841,
50856, 50873, 50889, 50905-6, 50915, 50936, 50976, 50999, 51023, 51057,
51091, 51133, 51182, 51205-6, 51211, 51217, 51271, 51283, 51307, 51371,
51400, 51456, 51496, B51531, 51569, 51591, 51632, 51646, 51653, 51661, 51673,
51720, 51735, 51826-7, 51846, 51870, 51881-2, 51890-1, B51897, 51959, 51988,
B52018, 52099, 52115, 52164, 52216, 52255, 52297

Growth analysis, specific leaf area, leaf area index, leaf area duration
48425, 48446, 48449, 48462, 48552, 48555, 48572, 48579, 48591, 48624, 48638,
48655, 48666, 48736, 48738, 48771, 48775, 48812, 48857, 48875, 48884, 48927,
48930, 48936, B48946, 48948, 48957-8, 48977, 49009, 49029, 49051, 49076,
49102, B49114, 49115, 49158, 49186-8, 49190, 49204, 49228, 49231,
49239, 49297, 49335, 49345, 49347, 49352, 49355-6, 49415, 49424, 49473,
49496, 49500-1, 49521, 49523, 49550, 49556, 49571, 49592, 49598, 49612-3,
49666, 49702, 49705, 49716, 49766, 49785, 49804, 49834, 49844-7, B49908,
49909, 49956, 49995, 50004, 50012, 50015, 50017, 50019, 50021, 50037, 50063,
50124, 50152, 50175, 50204, 50207, 50226, 50273-4, 50301, 50367, 50372,
50407, 50417, 50511, 50532-3, 50548, 50550, 50581-6, 50610, 50691, |
50753, 50757-8, 50791, 50812, 50841, 50843, 50862, 50905, 50915, 50936,
50976, 50985, 50989, 51002, 51023, 51057, 51061, 51082, 51091, 51133, 51153,
51182, 51248, 51254, 51296, 51306, 51373, 51400, 51409, 51412, 51433,
51496, 51511, 51565, 51569, 51575, 51591, 51619, 51661, 51689, 51735, 51822,
51854, 51864, 51870, 51895, B51897, 51920, 51982-3, 51988, 52000, 52003,
52020, 52051, 52067, 52099, 52138, 52154, 52198, 52215-6, 52228, 52235,
52246, 52255, 52260, 52264, 52297

Growth regulators and canopy functioning 50511

Growth regulators and carbon fixation pathways   49361, 49949, 50090, 50137, 51151-
-2, 51357

Growth regulators and carotenoids   48568, 49309, 49738, 50556, 50746, 51462

Growth regulators and chlorophyll
48479, 48521, 48551, 48568, 48725, 48897, 48995, 49121-2, 49161, 49174,
49276, 49395, 49421, 49539, 49660, 49696, 49738, 49804, 49818, 49980, 49983,
50079, 50085, 50154, 50191, 50292, 50419, 50453, 50556, B50659, 50746,
50920-1, 50950, 51151-2, 51199, 51357, 51713, 52177

Growth regulators and chloroplast (chromatophore) 49072, 49276, 49489, 50746,
51393, 51525, 52177, 52288

Growth regulators,and ecosystem and plant productivity
48538, 49229, 49704, 49804, 49818, 49855-7, 49918, 49980, 50158, 50165,
50511, B50659, 50747, 51066, 51713, B51897, 51969, 52016, 52129

Growth regulators and electron transport chain   50295, 50921, 51568, 52047

Growth regulators and gas exchange
48551, 48800, 48813, 48856, 48992, 49229, 49361, 49818, 49855, 49949, 49980,
50511, 50594, B50659, 51066, 51151-2, 51312, 51568, 51713, 51969

Growth regulators and photorespiration   50137, B50659

Growth regulators,and resistances to CO$_2$ and water vapour transfer   48424, 48733,
48813, 49145, 49229, 49387, 49797, 49818, 50008, 50594, B50659, 50821, 51212,
51251, 51568

Growth regulators and respiration   51312, 52224

Growth respiration see Respiration, growth and maintenance

H

H_2 and carbon fixation pathways   49619

H_2 and chloroplast (chromatophore)   49489

H_2 and gas exchange 48565, 49619

H_2, and resistances to CO_2 and water vapour transfer   49619, 50366

H_2 evolution, photoreduction
     48436-7, 48455, 48576, 48678, 48756, 48767, 48786, 49023-4, 59095-7, 49582,
     49640, 49695, 49699, 49888, 50024, 50053, 50131, 50233-4, 50248, 50252,
     50298, 50551, 50567, 50634, 50680, 50740, 50742, 50788, 50879, 51142, 51237,
     51239, 51621, 51682, 51759, 52104-5, 52319-20, 52380-2

H_2 isotopes see Deuterium ...

H^+ transport in chloroplast see Chemiosmotic hypothesis

Halobacterium photosynthesis (*cf*. Model of *Halobacterium* photosynthesis)
     48413, 48438, 48476, 48524, 48575, 48585-6, 48593, 48604, 48612, 48630-1,
     48644-5, 48741, 48759-60, 48816-7, 48905-6, 48931, 48985-7, 49046, 49053,
     49082, 49089-90, 49107, 49183-5, 49246-7, 49250-1, 49260, 49264, 49295-6,
     49306, 49313-5, 49323-4, 49353-4, 49414, 49459, 49522, 49540, 49585, 49654,
     49706, 49778, 49790, 49808, 49819, 49859, 49866, 49871, 49895-6, 49969,
     49994, 50009-10, 50058, 50061, 50068-9, 50081, 50084, 50087, 50091, 50119,
     50134-5, 50145, 50151, 50157, 50172, 50216, 50220-2, 50267, 50311-2, 50333,
     50349, 50379, 50399-401, 50403, 50420-2, 50460, 50484, 50503, 50513, 50543,
     50579, 50603, 50619, 50621, 50668, 50709-10, 50784, 50795, 50798, 50869,
     B50878, 50922-5, 50959, 51003, 51006-8, 51039, 51090, 51141, B51168, 51226,
     51300, 51318, 51325-8, 51460, 51465, 51523, 51554, 51562, 51633,
     51638-9, 51649, 51655, 51706, 51715, 51719, 51730, 51750, 51773, 51776,
     51792, 51797, 51878, 51912, 51939, 51955, 51957, 51965, 51989-92, 51999,
     52053, 52069, 52148, 52159-60, 52196-7, 52200, 52377, 52388

Hatch-Slack cycle see C_4 pathway ..

Herbicides see Pesticides, herbicides ...

Heterogeneity of leaf blade (organ), and carbon fixation pathways   48663, 49153,
     50480, 50999, 51588, 52147, 52211

Heterogeneity of leaf blade (organ), and carotenoids   49872, 52211

Heterogeneity of leaf blade (organ), and chlorophyll   48427, 48726, 49278, 49502,
     49515, 49872, 50076, 50324, 50414, 50480, 52211

Heterogeneity of leaf blade (organ), and chloroplast   48945, 49372, 49506, 49558,
     49755, 49886, 50414, 51243, 52066, 52153, 52211

Heterogeneity of leaf blade (organ), and electron transport chain   49506, 50324,
     50414, 52211

Heterogeneity of leaf blade (organ), and gas exchange   48629, 49109, 50198, 50999,
     51588, B52018

Heterogeneity of leaf blade (organ), and photosynthates   52122

Heterogeneity of leaf blade (organ), and resistances to CO_2 and water vapour transfer
     49724, B51897

Heterogeneity of leaf blade (organ), and respiration   50631

Heterotropy see Carbon metabolism types ...

Hill rection see Photosystem 2 activity measurement

Hill reaction, methods see Photosystem 2 activity measurement, methods

Humidity of air, and canopy functioning   48736, B51897

Humidity of air, and carbon fixation pathways   51618, 52265

Humidity of air, and carotenoids   49904

Humidity of air, and chlorophyll   49813, 49904, 51679

Humidity of air, and chloroplast   51322

Humidity of air, and ecosystem and plant productivity.   49831, 50447, 51618, 51714,
     51891, 52265

Humidity of air, and electron transport chain 48795, 51322

Humidity of air, and gas exchange   48883, 48885, 49356, 49398, B49621, 49747,
     49809, 50169, 50274, 50304, 50418, 50456, 50666, 50674, 50845, 50903, 51145,
     51227, 51418, 51497, 51652, 51657

Humidity of air, and resistances to CO_2 and water vapour transfer   48491, 48553,
     48590, 48627, 48733, 48883, 49077, 49281, 49363, 49398-9, 49478, B49527,
     B49621, 49798, 50095, 50304, 50456, 50494, 50666, 50674, 50676, 50821, 50905,
     51227, 51378, 51418, 51497, 51787, 51822, 51906

Humidity of air, and respiration   50903

Humidity of air, methods (*cf.* also Infra-red analyser for water vapour)   49045,
     50509, B50878

Hydration level of leaf, and canopy functioning
     48940, 49715, 49704, 49415, 49142, 49785, B51897

Hydration level of leaf, and carbon fixation pathways   48685, 48814, B49561, 49568,
     49704, 49717, 50688, 51104, 51566, B51897, 52065

Hydration level of leaf, and carotenoids   50514

Hydration level of leaf, and chlorophyll   48726, 48810, 49812-3, 49815, 49836,
     50514, 50628, 51051, 51165, 51864, 52064, 52077

Hydration level of leaf, and chloroplast   48685, 48810, 49812, 49815, 50760, 51384,
     52064

Hydration level of leaf, and electron transport chain   48702, 48726, 48810, 48881,
     49713, 49717, 49812, 49815, 50628, 51566, 52064

Hydration level of leaf, and gas exchange
     48473, 48548, 48564, 48578, 48685, 48692-3, 48702, 48775, 48806, 48814,
     48857, 49164, 49257, 49297, 49302, 49369, 49523, B49620-1, 49637-9, 49704,
     49713, 49717-8, 49806, 49809, 49922, 49927, 49968, 50021, 50165, 50167,
                                                                     (continued)

Hydration level of leaf, and gas exchange (continued)
        50175, 50241, 50266, 50273-4, 50291, 50347, 50610, 50662, B50687, 50758,
        50760, 50844, 50861, 51043, 51049, 51104, 51227, 51333, 51342, 51497, 51566,
        51570, 51596, 51689, 51720, 51865, B51897, 51936, 52004, B52018,
        52154, 52259, 52309, 52334

Hydration level of leaf, and photorespiration   49717, 51104

Hydration level of leaf, and productivity
        48473, 48564, 48702, 48775, 48806, 48814, 49102, 49785, 50021, 50165, 50344,
        50686, B50687, 51254, 51847, B51897

Hydration level of leaf, and resistances to CO_2 and water vapour transfer
        48424, 48627, 48775, 48815, 48940, 48975, 49142, 49281, 49387, 49399, B49527,
        49607, B49621, 49627, 49639, 49697, 49704, 49713, 49717, 49797-8, 49922,
        49927, 50008, 50021, 50096-7, 50165, 50175, 50266, 50291, 50331, 50494,
        50610, 50642, 50662, 51043, 51160, 51212, 51227, 51280, 51333, 51376-8,
        51497, 51570, 51596, 51720, 51787, 51822, 51865, B51897, 51898, 51906,
        51936, 51947, 51956, 52003-4, B52018, 52154, 52335

Hydration level of leaf, and respiration
        48702, 48814, 49297, 49717, 49836, 50266, 50347, 50760, 51689, B51897

Hydrogen   see H_2

Hydrogenase   see O_2 evolution mechanism and kinetics; H_2 evolution ...

Hygrometer   see   Humidity of air, methods

I

Ideotype   see   Model ...

Immobilization of chloroplasts and photosynthetic systems   see   Photosystems stabi-
        lization ...

Induction phenomena   see   Transient phenomena ...

Infra-red analyser for CO_2 48959, 50297, 50485, 50632, 51228, 51856, 52137, 52249

Infra-red radiation, effect on photosynthetic parameters   see   Irradiance, spectral
        composition ...; Temperature, high ...

Inhibitors of electron transport chain (*cf*. also Pesticides...; Antibiotics...)
        48488, 48607-8, 48659, 48663, 48681, 48749, 48757, 48787, 48860, 48934,
        49004, 49074, 49080, 49177, 49191, 49218, 49302, 49428, 49492, 49513, 49530,
        49556, 49559, 49590, 49687, 49745, 49816, 49848, 49858, 49892, 49920, 49939,
        49947, 49952, 49984, 49991, 50038, 50059, 50141, 50191, 50197, 50248, 50254,
        50319, 50323, 50335, 50354, 50361-2, 50368, 50411, 50435, 50534, 50559-60,
        50572-3, 50672, 50697, 50704, 50713, 50728, 50739, 50768, 50771, 50865,
        50892, 50927-9, 50975, 51011, 51016, 51031, 51048, 51070, 51093, 51102,
        51132, 51137, 51213, 51230, 51240, 51356, 51427, 51429, 51432, 51443, 51451,
        51458, 51488, 51509, 51521, 51556, 51559, 51588, 51592, 51604, 51609, 51615,
        51683, 51747, 51755, 51770, 51781, 51805, 51830, 51836, 51849, 51872, 52039,
        52041, 52046, 52071, 52206, 52224, 52283, 52314, 52336-7

Insertion level   see   Position of leaf on plant, insertion level ...

Intercellular spaces, CO_2 concentration inside
        48572, 48590-1, 48640, 48692, 48857, 48883, B48895, 48926, 49038, 49109-10,
        49150, 49162-3, 49249, 49368, 49398, 49455, 49469, 49702, 49821, 49854,
                                                                    (continued)

Intercellular spaces, CO_2 concentration inside (continued)
    50265, 50454, 50456, 50548, 50666, 50675, 50755, 50892, 50985, 51061,
    51071, 51178-9, 51186, 51498, 51515, 51566-8, 51905, 52352

Intracellular resistance see Resistance, intracellular (mesophyll)

Ionizing radiation (gamma, X, cosmic, *etc.*) and carbon fixation pathways 50979

Ionizing radiation (gamma, X, cosmic, *etc.*) and chlorophyll 50522, 51355

Ionizing radiation (gamma, X, cosmic, *etc.*) and chloroplast (chromatophore)
    50522, 51660

Ionizing radiation (gamma, X, cosmic, *etc.*) and electron transport chain 51993

Ionizing radiation (gamma, X, cosmic, *etc.*) and gas exchange 50198, 50979, B52018

Irradiance see also Model of radiation ...

Irradiance, compensation see Compensation irradiance

Irradiance, flash, and biliproteins 48846

Irradiance, flash, and carotenoids 51455

Irradiance, flash, and chlorophyll
    48792, 48853, 49177, 49274, 49286, 49531-3, 49872, 49945-6, 50038, 50628,
    50746, 50808, 50954, 51041, 51213, 51256, 51454, 51630, 51663, 51671,
    51869, 52043, 52090

Irradiance, flash, and chloroplast (chromatophore) 52111, 52209

Irradiance, flash, and electron transport chain .
    48488, 48545, 48594, 48789, 49074, 49078, 49172, 49238, 49507, 49594,
    49624, 49687, 49928, 49946, 49961, 50035-6, 50039, 50110, 50359-61, 50365,
    50580, 50624, 50672, 50955, 50965, 51177, 51213, 51256, 51277, 51279,
    51330, 51343, 51346, 51432, 51449, 51455, 51467, 51487, 51490, 51521,
    51651, 51980, 52057, 52084, 52090, 52092-3, 52299, 52354

Irradiance, flash, and gas exchange 48652, 49172, 49640, 50424-5, 51276, 52237-8,
    52387

Irradiance, flash, and respiration 52093

Irradiance (PhAR) and algae productivity
    48412, 48554, 48629, B48671, 48772, 48907, 48971, 49055, 49268-9, 49400,
    49425, 49504, 49581, 49591, 49603, 49728-9, 49735, 49800, 49824, 49874,
    49977, 50047, 50180, 50243, 50340, 50373, 50430, 50457-8, 50613, 50679,
    50766, 50880, 50899, 50934, 51139, 51184, 51210, 51313, B51350, 51387,
    51438, 51623, 51668, 51687, 51743, 51757, 51860, 52001, B52018, 52063,
    52085, 52102, 52156, 52165, 52245, 52324

Irradiance (PhAR) and biliproteins 49224, 49448

Irradiance (PhAR) and canopy functioning
    49000, 49125, 49556, 51895, 51941-2, 51962

Irradiance (PhAR) and carbon fixation pathways
    48477-8, 48500, 48508-9, 48623, 48768, 48807, 49214, 49262, 49454, B49561,
    49595, 49782, 49788, 49816, 49854, 49962, 50125-6, 50211, 50392, 50428,
    50571, 50626, 50734, 50752, 50767, 50840, 50999, 51122, 51167, 51451, 51605,
    51631, 51670, 51697, 52250, 52268, 52295

Irradiance (PhAR) and carotenoids
        48433, 48674, 48840, 48897, 49014, 49036, 49055, 49069, 49664, 50284, 50439,
        50478, 50514, 50556, 50618, 50939, 51148, 51172, 51292, 51451, 51462, 51974,
        52000, B52018, 52033, 52167

Irradiance (PhAR) and chlorophyll
        48427, 48433, 48447, 48507, 48526, 48528, 48535, 48554, 48569, 48591, 48660,
        48768, 48850, 48896-7, 48911, 48983, 49036, 49054-5, 49069, 49174, 49188,
        49248, 49263, 49273, 49283, 49291, 49362, 49379, 49381, 49435, 49442, 49448,
        49454, 49495, 49510, 49554, 49652, 49664, 49738, 49793, 49814, 49846, 49854,
        49864, 49878, 49910, 50007, B50052, 50060, 50094, 50138, 50173, 50211,
        50284, 50402, 50439, 50442, 50478, 50514, 50517, 50548, 50556, 50595, 50608,
        50752, 50766, 50798, 50856, 50916, 50939, 50974, 50984, 51021, 51080, 51110,
        51148, 51165-6, 51172, 51236, 51243, 51260, 51290, 51292, 51319, 51387,
        51428, 51431, 51451, 51479, 51589, 51712, 51830, 51869, 51903, 51941, 51974,
        52000, B52018, 52033, 52064, 52166-7, 52295, 52374, 52384

Irradiance (PhAR) and chloroplast (chromatophore)
        48426, 48514, 48535, 48546, 48750, 49032, 49072, 49357, 49411, 49595, 49762,
        49793, 49854, 49891, 49939-40, B50052, 50181, 50284, 50402, 50439, 50441-2,
        50592, 50774, 50802, 50939, 51148, 51243, 51303, 51353, 51361, 51460, 51945,
        52007, 52064, 52211, 52302, 52384-5

Irradiance (PhAR) and ecosystem and plant productivity
        48422, 48449, 48485-6, 48591, 48605, 48617, 48771, 49113, 49125, 49188,
        49241, 49329, 49355, 49510, 49574, 49651, 49730, 49844, 49846, 49861, 50007,
        50034, 50037, 50063, 50100, 50149, 50166, 50213, 50284, 50344, 50452, 50548,
        50574, 50581, 50822, 50856, 50984, 51109-10, 51126, 51148, 51191, 51448,
        51482, 51558, 51661, 51721, 51782, 51895, 51910, 51962, 52000, 52078, 52125,
        52198, 52300

Irradiance (PhAR) and electron transport chain
        48447, 48495, 48526, 48581, 48708, 48892, 48934, 49129, 49188, 49218, 49240,
        49463, 49544, 49551, 49709, 49722, 49782, 49854, 49864, 49888, 49925, 49961,
        B50052, 50101, 50211, 50232, 50276, 50353, 50402, 50439, 50552, 50580,
        50595, 50606, 50657, 50770, 50939, 50958, 50984, 51021, 51141, 51166, 51236,
        51301, 51330, 51337, 51343, 51347, 51427-8, 51440, 51467, 51500, 51556,
        51614, 51620, 51751, 51760, 51768, 51809, 51842, 51919, 51987, 52008, 52205,
        52302, 52354-5, 52374

Irradiance (PhAR) and gas exchange (cf. also Efficiency, photochemical; Irradiance
        (PhAR) and gas exchange, analysis of light curves; Irradiance (PhAR) and
        gas exchange, saturating irradiance)
        48428-9, 48435, 48441, 48453, 48491, 48505, 48508, 48514, 48520, 48528,
        48531, 48541, 48553, 48571, 48590-1, 48629, 48632, B48671, 48692, 48703,
        48733, 48754, 48787, 48809, 48857, 48882, 48959, 49000, 49030, 49036-7,
        49050, 49052, 49055, 49075, 49105, 49118, 49148, 49150, 49155, 49162-3,
        49187-8, 49229, 49241, 49252, 49297, 49302, 49331, 49356-8, 49362, 49364,
        49366, 49391, 49400, 49411, 49416-7, 49428, 49447-8, 49454, 49466, 49471,
        49504, 49513-4, 49520, B49527, 49545, 49601, 49618, 49638, 49699, 49701,
        49712, 49718, 49727-30, 49747, 49766, 49780, 49782, 49809, 49845-6, 49851,
        49854, 49878, 49938, 49941, 49957, 49993, 50007, 50018, 50024, 50037, 50050,
        50055, 50078, 50115, 50124, 50132, 50170, 50224, 50239, 50262, 50273-4,
        50329, 50347, 50372, 50378, 50406, 50418, 50428-30, 50441, 50454, 50456,
        50468, 50470, 50478, 50482, 50511, 50548, 50551, 50564, 50567, 50594, 50600,
        50604, 50606, 50609, 50618, 50631-2, 50675, 50692, 50755, 50758, 50776,
        50800-1, 50811, 50844, 50853, 50861, 50877, 50889, 50903, 50933, 50962,
        50970-1, 50984-5, 50999, 51009, 51030, 51061, 51108-10, 51139, 51142, 51148,
        51166, 51178-9, 51181, 51183, 51186, 51188, 51191, 51227, 51284, 51312,
        51354, 51366, 51386-7, 51418, 51447, 51491, 51495, 51497, 51517, 51534,
        51558, 51564, 51573, 51587, 51597, 51652, 51657, 51688-9, 51737-8, 51774,
        51846, 51903, 51941-2, 51956, 51962, 51974, 51981-2, 51987, 52009, B52018,
        52040, 52086, 52118, 52156, 52165, 52205, 52213, 52227, 52237-8, 52245,
        52247, 52252, 52266, 52297, 52309, 52320, 52334, 52346, 52352, 52354,
        52382, 52385

Irradiance (PhAR) and gas exchange, analysis of light curves
        48425, 48528, 48541, 48553, 48632, 48809, 48879, 48959, 49150, 49356-8,
        49364, 49658, 49718, 50011, 50511, 51089, 51148, 51181-2, 51903,
        52227, 52252, 52386-7

Irradiance (PhAR) and gas exchange, saturating irradiance
        48491, 48528, 48632, 49076, 49188, 49241, 49417, 49448, 49845, 49854, 49938,
        50018, 50108, 50566, 50889, 51061, 51067, 51109-10, 51587, 51652, 52040

Irradiance (PhAR) and photorespiration
        48509, 49065, 49214, 49391, 49513, 49528, 49957, 49981, 51303, 51631, 52106,
        52316, 52330

Irradiance (PhAR) and resistances to CO_2 and water vapour transfer
        48491, 48528, 48590-1, 48775, 48945, 48975, 49145, 49241, 49281, 49478,
        B49527, 49535, 49618, 49798, 49854, 50095-7, 50211, 50291, 50372, 50417,
        50456, 50494, 50507, 50548, 50675-6, 51030, 51160, 51166, 51186, 51227,
        51333, 51418, 51497, 51720, B51897, 51954, 51956, 52064, 52291

Irradiance (PhAR) and respiration
        48505, 49093, 49188, 49357, 50078, 50786, 50903, 51447

Irradiance (PhAR, total) measurement
        48944, 49283, 49490, 49644, 50269, 50509, 50650, 50852, 51765

Irradiance, spectral composition and algae productivity 48907, 49603, 50882, 51860

Irradiance, spectral composition and biliproteins 49134, 49224, 49565, 50159,
        50831, 50943, 51636

Irradiance, spectral composition and canopy functioning 51941-2

Irradiance, spectral composition and carbon fixation pathways
        48610, 50377, 50840, 51290, 51948, 52139-40

Irradiance, spectral composition and carotenoids
        49221, 49261, 49663, 50556, 50746, 50904, 51413, 51536, 52140, 52385

Irradiance, spectral composition and chlorophyll
        48521, 48602, 48725, 48755, 48920, 49134, 49161, 49221, 49274, 49495, 49515,
        49652, 49663, 49670, 49789, 49862, B50052, 50085, 50094, 50159, 50190,
        50295, 50377, 50453, 50556, 50746, 50831, 50881, 50904, 50920-1, 51240,
        51290, 51413, 51536, 51942, 52134-5, 52293, 52333

Irradiance, spectral composition and chloroplast (chromatophore)
        49134, 49223, 49263, 49411, 50159, 50439, 50917-8, 51290, 51345, 51353,
        51413, 51460, 52135, 52384-5

Irradiance, spectral composition and ecosystem and plant productivity
        48752, 48755, 49651, 49951, 51191, 51941, 52140

Irradiance, spectral composition and electron transport chain
        48785, 48836, 48920, 49224, 49240, 50190, 50377, 50633, 50766, 50918-9,
        50921, 51240, 51343, 51390, 51536, 52135, 52385

Irradiance, spectral composition and gas exchange
        48610, 48751-4, 48857, 48907, 48920, 49221, 49224, 49513, 49515, 49937,
        50274, 50377, 50650, 51089, B51168, 51312, 51366, 51942, 51971, 51981,
        52134, 52140, 52212, 52227, 52352, 52356, 52385

Irradiance, spectral composition and photorespiration 49513

Lipids, fatty acids and carotenoids 51732

Lipids, fatty acids, and chlorophyll 50612, 51226, 51273, 51732

Lipids, fatty acids, and chloroplast (chromatophore)
        48626, 48693, 48724, 48892, 48914, 49025, 49127, 49134, 49152, 49232-4,
        49242, 49265, 49292, 49437, 49538, 49558, 49615, 49633, 49794, 50032, 50098,
        50277, 50352, 50402, 50439, 50528, 50745, 50793, 50813-4, 50818, 50820,
        50939, 50946-7, B51168, 51216, 51329, 51365, 51394, 51396, 51398, 51434,
        51538, 51635, 51693, 51708-9, 51732, 51770, B51788, 51922, 52062, 52363

Lipids, fatty acids and electron transport chain 52323

Lutein   see Carotenoids ...; Xanthophylls ...

M

Magnetism and photosynthesis 49237, 49451, 49457

Maintenance respiration see Respiration, growth and maintenance

Malate dehydrogenase, methods see Malic enzyme, malate dehydrogenase, methods

Malic enzyme, malate dehydrogenase
        48500, 48542, 48569, 48622, 48693, 48856, 48861, 48919, 49299, 49392,
        49396, 49755-6, 49758, 49863, 49899, 50182, 50259, 50391, 50836, 50892,
        50981, 50999, 51004, 51054, 51120, 51150, 51167, B51168, 51175, 51223,
        51360, 51407, 51476, 51492, 51584, 51588, 51732, 52083, 52295

Malic enzyme, malate dehydrogenase, methods 49392, 52188

Mass culture of algae see Algae mass cultures ...

Maximum photosynthetic rate see Potential photosynthetic rate ...

Mehler reaction see Photosystem 1 activity ...

Membrane transport of CO_2 see CO_2 transfer across membranes

Mesophyll resistance   see Resistance, intracellular (mesophyll)

Microbody see Peroxisome ...

Microelements see Mineral elements (other than N,P,K) ...

Mineral elements (N,P,K) and algae productivity
        48629, 48648, 48745, 48808, 48937, 48971, 49059, 49416, 49591, 49735,
        49795-6, 49800, 49874, 50047, 50171, 50202-3, 50264, 50340, 50342, 50373,
        50375, 50613, 50730, 51668, 51686, 51843, 51863, 51904, 51913-4, 51916,
        52019, 52102, 52270, 52368

Mineral elements (N,P,K) and biliproteins 49396

Mineral elements (N,P,K) and canopy functioning 51082

Mineral elements (N,P,K) and carbon fixation pathways 48705, 48979, 49360, 49396,
        49469, B49561, 49595, 49900, 50041, 50043, 50428, 50999, 51113, 51158,
        51731, 52009, 52264, 52267

Mineral elements (N,P,K) and carotenoids   48674, 49108, 50536, 50721, 51033,
        51096

Mineral elements (N,P,K) and chlorophyll
        48674, 48752, 48995, 49208, 49395-6, 49498-9, 49517, 49708, 49738, 49993,
                                                                      (continued)

Mineral elements (other than N,P,K) and ecosystem and plant productivity
        48418, 48451, 48483, 48626, 49190, 49705, 49708, 49979, 50357, 50537, 50625,
        50654, B50687, 50942, 51033, 51196, 51206, 51583, 51714, 51735, 51828,
        52123

Mineral elements (other than N,P,K) and electron transport chain
        48432, 48581, 48606, 48608, 48625-6, 48647, 48795, 48900, 48909, 48984,
        49085, 49143, 49255, 49305, 49360, 49370, 49407, 49543, 49569, 49624, 49720,
        49739, 49770, 49877, 49897, 50147, 50160-1, 50177-8, 50186, 50318-9, 50434,
        50471, 50568, 50577, 50598, 50637, 50648, 50707, 50739, 50975, 50998, 51017,
        51117-8, 51132, 51301, 51440, 51443, 51556, 51581, 51879, 51918, 51938,
        52022, 52193, 52280, 52299, 52323, 52327, 52348, 52374

Mineral elements (other than N,P,K) and gas exchange
        48629, 48647, 48959, 49140, 49215, 49331, 49349, 49376, 49546, 49708, 50259,
        B50687, 50942, 51284, 51340, 51698, 51736-7, 51783, 51815, 52206

Mineral elements (other than N,P,K) and photorespiration 49331

Mineral elements (other than N,P,K) and resistances to CO_2 and water vapour transfer
        49546, 49708, 49881, 51815.

Mineral elements (other than N,P,K) and respiration 49349, 51698, 51736

Mixotrophy see Carbon metabolism types ...

Model of aquatic community production
        48441, 48652, B48671, 48911, 49037, 49059, 49258, 49268, 49364, 49400,
        49581, 49800, 49879, 50243, 50880, 51010, B52018, 52102, 52286

Model of canopy photosynthesis, prediction model
        48421-2, 48449, 48457, 48459, 48462, 48490, 48573, 48614, 48638, 48665,
        48829, 48838, 48871, 48879, B48946, 48958, 49118, 49186, 49228, 49231,
        49366, 49385, 49424, 49461, 49535, 49588, 49613, 49616-7, 49715, 49730,
        49733, 49852, 50012, 50016, 50063, 50067, 50305, 50384, 50447, 50511, 50533,
        50553, 50611, 50796, 50857, 50970, 51042, 51094, 51106, 51121, 51134, 51145,
        51157, 51203, 51412, 51546, 51551, 51575, 51619, 51754, 51891, 51903, 51908,
        B51961, 51962, 52017, B52018, 52183, 52219, 52239

Model of carbon fixation pathways 49944, 51071, 51507

Model of chlorophyll energetics (*cf*. also Chlorophyll in model systems)
        48463, 48583, 48783, 48801, 48873, 48963, 49294, 49435, 49438, 49583, 49770,
        50071, 50093, 50345, 50441, 50647, 50866, 50953, 51064, 51335, 51513, 51825,
        51939, 52303

Model of CO_2 in atmosphere, biosphere 49560, 51927

Model of electron transport chain
        48619, 49130, 49150, 49172, 49480, 49770, 49837, 49888, 49905, 50071, 50236,
        50249, 50345, 50471, 50518, 50598, 50867, 50953, 50955, 51064, 51208, 51250,
        51354, 51358, 51446, 51598, 51878, 51996, 52149, 52287, 52327, 52387

Model of *Halobacterium* photosynthesis 52159, 52388

Model of leaf gas exchange, ideotype of leaf photosynthesis
        48511, 48664, 48698, 48768, 48958, 48975, 49038, 49071, 49150, 49361, 49366,
        49368-9, 49398, B49527, 49658, 49689, 49718, 49759, 49782, 49878, 50050,
        50095-6, 50306, 50328, 50344, 50459, 50479, 50604, 50652, 50676, 50800-1,
        50817, 50876, 51038, 51119, 51180, 51233, 51354, 51675, 51935, 52020,
        52386

Mutants, chloroplast (chromatophore) in
         48514, 49035, 49234, 49463, 49505, 49614, 49682, 49891, 50326, 50414,
         50627, 51242-4, 51320, 51345, 51349, 51401, 51415, 51500-1, 51506, 51615,
         51654, 51660, 51923, 52221

Mutants, ecosystem and plant productivity of 50413, 51405

Mutants, electron transport chain in
         48569, 48595, 48681, 48811, 48851, 49079, 49342-4, 49597, 49682, 49954,
         49984, 50320-1, 50325, 50413-4, 50580, 50616-7, 50627, 50633, 50695, 50719,
         50914, 51320, 51500, 51615, 51628, 52037, 52055, 52221-2, 52241, 52279

Mutants, gas exchange in
         48514, 48811, 49514, 49682, 49957, 50024, 50413, 50719, 51383, 51405,
         51697, 51974, 52118

Mutants, photorespiration in 49957, 51695

Mutants, photosynthetic, isolation and selection 50719, 50882, 51696, 51717

Mutants, respiration in 49957

N

N_2, anaerobic atmosphere, and carbon fixation pathways 50391

N_2, anaerobic atmosphere, and chlorophyll 51427, 51712

N_2, anaerobic atmosphere, and chloroplast (chromatophore) 52052

N_2, anaerobic atmosphere, and ecosystem and plant productivity 50242

N_2, anaerobic atmosphere, and electron transport chain 51427

N_2, anaerobic atmosphere, and gas exchange 50242

N_2, anaerobic atmosphere, and resistances to CO_2 and water vapour transfer 50242

NAD see NADP, NAD ...

NADP, NAD
         48533, 48549, 48625, 48680-1, 48933, 48961, 48903, 49093, 49255, 49680,
         49692, 49782, 49883, 49930, 49970, 49973, 49991, 50101, 50235-6, 50277,
         50280, 50391, 50615, 50633, 50683, 50713, 50739, 50786, 50802, 50829,
         50847, 50935, 50981, 51068, 51263, 51276, 51301, 51320-1, 51343, 51352,
         51614, 51711, 51785, 51873, 51893, 51975, 52039, 52083, 52135, 52209,
         52314, B52321

NADP, NAD, methods 50500

Net assimilation rate see Growth analysis, net assimilation rate ...

Net photosynthetic rate see Gas exchange ...

Nitrogen see also N_2 ...; Mineral elements (N,P,K) ...

Nitrogen fixation and metabolism, nodule bacteria, and photosynthesis
         48587, 48720, 48745, 48787, 48919, 48942, 49070, 49225, 49266, 49360,
         49669, 49712, 49883, 50030, 50298, 50353, 50536, 50540, 50645, 50657,
         50896, 50945, 51229-31, 51240, 51458, 51492, 51625, 51778-9, 51808, 51913,
         51958, 52181, 52250, 52282

NMR see EPR, NMR

Nodule bacteria see Nitrogen fixation ...

Nuclear magnetic resonance see EPR, NMR

Nucleic acids see Proteins, amino acids, nucleic acids ...

O

O_2 and algae productivity 48932, 49581, 50690, 50731, B51350

O_2 and carbon fixation pathways
      48441, 48542, 48714, 49220, 50041, 50043, 50392, 50589, 50754, 50840, 51631,
      51975

O_2 and carotenoids 49014

O_2 and chlorophyll
      48569, 48982, 49209, 50183, 50795, 51166, 51260, 51425, 51428, 51712

O_2 and chloroplast (chromatophore) 50820, 50939

O_2 and ecosystem and plant productivity
      48857, 49065, 49293, 50183, 50382, 50405, 50733

O_2 and electron transport chain 48545, 48976, 49218, 50657, 51425, 51842

O_2 and gas exchange
      48429, 48436, 48441, 48571, 48659, 48692, 48714, 48723, 48730, 48809, 48857,
      48976, 49075, 49080, 49150, 49331, 49334, 49492, 49511, 49513, B49527,
      49699, 49780, 49860, 49938, 49957, 50011, 50043, 50115, 50117, 50183, 50273,
      50368, 50382, 50405, 50634, 50742, 50755, 50887, 50900, 50933, 51024, 51089,
      51108, 51477, 51701, 51759, 51851, 51974, 52106, 52227, 52266

O_2 and photorespiration 48443, 48723, 48976, 49403, 50041, 51104, 52106, 52330

O_2 and resistances to CO_2 and water vapour transfer 50183

O_2 and respiration B48895, 49075, 49403, 51249, 51851.

O_2 determination in water reservoirs see Algae, primary productivity, methods

O_2 electrode
      48716, 49022, 49888, 49920, 50390, 50404, B50878, 50927, 51105, 51108,
      51372, 51390, 51395, 51680-1, 51698

O_2 evolution mechanism and kinetic
      48445, 48463, 48488, 48532, 48608, 48811, 48865, 48909, 48922, 48984,
      49172, 49177, 49212, 49255, 49342, 49435-6, 49547-8, 49589, 49594, 49622,
      49631, 49640, 49720, 49722, 49883, 49946-7, 49970, 50035-6, 50039, 50147,
      50178, 50359, 50365, 50402, 50468, 50471, 50473, 50518-9, 50577, 50618,
      50620, 50624, 50628, 50699, 50842, 50914, 50958, 50975, 51016, 51213,
      51250, 51277-9, 51334, 51443, 51503, 51516, 51620-1, 51651, 51691, 51799-
      -800, 51877, 51967, 52057, 52090, 52135, 52193, 52299, B52321, 52327, 52354,
      52367, 52383, 52385

O_2 exchange see Gas exchange ...

O_2 isotopes, use in photosynthesis measurement
48659, 49109, 49491, 49528, 50430, 50940, 51685, 52367

Ontogeny of algae, and algae productivity
48937, 49833, 50613, 51308, 52244

Ontogeny of algae, and biliproteins 48978, 50289

Ontogeny of algae, and carbon fixation pathways
49673, 50341, 51464, 51626, 51850, 52247, 52319

Ontogeny of algae, and carotenoids 48513, 49036, 49108, 49986, 50613, 51462

Ontogeny of algae, and chlorophyll
48823, 48978, 49036, 49330, 49470, 49495, 49503, 49745, 49986, 50040, 50341,
50416, 50613, 51462, 51464, 51553, 51812, 51889, 52293

Ontogeny of algae, and chloroplast (chromatophore) 49741, 51503, 51553, 52244

Ontogeny of algae, and electron transport chain 48978, 50397, 51503, 51553

Ontogeny of algae, and gas exchange
48978, 49036, 49503, 49673, 50416, 50613, 50742, 50887, 51142, 51462, 51889,
52165, 52319-20

Ontogeny of algae, and respiration 48978, 49673, 50887, 51889

Ontogeny of canopy, and canopy functioning 48460, 48462, 50787, 50791

Ontogeny of canopy, and ecosystem and plant productivity
48449, 48459, 49000, 49187, 49617, 50152, 50787, 50791, 51131, 51157,
51306, 51371

Ontogeny of canopy, and gas exchange 49000, 49187, 51306-7

Ontogeny of canopy, and respiration 51306

Ontogeny of chloroplast, and carbon fixation pathways 51464

Ontogeny of chloroplast, and carotenoids 48561, 48840, 49243, 50133, 51462, 52209

Ontogeny of chloroplast, and chlorophyll
48561, 48878, 48943, 49243, 49665, 50133, 51245, 51259, 51351-2, 51464,
52209

Ontogeny of chloroplast, and gas exchange 49038, 52190

Ontogeny of leaf see also Position of leaf on plant, insertion level ...

Ontogeny of leaf, and carbon fixation pathways
48414, 48481-2, 48519, 48623, 48861, 48918, 49214, 49322, 49469, 49605,
49765, 49805, 50082, 50432, 50480, 50535, 50542, 50850, 50895, 51120, 51360,
51475, 51510, 52009, 52140, 52276, 52287

Ontogeny of leaf, and carotenoids 49570, 52140

Ontogeny of leaf, and chlorophyll
48414, 48553, 48725, 48834, 48918, 48995, 49123, 49278, 49322, 49395,
49554, 49570, 49738, 49805, 49983, 50029, B50052, 50079, 50099, B50144,
50441, 50454-5, 50474, 50542, 50856, 50895, 50985, 51360, 51475, 51779,
52009, 52140, 52353

Ontogeny of plant, and respiration
  B49.114, 49118, 49836, 50338, 50396, 50703, 51306, 51312, 51540, 51735,
  51837

Optical properties, leaf see Leaf optical properties

Oscillations, short-term fluctuations, steady and non-steady state in electron
  transport chain 48431

Oscillations, short-term fluctuations, steady and non-steady state in gas exchange
  49658, 49900, 50817, 50933, 50962

Oscillations, short-term fluctuations, steady and non-steady state in resistances
  to CO_2 and water vapour transfer 49478

Osmotically active substances and algae productivity 48673, 48675

Osmotically active substances and carbon fixation pathways
  48689, 49064, 49390, 49900, 50049

Osmotically active substances and chlorophyll
  48673, 49477, 49780, 51488

Osmotically active substances and chloroplast (chromatophore)
  48688-9, 49167, 50865, 51356, 52236

Osmotically active substances and ecosystem and plant productivity
  49065, 50123, 51252, 51570, 51815, 51817

Osmotically active substances and electron transport chain
  48688, 49714, 51578, 51691, 51853

Osmotically active substances and gas exchange
  48673, 48688-9, 48854, 49063, 50049, 51566, 51815-6

Osmotically active substances and photorespiration 49064

Osmotically active substances and resistances to CO_2 and water vapour transfer
  49708, 50642, 51392, 51815-6, 51947

Oxygen see O_2 ...

Ozone see Pollution of air ...

P

P680   48445, 48465, 48824, 49078, 49263, 49342-3, 49435-6, 49508, 49548, 49970,
        50038, 50088, 50106, 50110-2, 50121, 50179, 50435, 50468, 50562, 50616-7,
        50624, 50913-4, 50954, 51041, B51168, 51276, 51330, 51347, 51432, 51627-8,
        51674, 51830, 52057, 52089, 52229, 52253, 52281

P700, P750, P879, etc.
        48465, 48494-6, 48498, 48507, 48526, 48569, 48611-3, 48625, 48654,
        48707, 48717, 48735, 48740, 48768, 48773, 48777, 48785, 48788, 48802-3,
        48811, 48824, 48831, 48834, 48836, 48851, 48859, 48894, 48900, 48969,
        49026-7, 49062, 49157, 49182, 49235, 49256, 49286, 49305, 49342, 49348,
        49370, 49379, 49393, 49420, B49433, 49435-6, 49481, 49526, 49544, 49563-4,
        49586, 49597, 49634, 49682-3, 49687, 49781, 49812, 49837, 49848-9, 49854,
        49910, 49915, 49932, 49954, 49970, 50071, 50088, 50098, 50106, 50108, 50113,
        50118, 50197, 50201, 50223, 50227, 50268, 50271, 50350, 50427, 50468, 50526,
                                                                          (continued)

P700, P750, P879, etc. (continued)
      50561-2, 50580, 50616, 50628, 50661, 50682, 50708, 50859, 50866, B50878,
      50908, 50916, 50939, 50953-4, 50965-6, 50984-5, 51015, 51040-1, 51063-4,
      51114, 51141, B51168, 51177, 51208, 51256, 51268, 51276, 51292, 51330,
      51343, 51388, 51430, 51449, 51454, 51520, 51526, 51552, 51602-3, 51627-30,
      51726, 51729, B51788, 51809, 51849, 51931, 51937, 51984, 52034, 52037-8,
      52055, 52064, 52094, 52131, 52152, 52229, 52234, 52245, 52253, 52274,
      52281, 52289, 52310, 52312, 52345

Paramagnetic oxygen analyser see O_2 determination ...

Paramagnetic resonance see EPR, NMR ...

PEP carboxylase (PEPC) see Phosphoenolpyruvate carboxylase

Peroxisome, glyoxysome, microbody
      48589, 48658, 49397, 49542, 49893, 50155, 50205, 50437, 50528, 51473,
      51958, 52178, 52283

Pesticides see also Inhibitors of electron transport chain

Pesticides, herbicides and algae productivity 49189, 49311, 50323, 50451

Pesticides, herbicides and biliproteins 48506

Pesticides, herbicides and carbon fixation pathways
      49004, 49556, 49816, 49999, 50836, 52116, 52295

Pesticides, herbicides and carotenoids
      48506, 48561, 48840, 48880, 48896, 49013, 49200, 49218, 49261, B49422,
      49663, 50168, 50398, 50443, 50993, 51220, 51397, 51399, 51416, 51462, 51889,
      52116, 52331-2

Pesticides, herbicides and chlorophyll
      48506, 48561, 48662, 48700, 48749, 48792, 48819, 48880, 48892, 48896, 49218,
      49265, 49319, 49330, 49377, 49381, B49422, 49441, 49662-3, 49835, 49843,
      49926, 49946, 49984, 49989, 50038, 50136, 50168, 50254, 50293, 50364, 50397-
      -8, 50434, 50442-3, 50483, 50573, 50587, 50697, 50816, 50904, 50988, 51022,
      51127-8, 51220, 51292, 51346, 51397, 51399, 51416, 51429-30, 51488, 51522,
      51545, 51770, 51824, 51829, 51831, 51889, 52014, 52089-91, 52116-7, 52141,
      52221

Pesticides, herbicides and chloroplast (chromatophore)
      48534, 48880, 48892, 49265, B49422, 49643, 49683, 49793, 49843, 50323,
      50442-3, 50635, 50670, 50765, 51022, 51053, 51637, 51770, 51824, 51831,
      52117

Pesticides, herbicides, and ecosystem and plant productivity
      48446, 48662, 49608, 49843, 50136, 50483, 50573, 50587, 51127-8

Pesticides, herbicides and electron transport chain
      48506, 48518, 48595, 48700, 48757, 48785, 48792, 48836, 48880-1, 48892,
      48896, 48961, 48976, 49004, 49129, 49218, 49377, B49422, 49506, 49643,
      49653, 49662-3, 49685, 49688, 49720, 49843, 49850, 49902, 49911, 49984,
      50035, 50231, 50276, 50320, 50323, 50359-62, 50397-8, 50443, 50534, 50628,
      50633, 50665, 50696, 50704, 50765, 50768, 50788, 50865, 50928-30, 50965,
      51012, 51053, 51220, 51292, 51346, 51372, 51399, 51429, 51440, 51555, 51609,
      51702-3, 51747, 51770, 51805, 51824, 51831, 51879, 51944, 52008, 52041,
      52046, 52061, 52071, 52084, 52089, 52091, 52116, 52121, 52233, 52290

Pesticides, herbicides and gas exchange
      48446, 48506, 48571, 48742, 48749, 48882, 48976, 49004, 49173, 49189, 49218,
      49411, 49472, 49483, 49608, 49761, 49984, 50136, 50248, 50443, 50475, 50567,
                                                           (continued)

Pesticides, herbicides and gas exchange (continued)
      51292, 51372, 51391, 51831, 51889, 51963, 52370

Pesticides, herbicides and photorespiration 48976, 51458

Pesticides, herbicides and resistances to CO_2 and water vapour transfer 51997

Pesticides, herbicides and respiration 48506, 48742, 49843, 52093

Petiole see Stem, petiole, morphology, structure and physiological activity in

pH, effect on algae productivity
      48566, 48937, 49175, 49269, 50613, 52245

pH, effect on biliproteins 49924

pH, effect on carbon fixation pathways
      48477-8, 48481, 48542, 48704-5, 48787, 48863, 49230, 49516, 49816, 49863,
      49963, 50041, 50391, 50571, 50622, 50734, 50960, 51670, 52258, 52262

pH, effect on carotenoids 49822, 49924

pH, effect on chlorophyll
      49420, 49731, 49820, 49822, 49827, 49924, 50333, 50364, 50543, 50612, 50683,
      50898, 51509, 51875, 51911, 52206, 52365

pH, effect on chloroplast (chromatophore)
      48557, 49072, 49428, 49659, 49661, 50391, 51509, 51771

pH, effect on electron transport chain
      48557, 48597, 48789, 48794, 48933-4, 49026, 49084-5, 49097, 49383, 49450,
      49507, 49551, 49624, 49714, 49722, 49750, 49869-70, 49897, 49913, 49915,
      50036, 50110, 50276, 50314, 50362, 50365, 50562, 50648, 50713, 50729,
      50788, 50893, 50919, 50956, 51016, 51035, 51177, 51440, 51467-8, 51627,
      51691, 51751, 51755, 51760, 51853, 51879, 51918-9, 52039, 52046, 52092,
      52173, 52269, 52314, 52337, 52383

pH, effect on gas exchange
      48429, 48565, 48578, 49030, 49050, 49097, 49322, 49334, 49900, 50024,
      50248, 50376, 50567, 50742, 50971, 51180, 51477, 51499, 51539, 51564,
      51886, 51950, 52205-6, 52227, 52382

pH, effect on photorespiration 50041

pH, effect on photosynthates and plant productivity 49588, 50971

pH, effect on respiration 48578, 51886

PhAR, PAR see Irradiance ...; Canopy, radiation ...

Phosphoenolpyruvate carboxylase
      48481, 48508, 48515, 48539, 48542, 48621-3, 48657, 48693, 48811, 48856,
      48861-4, 48919, 48979, 49103, 49159, 49225, 49230, 49361, 49366, 49396,
      49446, 49530, B49561, 49568, 49734, 49755, 49863, 49916, 49982, 50043,
      50142-3, 50259, 50377-8, 50429, 50449, 50571-2, 50622, B50687, 50797,
      50836, 50847, 50853, 50870, 50911, 50960-1, 50999, 51004, 51031, 51054,
      51071-2, 51081, 51104, 51151-3, 51167, B51168, 51175, 51202, 51214, 51218,
      51222-3, 51234, 51274, 51324, 51357, 51360, 51380, 51447, 51485, 51492,
      51517, 51563, 51618, 51677-8, 51731-2, 51794-5, 51870, 51948, 51997, 52036,
      52083, 52101, 52139, 52147, 52227, 52247, 52262-4, 52266-7, 52276, 52292,
      52294-5, 52307

Phosphoenolpyruvate carboxylase, methods 49044, 50960, 52101

Phosphorus see Mineral elements (N,P,K) ...

Photoperiod and carbon fixation pathways 48861-3, 51107, 51201, B51897, 52262

Photoperiod and chlorophyll 48739, 48819, 49448, 49665

Photoperiod and chloroplast (chromatophore) 48546

Photoperiod and ecosystem and plant productivity
        48548, 49101, 49206, 49602, 49918, 49951, 50691, 51632, 51891

Photoperiod and electron transport chain 50728, 51084

Photoperiod and gas exchange 48862, 49326, 49448, 49534, 51089, 51632

Photoperiod and resistances to CO_2 and water vapour transfer 49898

Photoperiod and respiration 50815, 51089

Photophosphorylation, cyclic
        48474, 48518, 48606, 48677, 48688, 48762, 48870, 48904, 48978, 48980, 49004,
        49066, 49074, 49366, B49422, 49501, 49593, B49749, 49873, 50260, 50411,
        50471, 50562, B50687, 50719, 50766, 50824, 50834, B50878, B51168, 51240,
        51292, 51302, 51320, 51402, 51404, 51429, 51440, 51490, 51516, 51669, 51761,
        51781, 51806, 51879, 52008, 52045-6, 52084, 52098, 52126, 52173, 52194,
        52199, 52233, B52321, 52322, 52336, 52374

Photophosphorylation in photosynthetic bacteria see Photosynthetic bacteria phos-
        phorylation

Photophosphorylation, methods 48938, 49714, 49829, 50711, 50719, 51894

Photophosphorylation, model see Model ...

Photophosphorylation, non-cyclic
        48474, 48572, 48606, 48677, 48795, 48843, 48870, 48880, 48904, 48956, 48980,
        49004, 49143, 49348, 49366, B49422, 49507, 49551, 49556, 49684, B49749,
        49770, 49839, 49877, 49902, 50043, 50260, 50275, 50319, 50353, 50471, 50562,
        50683, B50687, 50719, 50765-6, 50770, 50799, 50824, B50878, 50883, 50921,
        50927, 50935, 50946, 51012, B51168, 51263, 51292, 51302, 51337, 51373,
        51516, 51604, 51645, 51761, 51943, 52008, 52022, 52045-6, 52108, 52121,
        52126, 52128, 52173, 52199, 52205, 52211, 52233, 52302, B52321

Photophosphorylation, pseudo-cyclic 48904, 50824, 51263, 51425

Photoreduction see H_2 evolution ...

Photorespiration enzymes
        48443, 48509, 48542, 48658, 48693, 48826, 48991, 49214, 49501, 49528,
        49542, 49554, 49737, 49801, 49906, 50000-1, 50041, 50043, 50126, 50137,
        50140, 50205, 50228, 50368, 50432, 50492, B50687, 50896, 50938, 50999,
        51047, 51054, 51071, 51238, 51606, 51631, 51692, 51975, 52065, 52073,
        52147, 52170, 52178, 52201, B52321

Photorespiration metabolic cycles
        48443-4, 48542, 48622, 48681, 48714, 48976, 49043, 49064-5, 49080, 49220,
        49366-7, 49390, 49528, 49554, 49622, 49737, 49890, 49893, 49981, 50059,
        50117, 50354, 50368, 50481, 50493, 50686, 50849, 50896, 50900, 50938, 50999,
        51093, 51238, 51303, 51599, 51695, 51870, 51975, 52025, 52170, 52201,
        52282-3, 52287, 52316, 52330

Photorespiration metabolic cycles enzymes, methods see Enzymes of glycollate
        cycle, methods

Photorespiration rate
        48415, 48444, 48659, 48723, 48809, 48826, 48902, 48945, 48976, 48992, 49052,
        49067, 49075, 49080, 49131, 49299, 49331, 49368, 49403, 49491-2, 49501,
        49513, B49561, 49606, 49638, 49860, 49957, 50011, 50041, 50126, 50339,
        50368, 50382, 50489, 50491, 50800, 50847, 50887, 50949, 51153, 51238, 51246,
        51249, 51457-8, 51678, 51692, 51696, 51921, B52018, 52040, 52106, 52216, 52254

Photosynthates and intermediates of carbon fixation pathways
        48649, 48656, 48705, 48714, 48947, 49004, 49043, 49056, 49070, 49220, 49390,
        49396, 49401-2, 49427, 49534, 49746, 49858, 49949, 49988, 50000-1, 50043,
        50116, 50218, 50234, 50343, 50430, 50490, 50589, 50622, 50838, 50870, 50940,
        51031, 51054, 51113, 51158, 51234, 51261, 51289, 51499, 51669, 51696, 51744,
        51850, 52247, 52304, 52307, 52351

Photosynthates and intermediates of carbon fixation pathways, and chloroplast
        (chromatophore)   50449, 52302

Photosynthates and intermediates of carbon fixation pathways, and electron trans-
        port chain   52108, 52302

Photosynthates and intermediates of carbon fixation pathways, and gas exchange
        49050, 51783, 52302

Photosynthates and intermediates of carbon fixation pathways, and photorespiration
        52170

Photosynthates and intermediates of carbon fixation pathways, and plant productivity
        49951

Photosynthates and intermediates of carbon fixation pathways, and respiration 49093

Photosynthates translocation and distribution (*cf.* also Model of photosynthates...)
        48442, 48480, 48512, 48517, 48540, 48548, 48564, 48570, 48609, 48665, 48728,
        48737, 48848, 48884, 48919, 48936, 48973, 48988, 48992, 49001, 49010, 49022,
        49042, 49101, 49112, B49114, 49136, 49193, 49202-3, 49206-7, 49293, 49301,
        49310, 49332, 49391, 49427, 49432, 49501, 49534, 49536-7, 49602, 49672,
        49716-7, 49719, 49725, 49736, B49749, 49785, 49792, 49833, 49840, 49846,
        49855, 49857-8, 49890, 49899-901, 49976, 49992, 50002, 50019, 50021-2,
        50034, 50082, 50126, 50158, 50165, 50194, 50285, 50338, 50344, 50355, 50450,
        50463, 50474, 50522, 50541, 50575-6, 50604, 50607, 50613, 50653, 50658,
        50676, B50687, 50712, 50733, 50747, 50782, 50812, 50823, 50863, 50872,
        50962, 50968, 50979, 50995, 50999, 51009, 51023, 51043, 51097, 51113, 51121,
        51130, 51153, 51171, 51224, 51266, 51283, 51374, 51400, 51405, 51412, 51433,
        51466, 51474, 51482, 51496, 51512, 51541, 51563, 51597, 51632, 51685, 51694,
        51701, 51710, 51744, 51808, B51897, 51902, 51920 30, 51969, 51971, 52004,
        52016, 52023, 52080, 52100, 52122, 52129, 52149, 52162, 52168, 52174-6,
        52179-81, 52195, 52209, 52214, 52254, 52275-6, 52300-2, 52329, 52342-3,
        52350-1, 52361

Photosynthetic bacteria carbon fixation pathways
        48542, 48670, 49298, 50944, 51171, 51584, 51794, 52351

Photosynthetic bacteria carotenoids see Carotenes; Xanthophylls of photosynthe-
        tic bacteria

Photosynthetic bacteria chlorophylls see Bacteriochlorophylls

Photosynthetic bacteria chromatophores see Chloroplast and chromatophore ...;
        Chromatophore ...

Photosynthetic bacteria electron transport chain
        48431, B48600, 48740, 48763, 48859, 48938, 49026, 49078, 49157, 49180-1,
        49343, 49393, 49409-10, 49547, 49739, 49781, 49848, 49873, 50071, 50106-7,
        50118, 50184, 50657, 50669, 50672-3, 50680, 50775, 50799, 50866, B50878,
                                                                    (continued)

Photosynthetic bacteria electron transport chain (continued)
        50908, 50954-6, 51015, 51021, 51041, 51084, 51141, 51163, B51168, 51176-7,
        51208-9, 51256, 51294, 51388, 51518, 51629-30, 51729, 51804, 51880, 51984,
        51986, 52024, 52058, 52092, 52186, 52289-90, 52312, 52380

Photosynthetic bacteria gas exchange 50141

Photosynthetic bacteria photophosphorylation
        48594, B48600, 49018, 49138, 50036, 50118, 50648, B50878, 50994, 51021,
        B51168, 51177, 51388, 51759, 52084, 52349

Photosynthetic bacteria reaction centres see P700 ...

Photosynthetic (chlorophyll) unit
        48526, 48654, 48783, 49188, 49263, 49283, 49532, 49622, 49674, 49770,
        B50052, 50098, 50320, 50424, 50545, 50580, 50641, 50939, 51012, 51276,
        51284, 51291-2, 51443, 51487, 51520, 51691, 51935, 52084, 52245, 52281,
        52312, 52354, 52386-7

Photosystem 1
        48423, 48431, 48447, B48448, 48495-8, 48501-3, 48506-7, 48518, 48536,
        48580-3, 48602, 48612-3, 48625, 48647, 48668, 48693, 48707-8, 48713, 48735,
        48773, 48783, 48787-8, 48811, 48823-4, 48832, 48834, 48836, 48851, 48896,
        48968-9, 48976, 48978, 49006, 49074, 49078-9, 49084, 49097, 49130, 49143,
        49212, 49235, 49240, 49248, 49255, 49266, 49318-9, 49330, 49342-3, 49348,
        49360, 49370, 49407, 49423, 49437, 49440-1, 49464, 49544, 49586, 49622,
        49634, 49636, 49671, 49674, 49682, 49685, 49687, 49722, B49749, 49770-2,
        49787, 49813, 49820, 49839, 49848-9, 49877, 49883, 49892, 49925, 49932,
        49940, 49946, 49954, 49984, 49997, 50071, 50106-8, 50113, 50120, 50201,
        50249, 50254, 50261, 50268, 50276-7, 50280, 50298, 50319-20, 50324,
        50336-7, 50393, 50398, 50403, 50424, 50427, 50441, 50444, 50468, 50471,
        50535-6, 50545, 50561-3, 50616, 50649, 50661, 50682, 50721, 50738, 50788,
        50804, 50835, 50865, 50917-9, 50926, 50931, 50939, 50953-4, 50965, 50975,
        50983, 50986-7, 50999, 51019, 51032, 51041, 51114, 51153, 51166, B51168,
        51236-7, 51292, 51302, 51308, 51320, 51330, 51375, 51425, 51427-30, 51443,
        51449, 51458, 51483, 51500, 51509, 51533, 51545, 51552, 51555, 51578,
        51609, 51627, 51629, 51634, 51650, 51663, 51682, 51746-7, 51756, 51761,
        51770, 51785, 51805-6, 51809, 51842, 51849, 51872, 51892, 51975, 52035,
        52041, 52047-8, 52055, 52058, 52082, 52091, 52096-7, 52135, 52150, 52194,
        52209, 52221, 52229, 52232-3, 52241, 52253, 52278-9, 52281, 52287, 52289-
        -90, 52310, 52320, 52323, 52345

Photosystem 1 activity measurement
        48466, 48572, 48581, 48625, 48681, 48688, 48757, 48834, 48978, 49004,
        49066, 49079, 49124, 49143, 49407, 49423, 49793-4, 49813, 49820, 49957,
        49997, 50197, 50320, 50325, 50411, 50536, 50590, 50665, B50687, 50719,
        50738, 50773, 50904, 50998, 51035, 51102, 51301-2, 51320, 51443, 51516,
        51578, 51592, 51682, 51756, 51785, 51805-6, 51842, 51975, 52045-6, 52096,
        52210, 52315

Photosystem 1 activity measurement, methods 50719, 50859, 51375

Photosystem 1, primary acceptor
        48542, 48968-9, 49097, 49837, 49848, 50107, 50471, 50561, 50661, B51168,
        51842, 52092

Photosystem 1 reaction centre see P700 ...

Photosystem 2
        48423, 48431, 48445, 48447, B48448, 48488, 48495-8, 48502-3, 48506, 48518,
        48526, 48532, 48536, 48580-3, 48595, 48602, 48607-8, 48613, 48647, 48681,
        48693, 48783, 48787-8, 48794, 48796-7, 48810-1, 48824, 48830, 48832, 48834,
        48846, 48851, 48877, 48896, 48976, 48978, 48984, 48991, 49066, 49074,
                                                                    (continued)

Photosystem 2 (continued)
      49078-9, 49084-5, 49177, 49213, 49240, 49255, 49263, 49318-9, 49330, 49342-4,
      49359, 49370, 49407, 49423, 49437, 49441, 49464, 49509, 49544, 49548, 49589,
      49594, 49622, 49634, 49636, 49643, 49662-3, 49671, 49674, 49676, 49685,
      49688, 49720, 49722, B49749, 49770-2, 49794, 49812-3, 49815, 49820, 49835,
      49839, 49848-9, 49877, 49891-2, 49905, 49925, 49939, 49946-7, 49954, 49970,
      49973, 49984, 49997, 50035, 50038-9, 50071, 50106-8, 50110-2, 50120-1,
      50177-9, 50197, 50206, 50254, 50261, 50276, 50314-5, 50319-21, 50323-4,
      50335-7, 50350, 50359-65, 50397-8, 50402-3, 50413-4, 50424-5, 50427, 50434-5,
      50441, 50468, 50471, 50534-6, 50545, 50562-3, 50606, 50616-7, 50624, 50627-8,
      50682, 50695-6, 50704, 50721, 50738-9, 50765, 50835, 50842, 50865, 50904,
      50913-4, 50917-9, 50926-8, 50930, 50939, 50954, 50958, 50965, 50975,
      50983-7, 50999, 51011-2, 51019, 51025, 51041, 51047, 51053, 51102,
      51114, 51132, 51153, 51166, B51168, 51213, 51216, 51220, 51236-7,
      51277-8, 51302, 51308, 51320, 51330, 51346-7, 51427-8, 51430, 51432,
      51443, 51455, 51458, 51483, 51488, 51493, 51500, 51509, 51533, 51545,
      51578, 51615, 51627-9, 51634, 51651, 51654, 51674, 51746, 51751, 51756,
      51785, 51799-800, 51805-6, 51824, 51830, 51877, 51892, 51918-9, 51926,
      51967, 51975, 52034-5, 52047-8, 52055-8, 52061, 52071, 52082, 52089-91,
      52096-7, 52121, 52135, 52194, 52205, 52209, 52221-2, 52229, 52241, 52253,
      52281, 52289-90, 52299, 52310, 52323, 52327

Photosystem 2 activity measurement
      48445, 48466, 48495, 48519, 48526, 48557, 48580-1, 48603, 48688, 48834,
      48847, 48904, 48978, 48980, 49004, 49024, 49050, 49066, 49079, 49085,
      49124, 49129, 49212-3, 49407, B49422, 49423, 49501, 49506, 49587,
      49710, 49713, 49793-4, 49813, 49820, 49842, 49854, 49911, 49946,
      49997, 50020, 50039, 50043, B50052, 50211, 50246, 50294-5,
      50314-5, 50320-1, 50325, 50350, 50411, 50435, 50530, 50536, 50590, 50627,
      50665, B50687, 50707, 50719, 50738, 50765, 50773, 50850, 50904, 50958, 50963,
      50975, 50998, 51016, 51034-5, 51079, 51102, 51117, 51132, 51153, B51168,
      51237, 51302, 51322, 51372, 51385, 51440, 51443, 51516, 51553, 51578, 51592,
      51650, 51690, 51702-3, 51756, 51767, 51770, 51777, 51785, 51805-6, 51853,
      51918, 51925, 51944, 51987, 52032, 52041, 52045-6, 52091, 52096, 52126-7,
      52193, 52205, 52210, 52233, 52302, 52322, 52360, 52374, 52383

Photosystem 2 activity measurement, methods   48768

Photosystem 2, primary acceptor
      48797, 49343-4, 49848, 49892, 50038-9, 50107, 50112, 50178-9, 50359-60,
      50471, 50617, 50682, B51168, 52071, 52090

Photosystem 2 reaction centre see P680

Photosystems stabilization, chloroplast immobilization, methods
      48428-9, 48436-7, 48603, 48847, 48937, 49023-4, 49107, 49582, 49667, 49695,
      49888, 49997, 50024, 50260, 50298, 5C314, 50634, 50774, 50826, 50834,
      50919, 51034-5, 51237, 51239, 51601, 51682, 51849, 51852-3, 51925-6, 52104-5

Phycobilins see Biliproteins ...

Phycobilisome
      49213, B49433, 49565-6, 49573, 49577, 49681, 49865, 49958, 49996, 50159,
      50230, 50299, 50464, 50531, 50777, 51074, B51168, 51174, 51264, 51344,
      51452, 51636, 52311, 52338, 52369

Phycocyanins
      48654, 48821, 48845-6, 48858, 48978, 49106, 49134, 49166, 49169, 49213,
      49224, 49448, 49452, 49565-6, 49573, 49577, 49681, 49865, 49958, 50159,
      50230, 50289, 50299, 50464, 50531, 50777, 50831, 50943, 51174, 51260, 51264,
      51276, 51338, 51344, 51452-3, 51460, 51470, 51590, 51622, 51636, B51788,
      51859, 51971-2, 52058, 52227, 52281, 52311, 52338, 52364, 52369

Phycoerythrins
        48654, 48845-6, 48867, 49106, 49213, 49452, 49565-6, 49573, 49577, 49681,
        49865, 50159, 50464, 50531, 50777, 50885, 50943, 50952, 50986, 51174,
        51260, 51276, 51319, 51452, 51460, 51542, 51636, B51788, 51857-9, 51972,
        52058, 52227, 52281, 52311

Phylogeny of biliproteins   48991, 51452, 51460-1, B51463, 51590, 51789

Phylogeny of carbon fixation pathways   48620, 48951, 49944, 50547, 50761, 51072,
        52050, 52243

Phylogeny of carotenoids   48991, 50515, 51460

Phylogeny of chlorophyll   48991, 50515, 51460, B51463, 51590, B51788, 52166

Phylogeny of chloroplast (chromatophore)   48529, 48543, 48620, 48761, 48877, 49031,
        49151, 49170, 49512, 49791, 49996, 50174, 50527, 50645, 50825, 51029, 51046,
        51401, 51460-1, B51463, 51972, 52243

Phylogeny of electron transport chain   48763, 48767, 48991, 50251, 50271, 50954,
        B51463, B51788, 51995, 52114, 52289, B52321

Phylogeny of gas exchange   52050

Phylogeny of photorespiration   51238

Phylogeny of respiration   48681, 51046

Phytochrome see Growth regulators ...

Phytoflavin see Ferredoxin ...

Phytopathological effects on carbon fixation pathways 48787, 48870, 50042

Phytopathological effects on carotenoids 50168, 51676

Phytopathological effects on chlorophyll 49675, 50168, 50200, 51059, 51297, 51676,
        51900

Phytopathological effects on chloroplast (chromatophore)   49031, 50200, 50833,
        51269, 51297

Phytopathological effects on ecosystem and plant productivity   48564, 48570, 49345,
        49493, 50013, 50168, 51710, 51753, 51782, 51787, 52051, 52168

Phytopathological effects on electron transport chain   48870, 51079

Phytopathological effects on gas exchange   48564, 48697, 48870, 49031, 49479,
        49605-6, 50013, 50200, 50332, 50833, 51059, 51186, 51193

Phytopathological effects on photorespiration   49606

Phytopathological effects on resistances to CO_2 and water vapour transfer   48697,
        49606-7, 50764, 51444

Phytopathological effects on respiration   48697, 49606, 51059

Pigments determination, sampling and extraction   48972, 49277, 49802, 50497, 50885,
        50898, 51161, 51293, B51662

Plastochron index see Leaf life span, plastochron index; *cf*. also Ontogeny of
        leaf...; Position of leaf on plant ...

Plastocyanin
        48544, 48601, 48625-6, 48681, 48707, 48717, 48894, 48994, 49073, 49078,
        49370, 49659, 49687, 49720, B49749, 49915, 50089, 50271, 50277, 50336-7,
        50471, 50562, 50637, 50965-6, 51047, 51176, 51232, 51320-1, 51358, 51467,
        51508, 51756, 51849, 51938, 52152, 52229

Plastocyanin, methods   50089, 52326

Plastoquinones see Quinones ...

Pollution of air, and algae productivity   48643, 49269, 50048, 50996

Pollution of air, and biliproteins   51858

Pollution of air, and canopy functioning   49217, 49545

Pollution of air, and carbon fixation pathways   48478, 48956, 49262, 49322, 49701,
        50140, 50343, 50490, 50739, 51120, 51447, 51873-4

Pollution of air, and carotenoids   50257, 51820

Pollution of air, and chlorophyll   49041, 49168, 49322, 49600, 49701, 49780, 50257,
        51820, 51858, 51886, 52359

Pollution of air, and chloroplast (chromatophore)   48733, 49322, 50792, 51571,
        51814, 51946, B52021

Pollution of air, and ecosystem and plant productivity   48666, 48733, 48812, 48930,
        49076, 49217, 49341, 49712, 49976, 50004, 50019, 50265, 50366, 50553, 50660,
        50990-1, 51885-7, 51902, B52021

Pollution of air, and electron transport chain   48956, 50739, 51070, 51806, 51872,
        B52021, 52210

Pollution of air, and gas exchange   48563, 48692, 48733-4, 48930, 48959, 49076,
        49322, 49349, 49545, 49701, 49968, 50122, 50195, 50265, 50376, 50489-91,
        50553, 50660, 50705-6, 50990-1, 51052, 51610, 51851, 51884-6, 51902, 51946,
        B52018, 52020, B52021, 52060, 52204, 52238, 52259-60

Pollution of air, and photorespiration   48733, 49513, 50140, 50489, 50491, B52021

Pollution of air, and resistances to CO_2 and water vapour transfer   48692, 48733,
        49076, 49147, 49262, 49881, 50265, 50366, 50905, 50969, 50990, 51946,
        52020, B52021, 52260

Pollution of air, and respiration   48733, 49322, 49349, 50195, 50489, 50491, 51052,
        51851, 51886, B52021, 52204

Porometer   49700, 50510, 51391, 51498

Position of leaf on plant, insertion level, and carbon fixation pathways
        48623, 48861, 49982, 50432, 50626, 50782, 51202, 51261, B51531, 51731, 52009,
        52264, 52267, 52276, 52292

Position of leaf on plant, insertion level, and carotenoids   50536

Position of leaf on plant, insertion level, and chlorophyll
        48465, 48768, 48809, 48827, 49208, 49813, 49982, 50083, 50138, B50144,
        50432, 50455, 50469, 50536, 50985, 51359, B51531, 51674, 51731, 51779,
        52009, 52264

Position of leaf on plant, insertion level, and chloroplast   48546, 49411, 49538,
        50536, 51538, 51641

Production modelling   see   Model ...

Production of dry matter   see   Biomass distribution ...; Ecosystem production ...

Proteins, amino acids, nucleic acids, and algae productivity   48772, 50937

Proteins, amino acids, nucleic acids, and carbon fixation pathways
        49220, 49222, 49225, 49401, 50430, 50702, 50750, 51031, 51149, 51374, 51605,
        52010

Proteins, amino acids, nucleic acids, and chlorophyll
        49771, 49983, 50072, 50079, 50573, 50702, 50750, 51051, 51586, 51868

Proteins, amino acids, nucleic acids, and chloroplast (chromatophore)
        48419, 48423, 48445, B48448, 48501-2, 48525-6, 48536, 48557, 48615-6, 48619,
        48628, 48653, 48668, 48707, 48750, 48761, 48766, 48824, 48831, 48835, 48877-
        -8, 48890-1, 48924-5, 48954, 49003, 49027, 49032, 49035, 49137, 49153,
        49170, 49209, 49212, 49240, 49405, 49412, B49422, 49437, 49541, 49557,
        49595, 49634, 49643, 49646, 49659, 49671, 49682-3, 49703, 49745,
        49771, 49812, 49850, 49867-8, 49891, 49929, 49943, 49996, 50033, 50071-2,
        50114, 50147, 50150, 50159, 50181, 50216, 50309-10, 50313, 50315, 50319,
        50371, 50381, 50410, 50414, 50521, 50590, 50627, 50635, 50695, 50707, 50715,
        50726, 50737, 50748-9, 50777, 50798, 50802, 50814, 50828-9, 50832, 50926,
        50951, 51011-2, 51025-7, 51029, 51045, 51047-8, 51050, 51103, 51115, B51168,
        51173, 51200, 51241-2, 51244, 51272, 51290, 51308-10, 51332, 51365, 51384,
        51401, 51426, 51435, 51443, 51501, 51506-7, 51538, 51553, 51605, 51615,
        51746-7, 51766, 51807, 51877, 51923-4, 51932, 51938, 51945, 51971-2, 52007,
        52010-1, 52037, 52062, 52111, 52161, 52190, 52202, 52220-1, 52234, 52243,
        52340, 52379

Proteins, amino acids, nucleic acids, and electron transport chain
        49143, 49643, 49771, 50186, 50590, 50804, 50829, 51190, 51615, 51804, 51967

Proteins, amino acids, nucleic acids, and gas exchange   48714, 48923, 50378, 50900

Proteins, amino acids, nucleic acids, and photorespiration   50354

Proteins, amino acids, nucleic acids, and respiration   52255

Protochlorophyll(ide)   see   Chlorophyll biosynthesis ...

Proton transport in chloroplast   see   Chemiosmotic hypothesis ...

Protoplasts, isolated   see   Tissue cultures ...

Pteridines   see   Ferredoxin-NADP reductase ...

Pyrenoid   50381, 51281-2

Q

Quantum yield and requirement
        48496, 48571, 48588, 48595, 48693, 48785, 48802-3, 48881, 48892, 48920,
        48922, 49060, 49155, 49172, 49224, 49435, 49463, 49465, 49518, 49544,
        49638, 49701, 49722, 49766, 49894, 49931-2, 50015, 50038, 50280, 50443,
        50456, 50606, 50641, 50650, 50754, 50756, 50866, 50877, 51074,
        51166, B51168, 51181, 51237, 51247, 51276-7, 51330, 51410, 51454, 51533,
        51564, 51651, 51760, 51987, 52055, 52354, 52387

Quantum yield and requirement, methods   see   Quantum yield and requirement

Quinones in photosynthesis
        48518, 48532-3, 48544, 48606, 48677, 48680, 48715, 48768, 48784, 48789,
        48831, 49026, 49074, 49078, 49130, 49157, 49211, 49305, 49342, 49397, 49563,
        49582, 49589, 49663, 49685, 49687, 49722, B49749, 49764, 49848-9, 49892,
        49914-5, 50032, 50036, 50104, 50106, 50108, 50110, 50118, 50268, 50321,
        50336, 50362, 50435, 50439-40, 50471, 50560, 50562-3, 50598, 50624, 50628,
        50672, 50683, 50695-6, 50775, 50867, B50878, 50908, 50914, 50926, 50929-31,
        50954, 50956, 50984, 51015, 51041, 51085, 51141, B51168, 51177, 51208,
        51346, 51431-2, 51454-5, 51519, 51609, 51630, 51651, 51669, 51747, 51799,
        51805, 51848, 52008, 52041, 52071, 52089-90, 52116, 52135, 52289-90, B52321,
        52349

Quinones, methods   51848

R

Radiation in canopy see Canopy, radiation ...

Radiation, light see Irradiance...; Model of radiation ...

Rain, precipitation, methods see Aerodynamic methods, bioclimatological methods ...

Reaction centres see $P680$; $P700$ ...

Recycling of CO_2 inside the cell and leaf   51563

Relative growth rate see Growth analysis, net assimilation rate ...

Relative water content see Water saturation deficit

Resistance, carboxylation and excitation   48945, 49150, 49369, 49618, 50511, 51136,
        51946, 51987

Resistance, cuticular   51481, 51704

Resistance, epidermal, leaf resistance
        48475, 48491, 48528, 48624, 48627, 48736, 48775, 48813, 48815, 48940, 48975,
        48977, 49020, 49038, 49051, 49077, 49142, 49162-3, 49210, 49281-2,
        49367, 49387, 49444, 49478, B49527, 49551, 49571, 49606-7, 49612, 49638-9, 49697,
        49704, 49723, 49727, 49747, 49881, 49922, 49927, 50021, 50095, 50165,
        50167, 50175, 50193, 50224, 50241, 50266, 50291, 50304, 50327, 50331,
        50348, 50372, 50417, 50454, 50494, 50507, 50548, 50594, 50610, 50642,
        50666, 50674-5, 50764, 50821, 50873, 51002, 51030, 51038, 51061, 51086,
        51133, 51136, 51153, 51160, 51212, 51228, 51280, 51296, 51376-8, 51392,
        51568, 51570, 51787, 51816, 51822, 51865, 51881-3, B51897, 51905-6, 51947,
        51954, 51956, 52003, 52065, 52124, 52284, 52291, 52372, 52375

Resistance, intercellular see Resistance, stomatal ...; Resistance, epidermal ...

Resistance, intracellular (mesophyll)
        48692, 48697, 48809, 48826, 48883-4, 48945, 48958, 49038, 49076-7, 49150,
        49162, 49163, 49241, 49369, 49469, 49605, 49606, 49618, 49704, 49717, 49766,
        49785, 49818, 49881, 50096-7, 50183, 50241, 50485, 50604, 50747, 50905,
        51030, 51061, 51136, 51181, 51186, 51246-9, 51497, 51570, 51815, B51897,
        51936, 51946, 51956, 51983, 51987, 52155, 52277

Resistance, leaf boundary layer
        48733, 48838, 48926, 48958, 49147, 49369, B49527, 49535, 49618, 49638,
        49717, 49723-4, 49881, 50485, 50969, 51253, 51675, 51946, 51987, 52020,
        52227

Resistance, mesophyll see Resistance, intracellular (mesophyll)

Resistance, stomatal (and intracellular) (*cf*. also Resistance, epidermal ...)
    48424, 48475, 48511, 48552-3, 48556, 48563, 48590-1, 48627, 48664,
    48679, 48692, 48697-8, 48733, 48775, 48813, 48826, 48838, 48857, 48883-4,
    48926, 48945, 48957-8, 48975, 48977, 49038, 49051, 49076-7, 49142, 49145,
    49147, 49210, 49241, 49249, 49281, 49341, 49363, 49369, 49398-9, 49444,
    49462, 49478, 49501, B49527, 49535, 49606-7, 49618-9, B49621, 49627, 49697,
    49702, 49704, 49708, 49713, 49717, 49719, 49723, 49726, 49785, 49797-8,
    49818, 49854, 49881, 49898, 49922, 49993, 50008, 50021, 50078, 50096-7,
    50165, 50175, 50183, 50241, 50265-6, 50291, 50296, 50304, 50366, 50417,
    50456, 50485, 50494, 50510, 50594, 50604, 50610, 50642, 50662, 50674-5,
    50755, 50783, 50821, 50843, 50862, 50873, 50905, 50969, 50985, 51002, 51014,
    51057, 51061, 51119, 51136, 51166, 51181, 51186, 51204, 51212, 51228, 51246-
    -8, 51280, 51296, 51333, 51378, 51392, 51444, 51497-8, 51570, 51596, 51642,
    51692, 51787, 51815-7, 51865-6, 51881-3, 51897-8, 51906, 51936, 51946-7,
    51956, 51983, 51987, 52003, 52020, 52060, 52065, 52124, 52154-5,
    52248, 52260, 52277, 52284, 52291, 52297, 52335, 52352

Resistances to CO_2 and water-vapour transfer at canopy level
    48736, 48838, 49469, 49785, 49881, 50843, 51296, 51720, B51897

Resistances to CO_2 transfer (*cf*. also Resistance, intracellular ...)
    48552, 48553, 48679, 48692, 48698, 49038, 49229, 49297, B49527, 49606,
    49638-9, 49854, 49881, 50241, 50454, 50861-2, 51038, 51061, 51227-8, 51418

Resistances to water vapour transfer see Resistance, epidermal, leaf resistance

Respiration and photosynthesis
    48506, 48541, 48578, 48649, 48680-1, 48693, 48702, 48717, 48731, 48768,
    48827, 48843, 48879, 48923, 48926, 48978, 49037, 49055-6, 49058, 49093,
    49148, 49150, 49215, 49228, 49239, 49297, 49331, 49404, 49416-7, 49423,
    49501, 49514, 49534, 49556, 49673, 49709, 49712, 49938, 49957, 50101,
    50115, 50180, 50234, 50273-4, 50281, 50416, 50418, 50478-9, 50489, 50493,
    50613, 50631, 50652, 50815, 50817, 50962, 50971, 50995-6, 51052, 51062,
    51085, B51168, 51169, 51301, 51457-8, 51527-8, 51558, 51587, 51681, 51688,
    51785, 51889, 51893, 51917, 51962, 52040, 52093, 52180, 52209, 52216,
    52227, 52254, 52346, 52367, 52383

Respiration, dark CO_2 efflux
    48435, 48505, 48548, 48632, 48718, 48742, 48800, 48809, 48923, 48957-8,
    49029, 49052, 49075, 49118, 49150, 49165, 49188, 49254, 49293, 49357-8,
    49368, 49403, 49417, 49424, 49471, 49491, 49513, 49562, 49572,
    49602, 49606, 49612, 49747, 49836, 50015, 50037, 50078, 50132, 50195,
    50266, 50328, 50347, 50372, 50396, 50406, 50479, 50491, 50594, 50651,
    50760, 50779, 50786, 50800, 50863, 50887, 50889, 50903, 51042, 51089,
    51094, 51186, 51246, 51249, 51306-7, 51312, 51441, 51447, 51527-8, B51531,
    51540, 51689, 51698, 51735-6, 51772, 51826, 51837, B51897, 51905,
    51921, 51935, 51962, 52040, 52078, 52224, 52250, 52255, 52277

Respiration, growth and maintenance
    48973, 50338, 50479, 50651-3, 50689, 50758, 50779, 50815, 51042, 51306,
    51826, 52255

Respiration of achlorophyllous tissues in light, light inhibition of respiration
    49366

Ribosome of chloroplast
    48468, 48547, 48616, 48628, 48791, 48876, 48924-5, 48928, 49136, 49153,
    49234, 49380, 49405, 49559, 49682, 49703, 49887, 49891, 50187, 50381,
    50410, 50694, 50704, 50963, 51970, 52011

Ribulose 1,5-bisphosphate carboxylase
          48414, 48441, 48499, 48508-9, 48515, 48519, 48539, 48542, 48549, 48569,
          48572, 48621-3, 48626, 48653, 48657, 48663, 48669-70, 48685-6, 48693,
          48704-5, 48722, 48761, 48766, 48768, 48788, 48807, 48811, 48826, 48877,
          48917-9, 48945, 48958, 48960, 48973, 48979, 48997, 49034, 49111, B49114,
          49153-4, 49214, 49249, 49259, 49322, 49334, 49361, 49366, 49396, 49401,
          49427-8, 49446, 49454, 49469, 49501, 49528, 49546, B49561, 49595, 49605,
          49619, 49632, 49648, 49673, 49680, 49701, 49704, 49710, 49734, 49782,
          49801, 49805, 49817, 49854, 49906-7, 49916, 49962, 49967, 49982, 50006,
          50033, 50041, 50043, 50049, 50126, 50137, 50140, 50142-3, 50228-9, 50258-9,
          50290, 50341, 50377, 50412, 50429, 50432, 50492, 50501, 50535, 50538,
          50542, 50645, B50687, 50702, 50723-4, 50739, 50743, 50750, 50755, 50836,
          50839, 50847-8, 50850, 50853, 50870, 50894-5, 50911, 50938, 50951, 50999,
          51004-5, 51025, 51047-8, 51054, 51071, 51076, 51103-4, 51122, 51149-
          -53, 51167, B51168, 51197, 51218, 51222, 51234-5, 51266, 51290, 51324,
          51332, 51357, 51360, 51380, 51382, 51437, 51447, 51451, 51458, 51464, 51475-
          -6, 51503, 51510, 51515, 51517, 51526, 51577, 51605-6, 51618, 51626, 51631,
          51678, 51697, 51716, 51718, 51732, 51762, 51783, 51790-1, 51870, 51873,
          B51897, 51923, 51958, 51972, 51975, 52009-10, 52036, 52065, 52073, 52076,
          52083, 52095, 52118-9, 52139-40, 52147, 52218, 52220, 52227, 52243, 52245,
          52247, 52258, 52263-4, 52266-7, 52271-2, 52276, 52292, 52307, B52321,
          52371

Ribulose 1,5-bisphosphate carboxylase, methods
          48670, 48686, 50006, 50389, 50492, 50538, 50743, 50894, 51076, 51164, 51197,
          51437, 51790-1, 52119

Ribulose 1,5-bisphosphate oxygenase see Ribulose 1,5-bisphosphate carboxylase ...;
          Photorespiration, enzymes ...

Root removal see Defoliation, decapitation, ear and root removal ...

Root, underground part, and carbon fixation pathways   48539, 50626, 51154

Root, underground part, and chlorophyll   51867

Root, underground part, and gas exchange   48540

Root, underground part, and plant production
          48548, 48683, 48687, 48775, 48893, 49112, 49310, 49345, 49352, 49385, 49523,
          49617, 49704, 49785, 49845, 49979, 50021, 50175, 50241, 50286, 50300, 50307,
          50636, 50658, 50757, 50915, 51157, 51204, 51412, B51897, 52003-4, 52100,
          52123, 52154

Root, underground part, and respiration   50078, 50338, 51673, 51826, 51867, 51921,
          52181

Rooted leaves, gas exchange in   51441

Rooted leaves, respiration in   51441

RuBP carboxylase, RuBPC see Ribulose 1,5-bisphosphate carboxylase

Rubredoxin see Ferredoxin ...

S

Saccharides and carbon fixation pathways
          48646, 48857, 49230, 49534, 49792, 49900, 49965, 50090, 50428, 50430,
          50490, 50837-9, 51004, 51031, 51170, 51561, 51563, 52116, 52301, 52319

Saccharides and chlorophyll   51526

Saccharides and chloroplast (chromatophore)   49301, 50386, 51404, 51526, 52347

Saccharides and electron transport chain   51402, 51404, 51526

Saccharides and gas exchange   49450, 50368, 51009, 51526, 51560

Saccharides and plant productivity   49523, 49619, 49951, 49980, 50747, B51897

Saccharides,and resistances to CO_2 and water vapour transfer   50747

Salinity of soil, and canopy functioning   48468

Salinity of soil, and carbon fixation pathways   49365, 49446, 49475, 49765, 49817,
        50056, 50073, 50999, 51205, 51618

Salinity of soil, and carotenoids   48672, 48710, 50478

Salinity of soil, and chlorophyll   48439, 48673, 48675, 48710, 49067, 49085,
        49215, 49446, 49517, 50057, 50478, 51583, 51674

Salinity of soil, and chloroplast (chromatophore)   50056, 50123, 52261

Salinity of soil, and ecosystem and plant productivity
        48439, B48701, 48747, 49020, 49067, 49091, 49193, 49267, 49517, 49523, 49704,
        49765, 50123, 50242, B50687, 50731, 51205, B51531, 51570, 51618, 51694,
        B51897

Salinity of soil, and electron transport chain   49084-5, 49517, 49562, 51578, 51674,
        51756, 51938

Salinity of soil, and gas exchange
        48415, 48439, 48673, 48692, 48745, 48854, 49063, 49148, 49215, 49267, 49475,
        49562, 50023, 50242, 50478, 50588, 50634, 50847, 50999, 51104, 51228, 51570,
        51618, 51815, 52155,

Salinity of soil, and photorespiration   48415, 50847, 51104

Salinity of soil, and resistances to CO_2 and water vapour transfer   50242, 51228,
        51392, 51570, 51815, 52155

Salinity of soil, and respiration   49562, 51618

Salinity of soil and water, and algae productivity
        48673, 48675, 48745, 49195, 49260, 49799, 50022, 50605, 50730-1, 51285,
        51624

Samples for pigment determination see Pigments ...

Seasonal changes see also Ontogeny of plant ...

Seasonal changes in algae productivity
        48470-1, 48541, 48550, 48562, 48567, 48592, 48629, 48641, 48643, 48712,
        48728, 48769-70, 48842, 48866, 48971, 48996, 49001, 49070, 49116, 49140,
        49197-8, 49269, 49311, 49317. 49338, 49350-1, 49416, 49425, 49445, 49485,
        49487-8, 49518-9, 49525, 49647, 49679, 49784, 49786, 49795, 49802, 49823-4,
        49874-6, 49879-80, 49933, 50022, 50047, 50064, 50077, 50171, 50180, 50203,
        50218-9, 50238-40, 50264, 50308, 50322, 50342, 50355, 50369, 50409, 50457-8,
        50544, 50566, 50605, 50972, 50988, 50992, 51060, 51069, 51124, 51129, 51143,
        51184, 51194, 51215, 51286-7, 51313, B51350, 51363, 51438, 51587, 51623-4,
        51699, 51742-3, 51862-3, 51914, 51916, 51973, 52001, 52019, 52042, 52044,
        52063, 52102, 52192, 52217, 52286, 52358

Soil moisture and carotenoids   48977, 49904

Soil moisture and chlorophyll   49711, 49904, 50105, 51380, 51834, 52064, 52077

Soil moisture and chloroplast (chromatophore)   51322, 51641

Soil moisture and ecosystem and plant productivity
    48473, 48548, 48564, 48573, 48596, 48721, 48829, 48884, 48977, 49042, 49142,
    49231, 49335, 49388, 49424, 49523, 49549, 49704, 49715, 49785, 49810, 49951,
    50021, 50105, 50165, 50175, 50241, 50272, 50574, 50586, 50636, 50906, 50936,
    50999, 51049, 51091, 51126, 51204, 51448, 51496, 51570, 51619, 51646, 51787,
    51834, 51866, 51895, B51897, 51909, B51961, 52003, 52017, B52018, 52154,
    52164, 52174, 52372

Soil moisture and electron transport chain   51322

Soil moisture and gas exchange
    48472-3, 48564, 48692, 48721, 48776, 48884, 49042, 49375, B49621, 49727,
    49934, 50105, 50182, 50241, 50262, 50266, 50610, 50662, 51075, 51380, 51497,
    51774, 51866, B51897, 51907, 52087, 52344

Soil moisture, and resistances to CO_2 and water vapour transfer
    48627, 48884, 49444, 49478, B49621, 49704, 50175, 50241, 50764, 51444,
    51497, 51866, B51897, 52003, 52064

Soil moisture and respiration   49010

Solar radiation and canopy see Canopy, radiation distribution ...; Canopy, radiat-
    ion profile ...

Specific leaf area see Growth analysis, specific leaf area ...

Spectral methods in photosynthesis research 48602, 48816, 49117, 49251, 49484, 49531,
    49533, 49685, 49750, 49778, 49849, 50084, 50189, 50227, 51015, 51796, 51912

Stabilization of photosystems see Photosystems stabilization ...

Stand see Canopy ...; Ecosystem ...

Steady state and non-steady state see Oscillations ...

Stem, petiole, morphology, structure and physiological activity in 50555

Stomata morphology and anatomy (number, dimensions, types, development, structure,
    etc.) (cf. also Leaf epidermis, stomata)
    48464, 48528, 48556, 48563, 48746, 48768, 48782, 48958, 48967, 49132, 49155,
    49355, 49473, 49478, B49527, 49627, 49638, 49666, 49708, 49763, 49818, 50037,
    50224, 50494, 50510, 50783, 51038, 51054, 51133, 51153, 51204, 51524, B51531,
    51725, 51787, 51933, B52018, 52353

Stomata physiology (mechanism of action, reactivity, etc.) (cf. also Leaf epidermis,
    stomata)
    48424, 48664, 48733, 48818, 48856, 48975, 49038, 49064, 49077, 49145, 49230,
    49369, 49387, 49391, 49510, 49523, 49710, 49763, 49785, 49798, 50021, 50165,
    50538, 50610, 50683, 50688, 50935, 51004, 51043, 51119, 51175, 51233, 51476,
    51592, 51720, 51787, B5189?, 52149, 52154, 52182, 52224, 52352-3,
    52375

Stomata role in photosynthesis
    48591, 48692, 48698, 48733, 48746, 48814, 48940, 49020, 49038, 49077, 49369,
    49398, B49527, 49713, 49719, 49727, 49818, 50021, 50126, 50175, 50241,
    50266, 50291, 50304, 50610, 50662, 50674, 50706, 50783, 50800, 50827, 50999,
    51024, 51119, 51204, 51497, 51652, 51815-6, B51897, 52182, 52372, 52375

Stomatal diffusive resistance see Resistance, stomatal ...

Stroma of chloroplast
        48478, 48674, 49093, 49366, 49397, 49558, 49659, 49709, 50033, 50391, 50520,
        50814, 51148, 51305, 51434, 51451, 51494, 51768, 51772, 51801, 52064, 52257

Sulphur oxides (and other sulphur compounds) see Pollution of air ...

Sunflecks in canopy see Canopy, radiation distribution ...

T

Taxons, algae productivity
        48541, 48592, B48671, 49149, 49339, 50467, 50934, 51129, 51215, 51914

Taxons, biliproteins in
        48846, 48867, 49169, 49577, 50777, 51460, 51529, 51643, 51789, 51859, 51971-
        -2, 52317

Taxons, canopy functioning of   51942

Taxons, carbon fixation pathways
        48657, 48861-2, 48917, 48950-1, 48953, 49034, 49064, 49337, 49758, 49807,
        49965-6, 49988, 50006, 50116, 50229, 50279, 50431, 50571-2, 50609, 50663,
        50744, 50752, 50797, 50819, 50836, 50850, 50894, 50940, 50960, 50999,
        51005, 51025, 51054, 51076, 51197, 51218, 51261, 51274, 51289, 51389, 51491,
        51515, 51625, 51762, 51997, 52010, 52083, 52112, 52220, 52243

Taxons, carotenoids in
        48433, 48672, 48674, 48855, 48868, 48899, 49280, 49336, 49380, 49458, 49580,
        49609, 50163, 50439, 50505, 50515, 50762, 50957, 51413, 51472, 51593, 51643,
        51658, B51788, 51820, 51952, 51972, 52000, 52130, 52305, 52317

Taxons, chlorophyll in
        48433, 48496, 48525, 48527, 48535, 48657, 48660-1, 48674, 48752, 48755,
        48941, 48972, 49235, 49248, 49265, 49280, 49330, 49336, 49373, 49380, 49458,
        49473, 49498, 49514, 49600, 49609, 49611, 49674, 49688, 49820, 49946, 49990,
        49997, 50043, 50138, B50144, 50163, 50296, 50317, 50348, 50439, 50505,
        50515, 50517, 50595, 50608, 50752, 50762, 50934, 50957, 51054, 51135, 51413,
        51460, 51522, 51643, 51658, 51672, 51712, B51788, 51799, 51820, 51900,
        51942, 51952, 51972, 51997, 52000, 52035, 52059, 52077, 52083, 52130, 52146,
        52227, 52244, 52305, 52317

Taxons, chloroplast (chromatophore) in
        48535, 48761, 48890, 49068, 49473, 49611, 49850, 49929, 50174, 50303, 50381,
        50521, 50773, 50793, 50939, 51026, 51028-9, 51272, 51322, 51364-5, 51396,
        51413, 51460, 51636, 52083, 52202, 52209, 52244

Taxons, ecosystem and plant productivity of
        48452, 48483, 48527, 48737, 48752, 48755, 48893, 48902, 48908, 49297, 49473,
        49514, 49571, 49820, 50103, 50139, 50300, 50348, 50367, 50554, 50591, 50638,
        50796, 50888, 50999, 51001, 51058, 51109-12, 51495, 51565, 51656, 51658,
        51688, 51753, 52000, 52099, 52138, 52346

Taxons, electron transport chain in
        48496, 48597, 48625, 48694, 48764, 48767, 48795, 48859, 48902, 49084, 49524,
        49698, 49742, 49764, 49820, 49830, 49919, 49997, 50043-4, 50271, 50439,
        50595, 50799, 51025, 51118, 51237, 51322, 51628, 51752, 51800, 51831, 51931,
        51986, 51995-7, 52289

Taxons, gas exchange in
       48436, 48441, 48541, 48590, 48629, 48649, 48751-2, 48857, 48902, 49252,
       49257, 49327, 49399, 49447, 49483, 49501, 49514, 49520, 49571, 49591,
       50043, 50049, B50144, 50273-4, 50296, 50298, 50331, 50348, 50588, 50609,
       50664, 50850, 50999, 51109-12, 51142, 51247, 51319, 51495-6, 51515, 51522,
       51534, 51642, 51682, 51688, 51831, 51917, 51942, 51956, 51971, 52165, 52227,
       52252, 52346, 52370

Taxons, photorespiration in   49064, 50938, 51054, 51246, 51457-8

Taxons, resistances to CO_2 and water vapour transfer   48590, 49399, 49571, 49628,
       50095, 50348, 51002, 51227, 51246-7, 51642, 51956

Taxons, respiration in   51246, 51457-8, 51496, 51688, 52346

Temperature see also Model of temperature ...

Temperature, high, and carbon fixation pathways   49648, 50838-9

Temperature, high, and carotenoids   49380

Temperature, high, and chlorophyll   48558, 49248, 49380, 49477, 49648, 49779,
       49814, 50330, 50756, 51080, 51216, 51252

Temperature, high, and chloroplast (chromatophore)
       48558, 49858, 50655, 50912, 51216, 51345, 51403, B52018

Temperature, high, and ecosystem and plant productivity   50597, 50655

Temperature, high, and electron transport chain
       48510, 49248, 49648, 50186, 50201, 50473, 50530, 50912, 51034, 51216, 51756,
       51761, 51917, 52336

Temperature, high, and gas exchange   48703, 48827, 50248, 50755, 50839, 51111-2,
       51917, 52205-6, 52227

Temperature, high, and respiration   51917

Temperature, leaf see Leaf temperature ...

Temperature, low, and algae productivity   49735

Temperature, low, and carbon fixation pathways   48649, 48657, 49906-7, 50082,
       51731

Temperature, low, and carotenoids   49570

Temperature, low, and chlorophyll   49381, 49477, 49503, 49570, 50528, 51731,
       51800, 51811, 52049

Temperature, low, and chloroplast (chromatophore)   49292, 49646, 50082, 50380,
       50655, 50707, 50982, 51322, 51403-4, 51538, B52018

Temperature, low, and ecosystem and plant productivity   50082, 50655, 52051, 52350

Temperature, low, and electron transport chain   48693, 49066, 49129, 51322, 51402,
       51404, 51800

Temperature, low, and gas exchange   48649, 48693, 48703, 49591, 50664, 50705-6,
       50827, 50864, 51086, 51111-2, 51497, 51522, 51701, B52018, 52227, 52350

Temperature, low, and photorespiration   49906

Temperature, low, and resistances to CO_2 and water vapour transfer   51497

                                                                                    (continued)

Thylakoid, granum (continued)
    51216, 51264, 51273, 51275, 51278, 51303, 51322, 51329-30, 51344, 51364,
    51379, 51381, 51413, 51451, 51471, 51478, 51483, 51487-8, 51494, 51509,
    51516, 51538, 51558, 51635, 51641, 51725, 51747, 51770, 51772, 51801, 51831,
    51853, 51901, 51924, 51943, 51967, 51977-8, 52006, 52062, 52064, 52111,
    52133, 52135, 52149, 52226, 52236, 52253, 52257, 52267, 52347, 52363, 52365

Thylakoid, localization of electron transport chain in see Electron transport
    chain, localization in thylakoid

Thylakoid polypeptides see Chloroplast proteins...; Proteins ...

Tissue cultures, carbon fixation pathways in   49103, 49782, 49916, 50046, 50750,
    50853, 50870, 51381, 51517, 52010, 52307

Tissue cultures, carotenoids in   48636, 49243, 51067, 51516

Tissue cultures, chlorophyll in   48636, 49103, 49122, 49165, 49243, 49916, 49923,
    50284, 50750, 51067, 51290, 51516, 52014

Tissue cultures, chloroplast in   49165, 50284, 51192, 51241, 51361, 51381, 51516,
    52223

Tissue cultures, electron transport chain in   49165, 49782, 49830, 51516

Tissue cultures, gas exchange in 49165, 49376, 49782, 49916, 49923, 50853, 51067,
    51517, 51667, 52307, 52370

Tissue cultures, growth of   48636, 49165, 49916

Tissue cultures, respiration in   49165

Transient phenomena in electron transport chain   48431, 49237, 50918

Transient phenomena in gas exchange   49658, 50169, 52354

Transpiration and photosynthesis
    48930, 49038, 49071, 49077, 49163, 49398, B49621, 49689, 49785, 50167,
    50182, 50304, 50332, 50348, 50456, 50548, 50638, 50666, 50674, B50687,
    50785, 50817, 50999, 51030, 51049, 51061, 51073, 51136, 51186, 51496-7,
    51570, 51642, 51657, B51897, 51905, 51956, 51983, 52003-4, B52018

Tritium oxide see Deuterium oxide, tritium oxide ...

U

Ubiquinones see Quinones in photosynthesis

Ultraviolet radiation see Irradiance, spectral composition ...

Uncouplers of electron transport chain (*cf.* also Antibiotics and electron transport
    chain)
    48430, 48474, 48533, 48580-1, 48595, 48795, 48865, 48892, 48898, 48934,
    49066, 49124, 49143, 49212, B49422, 49507, 49551, 49593, 49649, 49722,
    B49749, 49842, 49902, 49915, 50141, 50232, 50276, 50534, 50562, 50644,
    50770-1, 50865, B50878, 50927, 51035, 51166, 51301, 51429, 51467, 51489,
    51669, 51768, 51781, 51918-9, 52022, 52045, 52069, 52093, 52098, 52193-4,
    52205, 52315

V

Virus diseases   see   Phytopathological effects ...

Vitamin K$_3$ see   Quinones ...

Volume changes in chloroplast (chromatophore)   see   Chloroplast and chromatophore volume changes

Volume changes in leaf and other organs   48693, 50049, 50241, 51823, 52004

Volume of plant organs, measurement   see   Leaf volume ...

W

Warburg effect   see   O$_2$ and gas exchange

Water, heavy   see   Deuterium oxide, tritium oxide ...

Water saturation deficit
        48775, 48780, 49142, 49282, 49444, 49475, 49574, 49704, 49708, 50021, 50165,
        50175, 50241, 50674, B50687, 50760, 51153, B51531, 51570, 51596, 51720,
        51787, 51867-8, B51897

Water splitting mechanism   see   O$_2$ evolution mechanism and kinetic

Wind (air-flow rate) and canopy functioning   50936

Wind (air-flow rate), and ecosystem and plant productivity   49627, 50936, 51133

Wind (air-flow rate) and gas exchange   51169

Wind (air-flow rate), and resistances to CO$_2$ and water vapour transfer   49462,
        49627, 49723, 51133, 52020

Wind (air-flow rate) and respiration   51169

Wind measurement   see   Aerodynamic methods ...

X

Xanthophylls
        48410, 48654, 48674, 48727, B48841, 48855, 48964, 49006, 49108, 49117,
        49139, 49141, 49156, 49218, 49242, 49245, 49333, 49336, 49340, 49426, 49552,
        49570, 49655-7, 49664, 49986, 50032, 50162, 50278, 50398, 50440, 50444,
        50536, 50556, 50643, 50717, 50901, 51148, 51172, 51207, 51288, 51292, 51316,
        51516, 51676, 51838, 52000, 52033, 52075, 52130, 52211, 52227, 52274, 52331

Xanthophylls of algae
        48410, 48513, 48568, 48626, B48841, 48855, 48874, 49069, 49108, 49117,
        49128, 49141, 49280, 49283, 49325, 49333, 49336, 49552, 49752-4, 50477,
        50504-5, 50643, 50667, 50717, 50803, 50901, 50957, 51036, 51643, 51700,
        51743, B51788, 51818, 51972, 52130, 52227, 52244-5, 52274, 52378

Xanthophylls of photosynthetic bacteria
        48569, B48841, 48938, 49061, 50505, 50524-5, 50539, 50602, 50922, 51268,
        51388, 51472, B51788, 51810, 51835, 52274

Xerophytes   see   Drought ...; Temperature, high ...

X-ray diffraction determination   49051

X-rays   see   Ionizing radiation ...

Y

Yield   see   Ecosystem production, primary productivity (terrestrial); Biomass ...

Yield formation   see   Biomass distribution...; Photosynthate translocation ...

Z

Zeaxanthin   see   Carotenoids...; Xanthophylls

   This index contains a selection of plant genera and types interesting as
experimental material for physiological, ecological and agricultural studies. Latin
scientific names of plant genera and English names of plant groups and types are
the main items which present the reference numbers.

A

Abelmoschus see *Hibiscus*

Abies 48456-7, 48718, 48729, 48945, 49385, B49527, 49549, 49694, 50095-6, B50144,
       50241, 50272, 50631, 51086, 51087, 51658, 51933, 52086, 52123, 52138, 52182-
       -3, 52375

Abutilon 49478, 49763, B49921, 50546

Acacia 48692, B49691, 49990, 50241, 51227, 51498, 51507, 51570

Acer   48416, 48632, 48742, 48755, B49114, 49287, 49385, 49502, B49527, 49549,
       49717, 49844, B49921, 50003, B50144, 50241-2, 50439, 50507, 50534, 50611,
       50622, 50766, 51573, 51658, B51788, 51933, 51946, 51954, 52000, 52138, 52204

Acetabularia 48766, 48788, 48877, 49267, 49634-6, 49886, 50060, 50381, 50521-2,
       51503-4, 51971, 52047, 52161

Acorus 49599, 50206

Actinidia 49656

Aesculus 48632, 51658, B51788, 52138

Agathis 50537

Agave   48664, 49038, 50609, 51223, 51678, 51725

Agrostis 49091, 49297, 49496, 49500, 50226, 51933

Alder see *Alnus*

Alfalfa see *Medicago*

Algae (*cf.* also *Acetabularia*, A. blue-green, A.brown, A.green, A.red, *Anabaena*,
       Anacystis, *Ankistrodesmus*, *Chlamydomonas*, *Chlorella*, *Chrysophyta*, Diatoms,
       Dinoflagellatae, *Dunaliella*, *Euglena*, *Fucus*, *Laminaria*, *Nostoc*, *Oscillatoria*,
       Porphyridium, *Scenedesmus*, *Ulva*)
       48410-2, 48470, 48484, 48492, 48513, 48542-4, 48554, 48567, 48641-3, 48667,
       48673, 48694, 48710, 48712, 48728, 48769-70, 48808, 48819, 48825, 48842,
       48849, 48866, 48886, 48923, 48971, 48981, 48996, 49001, 49059, 49070, 49086,
       49099, 49106, 49116, 49120, 49140-1, 49149, 49168, 49176, 49189, 49194-6,
       49211, 49267-9, 49283, 49288-91, 49311, 49316-7, 49336-9, 49350, 49351,
       49371, 49400, 49416, 49418, 49425, 49445, 49468, 49483, 49485-6, 49488,
       49504, 49518, 49525, 49552, 49584, 49604, 49628-30, 49647, 49679, 49681,
       49690, 49694, 49729, 49735, 49741, 49753, 49768, 49786, 49795-6, 49799-800,
       49824, 49874-6, 49879-80, 49893, 49933, 49948, 50014, 50048, 50064, 50077,
       50080, 50092, 50171, 50180, 50185, 50195, 50203, 50219, 50238-40, 50243,
       50253, 50264, 50339-40, 50373, 50375, 50388, 50394, 50409, 50418, 50428,
       50430, 50451, 50475, 50504-6, 50544, 50596, 50605, 50671, 50679, 50698,
       50732, 50744, 50751, 50772, 50805-6, 50829, 50864, 50899, 50938, 50978,
       50992, 51013, 51060, 51069, 51108, 51110-1, 51123-4, 51138-9, 51143, 51161,
       B51168, 51184, 51194, 51215, 51285, 51287, 51313, B51350, 51363, 51438,
                                                                    (continued)

·Algae (continued)
51452, 51466, 51477, 51484, 51505, 51547, 51587, 51637, 51640, 51643, B51662,
51664, 51680-1, 51685-7, 51699, 51713, 51740-3, 51757-8, 51811-2, 51844-5,
51860, 51862-3, 51903-4, 51911, 51913, 51916, ·51966, 51968, 51973, 51979,
51995-6, 52001, 52019, 52044, 52063, 52103, 52130, 52142, 52188, 52192,
52209, 52218, 52225, 52242, 52274, 52286, 52308, 52318, 52324, 52357, 52378

Algae, blue-green (*cf*. also *Anabaena, Anacystis, Nostoc, Oscillatoria*)
48411, 48441, 48484, 48499, 48518, 48529, 48543, 48550, 48562, 48571, 48588,
48620, 48625, 48648, B48650, 48651, B48671, 48678, 48717, 48723, 48728,
48756, 48761-2, 48764, 48766, 48772, 48786-7, 48796-7, 48821, 48849, 48867,
48877, 48972, 48991, 49013-4, 49034, 49078, 49106, 49116, 49141, 49151,
49169, 49195-7, 49266-7, 49280, 49337, 49417, 49423, 49445, 49447, 49474,
49512, 49565-6, 49573, 49575, 49577, 49628, 49679, 49681, 49685, 49692,
49695, 49699, 49742-3, 49888,   49919, 49933, 49947, 49950, 49996,
50110-2, 50131, 50205, 50238-40, 50255, 50271, 50298, 50303, 50341, 50351,
50365, 50369, 50394, 50403, 50411, 50473, 50588, 50595, 50605, 50644, 50741,
50744, 50777, 50805, 50814, 50831, 50874, 50879, 50886, 50938, 50943, 50954,
50997, 51074, 51110-1, 51129, 51142-3, B51168, 51170-1, 51239, 51260, 51293-
-4, 51302, B51350, 51396, 51399, 51406, 51440, 51452, 51457-8, 51460, 51469,
51477-8, 51529, 51533, 51547, 51564, 51580, 51590, 51597, 51636, 51643-4,
51682, 51686, 51699, 51742, 51785, B51788, 51789, 51971-2, 51995-6, 52001,
52031, 52044, 52096, 52103, 52130, 52202, 52274, 52317, 52358, 52364,
52378

Algae, brown (*cf*. also *Fucus, Laminaria*)
48411, 48500, 48541, 48565, 48849, 49069, 49141, 49149, 49277, 49289, 49358,
49474, 49591, 49623, 49628, 49784, 50089, 50298, 50310, 50429, 50467, 50588,
50605, 50613, 51109, 51112, B51168, 51452, 51460, 51469, 51547, 51643,
B51788, 52042, 52130, 52227, 52274

Algae, green (*cf*. also *Acetabularia, Ankistrodesmus, Chlamydomonas, Chlorella, Dunaliella, Scenedesmus, Ulva*)
48411, 48441, 48500, 48541, 48566, 48648-9, B48671, 48723, 48728, 48756,
48766-7, 48820, 48823, 48849, 48854, 48877, 48937, 48991, 48993, 49108,
49141, 49149, 49195-6, 49198, 49235, 49254, 49263, 49267, 49277, 49289,
49337, 49417, 49470, 49472, 49474, 49494, 49628, 49667, 49679, 49684,
49695, 49743, 49762, 49787, 49802, 49838, 49888, 49950, 50022, 50100,
50116, 50174, 50180, 50239-40, 50255, 50271, 50298, ·50351, 50369, 50381,
50394, 50451, 50467, 50481, 50549, 50566, 50596, 50605, 50643, 50803,
50805-6, 50824, 50871, 50932, 50957, 50972, 51109-12, 51129, 51143, B51168,
51171, 51183, 51281-2, 51293, B51350, 51425-8, 51430-1, 51439, 51452,
51469, 51477, 51533, 51547, 51580, 51585, 51624, 51643, 51686, 51699,
B51788, 51916, 51952, 52001, 52044, 52161, 52165, 52225, 52227, 52244-5,
52270, 52274, 52371, 52378, 52385

Algae, red (*cf*. also *Porphyridium*)
48411, 48541, 48625, 48849, 48855, 48858, 48867, 48941, 48991, 49106, 49141,
49148-9, 49263, 49267, 49277, 49337, 49390, 49417, 49423, 49440, 49452,
49470, 49474, 49628, 49681, 49695, 49865, 49924, 49937, 49965-6, 50174,
50230, 50255, 50271, 50298, 50369, 50381, 50467, 50605, 50681, 50744, 50777,
50806, 50885, 50952, 50986, 51074, 51109, 51112, B51168, 51251, 51319,
B51350, 51452, 51460, 51533-4, 51539, 51547, 51585, 51590, 51643, 51679,
B51788, 51857-9, 51977-8, 51995-6, 52130, 52227, 52236, 52274

Allium 48665, 48877, 49091, 50103, 50439, 50636, 51004, 51481, 51745, B51788,
52013

Almond see *Amygdalus*

Alnus 48720, 49385, 50574, 50766, 51492, 51502, 52204

Aloe 49579, 50571, 51202, 51678

Alopecurus   50915

Alpine plants   49555, 50516, 50843, 51091, 51658, 52169

Amaranthus   B48560, 48664-5, 48700, 48883, 48930, 49309, 49367, 49413, 49524,
       B49561, 49628, 49758-9, B49921, 49999, 50183, 50279, 50572, 50609, 50625,
       50650, 50752, 50836, 50904, 50928, 50933, 51005, 51053, 51153, 51211, 51214,
       51227, 51353, B51531, 51678, 51747, 51905, 51933, 51944, 52276, 52294

Amygdalus   48940, 51678

Anabaena   48441, 48474, 48484, 48499, 49506, 48613, 48625, 48678, 48723, 48764,
       48847, 48867, 48877, 49034, 49036-7, 49106, 49169, 49195, 49213, 49220,
       49266-7, 49371, 49447, 49472, 49573, 49628, 49679, 49681, 49699, 49764,
       49849, 49933, 49958, 50070, 50109, 50113, 50120, 50174, 50268, 50271, 50289,
       50298-9, 50303, 50427, 50589, 50595, 50645, 50793, 50813, 50826, 50874,
       50879, 50945, 51017, 51089, 51110-1, 51129, 51396, 51399, 51452, 51457-8,
       51477, 51529, 51544-5, 51580, 51699, 51759, B51788, 51789, 52044, 52189,
       52194, 52220, 52293, 52338, 52378

Anacystis   48441, 48500, 48647, 48723, 48845-6, 48877, 49031, 49034, 49106, 49134,
       49166, 49169, 49196, 49222-4, 49267, 49337, 49441, 49573, 49575, 49577,
       49628, 49669, 49671, 49681, 49695, 49950, 49965, 50044, 50089, 50197,
       50201, 50298, 50393, 50415, 50595, 50740-1, 50744, 50793, 50813, 50831,
       50879, 50887, 50945, 50954, 51084-5, 51110-1, 51132, 51170, 51229-31,
       51396, 51399, 51401, 51429, 51457-8, 51469-70, 51477, 51533-4, 51547,
       51742-3, B51788, 51789, 51931, 51971-2, 52189, 52193, 52240, 52311

Ananas   48729, 50911, 51678, 51933, 51997

Andropogon   50754

Anethum   B51788

Angelica   49297

Anise   see *Pimpinella*

Ankistrodesmus
       48484, B48650, 49108, 49679, 50298, 50323, 50354, 50451, 50497, 50805,
       50945, 51110-1, 51129, 51284, 51572, 51580, 51686, 52044, 52298, 52378

Anthoxanthum   49500

Anthriscus   50285, B51788

Antirrhinum   48761, 48766, 48877, 49820, 50089, 50719, 51507, 51933, 52161

Apium   49091, 50593

Apple   see *Malus*

Apricot   see *Armeniaca*

Aquatic macrophytes (*cf.* also *Elodea, Nuphar, Phragmites, Typha*)
       48541, 48605, 48629, 48656-7, 48839, 48886, 48948, 49091, 49109, 49149,
       49188, 49267, 49417, 49474, 49520, 49599, 49628, 49690, 49694, 49768, 49788,
       49823, 49941, 49982-3, 50115-6, 50302, 50342, 50370, 50423, 50457-8, 50663,
       50741, 50871, 50915, 50938, 50971, 51078, 51129, 51178-80, 51205, 51251,
       51356, 51395, 51481, 51547, 51678, 51680-1, 51712, 51928, 51950, 51981,
       51988, 52036, 52040, 52146, 52212, 52217, 52227, 52252

Arabidopsis   50719, 51695-7

Arachis 48411, 48775, 48826, 48829, 49475, B49691, 49900, 49956, 50015, 50073, 50976, 51024, 51225, 51409, 51492, 51516-7, 51794, B51897, 51933

Arbor vitae see *Thuja*

Armeniaca 51678, B51788, 51832, 52359

Arrhenatherum 49496, 50915

Artemisia 48596, 49287, 49549, 50103, 50242, 50636, 50697, 50915, 51043, 51565, 52054, 52346

Asclepias B49527

Ash see *Fraxinus*

Asparagus 48800, 49091, 49334, 49450, 50280, 51067, 51933

Aspen see *Populus*

Asperula see *Galium*

Astragalus 50382, 51091

Atriplex 48495, 48622, 48692-3, 48719, 48733, 48766, 48857, 48945, 49091, 49248-9,49530,B49561, 49686, 49758-9, 50089, 50123, 50280, 50387, 50572, 50609, 50697, 50940, 50999, 51104, 51216, 51497, 51565, 51570, 51678, B51788, 51987, 52054

Atropa 48766, 48939

Avena   48451, 48483, 48499, 48539, B48560, 48664, 48682, 48761, 48766, 48806, 48821, 48836, 48877, 48902, 48951, 48954, 49091, 49126, 49262, 49380, 49477, 49539, 49558, 49576, 49593, 49628, 49709, 49855, 49878, 49929, 49998-9, 50099, 50133, 50208, 50280, 50434, 50440, 50523, 50534-5, 50558, 50570, 50650, 50654, 50802, 50808, 50829, 50946-7, 50964, B51168, 51204, 51292, 51411, 51492, 51525, 51589, 51780, B51788, 51802-3, 51933, 51946, 52015, 52243, 52287, 52373, 52385

Avocado see *Persea*

B

Bacteria, photosynthetic (*cf.* also *Chlorobium, Chromatium, Ectothiorhodospira, Halo-bacterium, Rhodopseudomonas, Rhodospirillum*)
        48543, 48574, 48576, 48620, 48670, 48678, 48699, 48756, 48831, 48859, 48869, 48877, 48991, 49016, 49098, 49151, 49178, 49211, 49256, 49307, 49379, 49408--10, 49481, 49533, 49547, 49721, 49848, 50036, 50054, 50106, 50179-80, 50184, 50203, 50223, 50233, 50245, 50247, 50280, 50298, 50415, 50526, 50540, 50741, 50908, 50938, 50944, 50954, 51041, 51063, 51065, 51100-1, 51110-1, B51168, 51170-1, 51183, 51436, 51452, 51460, 51469, 51472, 51547, 51584, 51590, 51629, 51742-3, 51749, B51788, 51794, 51810, 51880, 51913, 51964, 51966, 51985, 52038, 52274, 52289, 52312

Balsam pear see *Sechium*

Bamboo see *Bambussa*

Bambussa 50538, 52074

Banana see *Musa*

Barley see *Hordeum*

Barnyard grass see *Echinochloa*

Barnyard millet see *Echinochloa*

Bay laurel see *Laurel*

Bean see *Phaseolus*

Beard-grass see *Andropogon*

Bedstraw see *Galium*

Beech see *Fagus*

Belladonna see *Atropa*

Bentgrass see *Agrostis*

Bermuda grass see *Cynodon*

Beta   48499, B48560, 48582, 48625, 48722, 48729, 48756, 48761, 48766, 48806, 48848, 48857, 48870, 48936, 49025, 49091, B49114, 49156, 49287, 49301, 49360, 49432, 49546, 49605-7, 49628, 49677, 49690, 49694, 49708, 49766, 49817, 49820, 49845, 49851-2, 49980, 50168, 50262-3, 50270, 50274, 50280, 50295, 50463, 50574, 50610, 50692, 50721, 50745, 50747, 50766, 50850, 50999, 51020, 51096, 51170, 51181, 51412, 51496, 51524, 51661, 51667, 51674, 51712, 51777, B51788, 51896, B51897, 51915, 51933, 52010, 52149, 52161, 52198, 52243, 52300-1, 52361

Betula   48457, 48523, 48839, 48879, 49297, 49385, B49527, 49702, 49730, 49844, B50144, 50241-2, 50766, 50905, 50984-5, 51144, 51193, 51481, 51496, 51547, 51672, B51788, 51882, 51933, 52027, 52068, 52138, 52343

Birch see *Betula*

Bird's-foot trefoil see *Lotus*

Bitter cress see *Cardamine*

Blackberry see *Rubus*

Black salsify see *Scorzonera*

Blueberry see *Vaccinium*

Bluegrass see *Poa*

Box see *Buxus*

Brassica   48499, 48508, 48676, 48700, B48701, 48722, 48729, 48733, 48868, 48892, 48902, 48959, 49015, 49031-2, 49078, 49091, 49135, 49202, 49208, 49265, 49287, 49347, 49628, 49666, 49694, 49766, 49976, 50043, 50207, 50280, 50439, 50574, 50635, 50650, 50664, 50980, 51005, 51080, 51417, 51444, 51496, 51525, 51589, 51593, 51657, 51678, 51701-2, 51733, 51933, 51953, B51961, 52010, 52013, 52035, 52243

Brinjal see *Solanum*

Bristlegrass see *Setaria*

Broccoli   see   *Brassica*

Bromegrass   see   *Bromus*

Bromus   48747, 49091, 49217, 49850, 50207, 50207-8, 50226, 50574, 50635, 51933

Brussels sprouts   see   *Brassica*

Bryophyllum   48732, B49114, 50911, 51223

Buckwheat   see   *Fagopyrum*

Bulrush   see   *Scirpus*

Buxus   50439, 50517, 52138

C

Cabbage   see   *Brassica*

Cacti   see   Succulents, cacti

Cajanus   49193, 50301, 51492

Calluna   see   Heath plants and communities

Canna   49156, 49628, 52116-7

CAM plants (*cf.* also *Agave, Ananas, Bryophyllum,* Succulents, cacti)
        48610, 48622, 48729, 48732, 48861-3, 48898, 49038, 49091, 49446, 49756,
        49863, 50011, 50115, 50129, 50182, 50192, 50382, 50390, 50466, 50538, 50571,
        50607, 50609, 50783, 50797, 50891-2, 50911, 50938, 50999, 51005, 51223,
        51381, 51408, 51491, 51870, 51901, 51997, 52112, 52262-4

Camellia   see   *Thea*

Canarygrass   see   *Phalaris*

Cannabis   B49114, B51788, 51933

Capsicum   48598, 48913-4, 48945, 49091, 49219, 49628, 49976, 50366, 50404-5, 50439,
        50538, 51227, 51348, 51547, 51678, 51737, 51933, 52030

Cardamine   48716

Carex   48523, 48596, 48716, 48839, 48973, 49158, 49252, 49297, 49496, 49578, 49807,
        50224, 50554, 50754, 50756, 50915, 51144, 52000, 52068

Carica   49938, 49990, 50454-6, 51678

Carnivorous plants   49763, 52068

Carob   see   *Ceratonia*

Carpinus   48755, 49253, 51502, 51933, 52138

Carrot   see   *Daucus*

Carthamus   51678

Carya   50242, 50611, 51900

Cassava see *Manihot*

Cassia 49763, 49990, 51057

Castanea 50152, 50494, 50611, 50766

Castor bean see *Ricinus*

Catchweed see *Galium*

Cat's tail see *Typha*

Cattail flag see *Typha*

Cauliflower see *Brassica*

Cedar see *Cedrus*

Cedrus B49527, 50317, 51658

Celery see *Apium*

Cerasus 50576, 51678, B51788

Ceratonia 49204

Cereals see *Avena, Eleusine, Eragrostis, Hordeum, Oryza, Panicum, Secale, Sorgum, Triticum, Zea*

Chayote see *Sechium*

Chenopodium 48440, 48632, 48700, 48930, 48967, 49265, 49524, 49862, 49916, 49999, 50298, 50635, 50697, 50766, 50774, B51788, 51905

Cherry see *Cerasus*

Chervil see *Anthriscus*

Chestnut see *Castanea*

Chick pea see *Cicer*

Chinese cabbage see *Brassica*

Chinese gooseberry see *Actinidia*

Chlamydomonas
48484, 48499, 48595, 48625, 48663, 48677, 48680-1, 48723, 48761, 48766, 48788, 48819, 48925, 49003, 49006-8, 49078-9, 49171, 49259, 49267, 49342-4, 49366, 49377, 49506, 49544, 49582, 49625, 49628, 49682, 49688, 49793, 49850, 49867-8, 49954, 49984, 49986, 50060, 50089, 50111, 50174, 50186, 50255, 50298, 50309, 50326, 50381, 50403, 50415, 50521, 50562, 50616-7, 50629, 50690, 50694, 50742, 50805, 50887, 50913-4, 50927, 50929, 50945, 51011-2, 51028, 51110-1, 51118, 51284, 51308-9, 51374, 51455, 51460, 51477, 51507, 51533-4, 51615, 51640, 51663, 51705, 51716-7, 51762, 51971-2, 52011, 52103, 52118-9, 52161, 52165, 52209, 52220, 52225, 52243, 52278-9, 52298, 52319-20, 52378

Chlorella
48411, 48428-9, 48441, 48488, 48499, 48542, 48551, 48640, 48648, B48650, 48652, 48654, B48671, 48723, 48728, 48766, 48788-9, 48847, 48888, 48911-2, 48941, 48981, 49030, 49043, 49098, 49108, 49172, 49191, 49267, 49275, 49321,
(continued)

Chlorella (continued)
          49330, 49371, 49373, 49495, 49505, 49512, 49532, 49625, 49628, 49650, 49676,
          49688, 49690, 49694, 49780, 49793, 49833, 49858, 49946, 49950, 49965-6,
          50047, 50060, 50062, 50089, 50101, 50174, 50254, 50298, 50363, 50381, 50385,
          50394, 50397, 50415, 50424-5, 50497, 50534, 50551, 50596, 50618, 50633,
          50736, 50741, 50744, 50793, 50805, 50825, 50837-9, 50894, 50934, 50940,
          50945, 51074, 51102, 51110-1, B51168, 51170, 51251, 51265, 51276, 51293,
          51349, 51416, 51439, 51460, 51469, 51477, 51526, 51533-4, 51547, 51650-1,
          51690, B51788, 51824, 51850, 51863, 51889, 51971-2, 52041, 52058, 52072,
          52082, 52103, 52119, 52136, 52161, 52209, 52220, 52225, 52244-5, 52298,
          52354-6, 52378, 52383, 52385

Chloris 49091, B49561, 49758-9, 50511, 51618, 52083

Chlorobium
          48443, 48499, 48625, 48654, 48859, 48865, 49048-9, 49151, 49211, 49393-4,
          49692, 49781, 49942, 50054, 50071, 50106, 50141, 50280, 50298, 50488, 50740,
          50767, 51101, B51168, 51294, 51388, 51472, 51547, 51590, 51684, 51742-3,
          51961, 52186

Christophine see *Sechium*

Chromatium
          48441, 48443, 48455, 48499, 48597, 48706, 48767, 48795, 48859, 48942, 49016,
          49026, 49078, 49138, 49151, 49211, 49226-7, 49307, 49692, 49695, 49848-9,
          49935, 49942, 49965-6, 50054, 50071, 50107, 50280, 50298, 50415, 50740-1,
          50767, 50868, 50883, 50894, 50912, 50954, 51101, 51110, B51168, 51171,
          51294, 51469, 51472, 51590, 51742-3, B51788, 51986, 52058, 52290

Chrysanthemum 48755, 49496, 49712, 49934, 51933

Chrysophyta
          48411, 48728, 48749, 48758, 48764, 48849, 48996, 49078, 49267, 49591, 49802,
          49933, 50040, 50047, 50180, 50351, 50600, 50805, 51013, 51110-1, 51143,
          B51168, 51459, 51469, 51547, 51643, 51686, 51785, B51788, 52001, 52044,
          52103, 52130, 52161, 52274, 52378

Cicer   49091, 50143, 51492

Cinnamomum 51348, B51788

Cinnamon see *Cinnamomum*

Citrullus 49553, B51531, 52161, 52178

Citrus B49114, 49690, 49694, 49711, 49903-4, 50241, 50404, 50833, 51185, 51224,
          51424, 51481, 51497, 51547, 51595, 51652, 51822-3, 51933, 52029, 52155,
          52162, 52204, 52331-2

Cloud-grass see *Agrostis*

Clover see *Trifolium, Melilotus*

Club rush see *Scirpus*

Cocksfoot see *Dactylis*

Cocoa see *Theobroma*

Coconut palm see *Cocos*

Cocos   49091, 49331, 49345, B49691, 50465

Coco-yam see *Colocasia*

Coffea 48609, 49345, 49471, 49572, B49691, 50241, 50291

Coffee tree see *Coffea*

Colocasia 51678, B51788

Coniferous plants (*cf.* also *Abies, Cedrus, Cupressus, Juniperus, Larix, Picea, Pinus, Pseudotsuga, Sequoia, Taxus, Thuja, Tsuga*
   48722, 49327, 49385, 49493, B49527, 50138, 50181, 50241, 50705, 50762, 50766, 51369-71, 51447, 51658, 52138, 52182, 52209, 52225

Corchorus 51023, B51531, 51933

Corn see *Zea*

Cornelian cherry see *Cornus*

Corn lettuce see *Valerianella*

Corn salad see *Valerianella*

Cornus 48737, B50144, 50152, 50241-2, 50611, 51658, 52000, 52138, 52204

Corylus 50241, 50280, 51497, 52138

Cotton see *Gossypium*

Cottonwood see *Populus*

Cowberry see *Vaccinium*

Cowpea see *Vigna*

Crabgrass see *Digitaria, Eleusine*

Cranberry see *Oxycoccus*

Crataegus 50241, 50766, 52138

Cress see *Lepidium*

Crocus 51099

Crotalaria 51057, B51531

Cucumber see *Cucumis*

Cucumis B48560, 48658, 48660-1, 48703, 48766, 48915, 48951, 48954, 48982-3, 49091, 49105, B49114, 49121-2, 49161, 49271-3, 49421, 49426, 49476, 49542, 49628, 49712, 49738, 50042, B50052, 50082, 50090, 50169-70, 50280, 50328, 50419, 50446, 50597, 50647, 50650, 50763, 50812, 50950, 50993, 51026, 51028, 51145, 51232, 51259-61, 51297, 51507, B51531, 51547, 51819, 51869, 51933, 51946, 52060, 52161, 52288

Cucurbita 48535-6, 48870, 48903, 48951, 48954, 48995, 49031, 49091, B49114, 49628, 49792, 49929-30, 50089, 50538, 50541, 50650, 50715, 50766, 50808, 50895, 50982, 51295, 51547, 51607, 51678, B51788, 51933, 52035, 52120, 52161, 52195

Cupressus B49527, 51658, 52225

Curcuma 50301

Currant see *Ribes*

Cyanobacteria see Algae, blue-green

Cycas B51788

Cynodon B48560, 48721, 48729, 48857, 49091, 49287, 49628, 49645, 49759, 49999, 50572, 50999, 51678

Cyperus 48839, 48919, 49345, 49599, 49628, 49734, 49807, 49999, 50208, 50752, 51678

Cypress see *Cupressus*

D

Dactylis 48563, 48587, 48666, 48729, 48747, 48857, 49091, 49496-9, 49501, 49617, 49694, 50208, 50307, 50906, 50915, 50990, 51058, 51933, 52149

Dallis grass see *Paspalum*

Date palm see *Phoenix*

Datura 48632, 48930, 49367, B49921, 50538, 51051, 51137, 51656, 51933

Daucus 48899, 48902, 49091, 49976, 50407, 50439, 50870, 51368, 51678, B51788, 52013, 52383

Deadly nightshade see *Atropa*

Deciduous trees and shrubs (*cf.* also *Acer, Aesculus, Alnus, Amygdalus, Armeniaca, Betula, Carpinus, Carya, Castanea, Cerasus, Cornus, Corylus, Crataegus, Eucalyptus, Euonymus, Fagus, Fraxinus, Gingko, Hevea, Hibiscus, Juglans, Liriodendron, Malus, Mangifera, Morus, Persea, Persica, Pirus, Platanus, Populus, Prunus, Quercus, Ribes, Robinia, Rubus, Salix, Sambucus, Sorbus, Syringa, Tamarix, Tectona, Tilia, Ulmus, Viburnum, Vitis*)
48618, 48737, 48945, 48957, 49102, 49287, 49297, 49694, 49990, 50065, B50144, 50241-2, 50517, 50689, 51153, 51219, 51570, 51658, B51788, 51820, 51833, 52000, 52110, 52138, 52260

Desert plants and ecosystems
48572, 48692-3, 48729, 48780, 48857, 49287, 49443-4, 49555, 49686, 49694, 50103, 50129, 50241-2, 50382, 50418, 50638, 50844-5, 51026, 51333, B51531, 51565, 51827, 51949, 51956, 52248

Desmodium 49329, 50511, 52297

Dewberry see *Rubus*

Diatoms 48411, 48441, 48471, 48484, 48550, 48710, 48723, 48728, 48819, 48972, 48996, 49054-6, 49070, 49099, 49116, 49128, 49140-1, 49189, 49195-8, 49221, 49225, 49267, 49289, 49339, 49364, 49374, 49445, 49487, 49494, 49503, 49519, 49552, 49591, 49679, 49690, 49694, 49728, 49784, 49802, 49874, 49880, 49933, 49977, 50018, 50047, 50180, 50202, 50219, 50238-40, 50255, 50355, 50369, 50451, 50549, 50566, 50600, 50605, 50722, 50730, 50741, 50805, 50945, 51013, 51060, 51110-1, 51129, 51143, 51183, 51265, 51284, B51350, 51452, 51460, 51477, 51522, 51533, 51547, 51580, 51624, 51643, 51685, 51699, B51788, 51843, 51845, 51914, 51916, 51971, 52044, 52324, 52328, 52358, 52367, 52378

Digitalis 49083

Digitaria   48729, 48919, 49262, 49734, 49758-9, 49926, 49999, 50511, 50836,
        51162, 51588, 51632, 51678, 51908-10

Dill   see   *Anethum*

Dinoflagellatae, Dinophyceae
        48411, 48654, 48728, 48819, 48991, 48996, 49098-9, 49140-1, 49146, 49195-8,
        49289, 49339, 49417, 49445, 49494, 49752-4, 49784, 49802, 49880,
        50047, 50180, 50219, 50239, 50477, 50667, 50717, 50722, 50730, 50805, 51013,
        51110-1, 51123, B51350, 51387, 51452, 51460, 51547, 51643, 51668, 51699,
        B51788, 51818, 51845, 52358, 52378

Dioscorea   49783, 51547

Diospyros   50242, 52149

Dogwood   see   *Cornus*

Douglas fir   see   *Pseudotsuga*

Dunaliella
        48411, 48565, 48625, 48648, 48672, 48674-5, 48819, 49146, 49267, 49562,
        49591, 49743, 50478, 50528, 50588, 50741, 50945, 51013, 51110-1, 51183,
        51477, 51522, 51547, 51578, 51681, 51685, 51964, 51995-6, 52118-9

Dyer's plants   see   *Crocus, Indigofera, Phytolacca*

E

Ebony   see   *Diospyros*

Echinochloa   49491, 49758-9, 49911, 50697, 50796, 51002, 52052

Ectothiorhodospira
        48499, 49267, 49935, 49942, 50298, 50767, 51101, 51163, 51388, 51472, 51602,
        51748, 51986

Edible canna   see   *Canna*

Egg plant   see   *Solanum*

Elaeis   48729, B49691, 49694, 50241, 50462

Elder   see   *Sambucus*

Eleusine   49491, 49628, 49734, 49759, 50292-3, 51547, 51678

Elm   see   *Ulmus*

Elodea   48605, 48656, B49114, 49820, 50116, 50370, 50778, 51178-80, 51206, 52287,
        52383

Equisetum   48722, 49297, 49692, 49743, 50554, 50915, B51788, 52209, 52243

Eragrostis   49759, 51043, 51678, 51933

Ericaceae see   Heath plants and communities

Eucalyptus   48489, 48664, 48927, 49087, 49228, 49385, 49455, 49690, B49691, 49694,
        50241-2, 50574, 50646, 51227, 51497-8, 51502, 51568, 51933

Euglena   48411, 48441, 48499, 48529, 48625, 48639, B48650, 48727-8, 48733, 48758, 48761, 48766, 48788, 48822, 48849, 48876-8, 48891, 48941, 48991, 48996, 49003, 49141, 49170, 49191, 49263, 49319, 49429-30, 49557, 49628, 49703, 49745, 49784, 49827, 49965-6, 49981, 50060, 50094, 50121, 50174, 50195, 50271, 50298, 50381, 50394, 50403, 50521, 50549, 50690, 50726, 50805, 50881--2, 50931, 50966, 50990, 51018, 51028, 51047-8, 51110, 51122, B51168, 51173, 51260, 51345, 51364, 51401, 51451, 51460, 51462, 51464, 51469, 51477, 51479, 51493, 51499, 51547, 51580, 51626, 51643, 51762, B51788, 51794, 51945, 51971--2, 52011, 52044, 52130, 52161, 52189-90, 52202, 52209, 52274, 52329-30, 52378

Euonymus   48514, B50144, 50517, 52000, 52138

Euphorbia   48909-10, 49199, 49382, B49561, B49691, 50279, 50538, 50591, 50697, 51061, 51223, B51531, 51678, 52077

Evergreen grass see *Arrhenatherum*

Evergreen plants see Sempervirent plants

Evonymus see *Euonymus*

F

Fagopyrum   48589, 50948, 51153, 51576, 51933

Fagus   48755, 49252, 49287, 49385, 50122, 50343, 50439, 50630, 50632, 50766, 51001, 51495-6, 51497, 51933, 51954, 52138

Feather grass see *Stipa*

Fern-palm see *Cycas*

Ferns   48449, 48685, 48722, 48766, 48908, B49114, 49158, 49302, 49341, 49628, 49743, 49762, 50465, 51027, B51168, 51486, 51658, 51759, B51788, 51888, 52161, 52225, 52243, 52274, 52284

Fescue see *Festuca*

Festuca   B48560, 48563, 48666, 48721, 48729, 48747, 48857, 48893, 49029, 49091, 49297, 49382, 49387, 49496, 49498, 49501, 49578, 49627-8, 49694, 49717, 49740, 50208, 50276, 50307, 50779, 50906, 50915, 50999, 51133, 51353, 51642, B51897, 51923-4, 51933, 52004, 52115, 52124, 52287

Ficus   49355-7, 49990, 50007, 50439, 50494, 50517, 51092

Fig see *Ficus*

Finger-grass see *Chloris*

Finger-millet see *Eleusine*

Fir see *Abies*

Flatterdock see *Nymphaea*

Flax see *Linum*

Float grass see *Glyceria*

Forage crops (*cf.* also *Andropogon, Arrhenatherum, Brassica, Desmodium, Echinochloa,*
      Grasses, Leguminous plants, *Lotus, Lupinus, Medicago, Melilotus, Opuntia,*
      Trifolium, Vicia, Vigna, etc.)
      49077, 49102, 49147, 49329, 49580, 50226, 50465, 51359-60

Forest (including undergrowth) plants and ecosystems (*cf.* also Coniferous plants,
      Deciduous trees and shrubs, Ferns, *Fragaria,* Grasses, Heath plants and com-
      munities, Lichens, Liverworts, Mosses, *Oxycoccus, Salvia, Sphagnum, Vacci-*
      nium, etc.)
      48449, 48460-2, 48486-6, 48495, 48687, 48729, 48927, 49194, 49252-3, 49287,
      49385, 49555, 49574, 49609-11, 49690, B49691, 49694, 49990, 50065,
      50067, 50122, 50152-3, 50553, 50611, 50758, 51274, 51658, 52027, 52068,
      52074, 52087, 52109, 52138

Fountain-grass see *Pennisetum*

Foxglove see *Digitalis*

Foxtail millet see *Setaria*

Fragaria 49293, 49655, 49658, 50037, 50280, 50766, 51280, 51678, B51788, 52204

Fraxinus 48632, 48755, 49287, 50241-2, 50280, 50329, 51658, 51946, 52138

Fruit plants and trees see *Actinidia, Ananas, Armeniaca, Carica, Cerasus, Citrullus,*
      Citrus, Cocos, Cucumis, Cucurbita, Diospyros, Ficus, Fragaria, Hippophaë,
      Malus, Mangifera, Musa, Opuntia, Oxycoccus, Passiflora, Persea, Persica,
      Phoenix, Pirus, Prunus, Ribes, Rubus, Sorbus, Vaccinium, Vitis

Fucus   48702, 49257, 49267, 49277, 49470, 50467, 50806, 51109, 51189, B51788,
      52227

Fungi (parasitic)
      48564, 48570, 48697, 48870, 49606-7, 49675, 49893, 50013, 50136, 50200,
      50332, 50575, 51079, 51676, 51710, 52123, 52168

G

Galium 50443, 50697, 51933

Garden cress see *Lepidium*

Garlic see *Allium*

Gherkin see *Cucumis*

Gingko 48722, 50592, B51788, 52225, 52243

Glyceria 49520, 50226, 50915, 50999

Glycine 48442, 48477-8, 48499, 48542, 48548, B48560, 48579, 48611, 48659, 48711,
      48729, 48731, 48738, 48776, 48806, 48834, 48884-5, 48901-2, 48941, 48945,
      49000, 49080, 49101, 49165, 49326, 49333, 49341, 49378, 49469, 49534,
      49536-7, 49546, 49618, 49628, 49715, 49717, 49719, 49761, 49766, 49818,
      49822, 49834, 49860, 49893, 49901, B49921, 49993, 50016, 50167, 50204,
      50256, 50367, 50587, 50638, 50642, 50656, 50719, 50841, 50884, 50933, 50936,
      50942, 51026, 51057, 51062, 51066, 51076, 51131, 51153, 51156, 51187, 51227,
      51235, 51361, 51421, 51492, 51497, 51510-1, 51515, 51569, 51661, 51678,
      51807, 51851, 51895, B51897, 51905, 51921, 51930, 51933, 51946, 52004,
      52013, 52023, 52080-1, 52139-40, 52157-8, 52161, 52213, 52246, 52250, 52271,
      52287, 52301, 52361

Goat's-beard   see *Tragopogon*

Gossypium   48415, 48424, B48560, 48664, 48690, 48692, 48729, 48814, 48901-2, 48944-
        -5, 48992, 49042, 49091, 49367, 49387, 49469, 49529, 49678, B49691, 49694,
        49717, 49820, 49847, 49898-9, B49921, 49922, 49966, 49976, 50006, 50021,
        50241, 50638, 50655, 50662, 50664, 50719, 50847-8, 50850, 50999, 51204,
        51212, 51225, 51246-9, 51384-5, 51497, 51547, 51770, B51897, 51933, 51946,
        52004, 52010, 52204, 52243, 52304

Gourd   see *Cucurbita*

Gram chick pea see *Vigna*

Granadille see *Passiflora*

Grape fruit see *Citrus*

Grape vine see *Vitis*

Grasses (*cf*. also *Agrostis, Andropogon, Anthoxanthum, Arrhenatherum, Avena, Bromus,
        Carex, Chloris, Cynodon, Cyperus, Dactylis, Digitaria, Echinochloa, Eleusi-
        ne, Eragrostis, Festuca, Herdeum, Lolium, Oryza, Panicum, Paspalum, Penni-
        setum, Phalaris, Phleum, Poa, Saccharum, Secale, Setaria, Sorgum, Triticum,
        Zea*)
        48411, 48440, 48490, 48556, B48560, 48563, 48596, 48666, 48692, 48716,
        48721, 48729, 48747, 48768, 48839, 48857, 48908, 48919, 48945, 48973,
        48999, 49010, 49022, 49064-5, 49091, 49102, 49150, 49158, 49163, 49216-7,
        49287, 49297, 49329, 49382, 49490-1, 49496, 49501, 49520, 49546,
        49555, B49561, 49571, 49578, 49628, B49691, 49694, 49748, 49758-60, 49807,
        49860, 49889, 49995, 50148, 50163, ·50208, 50226, 50246, 50307, 50462,
        50511, 50636, 50658, 50718, 50754-6, 50836, 50850, .50915, 51058,
        51136, 51146, 51196, 51203, 51227, 51253, 51272, 51354, 51376-8, 51570,
        51596, 51677-8, 51707, 51733, 51761, 51933, 51951, 52068, 52083, 52109-10,
        52115, 52180, 52243, 52276

Groundnut see *Arachis*

Guayule see *Parthenium*

H

Halobacterium
        48438, 48476, 48524, 48575, 48585-6, 48588, 48593, 48604, 48612, 48630-1,
        48644-5, 48694, 48741, 48759-60, 48816-7, 48905-6, ~48931, 48986-7, 49053,
        49082, 49089-90, 49107, 49183-5, 49246-7, 49250-1, 49260, 49264, 49295-6,
        49306, 49313-5, 49323-4, 49353-4, 49459, 49522, 49540, 49585, 49654, 49706,
        49778, 49790, 49808, 49859, 49866, 49871, 49895-6, 49969, 49994, 50009-10,
        50058, 50061, 50068-9, 50081, 50084, 50087, 50091, 50119, 50134-5, 50151,
        50172, 50216, 50220-2, 50267, 50311-2, 50333, 50349, 50379, 50399-401,
        50403, 50421-2, 50460, 50484, 50503, 50513, 50543, 50579, 50603, 50621,·
        50668, 50710, 50717, 50795, 50869, 50878, 50922-5, 51003, 51006-7, 51039,
        51090, B51168, 51226, 51294, 51300, 51318, 51326-8, 51465, 51554, 51590,
        51633, 51638-9, 51649, 51715, 51719, 51730, 51773, 51797, 51912, 51955,
        51957, 51965, 51989, 51991-2, 51995-6, 51999, 52026, 52053, 52148-9, 52159-
        -60, 52196-7, 52200, 52377

Halophilous plants, salt marsh and strand plants
        48596, 48886, 49084, 49163, 49267, 49365, 49382, 49578, 49628, 49889, 50057,
        50382, 50609, 50638, 51043, B51531, 51570, 51678, 51949, 52146, 52346

Hawthorn   see   *Crataegus*

Hazel   see   *Corylus*

Heath plants and communities   49102, 50067, 52068

Hedera   48945, 49252, B49527, 49628, 49854, 50517, 51153, 51481, 51525, 51547, 52287

Helianthus
          48512, B48560, 48692, 48722, 48733, 48738, 48884, 48902, 49131, 49312,
          49341, B49527, 49628, 49675, 49793, B49921, B50052, 50207, 50574, 50585,
          50720, 50785, 50817, 50969, 50990, 50999, 51009, 51024, 51026, 51033,
          51096, 51127, 51198-9, 51227, 51234-5, 51254-5, 51497, 51583, 51674, 51678,
          51736, 51775, 51826, 51847, B51897, 51933, 52004, 52016, 52099

Hemlock   see   *Tsuga*

Hemp   see   *Cannabis*

Henbane   see   *Hyoscyamus*

Hepatics   see   Liverworts

Hevea   48945, B49691, 50241, 51551

Hibiscus   49091, 50538

Hickory   see   *Carya*

Hippophaë   51492

Holly   see   *Ilex*

Honey mesquite   see   *Prosopis*

Hop   see   *Humulus*

Hordeum   48419-20, 48426-7, 48433, 48451, 48483, 48495-6, 48521, B48560, 48564,
          48584, 48615, 48628, 48660, 48668, 48676, 48686, 48722, 48729, 48765,
          48788, 48806, 48850-1, 48853, 48857, 48902, 48941, 48950-1, 48953-4, 48979,
          49003, 49078, 49083, 49091, 49094, B49114, 49126, 49150, 49234, 49262,
          49270, 49278, 49320, 49343, 49373, 49380, 49387, 49469, 49514, 49534,
          49546, 49554, 49576, 49578, 49602, 49628, 49642, 49660, 49665, 49674,
          B49691, 49694, 49720, 49773, 49793, 49836, 49855, 49857, 49891, 49900,
          B49921, 49998, 50000, 50046, 50051, B50052, 50089, 50105, 50139, 50181,
          50196, 50208, 50215, 50294, 50320-1, 50357, 50398, 50439, 50536, 50599,
          50612, 50684, 50719, 50723-4, 50776, 50894, 50914, 51050, 51076, 51115,
          51146, 51151-3, 51164, 51170, 51197, 51200, 51220, 51228, 51245, 51292,
          51355, 51393, 51405, 51411, 51473, 51492, 51506, 51525, 51547, 51574, 51589,
          51613, 51628, 51631, 51675-6, 51678, 51707, 51712, 51770, 51780, B51788,
          51821, 51824, 51886, 51922, 51933, B51961, 51963, 52010, 52111, 52121,
          52135, 52163, 52168, 52204, 52209, 52211, 52220-2, 52243, 52287, 52339,
          52361

Hornbeam   see   *Carpinus*

Horse chestnut   see   *Aesculus*

Horsetail   see   *Equisetum*

Humulus   48483, B51788

Hyoscyamus   51933

I

Ilex 48737, B49527, 50241, 50452, 51658, 51933, 52287

Indian fig see *Opuntia*

Indian senna see *Cassia*

Indian shot see *Canna*

Indigo see *Indigofera*

Indigofera B51531, 51658

Inula 51658

Ipomoea 48434-5, 49091, 49345, 49367, 49723, 50609, 51024, B51531, 51933

Ivy see *Hedera*

J

Japanese persimmon see *Diospyros*

Jerusalem artichoke see *Helianthus*

Jointgrass see *Paspalum*

Jojoba see *Simmondsia*

Juglans 49155, B50144, 50241, 50507, 52138

Juniper see *Juniperus*

Juniperus B49527, 49549, 50374, 52138

Jute see *Corchorus*

K

Kaki see *Diospyros*

Kalanchoë 48481-2, 48500, 48610, 48861-4, 49038, 49446, 49951-2, 50044, 50182,
      50390, 50538, 50571, 50666, 50797, 50892, 50911, 50999, 51005, 51081, 51107,
      51202, 51389, 51408, 51516, 51678, 51683, 51794, 51799, 51933, 51997, 52264,
      52266

Kale see *Brassica*

Kenaf see *Hibiscus*

Kohlrabi see *Brassica*

L

Lactuca 48430, 48432, 48436, 48603, 48619, 48676, 48751-4, 48761, 48766, 48877,
        48901, 48905, 48944, 49091, 49156, 49172, 49212, 49287, 49325, 49370, 49437,
        49628, 49651-2, 49712, 49722, 49850, 49997, B50052, 50232, 50298, 50435,
        50643, 50648, 50725, 50859, 50966, 50998, 51116-7, 51191, 51366, 51391,
        51634, B51788, 51925-6, 51933, 52107-8, 52149, 52161

Lamb's lettuce see *Valerianella*

Laminaria 48541, 49257, 49277, 49744, 49966, 50045, 50259, 50467, 50806, 51109,
        51474, 52201, 52247

Larch see *Larix*

Larix 48729, 49100, 49385, B50144, 50241, 51221, 51289, 51497, 51714, B51788,
        51933, 52138, 52182, 52204

Lathyrus 48433, 49496, 50307, 51026, 51028

Laurel see *Laurus*

Laurus B49527, 50348, 50517, B51788

Lavandula 50316

Lavender see *Lavandula*

Leguminous plants (*cf.* also *Arachis, Cajanus, Cicer, Glycine, Lathyrus, Lens, Lupi-
        nus, Medicago, Melilotus, Phaseolus, Pisum, Trifolium, Vicia, Vigna*)
        48857, 49415, B49691, B49921, 50462, 51485, 51492, 52181

Lemna see *Lemnaceae*

Lemnaceae 48535, 48766, 48788, 48976, 49520, 49643, 49762, 49843, 49850, 50590,
        50719, 50763, 50971, 51401, 51671, 51875, 52035, 52161, 52243

Lemon see *Citrus*

Lens 50155, 51026, 51492, 51660

Lentil see *Lens*

Lepidium B48701, 49959

Lettuce see *Lactuca*

Lichens 48578, 48692, 48702, 48716, 49075, 49297, 49600, 49639, 50132, 50347,
        50376, 50844-5, 50987, 51342, 51439, 51571, 51689, 51786, B51788, 52113

Lilac see *Syringa*

Linden see *Tilia*

Linseed see *Linum*

Linum 49091, 49103, B49114, 49218, 49641, 50688

Liriodendron 50241-2, 50764, 52204

Liverworts 48536, 48691, 49638, 49878, 50439, 50766, 50951, 50987, 52161

Locust see *Robinia*

Lolium   48563, 48666, 48721-2, 48729, 48733, 48747, 48893, 48901, 48945, 49091,
          B49114, 49144, 49187, 49498, 49501, 49546, 49628, 49694, 49732-3, 49766,
          50017, 50055, 50226, 50286, 50474, 50574, 50906-7, 51042, 51058, 51181,
          51306-7, 51496, 51933, 52004, 52179, 52210, 52255-6, 52277, 52287

Lotus   49083, 49091, 49496

Lovegrass see *Eragrostis*

Lucerne see *Medicago*

Lupine see *Lupinus*

Lupinus 48929, B49114, 49443, 49628, 49661, 49966, 51091, 51154, 51353, 51492, 52257

Lycopersicon
          48475, 48538, B48560, 48665, 48815, 48902, 48915, 48917, 48926, 48951,
          48954, 48964, 49091, 49105, 49119, 49229, 49261, 49489, 49612, 49712, 49736,
          49840, B49921, 50006, B50052, 50156, 50280, 50366, 50404, 50438, 50486,
          50538, 50574, 50606, 50707, 50719, 50766, 50889, 50933, 50990, 51020, 51029,
          51130, 51169-70, 51207, 51261, 51322, 51481, 51547, 51570, 51678, 51703,
          51738, B51788, B51897, 51933, 51946, 52004, 52010, 52139, 52149, 52204, 52243,
          52316, 52370

Lycopodium B51788

M

Macereed see   *Typha*

Maidenhair tree see *Gingko*

Maize see *Zea*

Malus   48517, 48599, 48739, 48818, 48945, B49114, 49363, 49389, 49716, 50043,
          B50144, 50200, 50241-2, 50331, 50344, 50383, 50406, 50480, 50594, 50903,
          51020, 51082-3, 51097, 51147, 51186, 51357, 51383, 51678, 51744, 51815-7,
          51933

Mammoth tree see *Sequoia*

Mangifera 51678

Mango see *Mangifera*

Mangold see *Beta*

Mangrove communities
          48591, 48886, 49020-1, 49084, 49365, 50079, 51688, 52146-7

Manihot 48729, 48945, 49103, 49112, 49352, 49690, B49691, 49694, 49765, 49927,
          50999, 51030, 51077, 51400, 51678

Manioc see *Manihot*

Manna-grass see *Glyceria*

Maple see *Acer*

Marrow see *Cucurbita*

Meadow foxtail see *Alopecurus*

Meadowgrass see *Poa*

Medicago 48542, B48701, 48729, 48901, 48945, 49060, 49091, 49496, 49546, 49613,
      49628, 49694, 49697, 49715, 49785, 50006, 50130, 50207, 50270, 50307, 50529,
      50655, 50701-2, 50750, 50990, 50999, 51026, 51028, 51188, 51392, 51492,
      51507, 51678, B51788, B51897, 51933, 51946, B51961, 52067, 52204, 52207

Medicinal plants see *Artemisia, Asclepias, Atropa, Carica, Cynodon, Datura, Digi-*
      talis, Hibiscus, Hyoscyamus, Mentha, Papaver, Rhus, Ricinus, Ruta, Salvia,
      Thymus, etc.

Melilotus 49091, 49628

Melon see *Cucumis*

Mentha 50116, B51788

Mesquite see *Prosopis*

Milkweed see *Asclepias*

Millet see *Panicum*

Morus 48632, 48945, 50809-11, 51433

Mosses (*cf.* also *Sphagnum*)
      48449, 48535-6, 48543, 48546, 48716, 48766, 48849, 48941, 48991, 49158,
      49297, 49762, 49779, 50116, 50608, 50760, 50766, 51052, 51075, 51144,
      B51168, 51658, 51672, 51698, 51786, B51788, 52034-5, 52068, 52161, 52225,
      52259, 52274, 52288, 52383-4

Mulberry see *Morus*

Mung bean see *Vigna*

Musa 49345, 50241, 51933

Musk-melon see *Cucumis*

Mustard see *Sinapis*

N

Napier grass see *Pennisetum*

Needle grass see *Stipa*

Nerium 48572, 48692, 48726, 49249, B49527, 50049, 50348, 50494, 50517, 51165-6,
      51216

Nicotiana 48453, 48499, 48508, B48560, 48582, 48676, 48733, 48766, 48788, 48791,
      48877, 48881, 48902, 48920, 48925, 48944-5, 49002, B49114, 49387, B49527,
      49632, 49642, 49717, 49830, 49850, 49900, 49957, 50006, 50089, 50122, 50229,
      50242, 50296, 50298, 50381, 50390, 50410, 50439, 50483, 50501, 50538, 50542,
      50598-9, 50685, 50703, 50712, 50719, 50766, 50853, 50894, 50951, 50990,
      51005, 51027, 51055, 51093-4, 51118, 51170, 51237, 51241, 51261-2, 51273,
      51290, 51297, 51497, 51507, 51516, 51547, 51577, 51606, 51625, 51762, 51868,
      51933, 51946, 51972, 52010-1, 52055, 52095, 52161, 52220, 52243, 52287,
      52291, 52307, 52347, 52361, 52370

Nostoc 48474, 48499, 48625, 48649, 48745, 48764, 48849, 48867, 48877, 48978, 49075,
        49106, 49195, 49266-7, 49465, 49532, 49577, 49695, 49699, 49742-3, 49883,
        49919, 50131, 50159, 50298, 50562, 50879, 51074, 51085, 51110-1, 51344,
        51452, 51457-8, 51529, 51571, 51785, B51788, 51789, 51859, 51971, 51995-6,
        52058, 52369

Nuphar 48839, 49109-10, B49114, 50116, 51178

Nymphaea 49599

O

Oak see *Quercus*

Oat see *Avena*

Oil palm see *Elaeis*

Okra see *Hibiscus*

Olea   49152, 49204, B49527, 50241

Oleander see *Nerium*

Olive see *Olea*

Onion see *Allium*

Opuntia 48729, 48945, 49686, 49990, 50192, 50466, 50538, 50609, 50902, 51202, 51223

Orache see *Atriplex*

Orange see *Citrus*

Orchardgrass see *Dactylis*

Orchids 49475, 49587, 51933

Ornamental plants (*cf.* also *Agave, Antirrhinum, Asparagus, Canna, Chrysanthemum,*
        Coniferous plants, *Cyperus,* Deciduous trees and shrubs, *Eucalyptus, Euphor-*
        bia, Ficus, Hedera, Hibiscus, Ilex, Lathyrus, Lupinus, Orchids, *Passiflora,*
        Pelargonium, Perilla, Rosa, Tradescantia, Tulipa, etc.)
        48755, 48766-7, 48877, 48944, 48957, 48959, 48963, 49020, 49035, 49041,
        49085, 49091, B49114, 49204, 49340, 49399, 49458, 49568, 48628, 49712,
        49763, B49921, 49949, 49988, 50122, 50162, 50208, 50212, 50280, 50290,
        50439, 50452, 50517, 50538, 50609, 50611, 50683, 50691, 50716-7, 50719,
        50803, 50990-1, 51005, 51024, 51026, 51028, 51055, 51098, 51118, 51135,
        51223, 51270, 51481, 51507, 51658, B51788, 51933, 51946, 52010, 52066,
        52129, 52138, 52161, 52204, 52243

Oryza   48450, 48729, 48733, 48857, 48901-2, 48963, 49005, 49091, 49102, 49131-2,
        49239, 49241, 49287, 49328, 49362, 49511, 49599, 49628, B49691, 49694,
        49766, 49797, 49821, 49831-2, 49834, 49965, 50041, 50076, 50102, 50213,
        50431, 50469-70, 50719, 50752, 50785, 50822-3, 50995, 51002, 51170, 51372,
        51415, 51418-9, 51422-3, 51492, 51532, 51540-1, 51547, 51617, 51678, 51846,
        51854, 51867, 51886-7, B51897, 51933, 52003, 52009, 52143, 52172, 52175-6,
        52214-6, 52243, 52273

Oscillatoria 48441, 48592, 48678, 48728, 48867, 48981, 49034, 49195, 49258, 49266,
        49396, 49447-8, 49628, 49679, 49950, 50176, 50218, 50298, 50740, 50805,
        50880, 51110-1, 51129, 51340, 51529, 51547, 51580, 51590, 51686, B51788,
        51789, 52044, 52317, 52378

Oxycoccus   50554

P

Paddy   see   *Oryza*

Palms   see   *Cocos, Cycas, Elaeis, Phoenix*

Panicum   48622, 48729, 48839, 48857, 48919, 49091, 49262, 49299, 49491, 49520,
       49546, B49561, B49691, 49717, 49734, 49755, 49758-9, 49860, 49900, 50208, 50324,
       50754, 50836, 50949, 51043, 51181, 51576, 51678, 51933, 52083, 52100, 52153,
       52265, 52276, 52350, 52361

Papaver   49009, 52132

Papaya   see   *Carica*

Paper grass   see   *Stipa*

Paprika   see   *Capsicum*

Para-rubber tree   see   *Hevea*

Parasitic plants   50154, 51274

Parsley   see   *Petroselinum*

Parthenium   50638

Paspalum   48721, 48729, 49091, 49345, 49599, 49732-3, 49759, 49999, 50462, 50511,
       51678, 52100

Passiflora   49103, 49628, 49990

Pasture plants   see   Forage plants

Pea   see   *Pisum*

Peach   see   *Persica*

Peanut   see   *Arachis*

Pear   see   *Pirus*

Peavine   see   *Lathyrus*

Pecan   see   *Carya*

Pelargonium   48528, 48733, 48782, 49793, 50439, 50703, 50719, 51004, 51175, 51507,
       52106

Pennisetum   48665, 48721, 48729, 48766, 48919, 48945, 49132, 49287, B49561, B49691,
       49694, 49759, 49798, 50574, 50999, 51181, 51497, 51678, 51721, 52083, 52100,
       52276

Pepper   see   *Capsicum, Piper*

Pepperminth   see   *Mentha*

Perilla 48945, 49951, 50078, 52287

Persea 50241-2, 50280, 50327

Persica 49139, 49162, 49186, 49713, 50766, 51678, 51865-6, 52335

Persimmon see *Diospyros*

Petroselinum 48676, 48694, 48722, 49628, 50562

Phalaris 48747, 48839, 48945, 49628, 50915

Phaseolus 48425, 48439, 48447, 48483, 48499, 48507, 48525-6, 48539, 48542, B48560,
        48573, 48583, 48660-1, 48664-5, 48733, 48766, 48768, 48788, 48814, 48870,
        48877, 48890, 48902, 48921, 48941, 48943-5, 48951, 48954, 48959, 48975,
        48994, 49043, 49091, B49114, 49126, 49156, 49174, 49190, 49205, 49274,
        49322, 49341, 49346, 49349, 49365-7, 49370, 49453-4, 49469, 49541,
        49546, 49615, 49628, 49704, 49717, 49724, 49758, 49763, 49809-10, 49820,
        49846, 49884, 49900, 49990, 49997, 50012, 50025-9, 50043, 50059, 50089,
        50117, 50174, 50211, 50277, 50280, 50318, 50381, 50412, 50510, 50538,
        50570, 50573, 50637, 50664, 50686, 50736, 50749, 50765-6, 50841, 50856,
        50863, 50894, 50962, 50964, 50968, 50990, 50999, 51024, 51026, 51043, 51081,
        51120, 51153, 51160, 51167, B51168, 51170, 51181, 51227, 51260-1, 51295,
        51401, 51413, 51441, 51445, 51460, 51462, 51475, 51478, 51492, 51497,
        51507, 51537, 51589, 51610, 51658, 51674, 51676, 51678, 51712, 51735, 51779,
        51784, 51814, 51834, B51897, 51902, 51933-6, 51946-7, 51972, 52004, 52013,
        52048-9, 52161-2, 52202, 52209, 52226, 52233, 52287, 52361, 52384

Phleum 48666, 48733, 48747, 49498, 50019, 50208, 50307, 50915, 52115

Phoenix 49091

Photosynthetic bacteria see Bacteria, photosynthetic

Phragmites 49520, 49599, 50757, 50871, 50970, 51933, 52051

Phytolacca 49628, 50111

Picea   48449, 48456-7, 48718, 48729, 48754, 48839, 48879, 49385, 49461, B49527,
        49549, 49609, 49690, 49946, 50095-6, 50138, B50144, 50242, 50384, 50402,
        50417, 50452, 50574, 50706, 50766, 50958, 51001, 51086-7, 51195, 51221,
        51379, 51495-7, 51538, 51658, B51788, 51933, 52027, 52088, 52182-3, 52204,
        52237-8

Pigeon pea see *Cajanus*

Pimpinella 51933

Pine see *Pinus*

Pineapple see *Ananas*

Pinus   48452, 48456-9, 48486, 48552-3, 48638, 48729, 48737, 48871, 48879, 48945,
        49076, 49243, 49385, 49388, 49455, B49527, 49549-50, 49570, 49690, 49694,
        49701, 49730, 49747, 49826, 50067, 50095-6, 50122, 50124, 50138, 50140,
        B50144, 50152, 50181, 50241-2. 50287, 50317, 50489-91, 50564, 50574, 50660,
        50766, 50787, 50910, 50983,   51001, 51121, 51195, 51296, 51547, 51653,
        51658, 51672, 51704, 51753, B51788, 51799, 51933, 51982-3, 52086, 52088,
        52138, 52182-3, 52204, 52220, 52225, 52228, 52237-8, 52275, 52342-4

Piper   50735-6

Pirus   49479, 51481, 51658, 51678, 52309

Pistacia 49203-4, 49458, 50348

Pisum   48411, 48468, 48495-6, 48500, 48516, 48525, 48539, 48558, 48581-2, 48601-2,
           48612, 48615, 48636, 48653, 48662, 48676, 48715, 48761, 48766, 48785,
           48788, 48793-5, 48810-1, 48844, 48852,   48877, 48902, 48950-1,
           48954, 48956, 48965, 48988, 49003, 49091, 49111,
           B49114, 49124, 49210, 49244, 49255, 49348, 49361, 49370, 49376, 49401, 49411-
           -2, 49437, 49464, 49507-9, 49559, 49594, 49614, 49628, 49642, 49649, 49659,
           49672, 49688, 49694, 49695, 49758, 49763, 49769, 49772, 49816, 49850, 49869,
           49877, 49890, 49900, 49911, 49932, 49965, 49972-3, 49985, 49991, 50038-9,
           B50052, 50075, 50089, 50114, 50147, 50177-8, 50181, 50188, 50191, 50207,
           50225, 50248, 50260, 50319, 50325, 50356, 50359-60, 50364, 50371, 50381,
           50390, 50426, 50534, 50550, 50562, 50567-9, 50575, 50577-8, 50626-8, 50655-
           -6, 50682, 50719, 50728-9, 50773, 50804, 50828, 50859, 50894, 50990, 50998,
           51016, 51024, 51026, 51028, 51103, 51116, B51168, 51190-1, 51263, 51312,
           51315, 51320-2, 51332, 51337, 51343, 51347, 51492, 51500-1, 51507, 51509,
           51525, 51547, 51605, 51614, 51616, 51627-8, 51640, 51654, 51678, 51690-1,
           51710, 51733, 51746-7, 51752, 51892, 51894, B51897, 51918-9, 51931, 51933,
           51970, 51972, 52004, 52011, 52013, 52032, 52035, 52078, 52089, 52091, 52097,
           52126, 52128, 52135, 52139, 52161, 52163, 52170, 52220, 52241, 52268, 52287,
           52299, 52322, 52384

Plane tree see *Platanus*

Platanus 48632, B49921, B50144, 51881, 52005, 52138, 52204

Plum see *Prunus*

Poa   48666, 48700, 48716, 48747, 49217, 49297, 49496, 49731, 49758, 49909, 50208,
           50226, 50307, 50754, 50973, 50990, 52000, 52115, 52265

Pokeweed see *Phytolacca*

Poplar see *Populus*

Poppy see *Papaver*

Populus 48446, 48457, 48491, 48624-5, 48827, 48945, 48958, 48960, 48994, B49114,
           49206-7, 49287, 49341, 49385, B49527, 49690, 49694, 49844, B49921, 49955,
           50003, 50095-7, B50144, 50241-2, 50257, 50270, 50382, 50574, 50704, 50783,
           50861-2, 51358, 51390, 51502, 51508, 51672, 51787, B51788, 51820, 51946,
           51954, 52027, 52122, 52138, 52305

Porphyridium 48625, 48654, B48671, 48846, 48867, 49106, 49213, 49267, 49532, 49667,
           49681, 50089, 50271, 50298, 50378, 50415, 50464, 50777, 50831, 51174, 51264,
           51469, 51477, 51533-4, 51542, 52058

Portulaca 48883, B49561, 49758, 49999, 50193, 50279, 50572, 50609, 51005, 51516,
           B51531, 51678, 51908-10, 52276, 52294

Potato see *Solanum*

Prickly pear see *Opuntia*

Prosopis 49718, 49990, 50241, B51531, 51565

Prune see *Prunus*

Prunus (*cf.* also *Amygdalus, Armeniaca, Cerasus*)
           48632, 48692, 49103, 49162, 49287, 49686, 49844, 49934, B50144, 50241, 50517,
           50576, 50655, 51059, 51386, 51497, 51678, B51788, 51954, 52138, 52204

Pseudotsuga 48729, 49282, 49385, 49549, 49668, 50008, 50241-2, 50452, 50674-6,
           50821, 51086-7, 51496, 51933, 52138, 52182-3, 52204

Pumpkin see *Cucurbita*

Purple arrowroot see *Canna*

Purslane see *Portulaca*

Pyrus see *Pirus*

Q

Queensland arrowroot see *Canna*

Quercus 48416, 48457, 48632, 48879, 49203-4, 49253, 49341, 49385, 49473, B49527,
         49588, 49610, 49844, B50144, 50152-3, 50241-2, 50348, 50408, 50494, 50507,
         50565, 50607, 50611, 50766, 51502, 51658, 51883, 51933, 51946, 51954, 52000,
         52138, 52204, 52260, 52287

R

Radish see *Raphanus*

Rape see *Brassica*

Raphanus 48479, 48483, 48509, 48568, 48676, 48896-7, 49078, 49091, 49407, 49628,
         49662-4, B50052, 50280, 50441-4, 50670,50846,50998,51102,51118,51172,51191,
         51586, 51589, 51658, 51661, B51788, 51941-2, 52010, 52125, 52243

Raspberry see *Rubus*

Rattle-box see *Crotalaria*

Redwood see *Sequoia*

Reed see *Phragmites*

Rheum 49628

Rhodes grass see *Chloris*

Rhodopseudomonas
         48441, 48499, 48569, 48597, 48603, 48654, 48767, 48803-4, 48831, 48900,
         48938, 48941-2, 49016, 49018, 49026, 49028, 49040, 49057, 49061-2, 49078,
         49137, 49151, 49157, 49180-2, 49209, 49267, 49286, 49305, 49307, 49420,
         49456-7, 49526, 49532, 49698, 49764, 49774-6, 49789, 49848-9, 49885, 49910,
         49942, 49960, 49974-5, 50024, 50054, 50071-2, 50098, 50106-7, 50118, 50173,
         50223, 50237, 50298, 50415, 50524-5, 50539, 50602, 50623, 50634, 50648,
         50657, 50672-3, 50680, 50708, 50767, 50799, 50832, 50866, 50868, 50878,
         50937, 50939, 50954, 50955-6, 50974, 50994, 51015, 51041, 51063-4, 51101,
         51110, 51141, B51168, 51170-1 51176-7, 51209, 51260, 51268, 51365, 51435,
         51454, 51469, 51472, 51480, 51518-9, 51547, 51629-30, 51726, 51748-9, 51804,
         51835, 51848, 51852, 51937, 51980, 51986, 52024, 52037, 52055, 52058-9, 52084
         52084, 52092-4, 52104-5, 52131, 52186, 52289-90, 52340, 52349, 52380

Rhodospirillum
         48443, 48499, 48569, 48594, 48597, 48611-2, 48778-9, 48795, 48799, 48877,
                                                           (continued)

Rhodospirillum (continued)
48905, 48942, 48997, 49027, 49127, 49211, 49456, 49547, 49563-4, 49583,
49596-7, 49698, 49739, 49801, 49825, 49848-9, 49873, 49942-3, 50053-4,
50071, 50104, 50107, 50227, 50298, 50415, 50508, 50580, 50648, 50740, 50743,
50767, 50868, 50909, 50954, 50956, 51008, 51021, 51056, 51064, 51101, 51110,
B51168, 51170-1, 51197, 51256, 51260, 51294, 51388, 51401, 51437, 51454,
51469, 51472, 51480, 51547, 51559, 51585, 51590, B51788, 51790-1, 51871,
51984, 51996, 52058, 52092, 52105, 52186, 52234, 52274, 52351, 52380-2

Rhubarbe   see   *Rheum*

Rhus   48632, 48945, 50637

Ribes   49657, 52260

Rice   see   *Oryza*

Ricinus   48499, 48831, 49091, B49691, 49840, 49893, 50114, 50650, 50894, 50947;
51266, 51525

Robinia   49287, 50242, 51502, B51788, 51820, 51933, 52138

Rosa   48809, 49712, 51658, B51788, 52138

Rose   see   *Rosa*

Rosemary   see   *Rosmarinus*

Rosmarinus   50517

Rubber tree   see   *Hevea*

Rubus   48449, 51547, 51658, B51788, 52068

Rue   see   *Ruta*

Rumex   48751-5, 49535, 50285, 52370, 52372

Ruta   49923, 50284

Rye   see   *Secale*

Ryegrass   see   *Lolium*

S

Saccharum   B48560, 48729, 48901, 48963, 49060, 49287, 49523, 49628, 49690, B49691,
49694, 49766, 50463, 50532, 50538, 50574, 50999, 51560, 51591, 51678,
B51897, 52067

Safflower   see   *Carthamus*

Saffron   see   *Crocus*

Sage   see   *Salvia*

Sago-palm   see   *Cycas*

Salsify   see   *Tragopogon*

Salix   48523, 48716, 48722, 49297, 49844, B50144, 50574, 50888, 50916, 51038, 51144,
        51502, B51788, 52064-5, 52138

Salt marsh and strand plants   see   Halophilous plants, salt marsh and strand plants

Salvia   51833, 52260

Sambucus 52138

Scabwort see *Inula*

Scenedesmus
        48411, 48441, 48499, 48565, 48566, 48625, 48648, B48650, 48651, 48677,
        48696, 48728, 48758, 48764, 48766, 48847, 48941, 49078, 49108, 49123, 49195,
        49235, 49267, 49275, 49330, 49442, 49628, 49667, 49679, 49690, 49692, 49694,
        49743, 49793, 49946, 49950, 50060, 50089, 50298, 50353, 50381, 50451, 50497,
        50562, 50596, 50805, 50914, 50934, 50940-1, 51102, 51110-1, 51129, 51140,
        B51168, 51284, 51293, 51349, 51397-9, 51460, 51477, 51533-4, 51547, 51553,
        51560-1, 51580, 51624, 51640, 51686, B51788, 51995-6, 52044, 52186, 52354-6,
        52378, 52383

Scented vernal grass see *Anthoxanthum*

Scirpus 49403, 49578, 49694, 49807, 50116, 51678

Scorzonera   50439

Sea-buckthorn see *Hippophaë*

Secale   48570, 48714, 48766, 48821, 48951, 48954, 49091, 49380-1, 49514, 49646,
        49648, 49855, 49857, 49906-7, 50163, 50208, 50574, 50850, 50894, 50900,
        50963, 51146, 51339, 51380, 51411, 51473, 51492, 51678, 51733, 51770, B51788,
        51828, 51871, 52076

Sechium   49391

Sedge see *Carex*

Sempervirent plants (*cf.* also *Ceratonia, Citrus, Coffea,* Coniferous plants, *Hedera,*
        Ilex, Laurus, Olea, Pistacia, Simmondsia, Theobroma, etc.)
        48572, 49203-4, 49327, 49385, 49588, 50129, 50138, 50517, 50611, 50689

Senna see *Cassia*

Sequoia 49385, 52182-3

Service-tree see *Sorbus*

Sesamum 48692, B49691, 50387, 51497, B51531

Setaria 48930, 49091, 49131, 49759, B49921, 49989, 50511, 51678, 51933, 52292

Shrubs see Deciduous trees and shrubs; Sempervirent plants

Silk-grass see *Agrostis*

Silkweed see *Asclepias*

Simmondsia 50638

Sinapis B48560, 48676, 48725, 48755, 48766, 48945, 48949, 48951-4, 49911, 49988,
        50085, B50144, 50207, 50280, 50443, 50556, 50562, 50746, 50921, 51026, 51933,
        52010, 52161, 52243

Sisal   see *Agave*

Snowball   see *Viburnum*

Solanum (*cf.* also *Lycopersicon*)
   48537, 48539-40, 48542, B48560, 48700, 48722, 48729, 48806, 48899, 48902,
   48917, 48941, 48945, 49051, 49091, B49114, 49287, 49524, 49619, 49694,
   49766, 49820, 49850, 49893, 49965, 49990, 50006, 50089, 50149, 50175,
   50214, 50280, 50372, 50439, 50574, 50635, 50747, 50754, 51024, 51192, 51227,
   51394, 51420, 51481, 51525, B51531, 51547, 51678, 51707, B51788, 51794,
   51800, B51897, 51933, 51962, 52010, 52013-4, 52161, 52243, 52383

Sorbus   B50144, 50472, 52138, 52204

Sorghum   see *Sorgum*

Sorgum   48433, 48495, B48560, B48701, 48729, 48736, 48788, 48806, 48857, 48875,
   48902, 48919, 48945, 48951, 48954, 48966, 48977, 49091, 49103, 49131-2,
   49262, 49367, 49387, 49491, 49521, B49527, B49561, 49628, B49691, 49694,
   49715, 49717, 49734, 49758-9, B49921, 49966, 49999, 50044, 50066, 50142,
   50207-8, 50266, 50324, 50463, 50538, 50556, 50574, 50583-4, 50625, 50638,
   50650, 50791, 50857, 50920, 50999, 51014, 51081, 51137, 51170, 51181, 51227,
   51236, 51274, 51497, 51659, 51678, 51721-2, B51788, B51897, 51933, 52004,
   52010, 52101, 52179, 52204, 52243, 52295, 52361

Sorrel   see *Rumex*

Soybean   see *Glycine*

Spatterdock   see *Nuphar*

Spear-grass   see *Stipa*

Sphagnum   50083, 50554, 51052, 51144, 51478, 52068

Spice plants   see *Acorus, Anethum, Angelica, Anthriscus, Apium, Artemisia, Capsicum, Cinnamomum, Curcuma, Galium, Inula, Laurus, Lavandula, Melilotus, Mentha, Pimpinella, Piper, Portulaca, Rosmarinus, Ruta, Salvia, Thymus, Tropaeolum, etc.*

Spinach   see *Spinacia*

Spinach beet   see *Beta*

Spinacia   48414, 48423, 48436-7, 48445, 48454, 48494-6, 48498-502, 48510, 48522,
   48533, 48543, 48545, 48580, 48606-8, 48616, 48625-6, 48676, 48684, 48686,
   48688-9, 48704-5, 48707, 48722, 48748, 48757, 48761, 48766, 48773-4, 48788-
   -9, 48795, 48797, 48824, 48860, 48877, 48890, 48894, 48916, 48928, 48933-5,
   48941, 48950-1, 48955, 48961, 48963, 48965, 48968-9, 48974, 49003-4, 49023,
   49050, 49066-7, 49073, 49085, 49091-3, 49095, 49097, 49104, 49133, 49143,
   49171, 49191-2, 49232-3, 49235, 49241, 49248, 49259, 49263, 49279, 49322,
   49383, 49392, 49397, 49406, 49427-8, 49436, 49438-9, 49449-50, 49460, 49492,
   49516, 49532-3, 49543, 49548, 49551, 49569, 49584, 49586, 49624, 49628,
   49635, 49640, 49674, 49687-8, 49692, 49695, 49698, 49714, 49720, 49725,
   49743, 49750-1, 49764, 49782, 49794, 49820, 49842, 49877, 49883, 49888,
   49892-3, 49900, 49902, 49913, 49917, 49920, 49925, 49946, 49961, 49963-
   -4, 50006, 50031-3, 50044, 50049, 50089, 50111, 50127, 50160-1, 50190,
   50231, 50234-6, 50244, 50254, 50271, 50276, 50280, 50298, 50313-5, 50323,
   50330, 50335, 50337, 50350, 50358, 50361-5, 50368, 50381, 50390-1, 50394,
   50415, 50437, 50439, 50499, 50520-1, 50534, 50545, 50559-60, 50562, 50615,
   50648-9, 50661, 50700-1, 50707, 50734, 50739, 50743, 50745, 50765, 50768-71,
   50773-4, 50788, 50793, 50819-20, 50830, 50834-5, 50842, 50846, 50865, 50878,
   50894, 50896-8, 50919, 50927, 50929-31, 50954, 50960, 50965, 50966, 51025-8,
                                                                              (continued)

Spinacia (continued)
        51034, 51041, 51047, 51070, 51074, 51076, 51081, 51103, 51114-8, 51137,
        51149-50, 51158, B51168, 51170, 51173, 51176, 51197, 51213, 51216, 51237,
        51239, 51260, 51267, 51276-7, 51301, 51303-5, 51317, 51329, 51346, 51364,
        51375, 51399, 51401-2, 51404, 51426, 51432, 51434, 51443, 51455, 51468-9,
        51473, 51487-90, 51494, 51499, 51507, 51515, 51521, 51525, 51547, 51555-7,
        51600-1, 51604, 51620, 51635, 51640, 51647, 51669-70, 51678, 51693, 51708-9,
        51711-2, 51747, 51755-6, 51760, 51762, 51767, 51771, 51781, 51783, B51788,
        51794, 51799, 51806, 51813, 51829-31, 51836, 51838-42, 51849, 51853, 51871-4,
        51876-7, 51879, 51917, 51932, 51938, 51967, 51980, 51995-6, 52006-8, 52010-2,
        52022, 52039, 52045-6, 52069-70, 52073, 52096-8, 52150-2, 52161,
        52173, 52199, 52205-6, 52220, 52222, 52229, 52232, 52243, 52282-3, 52294,
        52299, 52302, 52310, 52315, 52326-7, 52336-7, 52341, 52348, 52365-6, 52379,
        52383

Spindle-tree see *Euonymus*

Spirodela see   *Lemnaceae*

Spruce see *Picea*

Squash see *Cucurbita*

Steppe plants 48596, 49287, 49424, 52203

Stipa 49382, 50103

St. John's Bread see *Ceratonia*

Stramony see *Datura*

Strawberry see *Fragaria*

Submersed plants see Aquatic macrophytes

Succulents, cacti (*cf*. also *Agave, Aloe, Bryophyllum,* CAM plants, *Opuntia, etc.*)
        48716, 48729, 48862, B49114, 49579, 49637, 49756, 49900, 50116, 50280, 50439,
        50466, 50538, 50609, 50640, 50797, 51005, 51113, 51407-8, 51491, 51678,
        51901, 51997

Sudan grass see *Sorgum*

Sugar beet see *Beta*

Sugar cane see *Saccharum*

Sumac see *Rhus*

Sunflower see *Helianthus*

Sun-hemp see *Crotalaria*

Sweet flag see *Acorus*

Sweet grass see *Glyceria*

Sweet potato see *Ipomoea*

Sweet root see *Acorus*

Sweet vernal grass see *Anthoxanthum*

Sweet woodruff see *Galium*

Synechococcus see *Anacystis*

Syringa B51788, 52138

T

Tall oat-grass see *Arrhenatherum*

Tamarisk see *Tamarix*

Tamarix 48491, 49091

Tapioca see *Manihot*

Taro see *Colocasia*

Tarragon see *Artemisia*

Taxus   B49527, 50241, 51658, B51788, 52138

Tea see *Thea*

Teak see *Tectona*

Tectona 49136

Telegraph-plant see *Desmodium*

Thea   48519-20, 48617, 50329, 50766, 51547, 51720, 51723, 51890-1, 51933

Theobroma 49345, 50462, 51252

Thuja   52138, 52182, 52225

Thyme see *Thymus*

Thymus 48596, 50591

Tilia   48632, 49473, B50144, 50439, 51954, 52138

Timothy see *Phleum*

Tobacco see *Nicotiana*

Tomato see *Lycopersicon*

Tradescantia 49628, 50049, 50538, 51005, 52384

Tragacanth see *Astragalus*

Tragopogon 49217

Trifolium 48411, 48587, 48729, 48747, 48767, 49091, B49114, 49164, 49187, 49496,
      49628, 49694, 49766, 50130, 50226, 50307, 50652-3, 50843, 51024, 51091,
      51182, 51353, 51392, 51492, 51496, 51547, 51678, B51897, 51933, 52004,
      52115, 52261, 52277

Triticum 48411, 48431, 48465-7, 48480, 48483, 48542, 48555-6, B48560, 48561, 48568,
      48584, 48651, 48655, 48664-5, 48676, 48697, 48722, 48729, 48743-4, 48750,
      48766, 48798, 48806-7, 48852, 48857, 48877, 48887, 48889, 48901-2, 48904,
      48918, 48929, 48945, 48951, 48954, 48980, 48992, 49060, 49064-5, 49091,
                                                          (continued)

Triticum (continued)
    49102, 49113, B49114, 49118, 49125-6, 49145, 49150, 49153-4, 49156, 49214,
    49231, 49262, 49292, 49335, 49380, 49395, 49431, 49449, 49463, 49469,
    49513, 49517, 49576, 49592, 49595, 49628, 49633, 49680, B49691, 49694,
    49696, 49707, 49715, 49748, 49758, 49766, 49803-5, 49820, 49851,
    49855-7, 49887, 49900, 49918, B49921, 49965, 49992, 49998, 50002, 50013,
    50034, B50052, 50128, 50158, 50163-5, 50208, 50246, 50270, 50273, 50280,
    50288, 50338, 50381, 50387, 50390, 50392, 50396, 50407, 50431, 50443,
    50449-50, 50453, 50530, 50552, 50638, 50650, 50733, 50737-8, 50752-3,
    50766, 50780, 50801, 50819, 50828, 50830, 50850, 50854, 50894, 50975,
    50989, 50999, 51043, 51068, 51076, 51079, 51096, 51116, 51136, 51146,
    51154, 51157, 51222, 51228, 51269, 51272, 51291-2, 51351-2, 51373, 51380,
    51448, 51456, 51473, 51492, 51507, 51520, 51525, 51547, 51589, 51596,
    51632, 51641, 51646, 51666, 51675, 51678, 51707, 51724, 51733, 51768-70,
    51780, 51829, 51831, 51864, 51886, 51893, B51897, 51898, 51929, 51933,
    51942, 51948, 51974, 51995-6, 52004, 52010, 52025, 52099, 52127, 52161,
    52164, 52174, 52177, 52191, 52198, 52240, 52243, 52249, 52265, 52272,
    52285, 52287, 52294, 52360-1, 52363

Tropaeolum 48766, B49114, 50783, 52287

Tropical plants 48495, 48687, 48729, 49329, 49345, 49555, 49990, 50065, 50462,
    50511

Tsuga   48449, 48729, 49385, B49527, 49549, 50241-2, B51788, 52183

Tulip see *Tulipa*

Tulipa 48766, 51004, 52149

Tulip-tree see *Liriodendron*

Tundra plants and ecosystems 48523, 48716, 48729, 48973, 49158, 49287, 49555,
    49609, 50482, 51144

Turmeric see *Curcuma*

Turnip see *Brassica*

Turpentine tree see *Pistacia*

Typha   48729, 48839, 49520, 49599, 49628, 50999, 51353, 51757

U

Ulmus   49473, B50144, 50242, 50565, 50873, 50990, 51153, 51502, B51788, 52138

Ulva   48541, 48592, 48702, 48767, 49267, 49277, 49417, 49673, 49917, 50022,
    50100, 50298, 50416, 50467, 50806, 51109, 51319, 51547, 52225, 52227,
    52386

V

Vaccinium 48449, 48523, 48530, 48899, 49142, 49297, 49609, 49611, 49861, 50067,
    50611, 51144, 51933, 52068

Valerianella 49287, 51917

Yellow pond-lily see *Nuphar*

Yew see *Taxus*

Yucca   50129, 50609, 51678

Z

Zea   48431, 48433, 48453, 48464, 48467, 48483, 48499-500, 48505, 48508, 48529,
48539, B48560, 48587, 48622-3, 48633-6, 48660, 48664, 48669, 48679,
48692-3, 48698, 48711, 48722, 48729, 48733, 48738, 48761, 48766, 48771,
48788, 48806, 48835, 48857, 48877, 48883, 48890-1, 48902, 48941, 48951,
48954, 49003, 49060, 49088, 49091, 49102, B49114, 49115, 49126, 49132,
49167, 49173, 49177, 49194, 49200, 49236-8, 49259, 49262, 49284, 49287,
49332, 49345, 49367, 49372, 49401-2, 49453, 49491, 49515, B49527, 49538,
49545-6, B49561, 49598, 49608, 49628, 49642, 49653, 49690, B49691, 49694,
49717, 49726-7, 49757-9, 49766, 49793, 49850, 49872, 49878, 49909, B49921,
49929, 49945, 49976, 49979, 50001, 50006, 50056, 50150, 50173-4, 50194,
50198-9, 50207-9, 50258, 50270, 50278, 50280, 50301, 50324, 50377, 50386,
50407, 50413-4, 50433, 50443, B50476, 50498, 50533, 50538, 50570, 50572-4,
50581-2, 50586, 50599, 50638, 50645, 50650, 50695, 50714, 50719, 50748-9,
50752, 50754, 50766, 50773, 50785, 50817, 50827, 50855, 50857, 50894,
50917-8, 50960-1, 50979, 50981, 50999, 51026-8, 51031, 51048, 51073, 51076,
51081, 51096, 51125-6, 51128, 51135, 51137, 51151-3, 51155, 51167, B51168,
51181, 51227-8, 51240, 51242-4, 51257, 51260-1, 51271, 51295, 51298-9,
51334, 51353, 51413, 51482, 51492, 51496-7, 51507, 51547, 51563, 51575,
51589, 51625, 51632, 51645, 51658, 51661, 51678, 51692, 51712, 51731-2,
51751-2, 51762, 51766, 51782, B51788, 51794, 51808, 51819, B51897, 51920,
51933, 51948, 51959, B51961, 52004, 52010, 52067, 52071, 52079, 52139,
52141, 52153-4, 52161, 52189, 52198, 52202, 52204, 52220, 52225, 52240,
52243, 52280, 52287, 52295, 52301, 52306, 52361, 52374, 52384

Zebrina see *Tradescantia*